D0272377

A Dictionary of the Bible

Consultant Editors

A Dictionary of the Bible

W. R. F. BROWNING

Oxford New York
OXFORD UNIVERSITY PRESS
1996

Oxford University Press, Walton Street, Oxford OX2 6DP

Oxford New York
Athens Auckland Bangkok Bogota Bombay
Buenos Aires Calcutta Cape Town Dar es Salaam
Delhi Florence Hong Kong Istanbul Karachi
Kuala Lumpur Madras Madrid Melbourne
Mexico City Nairobi Paris Singapore
Taipei Tokyo Toronto

and associated companies in
Berlin Ibadan

Oxford is a trade mark of Oxford University Press

British Library Cataloguing in Publication Data
Data available

Library of Congress Cataloging-in-Publication Data
Browning, W. R. F.
A dictionary of the Bible / W. R. F. Browning.
1. Bible—Dictionaries. I. Title.
BS440.B73 1996 220.3—dc20 96–13033
ISBN 0–19–211691–6

10 9 8 7 6 5 4 3 2 1

Typeset by Graphicraft Typesetters Ltd., Hong Kong
Printed in Great Britain
on acid-free paper by
Bookcraft (Bath) Ltd.
Midsomer Norton
Avon

PREFACE

THE invitation to compile this dictionary came to me from Angus Phillips of Oxford University Press and he has given constant encouragement. I am also grateful to members of my family and to academic colleagues for important assistance, to the Principal of Ripon College, Cuddesdon, for permission to use the library, and to Veronica Ions the copy editor for her meticulous examination of the typescript. The consultant editors gave immensely valuable advice. But for any errors that may remain, as also for controversial opinions expressed, I alone am responsible.

W. R. F. Browning

CONTENTS

INTRODUCTION

THIS is a dictionary of the books of the Christian Bible. Its aim is to expound the main ideas of the Bible and to give information about the principal people and places that readers encounter. The author's standpoint is that of a grateful indebtedness to the achievements of modern critical scholarship.

There are no entries for persons of whom little more than the mere name is recorded in the biblical text. On occasion, however, entries are allowed to stray beyond the biblical era, where a theme in the New Testament, for example that of 'ministry', has become more familiar as significantly developed over the centuries.

The dictionary is intended for students whether at school or university, and for a more general public of those who may be reading or studying the Bible at home or alongside others in a group from their local church or community. It is assumed that these students will use the NRSV or some other modern translation of the Bible, and therefore references to the Authorized (King James) Version are infrequent. The nomenclature follows, generally, that of modern translations; for example, those books in the New Testament which come after the four gospels and the Acts of the Apostles mostly consist of letters to individuals or to communities and they are usually called 'letters' in the dictionary, though the traditional 'epistles', still often heard in churches, is also used. This word perhaps reflects the fact that as well as similarities with other letters of the first century CE there are also marked differences in style and purpose.

The use of the neutral CE ('in the common era') in place of AD, and of BCE ('before the common era') rather than BC is increasingly preferred in biblical scholarship, to which so many important contributions have been made by Jewish scholars.

Sometimes one part of a syllabus for students of the Old Testament (the Hebrew Scriptures) is entitled 'The Early Traditions of Israel', and prescribed texts may include narratives about persons whose historical existence is doubtful. For example, Jubal (Gen. 4: 21), said to be the father of music, is mentioned as the narrator's way of indicating the cultural progress of mankind: Jubal was not an actual individual. There are other persons described in the dictionary where what is recorded of them in the biblical text is mentioned, but a note of caution must be assumed—that the material we are dealing with may well not be historical. Similarly, the narratives of the Exodus from Egypt and the occupation of Canaan were written in a much later era and have had that later interpretation stamped upon them. The task of disentangling the historical course of events from the interpretation has not led to a scholarly consensus. There are those who hold, for example, that the Israelites never 'entered the Promised Land' and were living in Canaan all the time. Areas of disagreement about interpretations have often been alluded to in the dictionary, and must be assumed in all the entries about the Exodus, the wanderings in the wilderness, and the occupation of Canaan.

The Land of the Bible

The disputes between the modern state of Israel and its neighbours demand an explanation of the territorial names used in the Bible and in this dictionary. They do not necessarily correspond with current political and geographical definitions.

Most of the events recorded in the OT and in the four gospels of the NT took place in an area about the size of Wales, never more extensive than about 500 km. (300 miles) from north to south. To the Hebrews it was 'the Promised Land' (Gen. 50: 24); to some modern travel agents it is (from Zech. 2: 12) 'the Holy Land'. 'Canaan' is the name given (Gen. 12: 5) by Hebrew writers to the country which they supposed was occupied by devotees of pagan gods until the arrival of the Israelite conquerors after the Exodus—a tradition of invasion not accepted as reliable history by modern scholars. The name 'Israel' is used in the OT records for the country after Hebrew hegemony (as in 1 Sam. 13: 19).

When the kingdom was divided (922 BCE) after the death of Solomon, the northern, and larger, part retained the name 'Israel', while the southern, with Jerusalem as its capital, was Judah. The independent northern kingdom existed for only 200 years. It was conquered by Assyria in 722 BCE, and many of the people deported. In a narrative (2 Kgs. 17) influenced by later Jewish hostility it is alleged that the country was repopulated and that the mixed race resulting became the Samaritans. When the kingdom of Israel disappeared, the name could then be applied again to the inhabitants in the still independent south—*they* were the 'children of Israel', and it was 'Israel' who went into exile in Babylon in 586 BCE. 'Judaea' is the Latin version of Judah and was adopted by the Romans for their province governed from Caesarea and forming part of Syria. It was not therefore coterminous with the ancient kingdom of Judah.

'Palestine' is probably the most suitable name for the whole country, even though it was only after 135 CE and the final Jewish revolt that the emperor Hadrian deleted 'Judaea' from his map and substituted 'Palestine'. But 'Palestine', which derives from the Hebrew for 'Philistine', does appear four times in the OT (e.g. AV of Joel 3: 4).

Names for God

The Hebrew *El* (or more commonly *Elohim*, the plural form of *Eloah*) is translated 'God' in English Bibles, and *Yahweh* by 'the Lord'—except that NJB prefers to retain 'Yahweh'. This dictionary uses the word which seems appropriate. In the NT the Greek *theos*, with or without the definite article, means 'God', but it is not always clear whether *kurios*, 'Lord', refers to Jesus or to God.

An asterisk before a word indicates that there is an article under that name in the dictionary. But words that recur repeatedly are not necessarily so marked, for example, the books of the Bible, Jerusalem, Jesus, Paul, and Peter.

ABBREVIATIONS

AV	The Authorized, or King James, Version of the English Bible, 1611
BCE	Before the Common Era (= BC, Before Christ)
CE	In the Common Era (= AD, *anno Domini*, in the Christian Era)
ET(T)	English Translation(s)
LXX	The Greek Septuagint
NEB	New English Bible, 1961 and later years; a new translation by British scholars
NJB	New Jerusalem Bible, 1985, a revision of the (English) Jerusalem Bible, 1966
NRSV	New Revised Standard Version, 1989, a revision by American scholars of the Revised Standard Version; Anglicized edition of NRSV, 1995
REB	Revised English Bible, 1989, a revision by British scholars of the New English Bible
RSV	Revised Standard Version, 1946 and later years; a revision by American scholars of the Standard Version
RV	Revised Version (NT, 1881; OT, 1885), a revision by British scholars of AV
marg.	Denotes a note in the margin of the English text

The titles of most books of the Bible are abbreviated; shorter titles are retained in full. The abbreviations are:

OT	Old Testament
NT	New Testament
Apoc.	Apocrypha/Deuterocanonical books
Gen.	Genesis
Exod.	Exodus
Lev.	Leviticus
Num.	Numbers
Deut.	Deuteronomy (or Deuteronomist)
Josh.	Joshua
Judg.	Judges
1 or 2 Sam.	1 or 2 Samuel
1 or 2 Kgs.	1 or 2 Kings
1 or 2 Chron.	1 or 2 Chronicles
Neh.	Nehemiah
Ps. (or Pss.)	Psalm (or Psalms)
Prov.	Proverbs
Eccles.	Ecclesiastes
S. of S.	Song of Solomon (or Song of Songs)
Isa.	Isaiah
Jer.	Jeremiah
Lam.	Lamentations
Ezek.	Ezekiel
Dan.	Daniel
Hos.	Hosea
Obad.	Obadiah

Mic.	Micah
Hab.	Habakkuk
Zeph.	Zephaniah
Hag.	Haggai
Zech.	Zechariah
Mal.	Malachi
Matt.	Matthew
Rom.	Romans
1 or 2 Cor.	1 or 2 Corinthians
Gal.	Galatians
Eph.	Ephesians
Phil.	Philippians
Col.	Colossians
1 or 2 Thess.	1 or 2 Thessalonians
1 or 2 Tim.	1 or 2 Timothy
Philem.	Philemon
Heb.	Hebrews
Jas.	James
1 or 2 Pet.	1 or 2 Peter
Rev.	Revelation
1 or 2 Esd.	1 or 2 Esdras
Rest of Esth.	The Additions to the book of Esther
Wisd.	The Wisdom of Solomon
Ecclus. [= Sir.]	Ecclesiasticus, or the Wisdom of Jesus the son of Sirach
S. of III Ch.	Song of the Three Children (or Song of the Three Jews)
Sus.	Susanna
Bel & Dr.	Bel and the Dragon
Pr. of Man.	Prayer of Manasseh
1/2/3/4 Macc.	1/2/3/4 Maccabees
Pss of Sol.	Psalms of Solomon

THE BOOKS OF THE BIBLE
according to the order of AV, RV, NRSV, REB

The Hebrew Scriptures

Genesis
Exodus
Leviticus
Numbers
Deuteronomy
Joshua
Judges
Ruth
1 Samuel (1 Kingdoms in LXX)
2 Samuel (2 ″ ″ ″)
1 Kings (3 ″ ″ ″)
2 Kings (4 ″ ″ ″)
1 Chronicles (1 Paralipomenon in LXX)
2 Chronicles (2 ″ ″ ″)
Ezra } (= 2 Esdras in LXX)
Nehemiah }
Esther
Job
Psalms
Proverbs
Ecclesiastes
Song of Solomon (or Song of Songs)
Isaiah
Jeremiah
Lamentations
Ezekiel
Daniel
Hosea
Joel
Amos
Obadiah
Jonah
Micah
Nahum
Habakkuk
Zephaniah
Haggai
Zechariah
Malachi

The Apocrypha/Deuterocanonical Books

Tobit
Judith
The Additions to the Book of Esther
The Wisdom of Solomon
Ecclesiasticus
Baruch
The Letter of Jeremiah (= Baruch ch. 6)
The Additions to the Book of Daniel
    The Prayer of Azariah and the Song of the Three Children
    Susanna
    Bel and the Dragon
1 Maccabees
2 Maccabees

Books in the Greek and Slavonic Bibles but not in the Roman Catholic Canon

(These four books, together with 4 Maccabees, and the Deutero canonical books, constitute the canon generally in use in Orthodox Christianity. They are approved, e.g., by the Holy Synod of the Greek Orthodox Church. All are included in NRSV.)

1 Esdras (= 2 Esdras in Slavonic and the Vulgate appendix)
Prayer of Manasseh (in Vulgate appendix)
Psalm 151
3 Maccabees

Books in the Slavonic Bible and the Vulgate Appendix

2 Esdras (= 3 Esdras in Slavonic; 4 Esdras in Vulgate)
(NB 1 and 2 Esdras in Vulgate = Ezra–Nehemiah, above.
3 Esdras in Vulgate is the Old Latin version of
1 Esdras.)

In Appendix to LXX

4 Maccabees

The New Testament Books

Matthew
Mark
Luke
John
The Acts of the Apostles
Romans
1 Corinthians
2 Corinthians
Galatians
Ephesians
Philippians
Colossians
1 Thessalonians
2 Thessalonians
1 Timothy
2 Timothy
Titus
Philemon
Hebrews
James
1 Peter
2 Peter
1 John
2 John
3 John
Jude
Revelation

THE BOOKS OF THE BIBLE
in the order of Catholic editions (e.g. NJB)

The deuterocanonical books are inserted into appropriate places within the OT canon. After Ezra–Nehemiah are placed:

Tobit
Judith
Esther
1 Maccabees
2 Maccabees

The Book of Wisdom and Ecclesiasticus follow the Song of Songs; Baruch is placed after Lamentations.

A The symbol employed in textual criticism for a MS of the New Testament in Greek written on parchment in Egypt (Alexandria? Hence known as the Codex Alexandrinus) in the 5th cent. A few leaves containing verses of Matthew, John, and 2 Cor. are missing, but the two epistles of *Clement are added at the end. There are 820 leaves. The MS was presented by the patriarch of Constantinople to the king of England in 1627 and is in the British Library in London.

Aaron Brother of Moses and *Miriam, of the tribe of Levi. Over the centuries Aaron's status in Israel's memory was allowed to grow. In the earliest tradition, Aaron must put into words what is the will of God (Exod. 4: 16) but he has no priestly functions. The account of Aaron collecting gold rings to be melted down in fire (Exod. 32: 24) seems to reflect what happened at the northern sanctuaries, with calves of gold, established by *Jeroboam (1 Kgs. 12: 28) in 922 BCE. In both passages idolatrous calf worship indicates independence of leadership (cf. Exod. 24: 14). But usually Aaron is subordinate to Moses: the uplifted hands of Moses brought victory over the Amalekites, and the role of Aaron (and Hur) was merely to hold Moses' hands up when he got tired. In the probably post-exilic (i.e. compiled after 500 BCE) source *P, Aaron becomes the ancestor of the Jerusalem priesthood who sacrifice and bless and exercise supreme authority (Num. 27: 21—Eleazar is a son of Aaron, 1 Chr. 24: 1) and officiates on the Day of *Atonement (Lev. 16: 32–4). Finally, in the tradition, Aaron is exalted *above* Moses (Ecclus. [= Sir.] 45: 20), which reflects the post-exilic situation of the division of a priesthood among twenty-four families who each performed Temple duty for a week; sixteen families claimed descent from Zadok (Eleazar's elder son) and eight from Ithamar (1 Chron. 24: 1–19).

*John the Baptist was said to be from a family of priests (Luke 1: 5).

Abaddon In Hebrew = destruction, and in Job 26: 6 and Prov. 27: 20 denotes the abode of the dead, also called **sheol* or *Hades. In Rev. 9: 11 it is personified to mean the destroyer of all life and peace.

Abba An Aramaic word for 'father'. Jesus used it in prayer in *Gethsemane (Mark 14: 36), and it is twice taken up by Paul as being an address used by Christians in prayer to God. It has been held that *Abba* represents an intimate form of speaking, especially by children speaking to fathers and corresponding to English 'Daddy'. However, the evidence is rather that *Abba* was equally an adult form of speech and was probably not the only term that Jesus used. Paul shows that the Gentile churches sometimes prayed *Abba* but this does not necessarily mean they were quoting Jesus' prayer; the Gentile Churches also prayed **Maranatha*—Aramaic for 'Lord, come!'—which was clearly not part of Jesus' prayer. Both *Abba* and *Maranatha* certainly do represent survivals into Paul's world of the early Aramaic speech of the Palestinian Church.

Abednego In Hebrew = servant of Nego, a corrupted form of 'Nebo'; it was the name given by King *Nebuchadnezzar to Azariah in Dan. 1: 7, one of Daniel's companions cast into the burning fiery furnace.

Abel In Gen. 4, the second son of *Adam and *Eve, whose offering to God was accepted, thereby incurring the wrath of his brother *Cain, who failed to win the divine approval and in jealousy murdered Abel. He is praised in Heb. 11: 4 for his faith which is a continuing encouragement.

Abiathar A priest in the army of King David (2 Sam. 20: 25), who later supported *Adonijah's unsuccessful bid for the throne against *Solomon. The king (Solomon) banished Abiathar to his estate at Anathoth (1 Kgs. 2: 26–7). *Jeremiah (Jer. 1: 1) may have been a descendant. Mark 2: 26 mistakenly regards Abiathar as already a priest in the time of *David's flights from *Saul (1 Sam. 21: 1–6) instead of his father *Ahimelech.

Abigail Wife of Nabal; she interceded with *David to save her husband's life (1 Sam. 25: 23–31). Nabal was so shocked by the successful intervention that ten days later he died. David then married the 'slender and beautiful' Abigail.

Abijah (or **Abijam**) The best-known of four OT characters of this name was the son and successor of *Rehoboam. He was king of Judah (911–908 BCE), after the disruption of the state into the separate kingdoms of Israel and *Judah. The editor of the book of Kings has no good word for him (1 Kgs. 15: 3) but the revisionist history in the book of Chronicles (2 Chr. 13: 19) records military successes against the apostate *Jeroboam. He 'grew strong' (13: 21)—and took fourteen wives.

Abilene An area NW of *Damascus round Abila. Luke 3: 1 mentions that when *John the Baptist began his ministry Abilene was governed by *Lysanias. An inscription, dated between 15 and 30 CE, at Abilene, refers to 'Lysanias the tetrarch'.

Abimelech Son of *Gideon (or Jerubbaal) and a concubine from *Shechem. Upon the death of Gideon, Abimelech got himself proclaimed king at Shechem (Judg. 9), contrary to the tradition in Israel which held that Yahweh alone was king. Before long the town rejected Abimelech but by way of reprisal he destroyed it; there is archaeological evidence of the destruction of Shechem between 1150 and 1125 BCE. He was killed while besieging *Thebez by a woman who heaved a millstone and broke his skull (Judg. 9: 53).

Abinadab (1) Man at whose house on a hill in Kiriath-jearim the ark of the *covenant was lodged after its return to Israel by the *Philistines; it remained there for twenty years (1 Sam. 7: 2) until *David brought it first to the house of Obed-edom (1 Chr. 13: 13) and later into Jerusalem itself. (2) The second son of Jesse also had that name (1 Sam. 16: 8).

Abishai A loyal soldier in *David's band who had to be restrained from killing *Saul when he was resting from the pursuit of his young rival (1 Sam. 26: 9). Abishai was later implicated with *Joab (2 Sam. 3: 30) in the tit-for-tat murder of *Abner, who had killed their brother Asahel (2 Sam. 2: 23); the biblical text claims that David had no part in this act of vengeance (2 Sam. 3: 37).

Abner A cousin of King *Saul, in whose army he was the commander (1 Sam. 14: 50). In the war 'between the house of Saul and the house of *David' (2 Sam. 3: 6) Abner continued with the forces of Saul's son for a time but later transferred his allegiance to David, who welcomed his defection and was dismayed and angered by his general *Joab's private feud (2 Sam. 3: 31); at *Hebron Joab had stabbed Abner in the stomach (2 Sam. 3: 27).

abomination Frequently in AV, and sometimes in NRSV, it describes an action or article incompatible with the true religion of Israel, as in Lev. 7: 18, or 1 Kgs. 11: 5 where REB prefers 'loathsome god'. In Deut. 27: 15 the word is rendered 'anything abhorrent to the Lord' by NRSV and 'anything abominable to the Lord' by REB and 'a thing detestable to Yahweh' by NJB. So the word embraces foreign gods, improper behaviour, and unclean foods.

Abomination of Desolation Used by AV, REB for '*the desolating sacrilege' of NRSV at Mark 13: 14 etc.

abortion Not an ethical issue in the Bible, for maintaining population growth was a Jewish priority.

Abraham Also known as Abram; first of the patriarchs of Israel. His kind of life, depicted in Gen. 11–25, might be that of about 2000 to 1500 BCE, but the Abraham of the stories is an individual merchant from the *Hittite realm typical of a people who migrated from the urban civilization of *Ur into Palestine. So also, the account of Abraham's agreement with *Lot (Gen. 13: 8) about their respective choice of land (Lot settled for the valley of Sodom) is meant to explain the origin of the *Moabites. This story is thought by some scholars to reflect the historical fact that there was a change of occupation when the decline in trade necessitated entry into stock-breeding. But throughout the events recorded of Abraham there runs the theological motif of God's promise of the land of *Canaan, and of Abraham's faith in response. This faith was put to the test when he was ordered to sacrifice his son *Isaac, who was essential for the fulfilment of the promise; but at the last moment a divine intervention saved Isaac and a ram trapped in an adjacent pile of scrub was offered instead. The Abraham patriarchal history ends with his buying a plot of ground for his burial and with his death (Gen. 25: 10). All three strands in Gen. contribute to the saga. For *J, he is a father of all nations; for *E, a model of faith; for *P, the pledge of the nation's survival. After Gen., Abraham is only mentioned in the later parts of the OT. But in the NT Abraham is venerated as a pioneer in the OT of trust in God (Rom. 4; Heb. 7: 1–10 etc.), second only to Moses.

Abraham was acceptable to God in virtue of his faith long before the *Law was given (Gal. 3: 7), so that his true sons are not the Jews who claim descent from Abraham (Matt. 3: 9, John 8: 33) and live by the Law, but Christians who live by faith in Christ.

Abraham's Bosom In Jesus' parable of *Dives and *Lazarus (Luke 16: 19–31) both characters die, and in the abode of the dead, which is divided into compartments separated by a chasm, Lazarus is imagined to be enjoying a place of honour with the patriarch Abraham. As is clear from the Beloved Disciple's position 'in the bosom' (AV) of Jesus at the *Last Supper (John 13: 23), this was where the chief guest would sit at a meal.

Absalom A son of *David; good-looking and ambitious; he gathered round himself a band of disaffected people who were prepared to overthrow David. He set up headquarters in *Hebron and marched towards Jerusalem, which David hastily abandoned. Across the River *Jordan, David arranged for secret intelligence from the capital through Hushai (2 Sam. 15: 34–5), who gave misleading advice to Absalom. This allowed time for David to regroup his forces which were still loyal, and Absalom was defeated. Explicit instructions to spare the life of David's son were ignored by *Joab, and Absalom, who had become entangled in an oak tree, was stabbed in the heart. David's lament for his rebellious son is one of the most poignant passages in the OT (2 Sam. 18: 33).

abyss The depths of the sea. Because the *Hebrews disliked the sea (cf. Rev. 21: 1), deep waters were regarded as the abode of *demons. It was the appropriate destiny for the Gadarene swine when the demons had entered into them (Luke 8: 31, 33), and from the abyss *antichrist will rise (Rev. 11: 7). In Rom. 10: 7 Paul refers to **sheol*, the abode of the dead, as 'the Abyss', which he seems to take from Ps. 71: 20 ('earth's watery depths', REB).

acceptance Abel's sacrificial offering of a lamb was acceptable to God, whereas God was not satisfied with *Cain's vegetarian offering. No reason for this discrimination is given in the narrative of Gen. 4. However, God 'will be gracious to whom he will be gracious and will show mercy on whom he will show mercy' (Exod. 33: 19). In the theology of Paul too God's choices are inscrutable (Rom. 9: 19–24), but those who are given faith are acceptable (Rom. 3: 9–25; 5: 17), and the practice of the early Church reflected the conviction that all are acceptable to God whatever their race or nationality (Acts 10: 34 f.).

access When visitors or ambassadors are granted an audience with the pope or a president, a trusted member of the staff shows them the way and the required etiquette and thus facilitates access to the presence. So it is by Christ (and 'by faith', NRSV marg., Rom. 5: 2) that believers have access to *grace—God's widespread kindness which welcomes both Jews and *Gentiles alike (Eph. 2: 18).

Aceldama (AV) Aramaic for 'Field of Blood' and rendered Hakeldama by NRSV; 'Aceldama which means "Blood Acre" ' by REB (Acts 1: 19): the field bought by *Judas Iscariot with the reward given to him by the priests for betraying Jesus and where he had a fatal accident. But according to Matt. 27: 7 the field, called the *Potter's Field, was bought by the priests with the money Judas returned, for use as a cemetery.

Possibly the latter story originated as an attempt to explain the adopted name of the place. Matt. is attempting to demonstrate that it has all been foretold, and so he uses a mangled sentence from Zech. (11: 13) and some isolated phrases in Jeremiah (18: 2 and 32: 9) as proof from prophecy.

Achaia The southern part of Greece made into a province by the Romans in 27 BCE. In the time (5th cent. BCE) of the historian Herodotus the name covered the area of Thessaly. When Paul visited the capital, *Corinth, the province was controlled by the Roman senate, who appointed Junius Gallio as the proconsul (Acts 18: 12).

Achan (or Achar) Son of Carmi (1 Chron. 2: 7). After the destruction of *Jericho (Josh. 6: 24) some of the booty reserved for the Lord was stolen by Achan; and when the Israelites suffered a reversal in failing to capture *Ai, it was suspected that someone had flouted the ban. A primitive process of investigation turned up Achan, and he and his family were stoned to death.

Achor The valley in which *Achan was stoned (Josh. 7: 25–6); the name in Hebrew = 'trouble'. But *Hosea (2: 15) prophesies that after deplorable apostasies the Lord will woo Israel back to its former allegiance: even the valley of Achor will become a door of hope.

Acts of the Apostles The fifth book of the NT. Without the Acts it would be impossible to write an account of the Christian Church of the first generation. With it, historians have interesting problems on their hands because of the apparent discrepancies both with the theology of Paul and with facts mentioned by Paul in his extant correspondence. And Paul is the central character in the second part of the book, as Peter is of the first.

Acts charts the course of events which took the gospel from Jerusalem, the capital of the Jewish world, and in the scriptures the place of revelation, to *Rome, the capital of the Empire. After a successful start with preaching to Jews (Acts 6: 7), the gospel moves out under the guidance of the *Spirit, ever wider, first to heterodox *Judaism (*Samaria, Acts 8: 5), to *proselytes (an Ethiopian *eunuch, 8: 27), God-fearing *Gentiles (a *centurion, 10: 2), and finally Gentile pagans (13: 46). From then on, the Christian mission goes on by land and by sea to Rome, not without internal dissent and external opposition; but, when Paul and his companions do finally reach Rome, the book surprisingly comes to an end (28: 16 ff.).

References to the Acts in early Christian literature regard it as the work of *Luke, author also of the third gospel. This tradition is unanimous from *Irenaeus (180 CE) onwards, and it is confirmed by evidence in the NT itself. Not only do both gospel and Acts share the same Greek style, they also share the dedication to *Theophilus (Luke 1: 3; Acts 1: 1). It also seems that the author slips into the first person plural ('we' and 'us') in four sections of Acts (16: 10–17; 20: 5–15; 21: 1–18; 27: 1–28: 16) which on the face of it looks as though he was himself present at those stages of the narrative, and this could fit in with the references to Luke by Paul himself (Col. 4: 14, Philem. 23–4; and cf. 2 Tim. 4: 11). If the writer was indeed recalling his own experiences in much of Acts, the book enjoys extra historical credibility.

But was he? The same hand that wrote the 'we-sections' seems to have been responsible for the remainder of the work, which is where the problems lie; for example, the Paul of Acts speaks much like Peter; the speeches are artificial compositions (like those of the Greek historian Thucydides (400 BCE), on his own admission); then, if the author of Acts was a companion of Paul, how strange that he seems unaware that Paul was a great writer of letters. He seems ignorant of Paul's characteristic doctrines—only one brief (and inaccurate) allusion to justification. The settlement agreed at Jerusalem, according to Acts 15, on the terms by which Gentiles might be admitted to the Church, is never quoted by Paul and the picture of harmony between Peter and Paul is different from the bitterness against Peter in Gal. 2: 11; there is also a savage invective possibly against the Pharisees (Phil. 3: 2 followed by 3: 5) which strangely contrasts with Paul's claim to be a Pharisee in Acts 23: 6. If Paul's account of the Jerusalem conversations in Gal. 2 refers to the conference described in Acts 15, then the latter must be a fabrication by the author of Acts, since the differences are obvious, and it is implausible that the Jewish Christian James should have misquoted the Hebrew of Amos 9: 11 in order to favour the mission to the Gentiles. The case against the author being a companion of Paul is formidable, and is all the greater if any weight is given to inconsistencies in the narrative of Acts (e.g. as between the three accounts of Paul's experience on the road to *Damascus) and neither the gospel of Luke nor the Acts contain any medical vocabulary that might have been expected from a physician (Col. 4: 14). Nevertheless, these arguments are not totally persuasive, provided that the different purposes of each author are taken into account. Paul was an evangelist writing in the white heat of controversy and using autobiographical details to support his conviction, whereas Acts comes from a later generation (perhaps about 80 CE) when the conflict is largely resolved; the author wishes to emphasize the unity of the Church and the evil of schism.

Now, given that the external evidence is unanimous for Luke as the author, can the book itself support this view?

1. Certainly if there was nothing in the tradition it is unlikely that the name of Luke would ever have come up. *Titus as a trusted spokesman for Paul (2 Cor. 12: 18) would have been a more likely choice.

2. Then the speeches; it is possible, as it was for Thucydides (he himself said so), that when he had reliable reports he used them.

3. As to Paul's claim to Jewishness in Acts, such as boasting of his Pharisaic upbringing and his support for men taking a Jewish vow (Acts 21: 24), this could be justified by his principle of becoming 'all things to all people' (1 Cor. 9: 22) and is not necessarily fictitious.

4. It is also held by some scholars that the visit to Jerusalem described by Paul as his second (Gal. 2) is also that mentioned in the Acts (9: 26) and that the visits of Acts 11: 30 and 15: 4 describe the same event, which took place after the epistle to the Galatians was written. At the conference of Acts 15 Paul was willing to accept Jewish food regulations for the sake of preserving the unity of the Church provided that he was authorized to continue his mission to the Gentiles. Doubtless Luke has made the most of this occasion in order to enhance the importance of Church unity, just as he records the apostles' laying on of hands to mark other great turning points in the expansion of the Church.

5. By ch. 15 the Church is theologically complete because all four types of mankind (Jews, Samaritans, proselytes, Gentiles) are within it, and now comes the geographical extension. Jerusalem, the place of revelation, is where Paul starts and finishes his long journey (15: 30–21: 15), all told in Acts with remarkable accuracy about the titles of officials and the perils of a shipwreck, and enlivened for readers of the age with graphic stories of Peter's miraculous escape from prison and Paul's wonder-working handkerchief.

While the main purpose of the Acts is to describe the triumphant (but not unmolested) progress of the gospel, there is a secondary motive in that Paul is presented as no less an apostle than Peter, though only in Acts 14 is Paul explicitly called an apostle. The author is motivated by trying to fasten opposition to Christians on the Jews rather than the Romans, whose authorities are described as friendly or helpful (e.g. Acts 18: 15; 22: 29). The book may even have been published in Rome.

AD Latin, *anno domini*, 'in the year of the Lord', commonly used by Christians to indicate dates, from the supposed date of the birth of Jesus in the year 1. The years before his birth are designated 'BC', Before Christ. However, Jesus was born, according to Matt. and Luke, while Herod was reigning, and he died in 4 BCE. Jesus' birth is therefore reckoned to have taken place in 4 BCE or just before, and the incorrect date is due to imprecise calculations by a Scythian monk, *Dionysius, in the 6th cent. CE.

Adam Hebrew for 'man'; and the first of his kind according to the OT. There are two accounts of the creation in Genesis. In the first (1: 1–2: 4a, *P), man and woman are created in the image of God, which means that human beings (*Adam* in Hebrew is a collective noun, meaning 'humanity') have a responsibility before God, unlike the animals: men and women are created together (1: 27) and are complementary to each other, and each have a freedom to obey or disobey God's will. In the second story (2: 4b–25, *J), 'Adam' is a proper name and he is regarded as the chief of all created beings by being placed first in order; woman is created later, for man is incomplete without her.

All is peace in the garden until Adam and *Eve disobey God's command, and then begin pain and toil and death, and the original relationship with God is broken. The myth of Gen. 3 is known as the *Fall.

In the NT Adam and Eve are regarded as historical persons and the *genealogy of Jesus is traced back to Adam by Luke (3: 38). Paul contrasts the first Adam of Genesis who was disobedient with the 'last Adam', Christ, who was obedient. All mankind, who are in solidarity with Adam, are corrupt and sinful; but through *faith in Christ they receive life and grace and the hope of *resurrection (1 Cor. 15: 21–2). In Paul's argument the universal human experience of new life has been made possible by the events of recent history, namely the life, death, and resurrection of Jesus.

Adonijah Son of *David, and his expected successor, but instead *Solomon was secretly anointed by the priest *Zadok. Adonijah was executed (1 Kgs. 2: 25).

adoption In the OT children are brought by adoption into families, as Moses (Exod. 2: 10) was by Pharaoh's daughter, and Esther by *Mordecai (Esth. 2: 7, 15). The whole nation was regarded by Jeremiah (3: 19) as adopted by Yahweh; it was the special relationship of *election of Israel, the chosen nation (Hos. 11: 1).

In the NT Christians have 'the spirit of adoption' (Rom. 8: 15) which enables them to say '**Abba*, Father'. For those who do the will of the Father are Jesus' brothers and sisters (Mark 3: 35). It is a matter for Paul of *grace, not of nature or of *law, though the legal process of adoption was well established in the Roman Empire and an adopted son became a full member of the adoptive family.

adoptionism A theory about the Person of Christ associated with the heretic Nestorius (d. 451 CE) that Jesus was a man gifted with divine powers. It was a view held by the *Ebionites and it has sometimes been suggested that the quotation of Ps. 2: 7 and Isa. 42: 1 ('You are my Son, the Beloved; with you I am well pleased') in Mark 1: 11 is a coronation or adoption formula: Jesus is at that moment being 'adopted' as the divine Son. However, there seems to be no implication here at all about Jesus' status before his *baptism; the words simply affirm God's act in Jesus; the beginning of the gospel is now and not at the end (the *Resurrection).

adoration The loving worship of God in response to his first loving us (1 John 4: 9–10). It is described as a 'sacrifice of praise to

God' (Heb. 13: 15) and is inconsistent with worship offered to any being less than God (Exod. 20: 2–4).

Adramyttium Port on the NW coast of modern Turkey. Paul was put on board a ship from Adramyttium at *Caesarea on his way to *Rome (Acts 27: 2).

Adria, Sea of In the 1st cent. CE the name was applied not only to the modern Adriatic (between Italy and the Balkan peninsula) but also southwards to the sea between the Peloponnesus in *Greece and Sicily. The ship carrying Paul and other prisoners drifted for a fortnight between *Crete and Sicily (Acts 27: 27).

Adullam A place in the hills of *Judah. When *David fled from *Saul he hid in a cave nearby and there gathered a supportive force of malcontents (1 Sam. 22: 1–2). Hence the term Adullamites applied by John Bright to MPs who refused to support Gladstone's Reform Bill in 1866 and deserted the Liberal Party.

adultery The seventh (or sixth, according to the Jewish reckoning, followed in Catholic Bibles) commandment (Exod. 20: 14) forbade adultery, and the prescribed penalty was death (Deut. 22: 22). The reference is not to any sort of extramarital sex but to that of a man having sex with a married woman, who is regarded as some other man's chattel. In the case of a betrothed virgin, she is to be stoned together with the man (Deut. 22: 23); but, if she is raped, the man alone is to be executed (Deut. 22: 25). Whether these ferocious penalties were ever carried out is uncertain; King *David at any rate suffered only prophetic rebuke (2 Sam. 11: 4; 12: 9). The episode recorded in some MSS of John 8: 1–11 has a group of scribes and *Pharisees testing Jesus' attitude to the *Law: was he in favour of the full rigour of the Law being applied to a proven adulteress? The story may have been incorporated by chance into these MSS of John from some non-canonical work and though it may not be genuine reminiscence it has the authenticity of Jesus' compassionate teaching and his encouragement to start life anew (cf. Luke 13: 1–5). The OT condemnation is stretched even to mental attitudes in the *Sermon on the Mount (Matt. 5: 28). Paul repeats the OT law (Rom. 13: 9), and elsewhere (e.g. Gal. 5: 19) it is presumably condemned under the general heading of 'works of the *flesh'.

advent An alternative translation of the Greek **parousia*, otherwise rendered 'coming' in the English versions, with reference to the eschatological expectation of Christ's coming as Judge of the world at the *End (Matt. 24: 39). Later, the Church regarded the life of Jesus in the *incarnation as the 'First Advent' and the 'Second Advent' was to be in the future. The Church season of Advent then developed (first with six *Sundays, later with four) into a time of preparation for *Christmas, and for the Last *Judgement.

adversary In Hebrew, '*Satan', or 'the Accuser' (Job 1: 6, NRSV marg.); his role is to find out about human frailties and provoke the innocent to sin. The sons of Zeruiah became an adversary to *David (2 Sam. 19: 22) in opposing the will of God. In the NT Satan is the adversary *par excellence* (1 Pet. 5: 8).

advocate A translation of the Greek '*Paraclete', used in the gospel and first epistle of John for the *Spirit (e.g. John 14: 16, NRSV, REB) who is to continue within the Church the teaching of Jesus—insights that were only possible to believers after the *Resurrection (John 16: 13). The term is derived from the lawcourts and means the counsel for the defence, who stand alongside the accused, and to that extent the AV 'comforter' (John 14: 16) conveys part of the meaning.

Aeneas The hero of the *Aeneid*, the epic poem in *Latin by Virgil. The name was indeed well known in the 1st cent. and not surprisingly therefore appears in the NT, where he is a paralytic at *Lydda cured by *Peter (Acts 9: 33–5).

Aenon A place near *Shechem in *Samaria, though not identified with certainty. When Jesus and his *disciples began a ministry of *baptism, John moved on to Aenon (John 3: 22–3). John was not yet in *prison (3: 24), whereas Mark (1: 14) appears to suggest that Jesus began his *teaching only after John's arrest.

aetiology An aetiological story professes to explain causes. The *Pentateuch, especially Genesis, is rich in stories designed to give explanations about the origin of natural phenomena and of customs in religious worship. How did the *rainbow come into existence, and what is its meaning? Gen. 9: 13 supplies the answer: it is a sign of the *covenant between God and his *creation that there shall never again be a universal *flood. Such a story is called aetiological, from the Greek word *aitia*, meaning 'a cause', as in Acts 28: 20. An example of a cultic aetiology is the account of the *Passover which answers the question in Exod. 12: 26. And if people asked why *Bethel was held to be a sacred place, the story of *Jacob (or Israel) having a *dream there (Gen. 28: 10 ff.) supplied an answer. While many aetiologies are *legends, some may be incorporated into narratives which have a basis in history, as in Josh. 4: 9, describing the existence of twelve stones in the *Jordan.

Africa Very little of the vast continent was known to writers of the Bible, but *Egypt and *Ethiopia are prominent in the narratives from Genesis onward and fundamental to the history of Israel. The *Exodus from Egypt is a vital part of Hebrew tradition, even though much of the biblical record is disputed by modern scholarship. In the NT there is an account of the infant Jesus being taken to Egypt out of range of *Herod's jealousy (Matt. 2: 13) and the cross of Jesus is borne by an African (Mark 15: 21), whose sons appear to be known to Mark's Church. *Apollos came to *Ephesus from *Alexandria (Acts 18: 24), and a Church was established early at Alexandria and in due course provided some of the greatest Christian scholars such as Clement and *Origen.

Agabus One of several Christian *prophets resident in Jerusalem who visited the Church at *Antioch and predicted a severe *famine (Acts 11: 28) during the reign of the emperor *Claudius (possibly 46 CE). Agabus also prophesied Paul's imminent *imprisonment (Acts 21: 10).

Agag King of the *Amalekites, whose life was spared by *Saul after Israel's victory. For this act of mercy he was denounced by the prophet Samuel, who then himself hacked Agag in pieces (1 Sam. 15).

agape Greek noun meaning '*love' not much used in secular writings but common in the NT for the gracious self-giving love of God shown in Christ; and correspondingly of unselfish human love (1 Cor. 13: 4).

The word came later to denote the shared *meal held by Christians in conjunction with the *Eucharist. It was the occasion for the exercise of mutual confidence and it would seem that 1 Cor. 11: 17–24 is a description of the *agape* in embryo and into which abuses such as over-indulgence in drink were creeping. By the time of Jude 12 abuses have worsened, and in the 4th cent. regulations to control the love-feasts were necessary. By the 8th cent. the *agape*, as a fellowship meal, was defunct.

age Human life was expected to end after seventy years, and anyone living for eighty years must have unusual strength (Ps. 90: 10). The days of Methuselah (969 years, Gen. 5: 27) are typical of the Hebrew belief that in the past people enjoyed an immense longevity. More realistically, the elderly are known to lack vigour (Job 30: 2), though their accumulated *wisdom was proverbial (Job 12: 12) and accounted for the use of 'elders' as an official title (Exod. 3: 16).

'Age' is also used to translate the Greek *aeon*, denoting a period of time—past, present, or future (1 Cor. 10: 11; Matt. 12: 32): thus, there are cares in the present age (Matt.

13: 22) and riches in the ages to come (Eph. 2: 7). It has sometimes been maintained that in the NT, especially in Luke–Acts, time since the *creation is regarded as divided into three ages—up to the coming of Jesus, the life and death of Jesus, and the continuing epoch of the Church.

agony From the Greek for a 'contest', meaning the nervous feeling experienced before such a contest. It is in that sense that Jesus was in an agony in the garden of *Gethsemane (Luke 22: 44).

agora The open space in the centre of a Greek city (Acts 17: 17), used for meetings, business, and political speeches.

agrapha A Greek word (in the plural) for sayings of Jesus which are not written in the gospels or found sometimes only in inferior MSS and not regarded as authentic parts of the text. The best-known example is the insertion by codex *D at Luke 6: 5; Jesus said to a man working on the sabbath, 'Man, if you know what you are doing you are blessed; but if you do not know, you are a transgressor of the law.' There are *agrapha* in the *Oxyrhynchus papyri and the *Gnostic documents discovered at *Nag Hammadi and in some writings of the Church Fathers. Since sayings of Jesus were in the first Christian generation passed on orally, it is thought that about a dozen of the *agrapha* may be authentic reminiscences which were not used by the four *evangelists.

agriculture From Genesis (2: 15) to Revelation (14: 14–20) the Bible is dominated by farmers and their endless labour as they struggled to feed the Palestinian population despite droughts (Jer. 14: 1–6), their main enemy, and pests and armed intruders. Nevertheless they did produce wheat, grapes, olives, and other crops and maintained herds of cattle and flocks of sheep. So vital was agriculture to the nation's survival that there were regulations governing it in the Law (Deut. 22: 9–10) and there were three religious festivals (the feast of Weeks at the time of first fruits and the feast of *Booths (*Tabernacles) at the harvest of fruits, and *Passover at the beginning of the corn harvest) associated with it. The processes of agriculture were frequently used in Jesus' *parables. Ploughing was done after the autumn rains, and then the seed was sown by hand. Harvesting took place in the months April to June. Olives and grapes were gathered in the early autumn and this was marked by the feast of *Tabernacles. This originally agricultural festival was provided with a historical interpretation by the *Holiness Code (Lev. 23: 43) and was said to commemorate the Israelites' wanderings in the desert when they possessed no permanent houses. (But surely 'booths' of greenery in the wilderness stretches the historical imagination.) After threshing and winnowing the grain had to be stored (Luke 12: 18) and the chaff could be burnt (Matt. 3: 12). Flax was available for clothes.

Agrippa (1) Herod Agrippa I, grandson of Herod the Great; the name 'Agrippa' was assumed on account of a friendship with Marcus Vipsanius Agrippa, son-in-law of the emperor *Augustus. In 37 CE he was given the former tetrarchies of *Philip and *Antipas, by the emperor Caligula, and in 41 the next emperor, *Claudius, added *Judaea and *Samaria, with the title 'king'. . He died in agony at *Caesarea in 44 CE. During his short reign (41–4 CE) he was able to impose the death penalty; James, the son of Zebedee, was executed (Acts 12: 2). (2) Herod Agrippa II succeeded his father at the age of 17 and from 50 CE assumed an increasing authority over territories in *Palestine; the emperor *Nero added parts of *Galilee and *Perea. Agrippa lived with his sister *Bernice after the death of her husband, and this notoriously incestuous relationship gives a piquant note to the scene of Acts 25: 13–23 when Paul as a prisoner testified before them after his appeal to *Caesar. Agrippa improved the amenities of Jerusalem and helped to complete the building of the Temple shortly before the revolt of 66–70 CE, when Agrippa took the side of the invading Romans. He died in *Rome in 93 CE.

Ahab King of Israel (*c*.869–850 BCE), son of *Omri and husband of *Jezebel, a Sidonian princess. He defeated the armies of *Damascus and came into conflict with *Assyria. The *Deuteronomic historian recounts his running battle with the prophet *Elijah over Jezebel's inclination to *Baal worship; the prophets of Baal were routed in a spectacular contest on Mount *Carmel (1 Kgs. 18). Ahab was also humiliated by Elijah for having concurred in the execution, by order of Jezebel, of *Naboth, whose vineyard Ahab wished to appropriate. For the biblical writer Ahab was the archetypal villain but on other criteria might be judged an able and successful ruler.

Ahasuerus Probably to be identified with *Xerxes I (486–465 BCE), who is mentioned by the Greek historian Herodotus. As portrayed in the book of Esther, Ahasuerus oscillated between the threat of a holocaust at the instigation of *Haman, followed by protection of the Jews through the intervention of his Jewish queen Esther. The story is fictitious and written to provide an account of the origin of the feast of *Purim; the book contains no references to the known historical events of the reign of Xerxes.

Ahaz King of *Judah, 735–715 BCE. The Assyrians were dominant, and *Syria and *Ephraim resolved to arrest their eastward expansion and tried to force Judah to join their coalition but Ahaz refused; Rezin of Syria and Pekah of Israel determined to replace Ahaz with Tabeel's son (2 Kgs. 16: 5; Isa. 7: 1–25). Isaiah supported Ahaz, but did not approve of his appeal to *Tiglath-Pileser to rescue him from the Syro-Ephraimite conspirators. When *Damascus, the capital of Syria, fell, Ahaz hastened to meet Tiglath-Pileser there. While he was in that city, Ahaz was attracted to a heathen *altar and had a replica installed in Jerusalem.

Ahimelech Priest at *Nob (1 Sam. 21: 1–6). During his flight from *Saul, *David, desperate for food, was given the *bread of the presence by Ahimelech. The gesture is used by Jesus (Matt. 12: 3–4) in defence of the *disciples who satisfied their hunger by plucking heads of corn on a *Sabbath.

Ahitophel A friend of *David who defected to *Absalom (2 Sam. 15: 12). When Absalom rejected his advice, he committed suicide.

Ai A small town or village captured by *Joshua probably about 1125 BCE; but the scale of the operation is much exaggerated in the account of Josh. 8 and seems to be an attempt to extol the military prowess of the Hebrews. It is also suggested, in the absence of any undisputed archaeological confirmation, that the account was compiled as an explanation of a well-known ruin—which is the meaning of the word in Hebrew—and has no historical basis.

alabaster Material used for containers of perfume. The long neck of the flask was broken to release the liquid (Mark 14: 3).

Albright, W. F. (1891–1971) American OT scholar who exercised great influence, both as the founder of a 'school' of interpreters and in his claim that archaeological discoveries could be correlated with the biblical text.

alcohol Both in the OT and NT 'wine' refers to fermented grape juice and always had some alcohol content.

aleph The first letter of the Hebrew alphabet (א). Textual critics use it as the symbol for a 4th-cent. MS known as *Codex *Sinaiticus, which was discovered by Constantin von *Tischendorf in 1844 on a visit to the Monastery of St Catherine on Mount Sinai; he rescued it from a pile of documents to be burnt as rubbish. Most of it, including the whole of the NT, was obtained by the Tsar of Russia and in 1917 it passed to the Bolsheviks, who sold it to Britain in 1933. It is a beautiful MS, with four columns to the page and is now held in the British Library (London). Some additional leaves were found in 1980.

Alexander the Great (356–323 BCE) King of Macedon from 336 who conquered the

Persian Empire. *Greek culture then permeated the Mediterranean region and Greek became the international language, leading to the Greek translation of the OT (*LXX), and the writing of the NT, and the early Christian liturgies, in Greek. The Seleucid dynasty derived from Alexander (from 275 BCE), which explains the interest in Alexander shown in 1 Macc. 1: 1–7.

Alexandria A city in *Egypt near the delta of the *Nile with a population of a million in the 1st cent. CE, including a large Jewish area. *Philo is the best-known Jewish scholar of the city. It is not known who established a Christian Church in the city (a legend attributes it to Mark) but Acts 18: 24 mentions that it had been the birthplace of *Apollos. Possibly the letter to the Hebrews was written from Alexandria; there are similarities in outlook and attitude to Philo.

alienation The broken relationship between God and mankind by reason of sin. Paul argues that the alienation is healed from the side of God by Jesus' self-emptying (Phil. 2: 7) and that by his *death Jesus effected *reconciliation (Col. 1: 21–2). It is also the main theme of the letter to the Hebrews (e.g. 2: 9).

aliens Aliens, or strangers sojourning in the land, were supposed to be treated generously (Exod. 22: 21; 23: 9). But neighbours across the borders were potential enemies. Antipathy to these states was increased by the reforms of *Josiah (621 BCE); religious practices tainted by beliefs of alien nations were purged and intermarriage with non-Israelites was prohibited (Deut. 7: 3). This nationalism was intensified for some by the experience of the Babylonian *Exile, but on the whole *Judaism of the Return integrated with a variety of alien cultures which only a minority resisted.

allegory A method of *interpretation or exposition where the words contain a secondary meaning, other than the straightforward one. John Bunyan's *Pilgrim's Progress* is an allegory. In the OT, Isa. (5: 1–6) has an allegory of the vine. In Gal. 4: 24 Paul uses the account in Gen. 16 and 17 of *Abraham's two sons in support of the opinion (which he already held) that *Isaac, the son of a free woman, *Sarah, represented the spiritual descendants of Abraham, whereas *Ishmael, the son of a bond woman, *Hagar, stands for those whose relation to Abraham is the inferior relationship of natural descent.

He means that the Genesis text bears a meaning over and above the literal, historical, meaning: Isaac stands for Christians, who came into a relationship with God as free sons. They have a faith like Abraham's: Isaac was born 'through promise'. Hagar bore children with the status of slaves—meaning Jews in bondage to the Law. The two women of Genesis are identified by Paul with two different covenants—the old and the new—and with two different Jerusalems—one present and another above, heavenly. Paul quotes Isaiah 54: 1 in Gal. 4: 27 about Jerusalem as it was before and after the Exile in Babylon: Sarah, who suffered the ordeal of being barren, will have more reason to rejoice than she who had a husband (Hagar). New Jerusalem, through the Church, offers so much more than the old: freedom, instead of slavery, the gospel instead of the Law.

At Alexandria *Philo used the allegorical method in order to diminish references and notions in the OT which would be offensive to pagans, and *Origen in Alexandria (*c.*200 CE) continued the method in the interests of Christianity. Beneath uncongenial details of ritual and history, Origen found timeless truths. By suggesting that the OT had layers of meaning beyond the literal, he made it acceptable and affirmed the unity of both OT and NT, against such as *Marcion.

Within the gospels, some of the *parables are already given an allegorical interpretation in which the details of the parable are said to contain a deeper meaning, as with the Sower (Mark 4: 3–8, explained by 4: 14–20). For those who incline to a rigorous view that Jesus' parables had invariably only a single meaning, such allegorizing is evidence of a development within the community after the time of Jesus. This is not necessarily illegitimate: it is open to readers of different cultures and generations to give their own interpretations of the texts. The original author does not enjoy a kind of copyright of interpretation. He does but send it on its way, and it then acquires new meanings, though not fanciful or arbitrary ones. However, many scholars today accept that Jesus himself probably used some allegory in his parables.

Taking Paul's distinction (2 Cor. 3: 6) between 'the letter and the spirit', Christians interpreted in an allegorical way those provisions in the Law which they no longer observed, such as the whole system of ritual purity; by this means they both respected the authorship of their received scriptures (the OT) and also made them relevant to their own faith and practice. Christian Fathers after Origen were fond of the allegorical method and applied it to the NT: Augustine regarded all the details in the parable of the Good Samaritan as having 'deeper' meanings; thus, the inn (Luke 10: 34) which provided succour designates the Church.

alleluia Sung by the multitude in *heaven ('Hallelujah', NRSV, REB, Rev. 19: 1 etc.), meaning 'Praise the Lord', as in Ps. 135: 3 etc. The cry was incorporated into Christian *worship and, as described by Augustine (d. 430 CE, the greatest of the early Fathers), formed the people's response to the Gradual sung after the first reading from scripture.

alliances In spite of the sense of a specially chosen people, foreign alliances were negotiated by *Solomon (1 Kgs. 5: 1–12) and *Ahab (1 Kgs. 16: 31). During Assyrian expansion both the northern and southern kingdoms made alliances to defend themselves, notwithstanding the opposition of the prophets (Isa. 20: 6; Jer. 2: 18) who urged the rulers to rely only on God. During the interval of independence under the *Maccabees, there was an alliance with *Rome (1 Macc. 8: 17).

almighty Frequently used in OT of God, especially in Job, and in the NT especially in Rev. *Paul uses the word once (= 'All-Sovereign', 2 Cor. 6: 18), where he quotes 2 Sam. 7: 14; interestingly, Paul adds to the quotation 'daughters' among the children of God, not merely 'sons'.

almond A fruit tree often mentioned in the OT and, because of the shape of the fruit, possibly originally associated with fertility rites. The white blossom is used in Eccles. 12: 5 as a symbol for old age.

alms In the Graeco-Roman world charity as the source of material support for the *poor and destitute was almost unknown. *Gifts of food and money were bestowed by the wealthy on those who could give something (e.g. political support) in return. But among the Hebrews there was an injunction encouraging compassion (Deut. 15: 11) in a concern for the helpless (Ps. 41: 1), especially *widows and orphans, and *Judaism developed this to the extent that righteousness was almost equated with almsgiving (Tob. 12: 9). The *synagogues had an organized system of aid to the poor and in the Temple there were six receptacles for receiving alms: these were in the shape of trumpets (cf. Mark 12: 41–2; Matt. 6: 2). Jesus condemned ostentatious almsgiving—as did some of the teachers among the *Pharisees.

The early Church appointed officers to administer charitable gifts (Acts 4: 32; 6: 3) and Paul urged his Churches to establish a proper practice of almsgiving for the impoverished Christians of Jerusalem (1 Cor. 16: 1 ff.). *Eternal life may depend on generosity and compassion (Matt. 25: 31–46).

aloes A fragrant spice (not a bitter plant) used as perfume (Ps. 45: 8 and S. of S. 4: 14) and on the linen wrapped round a corpse (John 19: 39).

alpha The first letter of the Greek alphabet; in Rev. 1: 8 God is called 'Alpha and *Omega' (the *last* letter of the Greek alphabet). The combination is used also of Jesus in Rev. 22: 13, meaning that Jesus was present from the beginning of *creation and will reign at the *End. Cf. Isa. 44: 6.

alphabet An alphabet was in use in the cities of *Palestine by 1200 BCE, including the Hebrew alphabet used in *Israel and *Judah. It was easier for traders than Egyptian hieroglyphs and Babylonian *cuneiform scripts, which were pictorial signs. The stele of Mesha, king of Moab, consisting of thirty-four lines, is an inscription from about 850 BCE in a script similar to that used for Hebrew. The Hebrews adopted the Phoenician alphabet of twenty-two consonants, but it was not until about 700 CE that Jewish scholars devised a system of vowel-signs, placed under the consonants, in order to preserve the tradition of pronunciation.

The Greek alphabet is ascribed by Herodotus to the Phoenician Cadmus and by the 9th cent. BCE there was a Greek alphabet with both consonants and vowels. There are twenty-one characters from *alpha to *omega.

Alphaeus In Matt. 10: 3 mentioned as the father of *James, after 'James the son of *Zebedee'. In Mark 2: 14 he is cited as the father of *Levi (usually identified as Matthew), and it is suggested that the mother of this James was *Mary (Mark 15: 40), whose husband was *Clopas (John 19: 25), which would mean that Clopas and Alphaeus are the same person; but this string of hazardous identifications beloved by the Fathers (e.g. Chrysostom) is dismissed by modern scholarship.

altar In the OT, a place of *sacrifice near which animals were slaughtered and on which oblations of corn, wine, and incense were burnt and offered, in the open air. The '*high places' (cf. e.g. 2 Kgs. 23: 5) were shrines with a natural kind of altar. *Solomon's Temple at Jerusalem had an altar in the courtyard overlaid with bronze (1 Kgs. 8: 64) and *Ahaz commissioned a larger altar with horns at the four corners. As places for sacrifice, altars were imbued with a sense of holiness where people would approach God and in some cases also seek refuge (1 Kgs. 2: 28). Altars might be of stone, either shaped or natural (Judg. 6: 19–23), or of earth (Exod. 20: 24), or of metal, as was the altar of incense (Exod. 30: 3). After the Jerusalem Temple was established as the national centre for worship, locally installed altars were officially no longer allowed.

In the NT there are eight references to an altar in the Temple or in the New Jerusalem according to the vision in Rev. (8: 5 etc.). The altar in the existing Temple at Jerusalem is referred to in Matt. 5: 23–4, for in Herod's Temple there was the new altar which the *Maccabees had restored (1 Macc. 4: 41 ff.) after the '*desolating sacrilege' of Antiochus Epiphanes had polluted the altar in 167 BCE. The reference in 1 Cor. 10: 21 to 'the Lord's table' is probably more a reference to the *Eucharist than to any structure used for its celebration, and in Heb. 13: 10 the writer claims that Christians have an altar 'from which those who officiate in the *tent (or *tabernacle) have no right to eat', which would appear to be a reference to the Eucharist contrasted with the sacrifices of *Judaism.

In the writings of the early Fathers of the Church structures on which the Eucharist was celebrated are sometimes called a 'table' and sometimes an 'altar'. They were made of wood, though stone altars were introduced when the Eucharist was celebrated at the tomb of a *martyr in the Roman catacombs in the 4th cent. CE.

Amalekites A nomadic people, whose ancestor was regarded by the Hebrews as *Esau (Gen. 36: 15–16), and who occupied part of the *Promised Land before the arrival of the *Israelites to dispossess them (Exod. 17: 8–16). They disappear from historical records after the time of *David (1 Sam. 15 and 30: 13–25).

amanuensis Latin for a secretary. Paul dictated letters to Tertius (Rom. 16: 22), who wrote on his behalf. It is sometimes held that Silvanus is responsible for the Greek style of 1 Peter (see 1 Pet. 5: 12).

Amarna Tell el- City on the eastern bank of the *Nile. For a time it was the capital of *Egypt, and in 1887 a cache of inscribed tablets was discovered which proved to be correspondence between Ikhnaton, the monotheist Pharaoh of Egypt (d. 1347 BCE), and rulers of neighbouring countries. They refer to the invasion of *Canaan by the *Habiru and several archaeologists argued attractively that these people were the Hebrews led by *Joshua of the biblical narratives. However, there are major problems which tell against any such identification.

Amaziah King of *Judah, 800–783 BCE. He slaughtered 10,000 *Edomites (2 Kgs. 14: 7) and his heart was so much lifted up by this feat that he rashly challenged Jehoash, king in *Samaria of the larger country Israel, and was humiliated. Later, he was murdered in *Lachish.

ambassadors 'Servants' (AV, NRSV), 'envoys' (REB), 'an embassy' (NJB) sent on a goodwill visit to another country (1 Kgs. 5: 1) or to make a formal protest (Judg. 11: 12). Paul describes himself as an ambassador for Christ (2 Cor. 5: 20; Eph. 6: 20), travelling abroad as his representative.

amen Hebrew, meaning 'certainly': it became a liturgical response by which worshippers identified themselves with the preceding *prayer of praise or petition (Ps. 41: 13; 1 Cor. 14: 16). The Hebrew has a connotation of *truth and steadfastness, and the AV translation is 'Verily, verily' (John 16: 23). The words often open a discourse of Jesus in the gospel of John and are an expression of his conviction of authority. The use in this gospel may be a reflection of Christian belief rather than a correct record of Jesus' own utterances. In Rev. 3: 4 Jesus is himself called 'The Amen'.

amethyst One of the precious stones in the high priest's *breastpiece (Exod. 28: 19) and in the walls of the New Jerusalem (Rev. 21: 20); red.

Ammonites Inhabitants of land east of the River *Jordan (= modern state of Jordan, whose capital is Amman): there was intermittent warfare between Israel and Ammon with oscillating fortunes. In spite of victories by *Saul (1 Sam. 11) and *David (2 Sam. 12) the Ammonites survived, and enjoyed a prosperous century (the 7th BCE), protected by the Assyrians.

Amon King of *Judah (642–640 BCE), father of Josiah. He was murdered apparently by opponents of his policy of appeasement with *Assyria, for which the Chronicler (2 Chr. 33: 23) sheds no tears.

Amorites Inhabitants of *Transjordan before the entry of the Israelites, perhaps settled by 1900 BCE. They resisted the newcomers but were expelled. Compiling the story in the 7th cent. BCE, the *Deuteronomist editor regards their expulsion as the proper reward for their *idolatry (Josh. 24: 18).

Amos A *prophet, but no professional in that role (Amos 7: 14). From *Judah in the south, he denounced evils in Israel, the northern kingdom, during the reign of *Jeroboam II (775–750 BCE). Coming from the countryside near *Bethlehem, Amos was appalled by the urban luxury he encountered (Amos 3: 15), and foresaw military catastrophe (Amos 6: 14) as the instrument of God's judgement. Earlier prophets (*Elijah, *Elisha) have their messages recorded in the historical books. Amos is the first prophet whose words are in his own book. His message of doom, rejected by the royal house (7: 10–17), is therefore reaffirmed in writing with a good deal of literary skill. The invasions under *Tiglath-Pileser III began in 734 BCE, not long after Amos has given his warning.

Amos, book of After the introduction (Amos 1 and 2), the book can be divided into the sermons (chs. 3–6) and the five visions (7: 1–9: 7), followed by the Promise (9: 8–15). It is particularly important for the rejection of the view that for each nation there was its own god—Yahweh being that of Israel, whose writ did not extend beyond the borders of the two kingdoms. On the contrary, the theme of Amos is that Yahweh is the universal God with moral demands on every nation (e.g. Amos 1: 5). Doubtless there was the special relationship between Yahweh and Israel, but the nation was warned that it would not be preserved by mere rituals. The priest *Amaziah did his utmost to silence this unwelcome message at *Bethel (7: 12), but Amos persisted. At any time the contract might be broken (Amos 5: 18); and yet a hope remained of ultimate restoration (9: 11–15); and even if this conclusion was added by an editor at a later date, it is how the book has been read within the context of the Bible as a whole by both Jews and Christians and as such it is not necessarily to be dismissed on the grounds of being inconsistent with the message of Amos in 750 BCE or treated as an unauthentic appendix. The book of Amos is quoted by *Stephen (Acts 7: 42) and by *James (Acts 15: 16–18), and in modern times has been much valued for its forthright appeal for social justice.

amphictyony A word borrowed from institutions in classical Greece and applied by some historians of Israel to its supposed organization before the monarchy as a confederation of twelve clans. It was suggested that there was a central shrine at which a cultic object was a shared responsibility among the twelve. But the amphictyony theory has now been generally abandoned.

Amphipolis A city and military post on the Egnatian Way. *Paul passed through it on his journey to *Thessalonica (Acts 17: 1). The Via Egnatia was constructed for rapid movement of soldiers moving from Italy to the east.

Amraphel The leader of an alliance of five kings who defeated a coalition of four rebels in the *Dead Sea area (Gen. 14). The identification, once popular, that this Amraphel was the famous *Hammurabi of *Babylon (1728–1686 BCE) is not tenable. The patriarch *Abraham is then brought by the narrative into this international conflict to rescue his nephew *Lot, who had been captured by the victorious Amraphel, and on his way home Abraham encountered *Melchizedek. Most scholars doubt whether Gen. 14 describes historical events.

anacoluthon A *Greek term for a sentence which is grammatically unfinished. It is possible that instances in Paul's letters are due to interruptions when he was dictating (e.g. Gal. 2: 4).

analogy Analogies, much employed in the Bible, are a means of reasoning by the use of parallel cases, and 'analogy' is therefore an umbrella term covering similes, metaphors, *typology, and *allegory. Thus Paul refers to Christ as a '*rock' (1 Cor. 10: 4), where he uses the account in Num. 20: 11 of the gushing of water out of a rock. The Jews of the 1st cent. CE believed that this supernatural rock somehow accompanied the Israelites on their journeyings through the *wilderness together with the numinous *cloud. So Paul applies this belief by analogy to Christ as the rock who gives life to his people. An analogy between sleep and death is found in 1 Thess. 5: 10.

For the medieval philosopher Thomas Aquinas the concept of analogy was fundamental as being an intelligible means of referring to the relationships between God and mankind; there are attributes of God which are similar to but not identical with those of human beings, as when we talk of the love of God.

Ananias (1) Husband of *Sapphira (Acts 5: 1–6), who collapsed when confronted by Peter with a false claim about his donation to the community funds. (2) A Christian in *Damascus who received Paul (Saul) into the Church (Acts 9: 10–20). (3) The high priestly president of the council taunted by Paul (Acts 23: 3).

anathema The Greek word is used in Luke 21: 5 of gifts in the Temple dedicated to God, but it came to have the opposite meaning—of a thing accursed. In Gal. 1: 9 Paul says of those preaching a different gospel: 'Let that one be accursed!' and this formula 'Let him be anathema' became a standard denunciation of heretics by Church councils, e.g. the Council of Trent in 1545–64.

ancestor worship The real or supposed ancestors of the Hebrews were venerated and were felt to be so close that their descendants even continued to suffer for their *sins (Exod. 20: 5). Not surprisingly it would seem that a cult of the dead infiltrated into Israel from neighbouring peoples (Deut. 26: 14) and former kings were worshipped (Ezek. 43: 7–9). The practice of calling up the dead, as when *Saul consulted *Samuel through the 'witch' of *Endor (1 Sam. 28: 7), is condemned by the *prophets (Isa. 8: 19–20).

anchor Vessels were anchored from the stern (Acts 27: 29). An anchor is used in Heb. 6: 19 as a metaphor for Christian hope.

Ancient of Days or **Ancient One** An Aramaic expression used in Dan. 7: 9 etc. for God as Judge. It conveys the notion of wisdom associated by the Jews with old age.

Andrew Brother of Simon Peter (Mark 1: 16), with whom he was fishing when invited by Jesus to become a *disciple. But John 1: 35–41 offers an alternative account; Andrew followed Jesus after hearing *John the Baptist's declaration; he then persuaded Peter to join them.

angels The Greek word *aggelos* means 'messenger' and as such angels are described as bearing messages from God to the *patriarchs (Gen. 22: 11 etc.). In the later (post-exilic) literature (perhaps influenced by contact with Zoroastrianism) angels are regarded as supernatural beings organized before God in a hierarchy (e.g. Dan. 7: 10; 9: 21). Against them are ranged evil angels under *Satan (Matt. 25: 41). Unlike the more conservative *Sadducees, the *Pharisees encouraged these developed beliefs (Acts 23: 9), which were needed to complement a doctrine of God's absolute transcendence. From the time of the book of Daniel (2nd cent. BCE) angels are given names (Dan. 8: 16; 10: 13) and specific tasks. In the gospels they are available to assist Jesus (Matt. 4: 11) and to be personal representatives of children before the Father (Matt. 18: 10). The *devil also has a troop of angelic assistants (Matt. 25: 41).

In several epistles (e.g. 1 Pet. 3: 22) the redemptive work of Christ is interpreted in terms of his triumph over supernatural celestial beings who may be angels. So also Paul in Col. 2: 8–15. The writers of the NT inhabited a different thought-world from us, but Paul's essential affirmation is that the *Law, which Jews believed was given to Israel by the mediation of angels, has been superseded and Christ's authority established.

angels of the Churches The letters to the seven Churches in Rev. 2–3 are addressed to their 'angels' because the author is assuming the Jewish belief that *communities on earth have their counterparts in *heaven among the angels. The angels share both in the achievements and in the failures of the Churches, for John does not conceive of heaven as supremely *good and earth as partly *evil; the whole created order is bound together in a unity.

anger The Bible describes the anger or *wrath of God as being his abiding reaction to human sin and an expression of his justice; it is not a sudden or passionate upsurge of emotion. He is 'slow to anger' (Ps. 103: 8). At Mark 1: 41, some MSS read that Jesus was moved not with 'pity' but 'anger'. Jesus was indignant not with the leper but with the example confronting him of the miserable condition of humanity.

animals According to the *Creation myth (Gen. 1: 28) humanity was given dominion over all sentient beings. This has often been accepted as divine authority for treating animals as property for food, religious sacrifices, and sport. Modern animal lovers in Protestant countries are inclined to reject this understanding of the text and respond with a moral generosity which believes that the rich and powerful have no absolute power but rather a duty to protect those parts of the Creation which are vulnerable and powerless. In the OT world wealth was measured by the size of herds (Job 1: 3) and religious zeal by *sacrifices (Deut. 15: 21). In Christian *worship animal *sacrifices are abolished (Heb. 10: 4) by reason of Christ's perfect sacrifice.

Anna An elderly and devout *prophetess who greeted the infant Jesus in the Temple (Luke 2: 36–8); a widow with whom many in Luke's own Church would identify (cf. 1 Tim. 5: 5).

Annas *High priest, 6–15 CE; father-in-law of *Caiaphas (John 18: 13). Although deposed, Annas continued to hold an influential position in the *Sanhedrin, and according to John 18 (but not mentioned in the synoptic gospels) Annas presided over a preliminary hearing of the case against Jesus before the trial by Caiaphas.

annunciation The message brought by the *angel *Gabriel to *Mary recorded only by Luke (1: 26–38) that she would be the mother of a son who would be 'great' and called 'the Son of the Most High'. The appearance of the angel has been a principal subject of Christian iconography from the earliest times.

anoint In the OT persons and things were consecrated by anointing with oil; kings were enthroned by anointing (Solomon, 1 Kgs. 1: 39); priests were anointed for the office (Aaron, Exod. 29: 7). *David refers to *Saul as 'the Lord's anointed' (1 Sam. 24: 6) and Cyrus, king of Persia, is God's anointed (Isa. 45: 1). In Hebrew this is 'Messiah' and in Greek 'Christos'. Jesus is said to be 'anointed with the *Holy Spirit and with power' (Acts 10: 38). Anointing was also used medicinally (Luke 10: 34) and as gestures of affection towards both the living (Luke 7: 38) and the dead (Mark 16: 1).

anthropology Usually the study of mankind's cultures, customs, and cults—but in Christian *theology it is used for the *doctrine* that men and women are made in the image of God, distorted by *sin. Some modern theologians have 'reduced' the doctrine of Christ (Christology) to an understanding of human beings (anthropology): for example, the *resurrection of Christ is interpreted not as Christ's dead body being raised to new life on *Easter Day, but as a multitude of individual believers' risings to new life when they responded to the Easter preaching.

anthropomorphism The attribution to the deity of human forms or behaviour, as when Moses and others went up the *mountain and saw God, and ate and drank (Exod. 24: 11), though there are numerous warnings against supposing that God is human (1 Sam. 15: 29). The *prophets proclaimed the living God and they invested him with a clutch of human emotions such as indignation, vengeance, and scorn, as well as joy and compassion. This is metaphorical language, rooted in human experience, and necessary in relation to God in order to use meaningful language of him at all.

antichrist The word occurs only in 1 John (2: 18, 22; 4: 3) and 2 John 7, where it stands for opponents of the Church, though the idea of the supreme enemy of Christ appears elsewhere—e.g. in 2 Thess. 2: 1–12, where 'that man of sin' (AV), 'the lawless one' (NRSV), is said to be expected before the **Parousia* could take place, and here has been identified with the Roman Empire or an emperor (*Nero?). Beyond the historical dimension, antichrist is a symbol for ongoing revolt against Christ until the Final *Judgement—though the revolt is embodied in particular historical persons, such as *Judas Iscariot.

antinomianism The doctrine held by a group of Christians taught by Paul that Christ has freed us from slavery under the *Law, indeed a freedom from any restraints at all. This travesty of his view is repudiated by Paul in Rom. 6; and Matt. 5: 17–18 also asserts the principle that laws and customs have a continuing validity in regulating private and public existence.

Antinomianism was not an intellectual justification of human frailty but a serious claim that the Christian *life was a life of the *spirit, not of the *body; the Christian was above ancient Mosaic scruples. It has frequently reappeared in the Church and as frequently been repudiated (as by John Wesley in 1740) as heretical, even by those who have been most clear that the gospel is not compatible with a legalistic ethic.

Antioch (1) A prosperous trading city on the River Orontes, in the north of the Roman province of *Syria, with a substantial and tolerated Jewish population. After the execution of *Stephen, Christians arrived from Jerusalem to begin missionary work (Acts 11: 19–20). *Hellenists from *Cyprus and *Cyrene admitted *Gentiles to the Church without undergoing the rite of *circumcision, and *Barnabas, himself a Cypriot, was dispatched to control developments. But so far from holding the line, Barnabas brought in Paul (Acts 11: 22–6). Their liberal policies brought them under suspicion and in 49 CE Paul and Barnabas met the Jerusalem leaders and (according to *Acts) a compromise was thrashed out (Acts 15: 1–19); but when Peter later visited Antioch he went back on the agreement and withdrew from sharing a meal with Gentiles (Gal. 2: 11–21). Paul soon left Antioch, travelling now without Barnabas, who had supported Peter. After the fall of Jerusalem (70 CE) there is no certain information about the Church in Antioch, but it is a reasonable supposition that the gospel of Matthew was compiled there or nearby. This gospel represents both a recognition of the Jewish roots of the Church, containing much typically Jewish ethical instruction, but also a firm commitment to the Gentile mission. The savage attacks on the *Pharisees, especially in ch. 23, suggest that the separation of Matthew's *community from the *synagogue is fairly recent. The Christian community is having to justify itself against a renascent Judaism. Matthew has the story of the pagan astrologers visiting the infant Jesus (Matt. 2) and at the end (Matt. 28: 19) he has the risen Christ authorizing the universal mission. In 117 CE *Ignatius of Antioch provides evidence that the Church there had acquired a clear structure with a leadership unified in the person of a presiding bishop. (2) Antioch of Pisidia, on the border of Pamphylia in modern Turkey, was made a colony by the emperor *Augustus. It was celebrated for the worship of the Phrygian moon-god, but the Romans suppressed the cult. There was a Jewish group in the city, which Paul and Barnabas visited (Acts 13: 14–52; 14: 19–21), but an initial welcome was followed by a campaign of persecution and their expulsion.

Antiochus Of the thirteen Seleucid kings of this name, the best-known is Antiochus IV Epiphanes, who ruled from 175 to 164 BCE. His efforts to Hellenize his dominions were resisted by Jews under the *Maccabees (1 Macc. 1: 10, 20; 2: 15 ff.). The book of Daniel reached its final form at this time, and the 'abomination' in 12: 11 refers to the desecration of the Temple by Antiochus.

Antipas (1) The son of *Herod the Great. After his father's death, Herod Antipas was assigned the territory of *Galilee and *Perea, and was ruler during the ministry of Jesus. Called 'that fox' (Luke 13: 32), he is said to have examined Jesus during the course of the trial by Pilate (Luke 23: 6–9). (2) The name of a Christian martyr mentioned in Rev. 2: 13.

Antipatris A town on the coast of Palestine. Paul spent a night in *prison here *en route* from Jerusalem to his trial in *Caesarea (Acts 23: 31). It was perhaps the OT Aphek (Josh. 12: 18) and had been rebuilt by *Herod the Great, and named after his father, Antipater.

antisemitism Both *John the Baptist and Jesus had condemned their Jewish contemporaries for failing to heed their messages, following the tradition of OT prophets; but the gospels were compiled when relations between Christians and Jews, Church and *Synagogue, had become strained, and the rebukes Jesus levelled at his fellow Jews (Matt. 12: 34) were in due course thrown by Gentile Christians at the non-believing Jews. This mutual suspicion is reflected in the gospels. No doubt Jesus encountered opposition from *scribes and *Pharisees from the earliest days of his ministry (Mark 3: 6), and in the Little *Apocalypse in Mark 13: 9 the disciples are warned to expect to be beaten in synagogues. This indeed became a threat in history. After the Fall of Jerusalem (70 CE) there was excommunication from the synagogue (John 9: 22; 16: 2) and in about 85–90 CE the rabbis at Javneh (or *Jamnia) included in the Eighteen Benedictions as number Twelve a curse against the Nazareans and the *Minim* (heretics) which may have sealed the rupture, though it is uncertain whether this addition was accepted everywhere and by all Jews. It is possible to trace in the Passion Narratives a tendency to lay more of the blame for the Roman penalty of *crucifixion on the insistence of the Jewish leaders and to diminish the responsibility of the Roman Pontius Pilate, who condemns Jesus with increasing reluctance. In Matt. 27: 25 the crowd exclaim, 'His blood be on us and on our children', which has been used as a particular provocation for antisemitism down the ages: but Matt.'s implausible allegation that 'the whole nation' was present (an impossibility!) in the **praetorium* is a good reason for regarding the cry as unhistorical. Luke continued the trend to diminish the Roman responsibility by recording the *centurion's confession when Jesus died that he was 'innocent' (Luke 23: 47). At its extreme, antisemitism is represented by John 8: 44, where the Jews are pilloried and demonized as children of *Satan—an attitude of bitterness and hostility rarely acquiesced in by the ex-Jew Paul (Rom. 9–11).

Antonia, tower of The name of the barracks (NRSV, REB), a fortress (NJB), or palace (AV), where Paul was arrested after the riot in Jerusalem (Acts 21: 27–36). It was built near the Temple of *Herod the Great and named after his friend Mark Antony.

Apelles Warmly greeted by Paul in Rom. 16: 10. It is known to be a name borne by members of *Caesar's domestic staff; and also of a well-known tragic actor. This Apelles may have been one of the household of *Aristobulus, a grandson of *Herod the Great, which probably means he was a slave.

apocalypse Though applied especially to the last book in the Bible, the Revelation to John, and called 'The Apocalypse', the word means more generally an 'unveiling' of divine secrets and in the OT books such as Daniel and parts of Isaiah and Zechariah there are apocalypses. In the NT, Mark 13 is often known as the 'Little Apocalypse'. In the OT *pseudepigrapha there are a number of documents, such as the Apocalypse of Abraham and the Apocalypse of Elijah. Fragments of 1 Enoch are among the *Dead Sea scrolls. The apocalypses traverse the whole range of human experience, and beyond, from references to the social setting of the author, such as the trials of persecution (Dan.), to revelations of events to occur at the end of time and disclosures under the guidance of an angel about the 'seventh *heaven' (2 Enoch). They were all published in *Judaism from about 250 BCE until about 200 CE, after which the *genre was to be continued by Christian commentators on Rev. such as Abbot Joachim (1132–1202) and Edward Irving (1792–1834) and by Jehovah's Witnesses.

apocalyptic Characteristic of OT apocalyptic is the fervent hope of a future salvation at the end of the current experience of national humiliation and suffering. Apocalyptic thought therefore takes the form of *eschatological literature and is usually pseudonymous; it consists of revelations attributed to *Abraham, *Enoch, *Baruch, and the like. Typically, apocalyptic thought might consist of historical narrative in which the story is told with tolerable accuracy from the time of the alleged author to the actual date of composition, after which the narrative (as in Dan. 11) becomes vague. It reflects a despair of historical process and predicts catastrophic cosmic upheavals and God's salvation of the nation of Israel: it is a literature of hope and consolation. What is at present an age of suffering will be transformed into the joys of *paradise; a land conquered by pagans will become a world fulfilled with the *glory of God.

Apocalyptic ideas included the notions of a general *resurrection and influenced the development of Christian doctrines, such as that of the Second Coming of Christ. Written between 170 BCE and 100 CE, several books in the *Apocrypha are apocalyptic, such as 2 Esdras, and there is apocalyptic doctrine among the *Dead Sea scrolls.

Apocrypha After the Fall of Jerusalem (70 CE) the future of *Judaism was maintained by rabbis of the Pharisaic tradition. They accepted as authoritative the twenty-four books of the Hebrew scriptures but rejected a number of Jewish works which were used in *Alexandria and which are known to us in MSS of the *LXX and called the Apocrypha (Greek for 'things hidden away'). Being composed after the time of *Ezra, when prophecy was held to have ceased, these Greek works, even if originally composed in Hebrew (e.g. 1 Macc.), were unacceptable. By and large, the Christians accepted the longer list—though when Jerome translated the OT into Latin for his *Vulgate, he treated the apocryphal additions as edifying but not part of the *canon. Eventually, however, the writings which Jerome had rejected were included from the Old Latin version which Jerome had worked so hard to supersede.

The books of the Apocrypha are called deuterocanonical (= at second-level) by Roman Catholics, to distinguish them from protocanonical (= first-level) books, but they are regarded as authoritative and included at appropriate places within the body of the OT. (But 3 and 4 Esdras were rejected as authoritative by the Council of Trent (1545–64) and relegated to an appendix.) At the Reformation Protestants reverted to the shorter canon of the Hebrew OT because they detected in 2 Macc. hints of the doctrine of *purgatory, which they repudiated; they also claimed to find in Tobit the unacceptable Catholic doctrine of justification by works. Luther's Bible of 1534 relegated these books to an appendix. The Church of England included the Apocrypha 'for example of life and instruction of manners' but not for the establishment of doctrine. The exact extent of the Apocrypha is not universally agreed, and some of the books are known by different titles. A list is given in the Introduction to this Dictionary.

Apocryphal New Testament Christian writings not included in the authoritative *canon of the NT which claim to be reminiscences of the life of the young Jesus and his miraculous powers, or supplements to the book of Acts about apostles who were there passed over. Several epistles and *apocalypses also exist. Although these works, deriving from the 2nd to the 9th cent., in *Greek, *Latin, Syriac, and other languages, give little reliable information about Christian origins, they do contribute to our understanding of Christian thought and the life and piety of these centuries. The discoveries at *Nag Hammadi have revealed many previously unknown apocryphal documents, some of which purport to give Christ's *teaching to the *disciples between the *resurrection and *ascension (the period is conveniently lengthened from 40 to 550 days).

apodictic law Divine law in the OT expressing absolute prohibitions or injunctions, as in the *Ten Commandments ('You shall not make an idol', Exod. 20: 4).

Apollos An Alexandrian Jew who is said to have 'taught accurately the things concerning Jesus' (Acts 18: 25) and yet, amazingly, 'knew only the baptism of John'. *Priscilla and *Aquila filled in his inadequate knowledge and in *Corinth Apollos soon became a leading teacher, revered by some above Paul (1 Cor. 1: 11–12; 4: 6–7). He returned to *Ephesus and declined Paul's generous invitation that he should pay another visit to Corinth (1 Cor. 16: 12). Perhaps they both suspected his absence would improve the chances of harmony in the Church there. Apollos has been proposed as the author of the epistle to the Hebrews in view of its apparent similarity in style and vocabulary to Alexandrian *Judaism. If so, the epistle might have been addressed primarily to the

Hebrew faction at Corinth (cf. 2 Cor. 11: 22) about 53 CE.

Apollyon Greek for 'the Destroyer' (so REB) in Rev. 9: 11, and therefore a subtle attack not only on the Greek God Apollo but also on the persecuting emperor *Domitian who regarded himself as Apollo on earth.

apologetic An 'apology' is, strictly, a defence—as Newman's *Apologia pro Vita Sua* (1864): A Defence of his Life—and there are many instances of apologetic in both OT and NT. Thus, Isa. 40–55 extols the transcendence of the God of Israel and compares him with the nonentities of the heathen. And both *Philo and *Josephus wrote to commend *Judaism to pagans; at Alexandria Philo maintained that the Hebrew tradition was as ancient and respectable as Hellenism, and in *Rome Josephus explained in his many works that the Jewish rebellion was instigated by a small group of fanatics.

The Acts, equally, has an apologetic purpose, though it may not be the primary aim of the book. But, as in the third gospel, where the centurion at the foot of the cross declares Jesus to be 'innocent', so in the Acts Christians are shown to be law-abiding citizens of the Roman Empire. If they are prosecuted, it is because there are Jews doing what they can to impede the progress of the gospel. Roman magistrates (like *Gallio and others) are shown to be friendly to Paul, who can boast proudly of his Roman citizenship, and make use of it (Acts 22: 27).

There are apologetic interests in the gospels, as when Matthew repeatedly adduces OT texts which have in his view been fulfilled in Jesus.

apophthegms A term alternative to 'paradigms' used in *Form Criticism of the gospels for short stories of an event. As used in sermons by early preachers, the terms of the stories may have been influenced by the current situation of their Church.

apostasy Repudiation of *faith in God, found in both the OT and NT. Israel is accused by *Jeremiah of apostasy (Jer. 2: 19) and apostates are said (Heb. 6: 6) to crucify again the Son of God. Paul says that the *day of the Lord will not come unless the 'apostasy' ('rebellion', NRSV; 'final rebellion', REB; 'Great Revolt', NJB) comes first (2 Thess. 2: 3).

apostle From the Greek *apostolos*, meaning one who is sent and enjoys the authority of the agent who instructs him. There was already a Jewish functionary, called a *shaliach*, who was trusted with the authority of his employer when taking messages. Thus, in the NT the apostles bear witness to the message of Jesus and continue his work.

The twelve *disciples in the gospels are called 'apostles' by Luke (6: 13; 22: 14; Acts 1: 2) and apparently by Paul (1 Cor. 15: 7) and only once each by Matt. 10: 2 and Mark 6.30); though the names vary slightly, the list is always of twelve, signifying that they are the leaders of an elect race, corresponding to the twelve *tribes of Israel. The betrayal of Jesus by *Judas created a vacancy in the number, and Matthias was elected to fill it. Subsequent deaths did not create vacancies, for in Luke's view apostles are unique. Yet the office was soon expanded to embrace Paul (unless it is Luke who was responsible for *restricting* a broader title to the twelve disciples), who claims the title on the basis of his vision of the Risen Christ (1 Cor. 15: 8) and his commissioning to go to the *Gentiles. A kind of second rank of apostles apparently included *Barnabas (Acts 14: 14) and *Junia (Rom. 16: 7—the latter name is feminine). The names of the Twelve are: Simon Peter, James and John, Andrew, Philip, Bartholomew (Nathanael?), Matthew (Levi?), Thomas, James the Less, Thaddaeus (Judas the son of James?), Simon the Cananean or Zealot; and Judas Iscariot—afterwards succeeded by Matthias.

Apostolic Council The Council of Jerusalem described in Acts 15 is sometimes called the Apostolic Council though it was presided over by *James, brother of the Lord and not himself one of the Twelve. The account in Acts was written much later than the event (which was probably held in 49 CE) and describes a far more temperate debate and harmonious decision than would be gathered from the violent language used by Paul in Gal. 2. The conflict was caused by the success of Paul's mission to the *Gentiles, whom he received into the Church by *baptism without their first undergoing admission to *Judaism by *circumcision. Jewish Christians feared an influx of pagan converts who had never known the discipline of the *Law or the heritage of God's ancient people of Israel. It followed that for strict Jewish Christians it was not acceptable to share a common meal with uncircumcised Gentile Christians. Although, to be sure, one important MS omits mention of the ban on eating flesh killed by strangulation (Gen. 9: 4) and so turns the decree into straightforward moral obligation, it is more likely that a compromise was reached: circumcision was not to be demanded of Gentile converts, but Gentiles were to respect Jewish Christian scruples about food so that full table fellowship should not be impaired. However, a grave difficulty about the historical accuracy of this decision is that Paul, himself said to be present at the debate, never once mentions the decree, even to settle the dispute at Corinth (1 Cor. 8).

Apostolic Fathers The Apostolic Fathers (*Clement of Rome, *Ignatius, *Hermas, Polycarp, and *Papias; and the authors of the epistle of Barnabas, the epistle to Diognetus, 2 Clement, and the *Didache) form the literary link between the NT period and later Christian generations. From about 95 to 150 CE a number of epistles and treatises give an important insight into how the tradition was both preserved and developed. On the one hand, there are frequent quotations from the OT (*LXX), but the doctrine of the ministry is in process of evolution: while the Didache suggests that sometimes prophets and teachers could preside at the *Eucharist, Ignatius emphatically asserts the authority of the bishop. Clement of Rome (about 95 CE) has a definitely Trinitarian concept of God. Polycarp of Smyrna, writing to the Church of Philippi, appears to suggest that parts of the NT (gospels, 1 Peter, Hebrews, and certain of Paul's epistles) were being read aloud in the assemblies. The epistle of Barnabas, written in Alexandria about 125, uses *allegory and *typology in the interest of *Christology. The Shepherd, written in Rome by Hermas about 125, consists of five 'visions' and at one time almost qualified for admission to the NT canon: both the Shepherd and the epistle of Barnabas are found in Codex *Sinaiticus.

Appian Way The road to *Rome from Puteoli where Paul disembarked (Acts 28: 13); 70 km. (40 miles) south of Rome on the road was the Forum of Appius, where a party of Roman Christians welcomed Paul.

apple of the eye A Hebrew expression used (e.g. Ps. 17: 8) for the eyeball, the part to be most carefully guarded.

aqeda A term used by rabbis for the elaborate theological development of the Genesis story of the sacrifice of *Isaac (Gen. 22: 1–19) with a doctrine of *atonement comparable with that of the NT. The rabbinic doctrine could have influenced Paul (e.g. in Gal. 3: 16–17). *Abraham's faith is held up as an example by *James (2: 21) because it issued in action.

Aquila A convert from Christianity to Judaism, who made a very literal translation (140 CE) into Greek from the Hebrew of the OT intended to replace the version of the *LXX used by the Christians.

Aquila and Prisc(ill)a Evicted from *Rome in 49 CE when the emperor *Claudius expelled Jews after disturbances about Christianity. They met Paul in *Corinth (Acts 18: 2) and afterwards he stayed with them in *Ephesus.

Arabah The desert valley which runs from the Sea of *Galilee to the *Red Sea was an important trade route from north to south. In Job 24: 5 it is translated '*desert' by NRSV and NJB and '*wilderness' in REB.

Arabia The vast *desert between Iraq in the east and the *Red Sea in the west. In the Bible the name was probably used for all the desert area in the east of *Palestine, including *Sinai. The climate is severe, with extremes of hot and cold and little rain except in the mountains, but caravans of camels kept open an important north–south trade route. There were Arabs in Jerusalem at *Pentecost (Acts 2: 11), and Paul explains that immediately after his call on the road to *Damascus he went off to Arabia (Gal. 1: 17), which was probably the Nabataean kingdom, not a great distance from Damascus, which was in the Roman province of *Syria.

Aramaeans According to Deut. 26: 5 'a wandering Aramaean was my ancestor'; a reference to *Jacob, and the Israelites' sojourn in *Egypt. It would seem that the claim was justified on the ground that *Abraham migrated from *Ur into Haran (Gen. 11: 28–32), in an area known as 'Aram of the two rivers' (the *Euphrates and the Tigris).

During the Israelite monarchy, Aramaeans fought against *David (2 Sam. 8: 3 ff.) and Benhadad II against *Ahab (1 Kgs. 20). The chief city was *Damascus; it remained powerful until defeated by the Assyrians in 732 BCE (2 Kgs. 16: 9).

Aramaic A language closely related to Hebrew and widely spoken in various dialects in *Palestine from the 8th cent. BCE up to and after the NT period. It was the language of diplomacy in 701 BCE (2 Kgs. 18: 26) but its use gradually spread right through the population, as did *Latin in Italy in the Roman Empire. Several words appear in Aramaic in the gospels ('*Abba*', Mark 14: 36; '*talitha cumi*', Mark 5: 41; '*golgotha*', Mark 15: 22) and it is likely that Jesus himself spoke Aramaic. The Aramaic *bar-* for 'son of' has replaced the Hebrew *ben-*, as in Barabbas, Bar-Jona. A few sections of the OT, notably part of Daniel, were written in Aramaic, and there were translations and interpretations of the Hebrew OT in Aramaic for use in *synagogues, called **targums*.

Ararat The region in modern SE Turkey where the sons of *Sennacharib sought refuge after murdering him (2 Kgs. 19: 37). *Noah's ark was made to settle in Ararat according to the legend of Gen. 8: 4, no doubt because it had the highest known mountains.

Araunah A Jebusite who owned a threshing floor (2 Sam. 24: 16). A pestilence which had ravaged the land to punish *David for having presumptuously held a census to ascertain the nation's strength did not get beyond this threshing floor. The greatly relieved king bought the premises as a suitable site for the future Temple (2 Chr. 3: 1 where Araunah is called Ornan), built by *Solomon after a similar census which had no dire consequences.

archaeology Discoveries at a multitude of sites in the Middle East under a succession of brilliant archaeologists—British, American, German, French, and, since 1948, Israeli—have made an enormous contribution to biblical, especially OT, studies and complemented the work of literary critics. As methods have been refined, so more accurate information has been derived about social life, military movements, and religious epochs. Numerous tells, or mounds, have been systematically excavated, stratum by stratum of human habitation, and texts, inscriptions, and *ostraca (potsherds) have been passed over to the experts in *epigraphy for dating. Places mentioned in the Bible have been definitely located by archaeology, and the historical records, which have reached us with the stamp of editorial interpretation, can be assessed. Excavations at *Jericho uncovered a mass of furniture in burial caves of the Middle Bronze Age (1900–1500 BCE) and so provided data for domestic life in the patriarchal period of the OT, though the patriarchs themselves continue to receive little archaeological verification. Some details in the Genesis stories may correspond with what excavations have revealed of the second millennium BCE, and to that extent OT studies are enriched. *Ur (Gen. 11: 31) is known to have enjoyed civilized urban amenities. Conditions described in the book of Judges have also been confirmed by excavations of small villages estabiished by the incoming Israelites and towns like *Shechem destroyed by them (Judg. 9: 45). It is possible that the millo built by *David (2 Sam. 5: 9) and *Solomon (1 Kgs. 9: 15) has been found, but there is no trace of Solomon's *Temple or palace. Shrines of the 8th cent. BCE have provided evidence of popular syncretistic religion; an inscription mentions 'Yahweh and Ashweh'—the kind of Canaanite religion which was *anathema to the prophets. But to what extent have archaeological studies confirmed or contradicted the OT narratives?

Articles for the general religious public have sometimes given the impression that archaeology proves the 'truth' of the Bible; but this is far from the case. Evidence which was once thought to demonstrate that the walls of *Jericho collapsed at the time of the Israelite invasion under *Joshua is now discounted on the basis of reliable dating of pottery. On the other hand, excavations at the cities mentioned in 1 Kgs. 9: 15 do suggest that *Hazor and Gezer were built in the age of Solomon. Other claims made for Solomon in the books of Kings and Chronicles seem exaggerated; for example, it was the dynasty of *Omri that contributed to the modest splendour of *Megiddo.

Much archaeological evidence is neutral in respect of corresponding literary data of the OT and is sometimes pressed too hard by conservative biblical scholars. But there is a wealth of material still waiting to yield its secrets, among which is the function of the *Qumran buildings and their relationship to the *Dead Sea scrolls.

Excavations in Jerusalem have revealed the great walls built around the Temple mount by *Herod the Great (37–4 BCE) as well as his buildings at Masada, Jericho, and elsewhere. Paintings and mosaics have also been found. John 19: 13 mentions that *Pilate sat on the judge's bench at a place called the Stone Pavement, or in Hebrew *Gabbatha, and the paved area on the site of the *Antonia fortress may have been in existence in the time of Jesus, though the date is disputed. There are scratch marks on the stones where soldiers played games. A notice threatening death to any *Gentile who entered the inner parts of the Temple has been discovered, though whether the Jewish leaders had authority under the Romans to inflict the death penalty is dubious (John 18: 31). In 1968 in Jerusalem the skeleton of a young victim of *crucifixion was found in a limestone box; a nail was still in place; his legs had been broken; and he had been fastened to the cross by the forearms (not the hands). The pools of Bethzatha (John 5: 2) and *Siloam (John 9: 7–11) have been identified.

Other archaeological discoveries relevant to the NT have been that of a decree of *Claudius found at Delphi in 1905 which makes it possible to date the proconsulship of *Gallio (Acts 18: 12) as 51 CE. At *Ephesus parts of the temple dedicated to the goddess *Artemis have been found. An inscription confirms the accuracy of Acts 17: 6 in giving the unusual name 'politarchs' (NRSV, marg.) to the magistrates of *Thessalonica. But on the whole archaeologists' main interests in places associated with Paul have been in classical studies, and NT scholars have had to search hard for relevant material.

archangel From the Greek, meaning a chief angel; seven are named in 1 Enoch 20, cf. Tobit 12: 15. In the two centuries BCE and the first CE the Jewish doctrine of *angels was developed by some (e.g. in the community of *Qumran) into a hierarchy of angels which included a rank of archangels (Test. of Levi 3: 5). Paul mentions the archangel's voice (1 Thess. 4: 16) and Michael the Archangel appears in Jude 9.

Archelaus Son of *Herod the Great, who inherited the territory of *Judaea, *Samaria, and *Idumea, but after popular disturbances was banished to Greece by the emperor in 6 CE.

architecture Israelite resources were concentrated on survival and there was little energy available for the kind of buildings that are the heritage of *Greece to civilization. Only in the short period between the reign of *David and the invasions of *Assyria was any substantial architecture possible, and this was achieved by *Solomon, who used the skills of Phoenician craftsmen. The construction of cities (1 Kgs. 9: 17) and palaces (1 Kgs. 7: 1) is recorded in addition to the massive Temple of Yahweh in Jerusalem. The Temple built by *Herod the Great reflected his enthusiasm for a Hellenistic style, and the immense fortification at Masada had columns and decorated capitals typical of the way Roman builders adapted Hellenistic motifs.

Domestic buildings in *Palestine were always modest in size and simple in style. There might be up to four rooms round a courtyard in which the cooking was done. Roofs were crude constructions and needed frequent repairs and renewals. Over a framework of timber there might be laid a matting of branches and baked mud. So when the bearers of the paralytic desperately tried to reach Jesus, they tore a hole in the roof in order to let down the man before Jesus in the house (Mark 2: 4). However, when Luke rewrote the story the stretcher-bearers are said to remove tiles (Luke 5: 19), for Luke has adapted the architecture to something more familiar to his sophisticated readers.

At *Jericho part of Herod's palace has been uncovered, and in Jerusalem excavations have revealed that the pool of *Siloam (John 9: 7) was a *basin with 4.5 m. (15 feet) sides and a surrounding portico.

Terms used by architects, such as '*corner-stone' and '*foundation', are appropriate metaphors in the NT for Christ (1 Cor. 3: 11; Eph. 2: 20) and the Church is called a *pillar (1 Tim. 3: 15) which supports *truth.

Areopagus A hill in *Athens where the city council met; Paul gave an address there (Acts 17) to the assembled *Gentiles, using a text about an *altar he had seen dedicated 'to an unknown god' (17: 23).

aretalogy In NT studies a narrative which is a recital of the qualities or virtues (Greek, *aretai*) of a worker of *miracles, such as Jesus and pagan philosophers like Apollonius of Tyana (d. 98 CE), whose biography was written by Philostratus in the 3rd cent. This would be the kind of discourse which the readers of 1 Pet. 2: 9 are summoned to proclaim. The Greek *aretas* (accusative) is used ('glorious deeds', REB).

Ariel In Isa. 29: 1–2 it stands for Jerusalem and has ominous overtones: the word means 'altar hearth'. The empty religious ritual will lead to the *sacrifice of the inhabitants themselves.

Arimathaea All the gospels mention that *Joseph, who sought permission to bury Jesus in his personal *tomb, lived in Arimathaea, 8 km. (5 miles) north of Jerusalem.

Aristarchus Accompanied Paul to Jerusalem and thence to *Rome (Acts 19: 29 etc.) and possibly at some time a fellow prisoner (Col. 4: 10).

Aristeas The fictitious name given to the Jewish author of a letter of unknown date purporting to describe how the Septuagint (*LXX) was miraculously translated from the Hebrew by seventy scholars in seventy days.

ark The English translates two unrelated Hebrew nouns: (1) The first is that of the vessel of three storeys in which *Noah and his *family and *animals were accommodated during the *Flood. This *legend of Gen. 6–9 is similar to, and was perhaps influenced by, the Babylonian Gilgamesh epic which also features an enormous boat. The same Hebrew word is used for the basket in which the infant Moses was placed (Exod. 2: 3). This ark is referred to by 1 Pet. 3: 20–1 in the context of Christian *baptism, which also saves through water. (2) The other noun denotes the Ark of the *Covenant, a kind of portable shrine with a lid carried in the wilderness and round *Jericho (Josh. 6) and supposedly containing the two tablets of the *Law (Deut. 10: 2, 5). It was regarded as a throne for the earthly presence of the invisible God, and those who desecrated it were punished (1 Sam. 6: 19). It was once captured by the *Philistines (1 Sam. 4) with disastrous consequences which led them to return it to Israel. The ark was established by *David in Jerusalem and installed by *Solomon in the Holy of Holies of his new Temple (1 Kgs. 8: 4–7). It was believed that its presence would protect the city from its foes, but in fact the ark was probably part of the loot when Jerusalem was destroyed by the Babylonians in 586 BCE. As the focus for the rite on the Day of *Atonement the ark was held by later generations to be overlaid with gold (Exod. 25: 10–16). This is referred to by Heb. 9: 4, which implies that there existed another tradition that the ark also contained a portion of *manna and *Aaron's rod (Exod. 16: 33; Num. 17: 10).

arm Used frequently in the OT to denote strength of a man or a nation, and so also of Yahweh—strong to create and to save (Job 40: 9; Isa. 59: 16) and also in the NT (Luke 1: 51). The plural arms, when used of a nation, does not have the modern meaning of weapons.

Armageddon The scene of the final *apocalyptic struggle between *good and *evil as depicted in Rev. 16: 16. The name (Harmagedon, NRSV) appears to be connected with *Megiddo, where many great battles were fought (e.g. 2 Kgs. 23: 29), in the NW of Palestine.

armour Israelite *soldiers were equipped with shields and spears (2 Chron. 14: 8), helmets (2 Chron. 26: 14), and coats of mail if they rode in chariots (Jer. 46: 4). Jews engaged in rebuilding the walls of Jerusalem under *Nehemiah also had coats of mail on account of the hostility of neighbouring Arabs. Pieces of armour are used metaphorically as equipment in Christian warfare against the powers of *evil (Eph. 6: 11 ff.).

army A small-scale army, or bodyguard, was established in the early days of the monarchy (1 Sam. 13: 2) and was enlarged by *David and *Solomon. After the defeat of Israel and *Judah by the *Assyrians and *Babylonians, Jewish armies were not operative until the Maccabean period (165 BCE). *Herod the Great organized an army, and in the NT there is mention of Roman armies (Luke 21: 20).

art In contrast with the genius of classical *Greece, Hebrew and Jewish creative talent was repressed by the strict prohibition of images of the kind that inspired the sculpture and painting of Israel's Egyptian and Mesopotamian neighbours. Decorated seals have been discovered, and at *Megiddo some statuettes of the '*Queen of Heaven' explain the wrath of the prophets (Jer. 7: 18) against popular fertility religion.

Artaxerxes In Ezra 4: 7 etc. the reference is probably to Artaxerxes I, king of Persia 465–424 BCE, whose appointment of *Nehemiah as governor in Jerusalem was probably to assist his ambitions against *Egypt.

Artemis *Diana of AV, NJB, the goddess worshipped in *Ephesus in a great temple made of marble. Miniatures of her were sold by silversmiths (Acts 19: 23–4) who feared for their livelihood if Paul's preaching was successful.

Asa King of *Judah, 911–870 BCE, who earns modified approval from 1 Kgs. 15: 14 for removing idols, but his alliance with *Damascus displeased the Chronicler (2 Chr. 16: 9), who promised divine retribution.

ascension The ascent of Jesus into *heaven (Acts 1: 2–11; a shorter version is in Luke 24: 50–3). There are two OT stories and also Ps. 110 which form the background of the accounts. When *Enoch had attained 365 years, he 'walked with God'; then he was no more, because God took him (Gen. 5: 24); the prophet *Elijah was taken to heaven by a *whirlwind (2 Kgs. 2: 11). Ascension stories (e.g. of Heracles) also circulated in the pagan Hellenistic world. The character of Jesus' appearances after Easter as recorded by Luke have a more 'solid' 'corporeal' and evidential nature than elsewhere in the NT, even though the risen body could pass through doors. It was necessary for Luke to explain that the appearances would cease and be replaced by the presence in the Church of the *Holy Spirit, and his solution was to use the OT notion of ascension after forty days (cf. Elijah's journey of forty days and nights to Horeb the Mount of God, 1 Kgs. 19: 8) to signify the termination of *resurrection appearances and the completion of Jesus' work of *redemption and his ascendancy in a Messianic kingdom over all things.

Traditionally, the Ascension has been regarded as a physical elevation and many famous pictures, e.g. by Tintoretto in Venice, show Jesus moving upwards in a *cloud, even leaving behind a footprint on the ground. However, given the OT background the Ascension should be regarded as Luke's way of expressing his understanding of the completion of one era and the inauguration of another, that of the Church, which at *Pentecost would shortly be endowed with the Holy Spirit which the heavenly Christ would send for the completion of his mission (John 20: 21). Although the Ascension has often been understood in the past in a literal way, there is a valid symbolical interpretation, for the spatial metaphor of height expresses the idea of transcendence.

Ascension of Isaiah An apocryphal work extant only in Ethiopic, probably Jewish in origin but containing extensive Christian additions. It describes the prophet's martyrdom and his subsequent elevation to *heaven, and is useful for the information it gives on *eschatological beliefs about *antichrist held in some Christian circles in the 2nd cent. CE.

asceticism The strict discipline of one's body has been a precept of many religions, and clearly Christians who expected an early return of the Lord to earth prepared for it with *prayer, *fasting, and abstention from sexual relations. There is strong approval of renunciation of *marriage in certain of the epistles (e.g. 1 Cor. 7: 7) and the gospels (e.g. Matt. 19: 12) and Rev. (14: 4). This was a minority opinion, but it became widely held that asceticism was a prerequisite for entering *eternal life. It was certainly also a proof of serious discipleship and in Luke 9: 23 the word 'daily' has been significantly added to Mark 8: 34, so making a practicable kind of asceticism acceptable in the continuing life of the Church which was not shortly to be terminated by the **Parousia*.

Ashdod A town under *Philistine control to which the *Ark of the Covenant was temporarily taken (1 Sam. 5: 6). Its walls were demolished by *Uzziah, king of *Judah (783–742 BCE) and later it was captured by the Assyrians. In the Maccabean period it was known as *Azotus.

Asher The name of one of the twelve tribes of Israel who occupied a fertile part of *Galilee (Deut. 33: 24–5). The inhabitants came to the help of *Gideon against the *Midianites (Judg. 6: 35). The tribe survived even after the conquest of *Assyria (1 Chron. 7: 30, 40; Luke 2: 36).

Asherah A goddess worshipped by the Canaanite predecessors of the Israelites to whom they were sometimes tempted to return. She was a mother-goddess and is mentioned in the *Ras Shamra tablets, and images possibly representing her on a sacred pole are ordered to be destroyed by force in Deut. 12: 3. Queen *Jezebel encouraged her cult in the northern kingdom (1 Kgs. 18: 19) as did *Manasseh in the south (2 Kgs. 21: 7). In popular religion denounced by the *prophets (Mic. 5: 12–13). Asherah was sometimes regarded as Yahweh's consort, whereas Yahweh of the official religion was not regarded as a sexual being.

Ashtoreth A goddess of love and motherhood worshipped among the *Canaanites (Judg. 2: 13) and called Astarte among the *Phoenicians, Aphrodite in Greek mythology. Her cult was encouraged by *Solomon (1 Kgs. 11: 5), for which he was condemned (1 Kgs. 11: 9). REB uses the form 'Ashtoreth'; NRSV and NJB prefer 'Astarte'.

Asia In the NT not the continent but the Roman province, governed from *Ephesus and now western Turkey. It was rich in resources and comprised *Syria, Mysia, *Lydia, and Caria; *Phrygia belonged partly to Asia and partly to *Galatia. If Paul's letter to the Galatians was addressed to Churches which he had visited (*Antioch, *Iconium, *Lystra, and *Derbe), then it was sent to the south of the area, within the province of Galatia; but *Colossae, to which a letter was sent, was in Asia, as also were the seven Churches of Rev.

asiarchs Officials (NRSV) in the Roman province of Asia. They were well disposed towards Paul at *Ephesus (Acts 19: 31—'dignitaries of the province', REB).

ass Both wild (Job 39: 5–8) and tamed (Deut. 22: 10) were known BCE; Jesus rode on a female ass (Matt. 21: 2), which the evangelist regards as a fulfilment of Zech. 9: 9.

assassins Men armed with daggers (Acts 21: 38), led by an Egyptian, threatened Jerusalem in 54 CE according to the historian *Josephus. The rebellion was put down, but the leader escaped. The term 'assassin' was used of Jewish nationalists (*Zealots).

assembly The people of Israel were summoned as a whole community to listen to Moses and *Aaron (Exod. 12: 3) and to *Solomon (1 Kgs. 8: 14). In the taunt song over *Babylon, the king of Babylon is mockingly made to refer to his expectation of taking a seat among the assembly of the gods (Isa. 14: 13), located on a mountain according to the *Ras Shamra texts found at *Ugarit.

Asshur The capital city of *Assyria until 900 BCE, when it became one only of several important cities in the greatly enlarged empire. The name was also used for the patron-god of the city and in the OT stands for the nation itself (Hos. 14: 3, AV).

assurance On the basis of Rom. 8: 16 some theologians of the Protestant tradition have held the doctrine that believers can be assured of their ultimate *salvation. The ground of this assurance and certainty is the work of Christ and his present exaltation; it confers a conviction (1 Thess. 1: 5) of being able to approach God with a true heart in 'full assurance of *faith' (Heb. 10: 22). The feeling of assurance must be validated by appropriate moral criteria.

Assyria One of the great empires of the ancient world, occupying a fertile area east of the River *Tigris, corresponding to modern northern Iraq. In the city of *Nineveh pottery which has been discovered is proof of habitation in the period 5000–3000 BCE, but its main importance for biblical history lay in the three centuries from about 900 BCE when its well-trained and equipped army, with chariots and infantry, terrified neighbouring countries (Jer. 1: 13). However, the northern kingdom was temporarily benefited by Assyria when it defeated Israel's Aramaean oppressors (2 Kgs. 13: 5). Nevertheless *Shalmaneser III, king of Assyria 858–824 BCE, records on the Black Obelisk, found in 1846 at Calah on the River Tigris and now in the British Museum in London, that he received gifts from *Jehu, king of Israel, in 841 BCE. Under *Tiglath-Pileser III (744–727 BCE) a period of aggressive expansion brought vast areas under Assyrian control and local neighbours were subject to it; many of the inhabitants were deported. But Menahem, king of Israel, secured immunity by forcing all the well-to-do inhabitants to surrender fifty shekels each (the current price of a slave) as tribute to Assyria (2 Kgs. 15: 20). The enlarged Assyrian cities were magnificently beautiful and in them the arts flourished.

When *Hoshea failed to send his annual tribute, Shalmaneser V (727–722 BCE) besieged *Samaria and deported 27,000 of its residents; he repopulated it with a variety of foreigners who brought their own religion, which was later amalgamated with the worship of Yahweh (2 Kgs. 17: 28). According to this slander generated by local Jewish enmity, the *syncretism initiated the impure *Samaritan religion. But this is a quite misleading view of Samaritanism.

*Judah long remained loyal to Assyria, until finally it joined the coalition against *Sennacherib (705–681 BCE), but during the siege of Jerusalem the Assyrians unexpectedly withdrew. In the late 7th cent. Assyria collapsed and *Medes and Chaldeans expelled Assyrians from Babylonia. Nineveh was captured in 612 BCE (cf. Nahum 3: 7). This gave a measure of space to Judah for Josiah's reforms.

astrology The alleged influence of stars and planets on human affairs has flourished at least from the time of the Assyrians and Babylonians and is mentioned in the OT (Job 38: 33) and condemned by *prophets (Isa. 47: 13–14). The *Magi who came to the infant Jesus with *gifts from the east (Matt. 2: 1–2) were apparently astrologers. In the narrative they were moved by an unusual phenomenon in the skies, but the gospel account that they offered gifts to Jesus and paid him homage (Matt. 2: 11) may be intended to signify that the astrologers surrendered their superstitious arts before the Lordship of Christ, very much as in the Acts (19: 19) magicians publicly burnt their books when they submitted to the Lord Jesus.

Assumption of Moses The story of the world, written about 90 CE, in terms of conflict between God and *Satan beginning with Moses and culminating in the *Judgement. The work survives only in fragments, but appears to be known by Jude 9.

Asyncritus Greeted by Paul in Rom. 16: 14; the name is found in a Roman inscription of the era of the emperor *Augustus (27 BCE–14 CE).

asyndeton A Greek term for two sentences put together without any appropriate conjunction.

Athaliah The only queen to rule *Judah (843–837 BCE); she organized a purge of male rivals—except one, *Joash, who murdered her and succeeded to the throne (2 Kgs. 11: 16).

Athens The capital city of Attica, *Greece. About the time when *Nehemiah was rebuilding the fortifications of Jerusalem, Athens, three miles from the Aegean Sea, was enjoying its era of unparalleled prosperity. The income from its empire and commercial enterprise made possible the construction of its magnificent public buildings. The Parthenon had marble friezes, now displayed, controversially, in the British Museum in London; Socrates taught philosophy in conversations in the *market-place; and Sophocles was winning prizes in the theatre. The cultural prestige of Athens continued into the age of *Augustus and Roman domination. It was a university town to which prominent Romans sent their sons to be educated; the emperor Hadrian lived in Athens from 123 to 126 CE.

In this self-assured city of learning Paul spoke on the hill called *Areopagus (Acts 17: 19, 22) and the author of the Acts catches the atmosphere well. The council members were not passionately aroused to riot; they assessed the context of the speech with academic impartiality, and a few were convinced (Acts 17: 34).

atonement *Sin destroys the relationship between God and people, and atonement is the means by which a *reconciliation is effected. In the OT, *sacrifices and *offerings are prescribed and were well developed in the post-exilic period with a view to removing the barrier caused by sin which cut a person off from God's favour. Sacrifices were appointed by a merciful God for the restoration of fellowship which an individual by his own personal act was incapable of achieving. In the time of the *Maccabees animal sacrifices were thought to be supplemented by the atoning value of suffering. In the stories of Maccabean martyrs their sufferings are held to be an atonement not only for the sufferers but also for others. So the youngest of the seven brothers appeals to God that through him and his brothers the wrath of the Almighty which has fallen on the nation may be brought to an end (2 Macc. 7: 38).

In the NT atonement is connected with the *life, *death, and *resurrection of Jesus and he may have understood his coming death in atonement terms (Mark 10: 45). In Paul's letter to the Romans the source of

God's mercy in reconciling mankind was Jesus Christ (Rom. 3: 25) and in the letter to the Hebrews the notion of sacrifice is employed; the defilement of *sin is cleansed by the blood of a sinless victim. The *blood of Christ's obedient and offered life is sprinkled on our sinful consciences so that we can then draw near to God with a similar obedience.

Atonement, Day of In Hebrew, *Yom Kippur*, a day of rest and *fasting to give space for atoning for the *sins of the past year. On this day, the tenth of the seventh month (Lev. 23: 27–32), and this day only, the high priest went with *incense to the Holy of Holies in the Temple. A bullock was slaughtered and its *blood was sprinkled on the mercy seat. Two goats were provided; one was chosen by lot for a *sacrifice on behalf of the nation. On the other the high priest laid his hands, thereby transferring symbolically the nation's sins on to the *scapegoat which was then released into the *wilderness. This symbolism is used to explain the sacrifice of Christ, in Heb. 8 and 9.

Attalia A port where Paul and *Barnabas embarked, in *Pamphylia, on the southern coast of Asia Minor, on their way to *Antioch (Acts 14: 25–6).

Augustan cohort A *cohort consisted of 600 soldiers, and to bear the name of an emperor was an honour.

Augustus The title, meaning 'revered', granted to Octavian when he became ruler of the empire in 27 BCE. He was therefore the emperor at the time of Jesus' birth. He initiated the process by which the Roman emperor was accorded deification; he closely associated his own imperial majesty with the State.

authority An authority which belongs absolutely to God (Ps. 29: 10) and exercised in the OT through *prophets, priests, and kings, is delegated to Jesus in the NT. He forgives *sins (Mark 2: 5 ff.) and exercises *judgement on behalf of the Father (John 5: 22). After the *resurrection the universal lordship of Christ is proclaimed (Matt. 28: 18) and his authority is then vested in the *apostles (2 Cor. 13: 10).

There are also human authorities recognized in the NT—the State (Rom. 13: 1–6; 1 Pet. 2: 17) when it acts responsibly, and husbands (Eph. 5: 22 f.), but authority is to be exercised in a spirit of service (Matt. 20: 25 ff.).

Authorized Version The English translation of the Bible published in 1611; though not formally 'authorized', it was generally adopted within a generation for public reading in church. The work was initiated at the Hampton Court Conference of 1604 called by King James I and so is known also as the King James Version. It was partly based on the translation from the original languages made by William *Tyndale (NT in 1526) and was completed in five years by six committees of fifty-four scholars in all, meeting in Oxford, Cambridge, and the Jerusalem Chamber of Westminster Abbey. Early reactions to the AV were hostile: it had 'all the disadvantages of an old prose translation' with 'uncouth and obsolete expressions'. But it began to be widely accepted from about 1760, and has been of enormous influence on English literature. A revision by Benjamin Blayney of Oxford in 1769 established some alterations in the text. (*See* translations of the Bible *and textus receptus*.)

autograph The original copy of a book or letter. None exists of any of the biblical writings.

avenger of blood In a case of murder, the victim's nearest relative was expected to avenge his death, but the killing was restricted to the perpetrator of the crime (Deut. 24: 16) and endless blood feuds were avoided. However, because there could be unintentional homicide, cities of refuge were provided where those who killed accidentally could live without fear (Josh. 20).

Azazel On the Day of *Atonement, a goat, on whom were laid the hands of the priest, was dispatched into the wilderness where Azazel, a *demon, dwelt (Lev. 16: 8). This was the ritual of the *scapegoat which was held to carry away the people's *sins. Later, the scapegoat was hurled over a cliff.

Azotus A city in northern Judah, just inland from the Mediterranean coast, formerly called *Ashdod. It was held by the *Hasmonean John Hyrcanus and later by *Herod the Great. *Philip is said to have visited the city (Acts 8: 40) after baptizing the Ethiopian *eunuch.

B

B Used in textual criticism to denote *Codex Vaticanus, a 4th-cent. parchment MS of the Greek OT and most of the NT, kept in the Vatican library since 1475 CE.

Baal Although the word simply means 'lord' or 'master', it was used as the proper noun for the principal object of Canaanite worship. Baal was a fertility and agricultural god, and when the Israelites settled down in the land of their adoption there was a constant temptation to transfer their religious allegiance from Yahweh, who they believed had led them out of *Egypt and through the *wilderness, to a god who could perhaps promote good yields of corn and grapes. Some personal names contain the word *baal*, even a son of *David (1 Chron. 14: 7). Baal was also thought to encourage erotic adventures which the prophets condemned, so that the *Deuteronomist historian is delighted when a Baal temple is destroyed (2 Kgs. 11: 18) and the followers of Baal routed (1 Kgs. 18: 20–40). However, Baalism survived, and again it was extirpated in the reforms of *Josiah (2 Kgs. 23: 4–5).

Much of the available information about Baal arises from the *Ras Shamra texts of the 15th cent. BCE, in which he is said to have a consort, Anat. In Palestine the consort was *Asherah (Judg. 3: 7) or Astarte (Judg. 10: 6)—distinct beings in the eyes of their worshippers.

The conflict between Yahweh and Baalism was for the people's allegiance: was it to a transcendent God with ethical demands? Or was it to a Canaanite god immanent in nature? But transcendent Yahweh could also be acceptably worshipped as the one who 'rides on the wings of the wind' (Ps. 104: 3); and the worship of Baal also had ethical implications.

Babel Hebrew for *Babylon. According to local tradition the city was founded by the god Marduk; in the OT it is the archetypal place of confused languages (Gen. 11: 1 ff.). Linguistic diversity was imposed by God, according to the *aetiological myth, as a punishment for human arrogance in building a lofty tower—possibly a *ziggurat, as developed by the Babylonians and the *Assyrians. Such a building had several platforms at various heights and the whole was crowned by a temple as much as 70 feet (approx. 20 m.) above ground level.

When the crowd on the day of *Pentecost understood the *apostles because they suddenly began to use a variety of foreign languages (Acts 2: 5–13), the story is probably intended by the author to be the reversal of the chaos at Babel: diversity and confusion are replaced through the *Spirit by unity and understanding in a common humanity. 'In our own languages we hear them speaking about God's deeds of power' (Acts 2: 11).

Babylon Ancient city and kingdom in southern *Mesopotamia. In the NT (1 Pet. 5: 13 and Rev. 18: 2 where 'Babylon' is an uncomplimentary reference to *Rome) and in Christian spirituality ('And Sion in her anguish with Babylon must cope', from St Bernard's hymn 'Brief life is here our portion') Babylon is synonymous with a state or place of exile, of longing for release, even therefore of a kind of hope. The *Exile of the Jews to Babylon after the capture of Jerusalem in 586 BCE dominates the Bible (cf. Ps. 137: 1; Isa. 14: 4), as Babylon itself dominated the Near East during her brief period of imperial expansion.

But there was a long history before, and after, that: King *Hammurabi, who died in 1750 BCE, formulated his famous Law Code and there were other cultural achievements, including an interest in astronomy. In the 8th cent. BCE the kings of *Assyria also ruled Babylonia (situated on the River *Euphrates, and corresponding to modern southern Iraq) and when *Hoshea the king of Israel joined *Egypt against King *Shalmaneser V of Assyria in 724 BCE, Samaria was besieged and captured in 722 BCE and some of the inhabitants (27,000 according to Assyrian claims) deported to Assyria after Shalmaneser's death by Sargon II (2 Kgs. 17: 6).

In 626 the Babylonians (known also as Chaldeans) threw off the Assyrian yoke and, with the *Medes, captured the Assyrian city *Nineveh in 612. Babylonian triumphs continued up to the borders of *Egypt, and during the conflict between Babylon and Egypt *Jehoiakim, king of *Judah, ignoring the advice of *Jeremiah (Jer. 27: 9–11), supported Egypt and suffered the consequences when *Nebuchadnezzar besieged Jerusalem in 597 BCE. The young King *Jehoiachin (Coniah in Jer. 22: 24), who had succeeded his father Jehoiakim, reigned only three months before the city was captured and the king and many leading citizens were taken into exile. *Zedekiah was installed as governor of Jerusalem, but when he rebelled, Nebuchadnezzar returned to the city, destroyed the Temple, and deported most of the remaining population to exile in Babylon (586 BCE, 2 Kgs. 24: 10–25: 21). In Babylon, where there were magnificent buildings and roads, the famous Hanging Gardens, and a *ziggurat to Marduk, the Jews eventually settled down with *prayer and *circumcision and engaged in trade and agriculture (Jer. 29: 7). Some of their number returned to Jerusalem by permission of Cyrus, the Persian conqueror of Babylon, in 538 BCE. Those who remained became increasingly independent of Palestinian Judaism. The Babylonian *Talmud is testimony to the strength of Jewish religious and social life in the country of the *Exile, and it is no surprise that when Jews in *Palestine revolted in 66–70 CE and again under Bar-Kochba in 132–5 CE, they got no support from Babylon.

baker *Pharaoh's chief baker was an important official (Gen. 40), and bakers' shops were a necessary part of city life (Jer. 37: 21). In an unflattering reference to the monarchy, *Samuel foretells that kings will abduct women to the royal bakeries (1 Sam. 8: 13).

Balaam A *prophet from *Transjordan, retained by Balak to utter threats against the enemies of *Moab. Although the legend of Num. 23–4 shows Balaam as obedient to the word of the Lord, elsewhere he was responsible for causing Israel to sin at Baal-peor (Num. 31: 16) and he becomes the very type of mercenary prophet who is hired to utter curses (2 Pet. 2: 15). Indeed the donkey on which he rode could see more clearly than the professional seer on his back (Num. 22: 33).

balm of Gilead Resin gathered from the mastic tree. *Joseph's brothers (Gen. 37: 25) sold him to merchants whose caravan of camels conveyed this commodity. The reference in Jer. 8: 22 is to a medicinal balm, perhaps the gum of the balsam tree.

ban Things or persons seized in warfare were placed under a ban, that is regarded as out of bounds for Israelites and wholly dedicated to God. Objects not necessarily taken in *battle might also be dedicated to God and might be destroyed by fire (Lev. 27: 28) or set aside for the priests (Lev. 27: 21 f.). Possibly it was feared that diseases could be caught from enemy possessions and the ban was designed to prevent infection, though a religious meaning was ascribed to it. *Achan was destroyed for having plundered forbidden booty (Josh. 7) and *Saul was rebuked by the prophet *Samuel for not destroying the most valuable possessions of the *Amalekites (1 Sam. 15: 23).

bank Banking institutions did not exist in Israel for safeguarding people's money or possessions or granting credit, but money could be deposited in the Temple treasuries (2 Macc. 3: 10 ff.). It was contrary to the law to exact interest on money lent to a fellow Israelite (Exod. 22: 25), though seemingly the practice was common in the time of Jesus (Luke 19: 11–27).

banquet A formal *meal for which invitations were sent. It was the custom for those invited to be told at the last moment when the *feast was actually ready (Luke 14: 17), and the seating arrangements corresponded to the host's assessment of his guests. One symbol of *eschatological joy in the Kingdom (Matt. 26: 29; Rev. 19: 9) is that of a heavenly banquet.

baptism The Greek word for the rite of baptism means 'to dip in' or 'to wash' but the classical meaning of drowning or overwhelming is found in the *LXX (Isa. 21: 4), and there is a suggestion of this sense in the NT when Jesus predicts his coming 'baptism' of death (Mark 10: 38–9) and perhaps when Paul refers to the Israelites being 'baptized' in the *Red Sea (1 Cor. 10: 2). Both these passages are interpretations of Christian baptism; for Paul baptism is compared to Israel's *Exodus through the sea; and in baptism Christians share symbolically in Jesus' *death and *resurrection; they are buried with him and rise to a new *life. So baptism is also regarded as the beginning of a new life (John 3: 4–5).

The Christian rite was not without some partial precedents in Judaism and in other religions. *Water is a natural and vivid source of purification, so *Gentiles who wished to become Jews baptized themselves before *circumcision. At *Qumran there were elaborate rites of *purification by water. *John the Baptist invited his hearers to repent and to be baptized in the River *Jordan, and Jesus accepted baptism at his hands, not for remission of sins (Matt. 3: 13–15) but to identify himself with his people. It was the moment when he was commissioned to proclaim the kingdom, the moment when he was adopted as Son of God (Mark 1: 10–11).

Baptism immediately assumed the role of rite of initiation into the *covenant which had been, and is, the purpose of *circumcision in *Judaism (Col. 2: 11–12). Peter exhorted his audience in Jerusalem to repent and accept baptism (Acts 2: 38). An Ethiopian *eunuch was baptized, without an elaborate course of instruction, by *Philip as soon as there was a handy pool of water (Acts 8: 38). It is not recorded that the Ethiopian became a member of a local Christian community.

It was held that in baptism the *gift of the *Spirit was conferred, as it had been with Jesus (Mark 1: 10), and the prerequisite for baptism was *faith (Gal. 3: 14). Whether the requirement of faith therefore excludes infants from the Christian *sacrament is much debated; there is no explicit evidence either way in the NT. It is argued that households who were baptized (Acts 16: 15, 33) would have included infants; and the parallel with circumcision as a rite of infant initiation points in the same direction.

Archaeological evidence from the early centuries shows that baptism was administered sometimes by submersion or immersion, in which the rite symbolically re-enacted the process of burial and resurrection, but also by affusion from a vessel when water was poured on the candidate's head, just as earth was sprinkled over the corpse at a funeral. There is a similarity of symbolism in both forms of ritual. Paul mentions a practice in his time of baptism 'on behalf of the dead' (1 Cor. 15: 29). He is referring, perhaps, to those at *Corinth who were baptized with a view to being united at the resurrection with their Christian friends who had died.

bar- Frequently a prefix in Aramaic to a personal name, meaning 'son of'.

Barabbas Aramaic for 'son of the father'. All the gospels name Barabbas as a prisoner held by the Romans at the time of Jesus' trial as a nationalist rebel (Mark 15: 7) but released. Mark and Matthew report that there was a custom to release a prisoner at *Passover (John calls it a Jewish custom which *Pilate accepted, John 18: 39) but there is no other evidence for such an arrangement. Some MSS of Matt. 27: 16 f. give the name as *Jesus* Barabbas (as NRSV)—in which case Pilate's question is 'Which Jesus do you want released?' The crowd is said to demand the *crucifixion of 'Jesus who is called the Messiah' and the release of Barabbas.

baraitha A saying included in the *Talmud but not recorded in the *Mishnah. One important baraitha is an instruction to be added as number 12 to the Eighteen Benedictions, condemning heretics. It was inspired by *Gamaliel II, the leader of the rabbinic assembly at *Jamnia, towards the end of the 1st cent. CE.

Barak An Israelite commander (Judg. 4) who routed Canaanites led by *Sisera in the late 12th cent. BCE.

barbarian The term used by the *Greeks of the classical period (from Herodotus, *c.*410 BCE) for those who spoke a foreign language, such as Egyptians. It continued to bear a pejorative meaning in the Hellenistic age and when Paul wished to refer to the entire world excluding Jews the nations were 'Greeks and barbarians' (Rom. 1: 14, NRSV, NJB, AV).

Bar-Kochba The name means 'son of a star' (from the prophecy in Num. 24: 17); he was the leader of guerrilla forces who fought the Romans from 132 to 135 CE when the emperor Hadrian prepared to erect a temple to the god *Jupiter on the site of the Jerusalem Temple. Bar-Kochba believed himself to be the Messiah, and was hailed as such by Rabbi Akiba. His enthusiastic supporters suffered (and inflicted) heavy losses before their final defeat. Christians in the area refused to join the revolt and were persecuted, according to Justin and Eusebius.

barley Known in Palestine before the arrival of the Israelites, the crop was regularly cultivated throughout the OT and NT periods. It was used in place of oats for feeding livestock and for *bread (John 6: 9) of cheaper quality than wheat bread. Harvesting barley took seven weeks and finished at the feast of *Pentecost in June (Ruth 2: 23).

Barnabas A Cypriot and a Levite who, on becoming a Christian, generously disposed of his real estate (Acts 4: 36–7) and was instrumental in persuading Paul to come to *Antioch (Acts 11: 25–6). Subsequently, Paul and Barnabas travelled together, accompanied by John Mark, a relative of Barnabas, and they are both reported to be at the Council of Jerusalem (Acts 15: 2; 12). A quarrel, however, ended the partnership; according to Acts 15: 38 this was because Paul doubted the reliability of Mark whom Barnabas again wanted with them, though the reason may well have been that Barnabas had sided with Peter against Paul in the confrontation at Antioch about sharing a common table with *Gentiles (Gal. 2: 13), which was a scene that the author of Acts probably found resistant to his theme of ecclesiastical harmony. However, Paul (or whoever wrote the epistle to the *Colossians) has a friendly reference later to Mark and Barnabas (Col. 4: 10).

Barnabas, epistle of A pseudonymous Christian tract written in *Alexandria at the end of the 1st cent. CE decrying the value of the literal observance of Jewish ritual laws. The work appears in full in Greek in the Codex *Sinaiticus, and in some early Christian circles seems to have hovered on the edge of the NT *canon though never actually accepted.

Barnabas, gospel of A medieval forgery by a convert to Islam purporting to give the secret teaching of Jesus. It consists of 222 chapters and claims that Jesus (who was not crucified—Judas took his place) prophesied the coming of Muhammad.

barrenness It was the worst of misfortunes for a Hebrew or Jewish wife not to bear *children (1 Sam. 1: 10 ff.). So the prediction that women who were childless will be thankful is an indication of how appalling will be the disaster to fall on Jerusalem (Luke 23: 29). The city was captured and destroyed by the Romans in 70 CE after the Jewish rebellion.

Barsabbas Joseph Barsabbas was one of the candidates for the place in the apostolic Twelve left vacant by the desertion of *Judas Iscariot (Acts 1: 23). A Judas Barsabbas travelled with *Silas to accompany Paul and *Barnabas when they returned to *Antioch to report the decision of the council at Jerusalem (Acts 15: 22), and it is suggested that Judas Barsabbas was a brother of Joseph Barsabbas.

Barth, Karl (1886–1968) Probably the greatest Protestant theologian of the 20th cent. Born in Switzerland, Barth became internationally known for publishing a commentary in 1919 on Paul's epistle to the Romans; brilliantly translated into English by the Cambridge NT scholar Edwyn *Hoskyns, it was not so much a commentary in the established sense as a manifesto recalling the Reformed Church back to the theology of Calvin and away from liberal concessions to human culture. God was known only as he made himself known in the revelation of Jesus Christ. Human experience, even religious activities, must not be identified with the activity and will of God. Primarily therefore Barth was a dogmatic theologian and not a biblical scholar; he rejected *Bultmann's programme of *demythologization and argued for a traditional understanding of the *Virgin Birth and the *Resurrection, though he was no *fundamentalist. He welcomed the procedures of biblical *criticism, holding that the word of God is the incarnate Word, Jesus Christ, to whom the NT bears witness but cannot express the fullness of the divine Word.

Bartholomew One of the Twelve (Mark 3: 18; Acts 1: 13; not mentioned by John) whom later tradition identified with *Nathanael (not mentioned by the synoptists), who was brought to Jesus by *Philip (John 1: 45). Perhaps the identification has been made because Bartholomew is named in the synoptic lists after Philip. Eusebius, historian and bishop of Caesarea (313 CE), claims him as a missionary to India.

Bartimaeus A blind beggar healed by Jesus outside *Jericho, according to *Mark (Mark 10: 46–52), though in Matt. 20: 29–34 the narrative is of two blind men healed. That the beggar followed Jesus 'on the way' (Mark 10: 52) indicates that he became a *disciple when his eyes were opened. Discipleship is one of Mark's themes and according to Acts 9: 2 the early Christian community knew itself as 'the Way'.

Baruch Son of Neriah (Jer. 36: 4); the names have been discovered on a clay seal of the 7th cent. BCE. Baruch was a friend and secretary of *Jeremiah who, in spite of possibly being an employee of King Jehoiakim, had the courage to read Jeremiah's subversive prophecies publicly in the Temple (winter, 604 BCE). The king burnt the scroll which contained advice to submit to the Babylonians. But the words were written out again (Jer. 36: 28). When Jeremiah was imprisoned, Baruch was entrusted with the legal deeds of property in Jeremiah's home town, Anathoth, a short distance NE of Jerusalem (Jer. 32: 12–16). This transaction was arranged by Jeremiah as a prophecy that after the Chaldeans (Babylonians) had gone, fields would again be bought and sold (Jer. 32: 44). However, it was not to be. After the siege and the Babylonian victory, Jeremiah and Baruch were taken by a group of anti-Babylonian conspirators to *Egypt (586 BCE).

It used to be thought that those sections in the book of Jeremiah which are biographical are the work of Baruch, though they have been edited in the interests of Deuteronomic history (Jer. 39: 11–12; 40: 1–6) with its characteristic refrain (40: 3) that evils had come upon the people 'because they have sinned against the Lord'.

Baruch, book of A miscellaneous collection of prayers and narrative from the 2nd cent. BCE attributed to *Jeremiah's scribe Baruch. It was contained in the Greek *LXX (though it may have been originally composed in Hebrew) and is therefore classified as part of scripture by Catholics, but its absence from the Hebrew OT puts it in the *Apocrypha for Protestants.

The work claims to be written during the Babylonian *Exile, advocates prayers for *Gentile rulers, and has affinity with the Wisdom literature (such as Job 28). (It is sometimes referred to as 1 Baruch to distinguish it from the two following.)

2 Baruch, 3 Baruch The Syriac Apocalypse of Baruch and the Greek Apocalypse of Baruch are translations of a pseudo-biographical work probably written in Hebrew in the first part of the 1st cent. CE. It contains a coherent account of Jewish *eschatological expectations, including the advent of a *Messiah.

Barzillai A man of wealth who gave support to *David during the time of *Absalom's rebellion (2 Sam. 17: 27–9). He later declined the invitation to join the royal court owing to his great age (2 Sam. 19: 34).

Bashan A fertile area east of the *Jordan whose fat cattle were proverbial (Amos 4: 1). Once ruled by *Og (Num. 21: 33), who was defeated at Edrei (Deut. 3: 1), it was taken over by the tribe *Manasseh.

basket Different words are used in Greek for the baskets which contained the fragments left over after the two feedings of the 5,000 and 4,000 (Mark 6: 43; 8: 8). At the latter, the baskets were large, such as the one in which Paul escaped over the wall at *Damascus (Acts 9: 25).

basin Half of the blood of animals sacrificed as burnt *offerings was poured into basins and Moses dashed it over the people (Exod. 24: 8). Basins (NRSV) were large bowls used for the service of the *altar in the *tent of meeting (Num. 4: 14).

bath A liquid measure of about 72 litres (16 gallons).

bath qol Hebrew for 'daughter of a voice'; the term used for the voice of God by the rabbis; cf. in the NT the heavenly voice at the *baptism (Mark 1: 11) and *transfiguration (Mark 9: 7) of Jesus.

baths The heat of *Palestine made frequent baths necessary, and in the NT period the Romans built public baths. It was long customary to provide water for guests to wash their feet on arrival at a house (Judg. 19: 21), and an oval tub for this has been excavated at *Samaria. Bathing was a means of ritual cleansing (e.g. of Aaron and sons in Exod. 29: 4) and for purification after *'leprosy' (Lev. 14: 8–9) and is a potent symbol of moral cleanliness (e.g. Jer. 4: 14) or of vindictive triumph (Ps. 58: 10).

Bathsheba The wife of *Uriah the Hittite, she was observed bathing by *David and much desired, and seduced. So he arranged for Uriah to be killed in battle and married Bathsheba (2 Sam. 11: 27). *Nathan the prophet rebuked David. Bathsheba gave birth to *Solomon, who, thanks largely to the intervention of his mother (1 Kings 1: 16), eventually succeeded David.

battalion Used in RSV for the whole 'company' (REB) or '*cohort' (NRSV), several hundred strong, which *Pilate brought up with him to Jerusalem from his residence in *Caesarea (Matt. 27: 27).

battle From beginning (Gen. 4: 23–4) to end (Rev. 20: 8) the Bible is a story of conflict, largely because Israel was situated between great powers to the north and south who therefore tended to clash on Israelite soil. God is himself frequently portrayed as a warrior, as in the Song of Deborah (Judg. 5: 4) and the psalms (e.g. Ps. 24: 8). Paul compares the Christian life to a battle (1 Cor. 14: 8; 2 Cor. 10: 4).

Baur, Ferdinand Christian (1792–1860) German NT scholar and Church historian of Tübingen University. He denied the authenticity of many of the letters ascribed to Paul and maintained that a bitter struggle in the apostolic Church between Peter and Paul was concealed by the author of the Acts in the interests of reconciliation. The conflict was resolved in the form of Catholic Christianity in the 2nd cent., when the fourth gospel was compiled (150 CE). It was an interpretation resembling the dialectical philosophy of history developed by Hegel (of thesis, antithesis, and synthesis). The principal English antagonist of Baur's view was J. B. *Lightfoot. Today it is widely conceded that NT literature should be understood in terms of its diversity and that the differences within the early Church are of a piece with the contradictions and tensions of universal history—and to that extent Baur has some modern support.

beam Squared timbers were used to support a flat Palestinian roof. In Jesus' parable of Luke 6: 41 REB uses 'plank' for AV 'beam', but 'log' of NRSV is misleading for English readers.

beard Full beards were worn by Jewish men, for whom it was a sign of vitality (unlike the Egyptians, Gen. 41: 14). It was an outrage when Hanun, king of the Ammonites, cut off the beards of David's envoys (2 Sam. 10: 4); and *Ezra pulled out his beard to show the depth of his distress (Ezra 9: 3). It has been suggested that the insult voluntarily suffered by the Servant (Isa. 50: 6) of plucking the beard is influenced by the ritual of the Babylonian New Year Festival when the king submitted to this cultic suffering to give reassurance to the people that in their time of distress they would also be succoured by God.

beast Mammals as distinct from birds or fish (Gen. 1: 30) either wild (as in Mark. 1: 13) or capable of being tamed; divided by the Jews into *clean and unclean. In Dan. 7 the four great oppressing nations are symbolized by beasts, and this is taken over by Revelation where the beast from the *abyss (sea) in 11: 7 is the anti-Christian power of *Rome, indicated by the 'seven hills' of Rev. 17: 9. Mark 1: 13 may be pointing the contrast between Jesus, with whom the beasts are at peace, and the enmity and disruption which has prevailed from the time of *Adam—a hint of paradise restored.

beatitudes There are numerous instances of the beatitude form in the OT, as in the psalms (1; 32; 119), but the word is particularly applied to part of Jesus' *Sermon on the Mount (Matt. 5–7) following his forty days' temptation in the *wilderness. It opens with a series of beatitudes (*blessings), which are essentially *eschatological; that is, they are related to the arrival in Jesus of the *kingdom of God. *Disciples are blessed when they recognize this intervention of God: they may be living in this world, but under the pressure of the Kingdom, they reject the usual values of this world. Thus, 'the poor in spirit' are blessed—those who are unattracted to, indifferent to, money and possessions are blessed. Behind Matthew's narrative is the recollection of Moses receiving the *Law on Mount Sinai. But it is placed firmly in a context of grace, not threats (Heb. 12: 20–1).

The beatitudes take a different form in Luke 6: 20–6, and the explicit mention that Jesus stood 'on a level place' (6: 17) seems to be making a deliberate contrast with Matthew 5: 1, since to be on level ground would not normally call for special notice. Luke loves to mention Jesus' compassion for the literally poor, the disadvantaged, the despised, and second-class citizens: so in Luke the first beatitude is 'Blessed are you who are poor', followed in 6: 24 by 'But woe to you who are rich'.

There are also isolated beatitudes scattered through the gospels, and one is also quoted at Acts 20: 35.

Beautiful Gate Peter and *John healed a lame beggar at the Beautiful Gate of Herod's Temple (Acts 3: 2), which was probably the bronze gate on the eastern side known as the Gate of Nicanor; it led from the Court of the Gentiles to the Court of the Women.

Beelzebul The *devil. 'Beelzebub' (= 'Lord of Flies'), in AV, following Latin Vulgate; Greek MSS have Beelzebul (= 'Lord of Heaven' or 'Lord of the House'), which is the more reliable spelling. It is a corruption of 'Baal-Zebul' and may refer to the god of Ekron (2 Kgs. 1: 2 f.). It is used insultingly by *Pharisees (Matt. 12: 24) to persuade the crowd that the powers which astounded them were not Messianic but demonic. Jesus' response is that demonic power does not effect exorcisms; *Satan would hardly be engaged in committing suicide.

The argument is couched in the thought-forms of the 1st cent. and was doubtless persuasive. For modern people who attribute blindness and dumbness to an explainable physical defect and who do not recognize the god of Ekron, or Zeus, or any other prince of demons, the discussion has no basis in reality.

Beeroth A Benjaminite city (Josh. 9: 17) where the murderers of Ishbaal, son of *Saul, lived (2 Sam. 4: 2–3), probably about 15 km. (9 miles) north of Jerusalem.

Beersheba The most southerly place of Israel, notable for its wells and therefore suitable for pasturing sheep. There are notes in Genesis that the ancestors *Abraham, *Isaac, and *Jacob stayed there (Gen. 22: 19; 28: 10; 46: 1). The phrase 'from *Dan to Beersheba' was a way of indicating the full extent of Israel (2 Sam. 17: 11).

beggar Illness or some physical disability such as blindness and lameness made it impossible to earn a living; recourse to people's charity was the only way of surviving. There are numerous accounts of beggars in the gospels and Acts (e.g. Mark 10: 46–52; Acts 3: 10); the only named character in a parable is the beggar *Lazarus in Luke 16: 19–30. Disinterested charity (Matt. 6: 3) would have seemed extraordinary in the Graeco-Roman world (*see* poor)—but it is commanded by Jesus, especially as recorded in the gospel of Luke (12: 33; 18: 22, etc.)

beggarly Paul asks the Galatians (Gal. 4: 9) how they could contemplate returning to the beggarly (AV, NRSV), 'bankrupt' (REB, NJB) *elemental spirits. He therefore contrasts the poverty of such a religious system with the riches of the gospel.

beheading *John the Baptist was beheaded by order of Herod Antipas (Mark 6: 27), and, probably, the apostle *James by Herod Agrippa (Acts 12: 2), but it was not the form of capital punishment permitted in the Jewish *Law, which was *stoning (Lev. 20: 27).

behemoth A great *beast whom God invites Job (Job 40: 15) to admire, usually thought to be either the hippopotamus, which frequented the *Nile though not the *Jordan, or a mythical creature from the era of primeval chaos.

Bel and the Dragon, book of An addition (ch. 14) found in the Greek OT to the book of Daniel; probably written in the 2nd cent. BCE, consisting of menacing satires against idolatry in the form of legends about the survival powers of *Daniel in the Babylonian era. The piece is regarded as a deuterocanonical part of the book of Daniel in Roman Catholic and Greek Orthodox editions of the Bible, but put into the *Apocrypha in Protestant versions.

Belial (or Beliar) In the OT the word in Hebrew denotes worthlessness (hence sons of Belial in AV of 1 Sam. 10: 27; 'worthless fellows', NRSV; 'scoundrels', REB). In the inter-testamental literature and *Dead Sea scrolls, Belial becomes the name of the wicked one, and so in the NT a synonym for *Satan (2 Cor. 6: 15).

beloved An adjective applied both to children (1 Cor. 4: 17) and to adult Christians (1 Pet. 2: 11) and even to Christ (Eph. 1: 6, though there the Greek uses the perfect passive participle of the verb as a noun). The 'Beloved Disciple' is mentioned in the gospel of John as especially close to Jesus (13: 23–5) and as the author of the gospel (21: 24). He has traditionally been regarded as John the disciple, but the identification is not made in the gospel.

Belshazzar Incorrectly asserted in Dan. 5: 2 to be the son of *Nebuchadnezzar and last king of *Babylon. *Cuneiform inscriptions found at *Ur, however, show that he was the son of Nabonidus of Babylon (556–539 BCE) and that he ruled as regent during the king's absence. The story in the book is of a feast given by Belshazzar during which a mysterious hand writes a legend of doom on the wall and he is slain that very night—a drama that has inspired many paintings and some great music (e.g. William Walton's oratorio).

Belteshazzar At the king's court *Daniel (Dan. 1: 7) accepted this Babylonian name without protest, though it appears to incorporate a prayer to the god Bel (Marduk, Isa. 46: 1), which seems inconsistent with his strict adherence to food regulations (Dan. 1: 8).

ben- Hebrew for 'son'.

benedictus The first word in Latin ('Blessed') of the oracle attributed in Luke 1: 68 to *Zechariah, who had been struck dumb for disbelieving the *angel (Luke 1: 20). With his power of speech restored he at once gives thanks to God for the birth of his son John. The canticle has been incorporated into Church liturgies.

benefactor flattering title accorded to *Gentile kings, e.g. of *Syria and *Egypt; but Jesus forbids its use among the *disciples (Luke 22: 25).

Benhadad The name of several kings of *Damascus, among them one who made war in Israel during the reign of *Ahab, who defeated Benhadad (1 Kgs. 20) but withdrew. The prophet *Elisha visited *Damascus and was consulted by Benhadad about his illness; but next day the king was suffocated with a blanket by his official, *Hazael (2 Kgs. 8: 15).

Benjamin The younger son of *Jacob and *Rachel (Gen. 35: 18), from whom the tribe of Benjamin claimed to be descended. He is mentioned as being close to *Joseph, and this suggests that historically the Benjaminites were associated with the northern (Joseph) tribes, which makes good sense as the name Benjamin means southerner. They occupied a strip of land which was a buffer between *Ephraim and *Judah. After an atrocity which led to an onslaught from the other tribes (Judg. 20), a weakened Benjamin became dependent on *Judah.

*Saul came from the tribe, but so also did Shimei and Sheba, opponents of *David (2 Sam. 16: 5; 20: 1). After the separation of Israel and Judah into two kingdoms, Benjamin became part of the southern kingdom (Neh. 11: 7) but descendants of the tribe still remembered their history, as did *Jeremiah (1: 1) and Paul (Phil. 3: 5).

Bernice Daughter of Herod *Agrippa I, and sister and mistress of Herod Agrippa II, with whom she listened to Paul in *Caesarea (Acts 25: 13 ff.) She had married her uncle Herod of Chalcis, who died in 48 CE.

Beroea A Macedonian city on the Egnatian Way which Paul and *Silas visited, there enjoying a favourable reception (Acts 17: 10–15).

beryl The green precious stone which was one of twelve on the high priest's *breastpiece (Exod. 28: 20), and in Rev. 21: 20 the mineral beryl forms the eighth of the foundation stones of the wall of the New Jerusalem.

beth- Hebrew for 'house of'.

Bethabara The scene of *John's baptisms (John 1: 28) as given in inferior MSS followed by AV. All English translations from RV prefer the more likely reading 'Bethany across the *Jordan', i.e. not the village of that name near Jerusalem.

Bethany A village over a mile east of Jerusalem on the way to *Jericho, and SE of the Mount of *Olives. Here Jesus raised *Lazarus from the dead according to John 11 and was entertained by *Simon the Leper (Mark 14: 3). It has been argued that the *Dead Sea scrolls contain evidence that Bethany was in fact a leper colony. Jesus spent his last night in the village before riding into Jerusalem (Mark 11: 1–11).

Bethel Formerly Luz, it was renamed by *Jacob (Gen. 28: 18–19). In the northern part of the divided kingdom, it became notorious when *Jeroboam established a rival sanctuary there with a *golden calf to discourage his people from travelling south to Jerusalem (1 Kgs. 13: 1–32; 2 Kgs. 10: 29). Bethel was therefore denounced as idolatrous by the prophets (Hosea 10: 15; Jer. 48: 13). When *Josiah, the reforming king of Judah (639–609 BCE), enjoyed a reign of prosperity, he was able to destroy the *sanctuary and kill its priests (2 Kgs. 23: 15–20), though the city itself, which seems to have been destroyed earlier by the Assyrians (722 BCE) and rebuilt, was spared. The Babylonians under Nabonidus were less generous than Josiah and they destroyed the city again. But once more it rose and began to thrive right through the Hellenistic and Roman eras and the Middle Ages, and today is known as Beitin.

Bethesda The name for the pool in Jerusalem by the *Sheep Gate where Jesus is recorded to have healed a paralysed man on a *Sabbath (John 5: 9). The reading in some MSS of Bethesda is followed by AV and REB, but NRSV prefers Beth-zatha, of other MSS, known as a northern extension of the city. The double pool has been excavated; there was also a subterranean bath nearby, which could have been in use for therapeutic purposes—so possibly John 5: 2–9 has fused two adjacent places into one.

Beth Horon A city situated on the road from the coast to Jerusalem and consequently often under attack (Josh. 10: 10; 1 Macc. 3: 16).

Bethlehem Small town about 8 km. (5 miles) south of Jerusalem. It was in existence certainly by 1250 BCE and is mentioned frequently in the OT especially in connection with King *David, where his great-grandmother *Ruth lived (Ruth 1: 19; 4: 11) and where he himself was born (1 Sam. 17: 12). There is a prophecy in Micah (5: 2) that the fortunes of *Judah will be renewed through a ruler in Israel born in Bethlehem of the house of David. This verse was taken by Christians as a prophecy of the birth of their Messiah, Jesus, who is described in the gospels (Matt. 2: 1; Luke 2: 4; John 7: 42) as being born in Bethlehem of *Judaea, although his home town was *Nazareth (Luke 4: 16). But the reasons given for the birth in Bethlehem are, historically, so unconvincing that they seem to be conceived to validate Mic. 5: 2 and 2 Sam. 5: 2. Luke describes an unbelievable universal *census which obliged the family to leave their home at Nazareth for the registration; while Matthew explains that on returning to Bethlehem from *Egypt (2: 22) they decided it would be safer after all to live at Nazareth rather than in their home at Bethlehem.

In 386 CE Jerome made his Latin translation of the Bible (the *Vulgate) while living in Bethlehem.

Bethphage A village on the Mount of *Olives from which Jesus obtained the *ass on which he rode into Jerusalem (Mark 11: 1).

Bethsaida A town on the NE of the Sea of *Galilee where Peter, *Andrew, and *Philip were born according to John 1: 44; 12: 21. Luke (9: 10–17) locates the feeding of the 5,000 in the neighbourhood of Bethsaida, though Mark's geography (Mark 6: 45) is unclear. Nevertheless the inhabitants were unreceptive to Jesus' message (Matt. 11: 21).

Bethshean (or **Bethshan**) An ancient site in a commanding position between the *Jezreel and *Jordan valleys where strong Egyptian influence has been detected by excavations. At the time of the entry into *Canaan under *Joshua the Israelites failed to capture the city (Judg. 1: 27) which was later taken by the *Philistines. But it became an Israelite town eventually during the conquests of *David, and the area, rich in grain, necessitated the appointment there of one of *Solomon's twelve commissaries (1 Kgs. 4: 12). In the Hellenistic and Roman periods the town was called Scythopolis and was the principal city of the Decapolis (cf. Mark 7: 31) and lay on the west of the *Jordan. Major excavations at the town continue.

Bethshemesh (1) A city, 25 km. (16 miles) SW of Jerusalem, for which Israelites and *Philistines contested. The *Ark of the Covenant was returned here after it had been captured and temporarily retained by the Philistines (1 Sam. 6: 1–16). Later, in the time of *Ahaz, the Philistines regained control. (2) Another city of the same name (but 'Heliopolis' in NRSV) in *Egypt, connected with a cult of the sun, is mentioned in Jer. 43: 13. The word *shemesh* means 'sun'.

Bethzatha The name in some MSS in John 5: 2 followed by NRSV etc. for *Bethesda.

beulah Hebrew for 'married' and to be applied to Jerusalem when the exiles return (Isa. 62: 4) and the city thus remarried to God.

Beza, Theodore (1519–1605). Swiss reformed theologian who succeeded John Calvin as leader in Geneva. He is remembered in biblical studies for having discovered a 5th-cent. MS of the gospels and Acts ('Codex Bezae'), which he presented in 1581 to the University library at Cambridge. In 1565 he brought out the first critical edition of the Greek NT by a study and comparison of available MSS and translations.

Bible The word derives from the Greek *biblia*, meaning 'books', and the plural witnesses to the fact that the Christian Bible is not a unity but a collection. The Old Testament consists of twenty-four books written in Hebrew (except for a few passages in Aramaic) and is often called the *Masoretic text. At the *Reformation the Hebrew books were rearranged and some were divided and so became thirty-nine in all, which is the number in the English AV and subsequent revisions. Roman Catholics, however, included as deuterocanonical works additional books from the *LXX, so the Catholic OT contains forty-three books.

The Hebrew Bible was traditionally divided into three parts—the Law, the Prophets, and the Writings:

The Law (Torah): the first five books, the Pentateuch: Genesis, Exodus, Leviticus, Numbers, Deuteronomy.

The Prophets (Nebi'im): Joshua, Judges, Samuel, Kings; the Latter Prophets: Isaiah, Jeremiah, and Ezekiel; and the Twelve Minor Prophets: Hosea, Joel, Amos, Obadiah, Jonah, Micah, Nahum, Habakkuk, Zephaniah, Haggai, Zechariah, Malachi.

The Writings (Ketubim): Psalms, Job, Proverbs, Ruth, Song of Songs, Ecclesiastes, Lamentations, Esther, Daniel, Ezra–Nehemiah, Chronicles.

Jews use the acronym Tanak for their scriptures: it is a made-up word from the initial letters of Torah, Nebi'im, and Ketubim. The Greek translation of the Hebrew Bible known as the *Septuagint (*LXX) adds: 1 Esdras, Wisdom of Solomon, Ecclesiasticus (or Sirach), Judith, Tobit, Baruch, the Letter of Jeremiah, 1, 2, 3, 4, Maccabees, and additions to Daniel and to Esther.

It is not entirely clear which books were accepted as authoritative in *Palestine in the time of Jesus, but by about 90 CE the rabbis restricted the canon to the Hebrew text, and this narrower list was preferred by several of the Fathers, though generally the Christian Church used and accepted the Greek OT. At the Reformation this larger list continued to have the authority of the Roman Catholic Church as defined at the Council of Trent, but for the most part Protestants preferred the Hebrew OT and regarded the other books as *Apocrypha ('hidden' or excluded), since they contained doctrines, like prayer for the dead, which were repudiated by Protestants.

The NT consists of twenty-seven books written in Greek, though the process of determining the *canon was tortuous: the four gospels of Matthew, Mark, Luke, and John; epistles to the Romans, two to the Corinthians, Galatians, Ephesians, Philippians, Colossians, two to the Thessalonians, two to Timothy, Titus, Philemon, Hebrews; the epistles from James, two from Peter, three from John, Jude; and the Revelation.

Chapters and verses were invented for convenience at quite late dates: the Jewish Masoretes—rabbinical scholars of the period 500–1000 CE dedicated to preserving the correct Hebrew text—divided the books into verses. The NT was divided into chapters by the English Archbishop Stephen Langton (died 1228) and verses were first introduced in 1551 in the Greek and Latin editions of the NT of Robert Stephanus in Geneva and reproduced in the English Geneva Bible of 1560.

Bible, English translations of The gospel of John was translated into Anglo-Saxon by the Venerable Bede (673–735 CE). An English translation of the entire Latin Bible was produced in 1380 under the influence of John

Wyclif, called 'the Morning Star of the *Reformation' but it was not well received by the Church authorities. William *Tyndale, in exile in Cologne, completed a translation of the Greek NT in 1525, in which Paul addresses the elders of *Ephesus (Acts 20: 20) as the 'congregation of the Lord' rather than 'Church'. Miles *Coverdale provided a revision (1539) which was called the 'Great Bible' (its psalter was incorporated into the Anglican Book of Common Prayer and is therefore still in use) but in 1560 exiles in Geneva, who had fled during the reign of Mary, published another Bible which became popular with the laity but not with the bishops, for it too used 'congregation' in place of Church. So they offered a revision of the Great Bible, known as the Bishops' Bible, read in the Church of England for forty-three years.

Roman Catholics also decided to make a translation; it was taken from the Latin and was the work of exiles at Douai and Reims in 1582 (NT) and 1609 (OT). Although wooden and literal it remained (though in a revision by Bishop Challoner in 1749) the official Roman Catholic translation in Great Britain until Mgr. Ronald Knox published a new translation (from the Latin) in 1955, which was elegant and clear but too different to win much approval.

What is known as the *Authorized Version or the King James Version (1611) was a revision of the Geneva Bible by three companies of scholars, divided into six panels. Their work became the standard translation for centuries after and part of English thought and speech; but by the 19th cent. it came to be regarded as unsatisfactory by reason of much new knowledge of Greek MSS and ancient versions in other languages. Accordingly a process of revision was set in motion and a Revised Version of the NT, based on the Greek text of *Westcott and *Hort, was published in 1881, followed by the OT in 1885 and the Apocrypha in 1895. There was consultation with American scholars, but a joint revision was not published. Instead the American revision (1901) was called the Standard Version and it had the English variations in an appendix—as the English RV also had the American variations in an appendix. These versions remained the usual text for students until the Revised Standard Version was published in the USA in 1946, and repeatedly, culminating in the New RSV of 1989. A popular modern translation of the Bible was that of the Scottish divine James Moffat, in 1935, and to supplement it a series of commentaries was published.

A completely new translation was available in Great Britain in 1961 called the New English Bible and this was revised to be more suitable for public reading and published in 1989 as the REB.

Roman Catholics (and others) have the Jerusalem Bible, which contains *exegetical notes translated from the French of the Jerusalem Dominicans in their Bible published in 1956. The English JB, and the New JB, are translations from the original languages, not the French, and one characteristic is the use of the Hebrew YAHWEH in the OT for the name of God rather than the LORD. The New American Bible, translated from the original languages, was published for Roman Catholics in 1970.

Both NRSV and REB use inclusivist language, so that where formerly a masculine word was used when it applied to people of both genders, men and women, now words such as 'community' are preferred to 'brethren' (Eph. 6: 23). NRSV and REB are used by Christians of most denominations except that in 1994 the Vatican forbade use of NRSV in Catholic public worship on account of inclusivist language.

Modern translations take into account important discoveries, such as the documents from *Qumran, and the resulting interpretations of Hebrew texts, and for the NT use the best available Greek text. Translators have been at pains to find dynamic equivalents for the original words, and have at times preferred paraphrases to literal translations.

See also translations of the Bible into modern English

biblical criticism A term covering a variety of methods used since the Enlightenment in the early 18th cent. for studying the books of the Bible. *See* criticism, biblical.

biblical theology Systematic theology derived from or based on the Bible. Many such theologies have been published, especially by Protestant scholars, since an early formulation of the task in 1787, by the German Johann Gabler. He maintained that it was necessary to expound the theology or theologies contained in the Bible and then to relate that theology to modern doctrines of God, the *Incarnation, the *Holy Spirit, *redemption, the Church, the *sacraments, and *resurrection life.

The second part of the task was undertaken throughout the 19th cent. by theologians who used biblical terminology with modern meanings in tune with current philosophy and culture. Rigorous historical examination of the Bible led to the view that God had revealed himself to human understanding gradually over the centuries; atrocities in the OT came to be regarded as primitive, and subsequently, it was thought, higher ethical standards were progressively recognized. Biblical categories such as 'sin' and 'redemption' were redefined by these theoreticians; the concept of miracle was rejected; the teaching of Jesus about the coming of the Kingdom was expounded in language congenial to liberal democratic, optimistic Western society.

After the optimism had been shattered by the world war of 1914–18, the biblical categories of human depravity became more believable and the Calvinist theology of K. Barth's interpretation of Paul's Epistle to the Romans (1919; ET 1933) heralded a new movement of 'Biblical Theology' in a specialized sense. This was dominant in America and Britain from about 1935 to 1965; it was fashionably pessimistic about human nature, opposed the idea of 'progressive revelation', and its slogan was 'God who acts'. NT *eschatology was proclaimed as a relevant interpretation of human history. The movement did not reject the methods of historical criticism of the Bible, and this openness was one of the factors which made it as acceptable in universities as its prophetic potential popularized it in pulpits.

Biblical themes of redemption, resurrection, freedom were held to have a validity of their own which made sense of human life and experience. NT concepts such as *kingdom, *salvation, *sacrifice, *grace were to be understood by their OT background; and the OT usage of such terms seemed to be straining forward to a fulfilment which was reached in the NT. Thus, Biblical Theology was a theology of unity rather than diversity, and as such could be presented as a coherent and attractive intellectual proposition. It had, too, ethical consequences. NT assertions about family and social life, such as the husband's precedence over his wife (Eph. 5: 22) or the authority of the State (1 Pet. 2: 13) were cited as a norm of Christian obedience: such hierarchies were regarded as definitive. Biblical Theology was therefore inclined to interpret categories like 'time' in a Hebraic rather than Greek way; everlasting life in the NT was said to be in terms of resurrection, not of *immortality—and this could be presented as attractive to modern Christians. For the 'resurrection of the body' implies that the richness and variety of human and social life will be fulfilled in eternity. It is a confident affirmation of the meaningfulness of historical existence.

As a movement in scholarship, Biblical Theology as expounded in its heyday faces many objections. Its emphasis on revelation and history, *exodus and *covenant, do not speak to burning issues of today, such as racism, the values of other religions, and the exploitation of natural resources. *Feminist theologies have exposed the patriarchal presupposition of much biblical literature; and it is difficult for many modern readers of it to be comfortable where women are regarded as subordinate and sometimes maligned. The Bible cannot be forced into a theological unity round one or two chosen themes; on the contrary, the sheer diversity of the material, even within the NT alone, is striking. And the insistence on the Hebraic background (and permanent validity) of NT concepts does not do justice to the proven penetration of Greek thought and culture even in Palestine of the 1st cent. Above all, *historical criticism has made categories unusable that depended upon history that has turned out to be non-history. The Biblical Theology movement did, however, influence ecumenical co-operation and also the worship of the Churches. 'Outlines of Biblical Services' for students were published, and since 1970 many of the official liturgical developments have shown the influence of Biblical Theology. It is assumed by the *lectionaries that events in the life of Jesus are anticipated by similar events in the OT and that the parallels somehow constitute the divine plan or pattern which can be celebrated in worship. Modern liturgies therefore contain many quotations and echoes of biblical words, assembled sometimes in a way seemingly innocent of modern NT scholarship, but which further intended revisions may well remedy.

Billerbeck, Paul (1853–1932) A German pastor whose encyclopaedic knowledge of the *Talmud and **midrash* was used for a four-volume commentary on the NT, referred to as Strack–Billerbeck, though his collaborator H. L. Strack wrote only a small portion of the work. Since the discovery of the *Dead Sea scrolls, the commentary's use of rabbinic literature compiled between 200 and 500 CE is out of date and the work has lost its former authority.

binding and loosing The expression occurs twice in Matt. (16: 19 and 18: 18) and is spoken first to Peter and then to the whole group of disciples. It is a rabbinical term, an authoritative declaration about what is permitted or forbidden in the *Law. It may have included the power of expulsion from the *synagogue. Taken in connection with John 20: 23 the expression in Matt. 18: 18 could imply the power to retain or forgive sins, and it has been so understood in some ordination rites.

biographies The gospels are not biographical in the modern sense; they give no information about Jesus' human personality or the events of his adult life before his *baptism by John. But they do resemble some of the *Lives* written in the Graeco-Roman world. Like them, the gospels aim to give credibility within a piece of history to the figure who is the principal subject of the biography.

birds Turtle-doves and young pigeons were allowed for *sacrifices (Lev. 1: 14; Luke 2: 24). Wild birds were hunted (Amos 3: 5; Ps. 91: 3). The migratory habits of birds were well known (Jer. 8: 7), and quails migrating from Africa to Europe fed the Israelites in the *wilderness (Exod. 16: 13). Birds regarded as *unclean are listed in Lev. 11: 13–19, together with bats. Birds are mentioned in the gospels as not having much financial value (Luke 12: 6–7, 24), yet are significant to God (Matt. 6: 26; 8: 20).

birth A mother was regarded as ceremonially *unclean after childbirth for seven days; and for thirty-three days was prohibited from entering a *sanctuary or touching anything hallowed. In the case of a female infant, the length of time required for *purification doubled (Lev. 12: 5).

The idea of a 'new birth' is used in the NT for the beginning of a life in Christ (1 Pet. 1: 3), and as a theological term called '*regeneration' (Tit. 3: 5, AV) was associated in the Church with *baptism.

birthday *Pharaoh's birthday was celebrated with a *banquet (Gen. 40: 20); Herod's with the severed head of *John the Baptist (Matt. 14: 8). Otherwise, birthdays do not figure in scripture.

birthright The first-born son possessed a right of *inheritance (Deut. 21: 17), and the first-born of both families and flocks were deemed to belong to God and had to be redeemed if they were to escape *sacrifice (Exod. 13: 11–16). It was a scandal that *Esau despised his birthright (Gen. 25: 34; Heb. 12: 16) and a paradox that *Jacob's guile in wresting it from Esau was an early stage in Israel's privileged status with God.

bishop Although 'bishops' are mentioned five times in the NT and the qualifications for the office defined (1 Tim. 3: 1–7), the translation of the Greek *episcopos* has been controversial: although used by AV, NRSV, and REB, in NJB the term is 'presiding *elder', which does justice to the fact that Tit. 1: 5–9 equates bishop and elder (Greek, *presbuteros*). It is clear that the later structure of the Church in which a single bishop presided over each local Church is not the situation in the NT. Certainly the AV suggestion that the traitor *Judas had been endowed with a 'bishopric' (Acts 1: 20) is misleading. One reasonable view is that single presiding bishops evolved out of a body of elders; one of the members showed a special aptitude for teaching. This arrangement established itself throughout the Church by the 2nd cent. CE and is accepted by *Ignatius of Antioch in the letters he wrote (107 CE) to Churches as he journeyed to Rome and martyrdom. It was the most effective means of securing community cohesion.

Bithynia Roman province, usually joined for convenience with neighbouring Pontus, in NW Asia Minor. Paul and *Silas had intended to preach there (Acts 16: 7) but were deterred either by opposition or by a prophetic message, which was afterwards regarded as divine intervention. However, by the time of 1 Pet. 1: 1, a Church had been established.

bittern Used in AV of Isa. 14: 23 etc. but NRSV and NJB give the more probable 'hedgehog' (following the *LXX). REB prefers 'bustard'.

blasphemy A deliberate assault on the majesty of God (Lev. 24: 16), punishable by death. In the NT his representatives are subjected to false accusations of blasphemy. Paul is said to have blasphemed when he had been a persecutor of the Church (1 Tim. 1: 13).

Jesus was accused of blasphemy because he forgave sins (Matt. 9: 3) which only God could do, and the charge of blasphemy before the Sanhedrin secured a unanimous verdict of guilty (Mark 14: 64) against Jesus, though the ground for this verdict is unclear since verbal abuse of the Temple or a claim to be the *Messiah were not themselves blasphemous. Possibly blasphemy lay in a claim to a unique relationship to God, or in 'leading Israel astray'.

Blasphemy against the *Son of Man, being humiliated and unrecognized for who he was, was forgivable (Luke 12: 10), because his words and deeds could be inadvertently misinterpreted. But blasphemy against the *Spirit (an exalted Jesus speaking through the Spirit?) cannot be reconciled with Christian fellowship (Heb. 6: 4–6) and the mutual forgiveness that the Spirit brings.

blessedness Paul argues that 'David' (meaning Ps. 32: 1–2) speaks of the blessedness ('happiness', REB) of those to whom God reckons *righteousness (Rom. 4: 6) independently of their observance of the *Law. It was a typical method of the rabbis to support a statement from the *Pentateuch (here about *Abraham, Rom. 4: 2–3) by a quotation from a psalm.

blessing Either words in praise of God or words used to make someone or something holy. Hence also the result of such words: in the OT a favour bestowed by God, as at *harvest (Deut. 28: 8). In the NT blessings are 'spiritual' (Eph. 1: 3), meaning all the *gifts promised by God to mankind through Christ (Acts 3: 25) and of Christ to his *disciples (Mark 14: 22; Luke 24: 50).

blindness *Animals that were blind were unfit for *sacrifices (Lev. 22: 22), and priests who were blind were disqualified from sacrificing (Lev. 21: 16–24), though the community was instructed to care for the blind (Lev. 19: 14). Blindness also becomes a symbolical term—the eyes of the blind opened by God is a vivid expression for the recognition of the signs of the age (Isa. 29: 18). In the NT, where the healing of blind persons is one of the indications of the coming of the *Kingdom, the gradual healing of the blind man at *Bethsaida (Mark 8: 22–6) is significant; it is what is happening to the disciples, and at *Caesarea Philippi a partial insight is reached by Peter (Mark 8: 29). Similarly *Bartimaeus' healing (Mark 10: 46–52) is a vivid illustration of the new insights of all disciples who follow Jesus on the way (10: 52), that is, of disciples who follow him on the Christian way (Acts 9: 2). A man blind from birth whose sight was restored by Jesus (John 9: 6–7) became a believer (9: 38); he is converted, which implies rejection of Judaism and expulsion from the synagogue (9: 22).

blood In Hebrew thought blood was the seat of *life, or even identified with life, and it therefore had an essential role in *sacrifices, which were themselves fundamental in Hebrew society. The priests were consecrated by blood (Exod. 29: 19–21); blood sprinkled on the *altar expiated sins (Lev. 17: 6); blood dashed over the people established the *covenant with the Lord (Exod. 24: 8).

In the NT the blood of Jesus signifies the atoning power of his *death. So the obedience of his life and death becomes the basis for the new covenant (1 Cor. 11: 23–9); and the words of Jesus in verse 25, 'This cup is the new covenant in my blood', are the oldest record in the NT of what he is doing and explaining at the *Last Supper. Paul's account then appends Jesus' injunction that the rite should be repeated, and this became accepted as instruction for the regular celebration of the Church's *Eucharist. In Catholic doctrine the *wine in the cup, as with the *bread, when blessed in the appropriate form by a duly authorized priest or bishop having the right intention, becomes the whole substance of the Body and Blood of Christ, only the appearances (the 'accidents') of bread and wine remaining. Other doctrines of the Eucharist, deriving from the Reformation, hold that the word 'is' means no more than 'represents': the Eucharist is a commemoration of the sacrifice of Christ. But possibly the language ('is') is performative rather than descriptive. It changes the status of the bread and the wine. Drinking from the cup of the new covenant is a means of fellowship with the Paschal (*Passover) lamb slain for us (1 Cor. 5: 7): and is given by Jesus for his followers to keep in touch with him who died on the cross and who was raised to new life at *Easter.

Boanerges Jesus gave this name, 'sons of thunder', to *James and *John, sons of *Zebedee (Mark 3: 17), perhaps on account of their tempestuous tempers.

boats There are not many references to boats in the OT since the Hebrews were not a seafaring people; indeed the sea had connotations of *evil (Ps. 107: 26; cf. Rev. 21: 1). However, in the NT boats are familiar parts of the scenery since Jesus called fishermen to be *disciples and sometimes addressed people on the shore from a boat (Mark 4: 1). A boat soon became a symbol in Christian art; both the cradle and the coffin are shaped like boats, so the boat is an apt symbol of *birth and *death, of new birth and *resurrection (a journey by boat into the next world).

Boaz (1) A landowner in Bethlehem in the time of the judges who acted generously at harvest time to *Ruth, a *Moabite widow. After marriage, they had a son Obed, who became the grandfather of *David (Ruth 4: 17), an ancestor of Jesus (Matt. 1). (2) One of the two pillars situated in front of Solomon's Temple (1 Kgs. 7: 21). *See* Jachin.

body Although the word 'body' in the Bible can denote a corpse (Mark 15: 43), it normally refers to the whole physical being of a human person active in the world (1 Cor. 6: 20), and the limbs, the mind, the spirit are various parts of the whole organism (Rom. 6: 12–13). Human bodies, as made by God, are 'very good' (Gen. 1: 31), and the *resurrection of Jesus conveys the promise of a resurrection of those who belong to him. This belief is to be understood not as the reassembling of human bones and flesh out of the graves, as it has often been pictured in the past (1 Thess. 4: 13–17), but as signifying that all that is good in human life will be taken up into *eternal life. It affirms a continuity of this bodily life with that which is to come.

Body of Christ It was not unknown but it was unusual in the 1st cent. to use 'body' for a collection of people as Paul does for his important phrase 'You are the body of Christ' (e.g. 1 Cor. 12: 27). He was probably attracted to the idea because of the Eucharistic words; the oneness of the loaf (called Christ's body) shared by the *congregation led naturally to the application of the title 'Body of Christ'.

There has been disagreement about the interpretation of the phrase—is the Church metaphorically or really the Body of Christ? Does it resemble the Body of Christ? Or is it in fact, in some way (but plainly not literally), the same body that lived and died and was raised in Palestine? On the whole Protestants have preferred the metaphorical interpretation. Catholics hold that the Church is sacramentally or mystically united to Christ—the instrument by which the heavenly Lord continues in the world what he had begun: he is the Head, the Church is his Body. For Protestants the Church is not the extension of the *Incarnation any more than it is the extension of Christ's *atonement, which is once and for all. Head and Body are separate from each other. Protestants point out that monstrous errors have been perpetrated by ecclesiastics who have identified their decisions with those of Christ himself.

Nevertheless Head and Body are inseparable, necessary to each other. Paul on the *Damascus road is reported (Acts 9: 4) to have heard Jesus accusing him 'Saul, why are you persecuting me?', when he was persecuting Christians: there was a very close relationship between Church and Christ, and the Body is more than a mere metaphor. Perhaps Body language should be explained as similar to the Bride and Bridegroom imagery of Eph. 5; both are necessary to each other but are not identical.

boil (1) Festering boils on humans and on animals were the ninth *plague on *Egypt (Exod. 9: 9) but failed to cause Pharaoh's heart to soften and release the Hebrews from their slavery. Moses threatened further disasters. (2) King *Hezekiah suffered the same affliction and Isaiah recommended a poultice of *figs (2 Kgs. 20: 7).

bones In Luke 24: 39 the risen Jesus invites the eleven *disciples and their companions to touch him and assure themselves that he is no mere ghost but possesses flesh and bones. It was, however, a body which could pass through the door (24: 36) and this is presumably why in the narrative the body, with its essential constitution of bones, could be mistaken for a ghost. Luke's narrative is to be complemented by Paul's belief (1 Cor. 15: 42–53; Phil. 3: 21) that flesh and blood cannot inherit the kingdom of God; Luke's story is perhaps to be regarded as an assertion that the risen Jesus is one and the same person continuous with Jesus who had ministered in *Palestine.

Book of Life The Hebrews believed that Yahweh kept a register of living persons, from whom the ungodly could be blotted out (Ps. 69: 28). In the NT Paul mentions a book which records the names of believers (Phil. 4: 3). At the *judgement of the dead (Rev. 20: 12) some are entered in the '*book of life', according to their works. The seer seems to envisage a kind of accounts book (or scroll, Rev. 5: 1) with debit and credit columns.

books Books of the Hebrew Bible were scrolls, wrapped round rollers (Jer. 36: 2) and unrolled for the appropriate place (Luke 4: 17). Two hands were therefore required for reading them. The Christians' bibles by the 2nd cent. CE were in the form of papyrus *codices with pages fastened on the spine. These were not only more convenient than rolls but also served to point the difference from practice in the *synagogues.

booth A shelter (so REB) used by farmers when watching over their crops (Isa. 1: 8). Because they were only temporary structures, they were reminiscent of Israel's journeying through the *wilderness, so at the feast of *Tabernacles (Booths), when the harvest was gathered, every Israelite was required to live in a temporary booth (Lev. 23: 42).

border of garment Israelites were ordered to have fringes (NRSV) on their garments (Num. 15: 38), and the gospels record them on Jesus' cloak (Mark 6: 56; Luke 8: 44). The translations 'hem' and 'tassels' have also been used.

Bornkamm, Günter (1905–90) A German NT scholar who had been a pupil of Rudolf *Bultmann, but his *Jesus of Nazareth* (1960) and his *Paul* (1975) allowed for more historical reliability than Bultmann did, even though he was a pioneer in emphasizing the various theological and apologetic interests of the evangelists (*Redaction Criticism).

bottles Vessels to hold *water or *wine were made of leather—hence the translation 'skins' (of goats or pigs; Matt. 9: 17). The animal's shape was retained for the bottle, and the neck was tied. The gases of fermenting wine could destroy old wine skins (Mark 2: 22)—a reference to the incompatibility, as Mark understood it, of the new institution of the Church with old *Judaism. Vessels of earthenware were also available (Jer. 19: 1).

Bousset, Wilhelm (1865–1920) A German scholar of Göttingen, frequently mentioned in NT commentaries for his comparisons of ideas and events in the gospels and epistles with those of Hellenistic religions. The implication was that Christian thought was influenced by the surrounding cultures.

bow A weapon sometimes made of bronze (Ps. 18: 34), wielded by both soldiers and huntsmen. Arrows were made from reeds or polished wood (Isa. 49: 2), occasionally with poisoned tips (Job 6: 4). They were carried in a quiver (Lam. 3: 13) hung over the right shoulder.

bowels Human emotions were thought to be generated in the entrails; hence the AV translation 'bowels' in Phil. 2: 1 etc. Modern custom speaks of the heart in a similar way. There is a literal physical usage in 2 Sam. 20: 10 and Acts 1: 18, but modern versions prefer 'compassion' (NRSV) or 'warmth of affection' for the metaphorical use.

bramble Used by Jotham as a sarcastic description in a parable for *Abimelech (Judg. 9: 14–15). Jesus uses the word in a parable (Luke 6: 44). There were many varieties of prickly thorn bushes, sometimes used as *hedges.

branch A great crowd in Jerusalem for the *Passover went out to meet Jesus carrying branches of palm trees (John 12: 13). According to Mark 11: 8 people accompanying Jesus carpeted the road with leafy branches (rushes?) which they had gathered in the fields.

brass The frequent AV rendering (e.g. Deut. 8: 9) is incorrect and probably copper is meant, which became bronze (as 1 Sam. 17: 5) when alloyed with tin. NRSV and REB use 'brass' for the metaphorical sense (of unyielding hardness) in Isa. 48: 4, while NJB prefers 'bronze'.

bread Barley was generally used for making bread, though wheat was not unknown, and even lentils and beans (Ezek. 4: 9) combined with grain. A little of old fermented dough was added to the new mixture as leaven to make it rise. The 'showbread' or 'bread of the presence' consisted of twelve loaves in two rows in the holy place of the *tabernacle (Exod. 25: 30) and it was these which were given to *David and his men by the priest at *Nob (1 Sam. 21: 6), an act of humanity which was strictly illegal. This incident was once sympathetically recalled by Jesus (Mark 2: 25). *Unleavened bread, i.e. bread made without yeast, was eaten by Jews at the Feast of Unleavened Bread, which was originally a solar agricultural festival but was later amalgamated with Passover.

Bread was vital both for existence and for religion. The prospect of 'bread without scarcity' (Deut. 8: 9) was no mean promise. And Jesus as 'the bread of life' was a powerful image (John 6: 33).

breaking of bread Early Christian fellowship was expressed in shared property and shared *meals (Acts 2: 42); they broke bread in various houses in turn (Acts 2: 46). The 'breaking' was only part of the preparation for the meal, but the name of the part was applied to the whole. In the action the group was conscious of Jesus' presence. Some may have had memories of the Feeding of the Multitudes, and the image of the future Messianic Banquet was well known. So usually in the NT (probably not at Acts 27: 35) the breaking of the bread denotes the *Lord's Supper—the Christians' covenant-rite, which Paul had received as a tradition and which he passed on (1 Cor. 11: 17 ff.). Probably the breaking of bread was from earliest days a weekly celebration.

breastpiece 'Breastplate' in AV and sometimes in NRSV, REB, NJB. It was a dramatic part of the ceremonial garb worn by *Aaron (Exod. 28: 15) when he ministered before the Holy of Holies in the *tabernacle. Its four rows of three precious stones represented the twelve tribes which Aaron therefore symbolically brought before the Lord.

A military breastplate was worn as protective armour (1 Kgs. 22: 34) and as such the word becomes a metaphor in the NT (e.g. Eph. 6: 14).

bribery The strict impartiality of the judicial system was intended even in the earliest codes, and bribery was severely punished. Incorruptibility, however, was not totally enforceable (Amos 5: 12).

brick In *Egypt, the oppressed Israelites were forced to build cities with bricks made of mud or clay and dried in the sun. As an additional hardship they were obliged to collect their own straw (Exod. 5: 7), which was needed to prevent the bricks from cracking.

bride, bridegroom In the OT the *marriage concept is used to describe the relationship of God to his covenant people, Israel, which is regarded as his 'bride' (Isa. 62: 5). In the NT the image is applied to Christ and the Church (Eph. 5: 25 ff.) and implies a relationship of inseparability but not of identification. This image was anticipated by Jesus (Mark 2: 19), though the following verse (2: 20) reflects the situation in the Church and therefore casts doubt on 2: 19 as being from the lips of Jesus himself.

brimstone The colourful word of AV retained by REB and NJB but changed to 'sulphur' in NRSV (Gen. 19: 24) for the burning yellow stones hurled down on Sodom and Gomorrah: just the sort of volcanic material found in the Dead Sea valley.

bronze Bronze was harder than copper and came to replace it in weaponry. It also had a bright, shiny surface—hence the simile in Rev. 1: 15.

brother Used of sons of the same parent(s) (e.g. Judg. 8: 19), of related people (e.g. Deut. 23: 8, NJB), or in reference to a forgiven foe (1 Kgs. 20: 33, NRSV). In the NT it is used of a fellow Christian (1 Cor. 6: 6, NRSV marg.) and several times in Acts 15.

brotherly love The relationship which should prevail amongst members of the Church (Rom. 12: 10). Christ's *love for his *children is the example to be followed (Eph. 5: 1–2). In practice this mutual love is a proof of genuine *faith (1 John 2: 9–11).

brothers of Jesus The natural understanding of Mark 6: 3 is that *James, *Joses, *Simon, and *Judas were the younger sons of *Mary and *Joseph. But the wider meaning of '*brother' enables those who believe in the perpetual virginity of Mary to hold that James (Gal. 1: 19) and the others were children of Joseph (who disappears from the gospels, perhaps because of old age, after the infancy narratives) by a former marriage, or cousins (cf. 1 Kgs. 20: 32, REB).

buildings Although there was no lack of stone in Palestine, the surplus wealth and technical skills which existed in classical Greece of the 5th cent. BCE, and in the era of *Alexander the Great, were not available to the Hebrews. Wood was used extensively, transported from Lebanon (1 Kgs. 5: 6 ff.).

Surveyors used measuring-lines (Ezek. 40: 3) and could check building progress with a plumb line. Foundation stones were important and were laid with great care, and it is not surprising that in the NT Christ is said to be the foundation or *corner-stone of the Church (Eph. 2: 20, 1 Pet. 2: 7). For the Church is the building itself, though the metaphor changes in 2 Tim. 2: 19, where the Church is described as an immovable foundation. In yet another metaphor it resembles the pillar of a building; for it upholds the *truth of God (1 Tim. 3: 15).

bulls In the ancient Near East bulls were often prominent as cult objects and their strength and procreative power were not lost on the Israelites. When *Jeroboam I tried to wean the divided kingdom away from Jerusalem, he set up idols of two young bulls in *Dan and *Bethel. Bull statues were, however, legitimate when they were regarded as thrones for the deity or used as supports for the *molten sea (2 Chron. 4: 4).

Bultmann, Rudolf (1884–1976) German NT scholar. Professor at Marburg University and generally regarded as the foremost NT scholar of the 20th cent. His commentary on the gospel of John has been fundamental for all subsequent study of it. He was a leading exponent of *Form Criticism and proposed a minimizing view of the value of the synoptic gospels as historical records. He held that John's gospel had been compiled from several sources, and was influenced by *Gnosticism. Bultmann associated the NT with Hellenistic culture and religions (he wrote before the discovery of the *Dead Sea scrolls) and argued that just as this environment became a vehicle for the gospel in the 1st cent., so contemporary thought could be used by theologians in the 20th cent. For us the supernatural elements in the gospels are no longer credible as described. But as a theologian Bultmann maintained that instead of being discarded they could be acceptably reinterpreted by preachers in terms of the modern philosophy of existentialism. The speculative mythological language of the NT could be re-expressed so that Christ reveals to us a new and authentic mode of existence. Bultmann's programme of *demythologization became the main issue debated in NT studies from about 1945 to 1965. It was repudiated by Catholics and others who argued that *myth and *poetry continue to be a way of speaking, by *analogy, of God, and that Bultmann's version of the Christian faith was altogether too thin.

burden Although used literally (Exod. 23: 5, NRSV) for a heavy load (REB) it is also found in a metaphorical sense for something hard to bear (e.g. royal responsibilities, Hos. 8: 10). Jesus does not inflict a heavy burden (Matt. 11: 28–30). The word is used by AV (e.g. Isa. 13: 1) for a prophetic *oracle.

burial Because of the high temperatures in Palestine, it was desirable to dispose of dead bodies as soon as the appropriate customs could be organized. The corpse was washed, anointed, and wrapped in linen cloth and carried in procession on a bier to the *grave or *tomb. There could be professional mourners and even flute-players (Matt. 9: 23). For some, graves were dug, but *caves were also convenient, and *Abraham's family are said to have been buried in the cave of *Machpelah at *Hebron (Gen. 23: 19). Wealthy individuals arranged for their tombs to be cut out of the rock in a garden, as did *Joseph of Arimathaea (Matt. 27: 60), and sealed with a stone which rolled on a kind of track. Cremation was not practised by the Hebrews. In the NT era ossuaries made of limestone were used in Palestine; after bodies had disintegrated, the bones were collected and stored in chests, sometimes giving the name of the deceased.

burnt offering A *sacrifice to the God of Israel in which a whole *animal, except for the hide, which was spared for the priests, was burnt on the *altar. It was supposed that Yahweh enjoyed the smell, which was pungent (Lev. 1) and bestowed his favours in return. The animal was always male (Num. 29: 2–38).

bush, burning Moses experienced his call to lead the enslaved Israelites out of *Egypt when he saw on Mount Sinai a bush apparently on fire but never becoming burnt up (Exod. 3: 2). He accepted the vision as a promise from God to be faithful; and in a gesture of reverence in the presence of the holy he removed his sandals, as Muslims do today when they enter a mosque. The natural cause of the burning phenomenon may have been a discharge of atmospheric electricity on pointed objects during a storm.

Byzantine text The Greek text which was the basis of Erasmus' NT (1516) and so of the translators of the English AV of 1611. The name derives from Byzantium (Constantinople). Being the text in current use in the capital of the Eastern Empire (330–1453 CE), it enjoyed a certain primacy.

C

Caesar Originally the name of a Roman aristocratic family and assumed by *Augustus as the adopted son of the dictator C. Julius Caesar, and passed on by Augustus to his adopted son *Tiberius. Then the title continued to be used by Caligula, *Claudius, and *Nero. After that, succeeding emperors retained the name not only for themselves but also for the heir presumptive to the throne. Their power was exercised in the empire by a remarkable combination of legal authority and political acumen. The names of the ruling Caesars in the NT era are:

27 BCE–14 CE	Augustus
CE	
14–37	Tiberius
37–41	Gaius (given the surname Caligula by his troops)
41–54	Claudius
54–68	Nero
68–9	Galba
69	Otho; Vitellius
69–79	Vespasian
79–81	Titus
81–96	Domitian
98–117	Trajan
117–38	Hadrian

Caesarea Built on the coast by *Herod the Great in honour of *Caesar *Augustus; it became the official residence of Roman prefects. Its inhabitants were both Jews and *Gentiles, and the Christian way was brought there by *Philip (Acts 8: 40; 21: 8). Peter became aware of the equality of Jews and Gentiles in the Church by being summoned to Caesarea by *Cornelius, a *centurion and a Gentile God-fearer. Paul was taken to Caesarea for trial (Acts 23: 23).

Caesarea Philippi An inland city on the River *Jordan at the foot of Mount *Hermon where *Herod the Great built a temple to Caesar *Augustus, who had given him the city. Formerly called Panion, the city was named after the emperor by Herod's son, *Philip the tetrarch, who egotistically added 'Philippi'—to avoid confusion with the city on the Mediterranean coast.

Caiaphas Son-in law and successor of *Annas, high priest in Jerusalem. He held office from 18 to 37 CE, but the statement in John (18: 13) that he was high priest 'that year', combined with the reference to a preliminary investigation by Annas (John 18: 13–24) has suggested the possibility of the high priesthood being held for one year only at a time. But, more probably, John's meaning is that Caiaphas was the high priest 'in that memorable year of the *crucifixion'. An ossuary with an Aramaic inscription thought to mean 'Joseph, son of Caiaphas' was found in 1990 in the Caiaphas family *tomb in Jerusalem; but the translation is by no means certain.

Cain The elder son of *Adam and *Eve. Cain offered fruits of the earth (he was a farmer) to the Lord, but they were rejected; whereas his brother *Abel offered meat, which was accepted (Gen. 4). How the verdict was signalled is not described, or why. Possibly the author has in mind a preference for a nomadic to a settled way of life. In Cain's subsequent murder of Abel some commentators have discerned a hint of an ancient belief that a human *sacrifice enhanced the fertility of the soil. Heb. 11: 4 suggests that Abel was favoured because of his superior goodness and *faith.

Caleb One of the spies sent ahead by Moses to explore *Canaan (Num. 13: 6) and permitted to enter it (Num. 14: 24) unlike Moses himself (Deut. 34: 4). He settled at *Hebron (Josh. 14: 14).

calendar The waxing and waning of the *moon prescribed the division of the year into twelve months, usually designated during the *Exile merely by numbers instead of the old Canaanite names formerly in use. But the Babylonian *Nisan (corresponding to April) is found at Neh. 2: 1 and Esther 3: 7 as the first month, and on the 14th of this month preparations for *Passover began. Other Babylonian names were also adopted. *See* months.

In the early part of the 2nd cent. BCE the book of Jubilees, a commentary on Genesis, rejected the lunar calendar, and it is possible that this solar calendar of 364 days to a year, alternative to the official calendar, was adopted by the Dead Sea community at *Qumran, where copies of part of Jubilees have been found amongst the scrolls.

calf, golden (or **molten**) The archetypal image of apostasy, in Hebrew tradition. There is much evidence of the worship of cattle in the Near East, certainly from the 6th millennium BCE, in Anatolia, but the classical account of this *apostasy in the OT is when Moses was so long absent, engaged in conversation with the Lord on Mount Sinai, that the Israelites down below began to be mutinous. So *Aaron ordered a mass collection of golden earrings (Exod. 32: 2), reshaped them in the fire, and 'there came out this calf'! On his return Moses had the idol melted down again. Cf. 1 Kgs. 12: 28. *See also* Jeroboam (1).

call, calling As part of his redemptive action, God 'calls' Israel (Isa. 49: 1) and individuals (Isa. 41: 25), and because the calling is sometimes to new or unexpected work an unexpected (Luke 1: 59–63) or new (John 1: 42) name is bestowed. The calling of God to people goes on in the apostolic Church (Gal. 1: 15); though all are called to share as equals in the Christian life, some have a calling or vocation to particular functions within it, as was the case with *Matthias (Acts 1: 24) and Paul (Gal. 1: 15).

Martin Luther, and many since the *Reformation, have maintained a high doctrine of vocation or calling to one's secular occupation, but this sense of a calling is unknown in the NT.

Calvary 'The place of the skull' (Mark 15: 22) or simply 'Skull' (Luke 23: 33), **golgotha* in Aramaic, is rendered *calvaria* in Latin. It was situated outside the city wall and adjacent to an important road.

camel Regarded by the Hebrews as an *unclean *animal, not to be eaten (Lev. 11: 4), but useful for carrying goods (2 Kgs. 8: 9).

Cana A village in *Galilee. The gospel of John is the only writing in the NT to mention it. Cana is the scene of Jesus' *miracle at a wedding (John 2: 1–11), and of the healing of a nobleman's son (John 4: 46–54); and it is where *Nathanael lived (John 21: 2).

The miracle of water turned into *wine (John 2: 1–11) is the first of Jesus' 'signs' in this gospel. The story is intended to demonstrate that Jesus is beginning to fulfil and transcend the expectations of *Judaism; it is carefully noted that the water in the jars was required for Jewish rites of *purification. John may be elaborating some well-known words of Jesus, preserved at Mark 2: 21–2, about the virtues of new wine, and he may also have been influenced by Hellenistic stories of the transformation of water into wine.

Canaan(ites) The *'Promised Land'. The extent of Canaan, composed of independent towns and villages before the arrival of the Israelites, comprised what is now modern Israel and Lebanon. *Agriculture was the main occupation, and this was reflected in the *harvest festivals, which Israel inherited (Exod. 23: 14–17). But much of Canaanite religion and morality was decisively repudiated by the conquering Israelites according to the Deuteronomist historian (Deut. 20: 16–18), whose narrative was prejudiced, polemical, and partial.

In the NT the land is mentioned in *Stephen's speech (Acts 7: 11), and where Mark (7: 26) refers to a Syro-Phoenician woman, Matthew (15: 22) calls her a Canaanite, which is a simplification of the more precise

Marcan term, and also carries rather derogatory overtones.

Candace The title of queen mothers who ruled Ethiopia (modern Sudan). It is mentioned (Acts 8: 27) to illustrate the progress of the Gospel beyond Palestinian *Judaism: a court chamberlain sympathetic to Judaism was returning from Jerusalem in a covered wagon when he encountered *Philip, who converted and baptized him.

canon The word in Greek means a 'rule' or 'standard' and is applied to those books in the Bible which have been recognized as authoritative, but in both OT and NT the process of finalizing the canon was gradual and controversial.

Jews agreed that the *Pentateuch in its final form after the *Exile was fundamental. After that, the Prophets, which included Joshua and the historical books, as well as the major and minor prophets, became accepted but not by the *Samaritans. Finally, the section called the Writings (*Ketubim*) (consisting of eleven other books in Hebrew), was eventually settled in the 1st cent. CE, though not always in the same order. The rabbis who taught at *Jamnia are regarded as having a responsibility for finalizing the collection, and what they did above all was to exclude those books which were written in Greek and were widely read amongst Greek-speaking Jews. These came to be known as the *Apocrypha. But the notion of a fixed OT canon was very loose in the 1st cent. and it is noticeable that the epistle of Jude (verses 14–16) quotes from the book of 1 Enoch as though it was scripture, although it was not accepted as canonical.

Christians, living in the Hellenistic world, naturally accepted the Greek OT, the *Septuagint (LXX), as their scriptures. This contained the books excluded from the Hebrew canon such as the Wisdom of Solomon, Ecclesiasticus, 1 and 2 Maccabees; and when Jerome was translating into Latin he learnt from Jewish scholars that these were regarded as secondary (even by Hellenistic Jews). So he refused to include them in his Latin Bible. However, Jerome's view did not prevail, and, called 'deuterocanonical' (second-level), they were incorporated into the *Vulgate from the Old Latin version and retained their authority, as has always continued to be the doctrine of the Roman Catholic Church. At the *Reformation the books of the Apocrypha were excluded by Protestants from authoritative scripture; so, when Article 6 of the Anglican Thirty-Nine Articles declares that the OT (Hebrew books) and the NT contain all things necessary to *salvation, it is asserting that these books contain the maximum number of scriptural writings that can be regarded as having authority for Anglican Christians. The books of the Apocrypha are, however, usually printed between the two Testaments and passages from them are included in *lectionaries for optional use in public worship.

But the fact is that it took time also for the NT list to be finalized, and many occasional sayings of Jesus (known as **agrapha*) and several whole books are sometimes quoted by the early Fathers as authorities even though they are not within the final canon of twenty-seven. The main stages of the process of recognition were:

1. Epistles addressed to various Churches were preserved by them, and in due course the letters of Paul formed a collection. Conceivably, this was the contribution of *Onesimus; and the epistle to the Ephesians could have been an editorial introduction to the corpus, summarizing Paul's essential thoughts. Oral traditions about Jesus were constantly repeated and much valued as the 'living tradition' (*Papias) but in due course this was cast into the written form of the four *gospels, which are regularly quoted from the middle of the 2nd cent. CE onward. The adoption by Christians of the *codex (book), in place of the roll, was probably an influence now on the composition of the canon. Four gospels conveniently contained in a single codex tended to sideline alternative rivals.

2. The heretic *Marcion, about 140 CE, issued his own NT consisting of most of Luke and ten epistles of Paul, apparently in opposition to a larger collection already circulating in the Catholic Church.

3. *Irenaeus, about 180 CE, quotes most of the NT books as having an authority equal to that of the OT.

4. The *Muratorian Fragment, probably from about 190 CE, gives a list which includes four gospels (Matthew by inference—the beginning of the document is missing) Acts, thirteen epistles of Paul, the Johannine epistles, Jude, and the Revelation, but not Hebrews, James, or 1 and 2 Peter. The Shepherd of Hermas is allowed for private reading, and it is conceded that the Church is divided about the Apocalypse of Peter.

5. Clement of Alexandria (d. 215 CE) admits fourteen epistles of Paul (Hebrews is included). There is no mention of James, 2 Peter, 3 John; but he accepts the Shepherd of Hermas.

6. Eusebius (d. 340 CE) has a threefold classification; he notes the accepted, the disputed, and the rejected books. His first category includes the four gospels, Acts, fourteen epistles of Paul, 1 Peter, 1 John, and (with hesitation) Rev. In the second category, James, Jude, 2 Peter, and 2 and 3 John are regarded as widely approved, but much less so are the *Didache, Acts of Paul, the Apocalypse of Peter, the Shepherd of Hermas, and the epistle of Barnabas. The totally rejected books are the gospels of Peter, Thomas, and Matthias.

7. Athanasius of Alexandria wrote a letter in 367 CE giving the list of twenty-seven NT books (and some others for private reading only) and this provides the earliest evidence of the final canon for Eastern Christendom.

8. Augustine's criterion was universal acceptance and his North African Church followed Athanasius in 393. Jerome published his Vulgate in 405 and this was decisive in the West for establishing the content of the canon. But Jerome admitted the epistle to the Hebrews and Revelation reluctantly and only on the ground that they had been recognized by early Fathers.

9. The Council of Carthage (397 CE) forbade non-canonical books to be read publicly, and appended the authorized list, of twenty-seven books.

The provision of the authorized canon was a defence of orthodox Catholicism against heresies such as *Gnosticism and Montanism. Among the criteria for gaining admission to it were: authorship by an apostle; the reliability of the witness to Jesus Christ; and wide consensus of the Churches.

At the Reformation, Martin Luther translated all the books of the OT and the NT into German, but relegated the OT Apocryphal books to an appendix—and also James, Hebrews, Jude, and Revelation (not a precedent followed by modern Lutheran Churches).

Canonical Criticism This term, though not welcomed by its exponents, has been applied to a kind of biblical scholarship developed by a small group of university teachers in North America. It began with the breakdown in the 1960s of the attempt by *Biblical Theology to use historical criticism in the service of the Church's theology; those events judged to be historical could be regarded as the revelation by which God saves his people—hence the term 'salvation history'. But the ongoing work of archaeology and literary criticism demonstrated the fragility of the historical evidence.

What is offered by Canonical Criticism to the Church is an invitation, certainly not to reject historical criticism, but to build its theology on the books of the Bible as wholes and not to separate the 'authentic' from the 'non- authentic' sections, nor to concentrate on the stages of a book rather than the finished work. The suggestion is that the Bible should be respected as it traditionally has been in the life and worship of the Church. Historical criticism and sociological analysis are not rejected but they are given a subordinate status. The text must be heard as scripture. Books of the Bible must be studied as wholes.

It was the complete books that were accepted after the *Exile in *Judaism and which therefore shaped the life and beliefs of the community and established its self-identity.

Canonical Criticism is the work of Christian scholars, but they have devoted most of their attention to the OT, and within the OT particularly to the *Torah, which is regarded as the centre of the OT, in the sense that it determines the interpretation of the rest of the OT. This, however, is a problem;

for example, in the Torah God's will is made known; his work is clearly understood in Israel—but in the OT book Ecclesiastes the work of God defies a man's ability to comprehend. A further problem for Canonical Criticism is the wide disagreement about the content of the canon, not only between Jews (for whom the OT *is* the Bible) and Christians, but between Jews and Jews, and between Roman Catholics, Orthodox Christians, and Protestants—disagreements reflected even in the order in which the OT books are placed.

The advocates of Canonical Criticism are as much theologians, concerned to establish a firm foundation for preaching, as biblical scholars wrestling with the traditional apparatus prescribed in a secular university. Among the contributions within a Religious Studies faculty are those made without any commitment to a Church or to theism, whereas canonical critics regard the Bible as a unity, and for them not only are individual books viewed as wholes, but both Testaments are complementary; and the OT is read in a Christian way, for the *day of the Lord predicted at the end of the OT (Mal. 4: 5–6) has arrived.

Although the origin, content, and status of the canon of scripture is now very much part of scholarship's agenda, few are prepared to align themselves with the Canonical Criticism movement.

Capernaum One of the most frequently mentioned towns in the gospels, on the NW shore of the Sea of *Galilee, and the centre of a fishing industry. Here Jesus healed (e.g. Matt. 8: 5–13), taught (Mark 1: 21), and called *disciples (Mark 2: 1, 14). A *synagogue has been excavated which belongs to the 4th cent. CE, but beneath it are the ruins of one from the 1st cent. CE. Other excavations have revealed a 1st cent. CE private house beneath a 5th-cent. CE church which the Franciscan friars responsible for the work identify as that of Simon Peter.

Cappadocia Roman province in E. Asia Minor, where a settlement of Jews was established (Acts 2: 9) and a Church founded which is addressed in 1 Pet. 1: 1.

captain A title in the Bible for military officers of varying duties and ranks, sometimes translated 'commander' or even 'comptroller' (1 Kgs. 16: 9, REB). In Acts 4: 1, there is a captain of the Temple (NRSV) who may have been in charge of the Temple police. 'The captain of our *salvation' (Heb. 2: 10, AV) is better translated 'pioneer of their salvation' (REB, NRSV) or 'leader of their salvation' (NJB).

captivity epistles Traditionally the letters to the Philippians, Colossians, Philemon, and Ephesians; they are grouped together because they all mention Paul's current *imprisonment. But where was the prison? It is probable that the first three of these epistles were written in Paul's relatively relaxed regime in prison at Rome (Acts 28: 30 f.), but it is all too likely that this final imprisonment was not the only one (cf. 2 Cor. 6: 5), and accordingly some modern scholars have argued for their composition at *Caesarea or at *Ephesus. For the former suggestion, there is little or no evidence; it is just that it is perfectly possible. For Ephesus, there is support from 1 Cor. 15: 32: 'I fought with wild animals at Ephesus'—but this is not a reliable support, since from such an ordeal in the amphitheatre Paul would hardly have survived. The words are surely metaphorical, and refer to human opposition. Probably therefore these three epistles were written in Rome about 58–60 CE, and then the references to the praetorian guard (Phil. 1: 13) and *Caesar's household (Phil. 4: 22) are natural enough, though admittedly it would not be impossible to associate them with Ephesus. The letter was presumably taken to Philippi by Epaphroditus (Phil. 2: 25) as a visible reassurance that he had recovered from his illness (2: 26). The journeys required for these exchanges might have taken some four months, but such a delay does not rule out the Roman hypothesis. The distance on the other hand from *Colossae to Rome, and so the problem of the journeys of *Onesimus, the absconding slave, give Ephesus more plausibility as the place of composition for the epistles to the Colossians and to Philemon; but the advanced theological thought of Col. argues a late, i.e. Roman, date for these two epistles also.

There is a fairly wide agreement that the epistle to the Ephesians comes from the Pauline circle rather than from Paul himself. The Church is predominantly *Gentile (Eph. 2: 11–22), and this composition is accepted without having to defend it against Jewish Christian opposition, as in earlier days; and the Church has received a structure of leadership and administration (Eph. 4: 8–12) rather different from the looser conglomeration of diverse gifts which threatened anarchy at Corinth (1 Cor. 12).

Carchemish A Syrian city on the River Euphrates which was bitterly fought over and passed in turn to *Assyria, *Egypt, and *Babylonia (605 BCE; Jer. 46: 2).

Carmel, Mount A hill of 168m. (556 feet) overlooking the sea. In its region *Elijah and the prophets of *Baal engaged in a struggle for the nation's allegiance (1 Kgs. 18). Elijah's triumph eliminated the threat of the religion of Melkart which the queen *Jezebel was promoting.

carnelian The sixth precious stone in the foundation of the wall of the New Jerusalem (Rev. 21: 20, NRSV, REB); a red stone, a ruby (NJB) rather than sardius (AV).

carpenter Worker in timber. Carpenters from abroad were employed in the building of *Solomon's Temple (1 Kgs. 5: 6) but by the *Exile there were Jews who had learnt the craft (Jer. 24: 1) and would be available for the restoration. Their tools are listed in Isa. 44: 13–17. Jesus worked as a carpenter (Mark 6: 3), and was the son of a carpenter (Matt. 13: 55), probably making agricultural implements. But in Talmudic sayings there is a noun for 'carpenter' which is a metaphor for a 'learned man', though it does not seem to be the meaning at Mark 6: 3. Sociological studies of the country put carpenters into the artisan class which formed 5 per cent of the population, below peasants but above beggars and day labourers.

Carpus A resident of *Troas. *Timothy is instructed to bring the cloak and books which Paul left with Carpus (2 Tim. 4: 13), which seems to imply that the writer hoped for a period of activity longer than the imminent *martyrdom foreshadowed in 4: 6–8.

Castor and Pollux The name (REB) of the ship of Alexandria in which Paul sailed from *Malta (Acts 28: 11) with these *twin brothers (NRSV) as figurehead. In Greek mythology they were the sons of the god Zeus and brothers of Helen.

catechetical material The Greek word *catechesis* means 'an echo', and in NT studies it denotes sections of the epistles which are believed to represent oral instruction for those being baptized. An example would be 1 Pet. 1: 2–4: 11. Catechetical usage may also have shaped certain sections of the gospels, as in the accounts of the *Lord's Supper (e.g. Matt. 26: 14–30).

catena (plural: ***catenae***) A Latin term used in biblical studies for a series of connected items, such as miracle stories or short sayings.

catholic epistles A term used for those NT epistles that are not addressed to particular Churches or individuals: James, 1 and 2 Peter, 1 and 2 John, and Jude. These epistles all take their titles from the supposed writer rather than the groups addressed (as with the epistles of Paul). 'Catholic' here means 'general' and has no doctrinal or ecclesiastical implication.

caves Palestine abounds in caves, some so large that in them a contingent of 400 armed men could take refuge (1 Sam. 22: 1–2). Most famous of all are the caves in which the community at *Qumran hid their MSS before the Roman invaders (66–70 CE) overran the place.

cedars Great trees which grew in *Lebanon were used for the construction of palaces and temples; they were long-lasting and had an attractive fragrance. *David and *Solomon bought them from *Tyre (2 Sam. 5: 11; 1 Kgs. 5: 8), whence they were floated on the sea to *Joppa and then hauled up to Jerusalem. The splendour of cedars (Ps. 92: 12) was proverbial; their height was between 80 and 100 feet (up to 30 m.).

celibacy The unmarried state which is commended by Paul (1 Cor. 7: 20–7) but without insisting on it as a discipline. There is no **sin* in getting married and those who are married need not obtain separations. Paul's advice is to avoid any personal unsettlement and trouble in view of the *eschatological crisis and *judgement on the world which he thought, at that time, to be imminent. He confirms, however (1 Cor. 9: 5), the picture in the gospels of Peter and other apostles being married men (Mark 1: 30).

Cenchreae A town near *Corinth visited by Paul (Acts 18: 18–20). A Church was established there; *Phoebe was one of its *deacons (Rom. 16: 1).

censer A vessel in which *incense was sprinkled on hot coals (2 Chr. 26: 19); probably a decorated rod with a bowl at the end (Heb. 9: 4). In Rev. 8: 3 an *angel swings a golden censer and the ascending smoke symbolizes people's prayers.

census Governing authorities needed to know the exact number and distribution of a population for the purposes of taxation and conscription, but the enquiry was resented by those in Israel who were opposed to the despotic claims of the monarchy. *David is said to have been punished for ordering a census—though the punishment was inflicted on the very people being counted (2 Sam. 24: 15).

The census at the time (4 BCE) of Jesus' birth (Luke 2: 1–5) involves a historical problem in that *Quirinius, said to be governor at the time, did not become governor of the province of *Syria until the year 6 CE—when indeed a census *was* held. A universal census, with such social chaos and movement as described in Luke 2: 1, is in any case most unlikely. One way of reconciling the data is to suppose that Quirinius had a minor post in Syria in 4 BCE and that Luke calls him governor because that is how he was generally remembered. Another suggestion is that Luke's account of the journey of Joseph and Mary from their home in *Nazareth in *Galilee to *Bethlehem in *Judaea is an editorial device in order to fit Jesus' birth into plausible OT prophecies. For example, Micah (5: 2) appears to associate a Messiah with the house of David who came from Bethlehem (1 Sam. 16: 1).

centurion A soldier in a Roman legion who was in charge of a hundred men—perhaps equivalent to a modern sergeant, though there were various grades and those in the highest were important figures in the province in which they served (Luke 7: 2). *Cornelius was such a one (Acts 10: 1–2).

Cephas The surname given to Peter the disciple by Jesus when his brother *Andrew introduced him (John 1: 42). In Aramaic the word means 'rock' and is therefore equivalent to the Greek 'Peter', which also means 'rock', as shown by Matt. 16: 18. His original name was Simon.

chalcedony The third precious stone used in the foundation of the walls of the New Jerusalem (Rev. 21: 19), 'agate' (NRSV); perhaps crystallized quartz and silica.

Chaldea Part of *Babylonia, on the Persian Gulf, which eventually assumed control of the whole country, so that Chaldea and Babylon were interchangeable. The rulers of Babylonia best known through the OT were Merodach-baladan and *Nebuchadnezzar, the Chaldean kings (2 Kgs. 20: 12; 24: 1). The country had a reputation for its astrologers (Dan. 2: 2).

chapters and verses *See* Bible.

charisma Greek word for a spiritual *gift, as in 1 Cor. 12: 4.

charity Although often used, especially in 1 Cor. 13, by the AV to translate the Greek *agape*, modern English translations have rejected it on the ground that it has now been appropriated for the more limited reference to almsgiving. *Agape* means disinterested '*love' which is a fruit of the *Spirit (Gal. 5: 22). When used of God in relation to mankind, AV too translates *agape* by 'love' (1 John 3: 1).

Chebar A canal in *Babylonia near which Jewish exiles were settled (Ezek. 1: 3). Excavations have confirmed this site at Tel-abib (Ezek. 3: 15).

Chemosh The god of *Moab for whom *Solomon built a high place (1 Kgs. 11: 7) which *Josiah later destroyed (2 Kgs. 23: 13). A stela dedicated to Chemosh by King Moab was discovered in 1868 with cultic language similar to that of Israelites for Yahweh.

Chenoboskion Site of early Christian monastery on the River *Nile in *Egypt where *Gnostic MSS were found in 1945–6. The place is near the modern town of *Nag Hammadi, which has given the name to the MSS.

cherubim Winged creatures ('cherubim' is the Hebrew plural of 'cherub') which were frequently represented in the art of ancient Assyria. Two may be seen in the British Museum in London. They are not human and are very different from the domesticated, if ethereal, cherubim associated with deity in Western art.

In the OT cherubim are described as being golden figures with a single face (Exod. 25: 18) on the mercy seat on the top of the *Ark of the Covenant, but in *Ezekiel's vision (Ezek. 10: 21) they are given four faces. Clearly the cherubim are regarded as unnatural creatures in the OT and the descriptions are inspired by the representations of their Near Eastern neighbours.

In the NT cherubim are probably to be identified with the four living creatures, with six wings apiece, of Rev. 4: 8 who worship God without ceasing, and as such are met in Christian hymns, where with the seraphim they give glory to the Holy Trinity.

chiasmus A literary device used in rhetoric and in *poetry to designate the inversion of an order of words, phrases, or rhyming which follow subsequently in the material. It can be represented by the Greek letter χ (khi, ch) or the symbols a, b, c; c, b, a. In the OT it is found in the psalms. In the NT chiasmus is used by Paul; for example Rom. 10: 9 f., 1 Cor. 7: 3, and in 1 Cor. 15: 12 ff. And when the gospels are seen as sophisticated literary constructions, chiasmus has been detected in the order of narratives, as when the *temptations of Jesus in Luke 4: 3, 7, 9 seem to be taken up and somehow answered by the *miracles in Luke 4: 29–30; 4: 35, 36; and 5: 6. Those of Luke's readers who have eyes to see thus perceive that Jesus is the Son of God. It is suggested that the structure of Matt. has a chiasmic form: the *Sermon on the Mount at the beginning is answered by the final discourse (24–5); 10 is related to 18; and 13 is in the middle.

children Childlessness was such a disaster that the Hebrews could regard it only as a punishment from God (Gen. 20: 18); but children in abundance (Ps. 127: 5) were a mark of divine blessing, and, in the case of sons, the means of perpetuating the family line and taking on the father's trade to ensure prosperity. Although mothers usually chose a child's name (1 Sam. 1: 20) it was the father who gave religious and practical instruction (Exod. 10: 2; Ecclus. [= Sir.] 30: 1–13), and discipline was severe (Prov. 13: 24). Christians are sometimes addressed as 'children' (Gal. 4: 19; 1 John 2: 1), a practice they perhaps connected with Jesus (John 21: 5). And they referred to themselves as 'children of God' (Rom. 8: 16) to denote the special relationship with God which they inherited as the people of the new *covenant.

Chios A mountainous island off the west coast of Asia Minor (Acts 20: 15). It has been famous for its wines, and by ancient Greeks was celebrated as the birthplace of Homer.

Chloe A woman of substance, possibly a Christian, who lived in *Corinth. 1 Cor. was written when Paul learnt of the unsatisfactory state of that Church from slaves who belonged to Chloe (1 Cor. 1: 11).

Chorazin A town on the Sea of *Galilee, north of *Capernaum, rebuked by Jesus for its lack of response (Matt. 11: 21), but otherwise not mentioned in the gospels.

chosen In the OT certain individuals are 'chosen' (e.g. Moses, *David) and certain groups (priests), but also the whole people of Israel are the chosen, the elect, people of the *covenant (Deut. 7: 6–9); in the NT the 'chosen' are members of the new Messianic community (Eph. 1: 4), though Jesus warns his disciples that the Messianic invitation is not the same thing as being accepted (Matt. 22: 14). Christians addressed by 1 Pet. 1: 2 are said to have been chosen in the foreknowledge of God.

***chriae* (or *chreia*)** The latter spelling is now more usual. The term was applied to a popular literary form in the ancient world where it was defined as 'a concise reminiscence aptly attributed to some character'. Chreiai might be jokes or maxims or examples, and it was part of the education of students to compose them, or to expand a concise *chreia* into a whole paragraph.

*Form Critics of the NT (e.g. Martin Dibelius) in the first part of the 20th cent. used the term for the category of short pointed sayings of general significance but originally spoken to a particular person or within a definite situation. No circumstantial details are included. An example is: 'Let the children come to me; do not stop them; for it is to such as these that the kingdom of God belongs' (Mark 10: 14). Other examples are Mark 2: 4 f. and Luke 11: 27 f.

A concise *chreia* could, however, be expanded, as in Mark's account of Jesus' Cleansing of the Temple (11: 11–15), while Luke (who had previously mentioned Jesus' first entries into the Temple at 1: 22 and 2: 41) condenses Mark's expanded *chreia* into a concise one (Luke 19: 45–6).

Christ The translation *Christos* is employed in the *LXX for the Hebrew Messiah, meaning 'anointed'; originally it applied to the king, who was anointed, as *David by *Samuel (1 Sam. 16: 13), and his successors (Ps. 2: 2; Dan. 9: 25), and in the NT to Jesus as the one who fulfilled the OT expectations (Luke 2: 11).

'Christ' was so often applied to Jesus, as the Messiah, that it virtually became a surname attached to Jesus (John 1: 17).

Christian The name may well have been applied to followers of Christ in the first instance by opponents. According to Acts 11: 26 it was first heard in *Antioch of Syria. By the time that 1 Pet. 4: 16 was circulating, it was the customary usage in Asia Minor, and the author seems quite comfortable with it.

Christmas The early Church had no celebration for the *birth of Jesus. The date was unknown. However, with the growth of a calendar by which events in the life of Jesus could be commemorated in turn, the birth of Jesus was inevitably given a day, just as the single-day celebration at *Easter was eventually spread through Holy Week. At first, 6 January was observed and a whole sequence of stories was rehearsed—*birth, the visit of the *Magi, and *baptism. But in Rome, where 6 January was set aside for the Magi, it seems that in the 4th cent. CE the Church designated 25 December as the appropriate day for celebrating the Nativity because it had been the day of the pagan festival of Sol Invictus, when the unconquerable sun triumphed annually over the darkness of winter and the days grew longer again. There was a verse in Malachi (4: 2) about the 'Sun of Righteousness' which could then seem to be especially apposite.

Christology A doctrine of the Person of Christ, and the study of it, has been an essential part of rational thought about their beliefs once monotheistic Jews found that they were worshipping Jesus Christ as God. It is the work of interpreting the significance of Jesus Christ for *faith. Down the ages it has taken many shapes, beginning with the application of 'Christ' (= '*Messiah') to Jesus; this asserted the connection of Jesus with the aspirations and beliefs of the OT and the people of Israel, however interpreted.

After long and sometimes acrimonious debates the Church gave a final definition of its Christology at the Council of Chalcedon in Asia Minor in 451 CE, affirming belief in Jesus Christ as One Person in Two Natures, which are united without confusion. Much subsequent thinking started with the premiss that Jesus was the second Person of the Trinity and then speculated how he could have been man. An early suggestion had been that Jesus only *appeared* to have had a physical body (this became known as the heresy of *docetism and was ruled out by stressing the genuine humanity of Jesus, descendant of *David, 2 Tim. 2: 8). Nevertheless there continued a long tradition in the Church which emphasized the divine nature of Christ at the expense of his humanity.

Much recent Christology claims to work 'from below up', that is to say, to begin with the humanity of Jesus and go on to show that the evidence leads to a recognition also of his divinity. It is a procedure beset with problems; for it necessarily depends upon controversial assessment of the historical value of the gospels which recount words and deeds of Jesus within the context of an interpretative apparatus. Examples are the messages of *angels at the beginning and end, voices from *heaven (at the *baptism and *transfiguration), various theological reflections (e.g. Luke 23: 44–5), and constant references (especially in Matthew) to OT prophecies.

On the other hand, there is much in the narratives of the gospels which can be accepted as historically trustworthy and which provides a foundation for a modern Christology. Events which are recorded in spite of being obviously embarrassing to the early Church and which might therefore have been understandably omitted impress for their veracity; an example is the *baptism of Jesus and the history of its treatment from Mark, through Matthew (3: 14); Luke (3: 21), where the baptism by John is not mentioned explicitly; and John (1: 33), which does not record the baptism of Jesus at all. Such a progress reveals the embarrassment felt that Jesus should have undergone a 'baptism of repentance for the forgiveness of sins' (Mark 1: 4). It was his self-identification with his people, but theological interpretation is evident: there is the mention of the heavenly voice, where Matthew uses a rabbinic image, the **bath qol*, to confirm the readers' inference that there is a pattern of events similar to that of the *Exodus (of Israel). There can be no doubt about the *crucifixion and little doubt that Jesus at the end cried, 'My God, why have you forsaken me?' (Mark 15: 34), where Luke (23: 46) prefers the cry, 'Into your hands I commend my spirit'. Equally authentic are the stories of Jesus consorting with disadvantaged, unpopular, and despised people, and the accounts of *miracles of healing and exorcizing.

Given the recognition that the gospels portray a genuinely human Jesus whose mode of life and teaching provoked opposition almost from the start (Mark 3: 6), we are entitled to take a further step by noting the theological interpretation given to some of these events. That Jesus was born somewhere in Palestine at the end of the reign of *Herod the Great is a fact of history, but the interpretation of Matthew and Luke that the birth by *Mary without the agency of a human father is an assertion that the *man* Jesus came from God to inaugurate a new relationship between God and humanity. The *Transfiguration (Mark 9: 2–8), often dismissed by critics as a post-resurrection narrative which has been lifted out of the oral tradition into its present significant location within the ministry, need not be regarded as theological fiction. The account of Jesus' shining garments is paralleled by authentic accounts of the luminosity of saints. But the theological apparatus of the *cloud and the arrival of Moses and *Elijah are pointers to the belief that this wholly human figure can be discerned by faith to be invested with a transcendental '*glory'.

Beginning therefore with clear expressions of the humanity of Jesus, it is discernible that the gospels also bestow on it a depth of glory which goes beyond our own humanity. In him there is a combination of the authority in his teaching, an outreach of unselfish love, and a total obedience, with humility, to God's calling.

The NT account of the Person of Christ is also expressed in the use of an extraordinary range of titles, or names, or images, applied to Jesus.

Among these 'titles' the best known are 'the *Son of Man' and 'the *Son of God', which have often been seen as straight references to Jesus' humanity and divinity respectively. But it is not so simple. Both terms have a long history in the OT and later literature and have been the subject of a whole industry of scholarship. Was 'Son of Man' used by Jesus himself with or without the definite article? Are the sayings in the gospels authentic utterances of Jesus? If so, do they refer to himself or to some other figure? Are they references to an individual or to a collection of people? Does the Aramaic simply mean 'man' in general, or does the background of the term denote a supernatural, apocalyptic figure that was taken over by the Palestinian Christian community and applied in the first generation of the Church to Jesus?

There is wide agreement that at least certain sayings in the gospels cannot be the words of Jesus himself, such as Matt. 16: 13. There is also considerable agreement about the almost complete absence of the term 'Son of Man' outside the gospels: that the phrase was so consecrated to the lips of Jesus that others in the NT did not use it of him. As to the meaning, it is probable that, in Aramaic, it was an oblique, reticent way of referring to himself—'I, being the man I am . . .' 'The Son of Man' was not a Messianic title and in using it Jesus was not laying claim to Messiahship; but neither was he denying such a claim.

'Son of God' was a familiar expression in Hebrew and Jewish thought: it does not characterize a divine being but is used of male persons who are believed to stand in a close relationship to God. The gospels maintain that Jesus was Son of God in a preeminent sense, as is apparent from the application of the expression to him by the Voice at Jesus' baptism and transfiguration. In the Hellenistic world into which the Church advanced, 'Son of God' was being accorded to the Roman emperor *Augustus and his successors, so that 'Son of God', an early use in the Church (1 Thess. 1: 10), made possible an easy transition from Jewish to Gentile understanding. The early and authentic origins of the phrase as applied to Jesus is indicated by the unconcealed confession of the ignorance of the Son (Mark 13: 32), though the Son does enjoy a unique knowledge of the Father (Matt. 11: 27).

Another title which made an easy transition to the Hellenistic world was 'Lord' (*kurios*, in Greek; used in the *LXX with the definite article for 'The Lord' or Yahweh) which could mean merely 'sir' but to *Gentile converts would imply supernatural status. It was already acquiring this meaning in the third gospel, written for Gentile readers (Luke 22: 61). Cf. Phil. 2: 10–11.

There is therefore in the NT, especially in the gospel of John, material which led to the classical Christological definitions. Some forty-two names or titles of Jesus are used in the NT; his humanity is clear—he is a 'son of *Joseph', '*rabbi', '*prophet'—yet the gospels were compiled 'that you may come to believe that Jesus is the Messiah, the Son of God' (John 20: 21, making explicit what was unformulated in the synoptic gospels). Jesus was tempted to escape the cup of suffering (Mark 14: 36) but through death was vindicated by *resurrection and then bestowed the *Spirit on his disciples (Luke 24: 49; John 20: 22).

Chronicles, books of A revised version of OT history. The two books of Chronicles derive their name from a remark of the Latin Father Jerome (d. 420 CE) that they are 'a chronicle of the whole of sacred history', and indeed they are a kind of commentary on the books of the OT from Genesis to the Fall of Jerusalem—and beyond, if the books of Nehemiah and Ezra, which come from the same circle, are to be included. In that case the chronicler goes on beyond the Exile up to the rebuilding of the Temple under

*Zerubbabel in the 6th cent. BCE and the laws, a century later, of Nehemiah and Ezra.

In the Hebrew Bible 1 and 2 Chronicles form a single volume and are called 'the things of the days' (= history), but in the *LXX, where they are now two volumes, the title is the 'paralipomena', 'the things left out', and '1 and 2 Paralipomena' is the title given to the books in older Roman Catholic translations. The implication is that 1 and 2 Chron. are giving supplementary information not contained in the other books. But the main purpose of the work is to encourage Jews of the Persian period (4th cent. BCE) to maintain faithful worship in the Temple.

The genealogies of the first nine chapters (Adam to Saul) are important as a theological demonstration of how God's purpose has been carried on from generation to generation. *Abraham and Moses in this story receive scant attention compared with *David and *Solomon. For they are the kings responsible for the Temple and its glory—David for conceiving the idea and leaving the finance, Solomon for the execution. These two kings are celebrated as beyond criticism, and their less attractive deeds as recorded in the books of Samuel and Kings are suppressed by this revisionist. Because so much of Samuel and Kings is repeated in Chronicles the author's point of view is all the clearer by reason of the divergence from his predecessors. He is interested above all in the religious cult at Jerusalem (he ignores the apostasy of the northern kingdom) and expands the earlier accounts of the reforming kings of Judah (such as *Josiah). Although 2 Chron. ends with the capture of Jerusalem by the Babylonians (2 Chron. 36: 15–21), there is appended an assurance (36: 22–3): all is not lost! And of this the continuation in the books of Ezra and Nehemiah is the proof.

chronology of the New Testament The life of Jesus and the travels of Paul are fixed by the NT within the stream of secular history. Calculations of the monk *Dionysius in Rome in 532 CE resulted in the *birth of Jesus being put at year 1 of the new Christian Era, corresponding to year 754 of the Roman calendar, which began with the foundation of the city. Collating various subsequent dates then puts the reign of *Herod the Great in Jerusalem as from 37 BCE to 4 BCE. But Jesus was born shortly before the death of Herod the Great, i.e. in or before 4 BCE (Matt. 2: 1, 19). He was crucified under *Pontius Pilate, who was *prefect of Judaea from 26 to 36 CE. Precise dates for the ministry of Jesus are beset with uncertainty, since the evangelists are not much interested in chronology; Luke is the exception, but his efforts in 3: 1, as in 2: 1, have left problems for his readers. A *universal* *census (2: 1) is highly improbable: how could thousands of people throughout the Roman Empire have been uprooted? Moreover, the census of *Quirinius was held in 6 or 7 CE, which does not coincide with the reign of Herod the Great, and in any case did not apply to residents of *Nazareth in *Galilee. And why should *Joseph have been required to travel to Bethlehem on the ground of his ancestor being born there many generations previously? It looks very much as though Luke was providing a story to fulfil the prophecy of Micah (5: 2–4). The 'star' followed by the *Magi (Halley's comet was seen in 12 BCE) is another feature of the *infancy narratives which support a view that they are more theological interpretation than precise historical records.

The length of the ministry of Jesus may have been one year, as the synoptists appear to imply, or two years (at least), as the mention by John of three Passovers indicates (2: 13; 6: 4; 11: 55); but as the synoptists and John disagree about the date of the *Lord's Supper and the *Crucifixion, and because the methods of establishing the details of the Jewish *calendar were unreliable, it is impossible to be sure of the year. Even the note in John 2: 20 that the Temple had so far taken forty-six years to build is not a safe guide, since Herod did not necessarily order construction to begin immediately upon taking his decision. Perhaps April in 30 CE could be accepted as a reasonable date for the crucifixion.

There are some fixed dates for Paul.

*Aretas, who governed *Damascus (through an official) when Paul escaped from it, ruled the city from 37 CE (2 Cor. 11: 32). This puts Paul's experience on the Damascus road at about 34 CE. *Gallio was proconsul of Achaia (Acts 18: 12) from 51 to 52 CE, which gives a date for Paul in *Corinth. The edict of the emperor *Claudius which expelled Jews from Rome was probably issued in 49 CE; hence Paul met *Aquila and *Priscilla soon afterwards in Corinth (Acts 18: 1–3). He had probably written the epistles to the Thessalonians by that time. The procurator *Felix (Acts 24: 27) was deposed in 60 CE, so Paul's trial under *Festus and the voyage to Rome probably took place in that year. *James the Lord's brother was stoned in 62. The martyrdom of Peter and Paul probably took place soon after the great fire in Rome in 64.

chronology of the Old Testament A great deal of the OT is in the form of historical narrative and occasionally its dates can be checked against external evidence. The Creation stories are followed by accounts of the ancestors—*Abraham, *Isaac, *Jacob, *Joseph—and then by the *Exodus and the conquest of *Canaan under Moses and *Joshua. There is no means of verifying dates or history until the period of the Judges, when village life depicted in Palestine seems to be that of the Iron Age. The first three kings of the undivided kingdom lived from about 1020 to 922 BCE. Thereafter the two books of Kings provide more reliable information, and the reigns of the kings of Israel and Judah are dated in relation to each other, from about 925 to 596 BCE, though precise dating is often impossible. *Babylonian records establish the fall of Jerusalem as in March 597 BCE, and its second fall (2 Kgs. 25: 8, 21; Jer. 52: 12, 31–4) was then in 587 or 586 BCE.

After the Exile some of the biblical dates are given in terms of the reigning kings of Persia, Greece, and Syria. *Nehemiah and *Ezra were in Jerusalem in the middle of the 5th cent. BCE; *Alexander the Great conquered Palestine in 332 BCE, and from 320 to 200, when the empire was divided, Palestine was under the *Ptolemies, and there was local government by high priests. The Syrian period lasted from 200 BCE until Pompey took Jerusalem for the Romans in 63 BCE. The desecration of the Temple by *Antiochus Epiphanes (1 Macc. 1: 54) took place in 167 BCE.

The following is a list of key dates in OT history, though until the establishment of the monarchy they can only be possibilities:

BCE	
1700	Abraham
1250	The Exodus
1100	The Judges
1020	The monarchy of Saul and David
962	Accession of Solomon
924	The kingdom divided
842	Jehu's revolt
722	The fall of Samaria
597, 586	The fall of Jerusalem and the Exile
520	The rebuilding of the Temple begins
332	Conquest of Palestine by Alexander the Great
166–160	Judas Maccabaeus
37–4	Herod the Great

Church The Christian *community. In the Graeco-Roman world of the 1st cent., there were many religious guilds and societies, but the Christians took over the word *ecclesia* from the *LXX, where it denotes the assembly of the people of Israel; but unlike the word *synagogue*, also used in the *LXX, *ecclesia* was not a peculiarly Jewish word: it was the ordinary word in classical Greek for a gathering of the people at the call of a herald, and is indeed used in Acts 19: 32 of a secular assembly. It was thus a fitting noun to apply to a society which included many *Gentiles. But 'Church' is never used as today for a building, or for a denomination (e.g. the Methodist Church).

In Acts the word is sometimes used in the singular of a local Christian community but in 9: 31, it refers to the whole Church so far as it had then extended. In Acts 20: 28 Paul charges the elders of *Ephesus 'to shepherd the Church of God that he obtained with the blood of his own Son', which must surely mean a body wider than the community at Ephesus: it points to the universal Church, as in 1 Cor. 16: 19, where the local Church which meets in a house seems to be regarded as one unit in a greater whole. In 1 Cor. 12: 28 Paul refers to those whom God has appointed in the Church—surely the universal Church—'apostles, prophets, etc.' So too Eph. 1: 22.

The universal Church is also mentioned under different names: 'the Israel of God' (Gal. 6: 16), 'the true people of the circumcision' (Phil. 3: 3, NJB), asserting the claim that the Church was both the continuation and consummation of the chosen people of the OT. The fullest description of the *ecclesia* in the NT is provided in 1 Pet. 2: 9—'a chosen race, a royal priesthood, a holy nation, God's own people'.

The English word 'Church' is therefore a very satisfactory rendering of *ecclesia*. The word, like the Scottish *kirk* and German *Kirche* (which Martin Luther detested and used mostly to denote the heathen shrines of the OT), is from the Greek *kuriakos*, 'belonging to the Lord'. Possibly the word *Kirche* was brought by traders up the Danube and down the Rhine. Wyclif was responsible for the word 'Church', and though *Tyndale and Cranmer substituted '*congregation', this was changed back to 'Church' in AV, possibly because of Tyndale's absurd version of Acts 8: 1: 'There was a great persecution against the congregation which was at Jerusalem and they were all scattered', where *ecclesia* must refer not only to the body of Christians gathered as a congregation, but also to a corporate body, whether assembled or not.

The southern European nations derive their words (*église* etc.) directly from the Greek (and Latin) *ecclesia*.

Whether Jesus intended this Church continues to be debated. It is said, for example, that because Jesus expected the world to end quite soon he could not possibly have contemplated anything so long-term as founding a Church. This, however, is disputable. Mark ends his *apocalyptic chapter (13: 32) with a warning to any Christians who supposed the end to be near; not even Jesus knows when the end will come. There is indeed to be a future *judgement, but Jesus says that God's saving power is to be experienced in the present (Luke 11: 20); the kingdom is within the disciples' grasp (Luke 17: 20–1); it is present when evil is met with love and mercy rather than with hate and violence, and it is to be present above all in Jesus' suffering and *death. Jesus has necessarily employed the concepts that were available to him, and then given them new content. The Reign of God, he said, is breaking into human affairs, but there will be an interval before the Reign is wholly and unconditionally present. It is in this interim period that the Church must work, and Jesus made provision for a community of disciples. The term '*Son of Man', derived from Daniel 7, where he represents 'the saints' (AV), involved the notion of community; but it is not possible to assign the foundation of the Church to a particular day or hour, such as after Peter's confession. Rather, the Church was created by the totality of the *life and death and *resurrection of Jesus. The words of Jesus reported in Matt. 16: 18 and 18: 17 are the only instances in the gospels of the word *ecclesia*, and are unlikely to be authentic words of Jesus, since the kind of authority there promised to Peter (16: 18) was never in fact enjoyed by him, so far as the NT evidence suggests. In Matt. 18: 17 it is the local Church community that is referred to—so that at the time of the composition of Matthew the Church was seen as a universal body with local manifestations.

Admission to the Church was by *baptism, but the ideals of holiness were far from being achieved. Although membership of the Church was open to all, Jews and Gentiles, slaves and free, rich and poor, men and women, in practice social differentiations sometimes persisted (1 Cor. 11: 21), as did typically Greek fondness for lawsuits (1 Cor. 6: 1–11). False teachings and unseemly behaviour crept in (1 John 4: 1–6; 1 Tim. 4: 1–5) but nevertheless the quality of corporate Christian life did attract converts, and it seems that outsiders might be present at worship (1 Cor. 14: 16), which would consist of prophecies and teaching, singing and readings from the OT, and *Eucharistic celebrations on the first day of the week (1 Cor. 16: 2). At some meetings, Paul had asked for a collection to be taken to assist Christians in Jerusalem who were destitute in a time of famine as a result of disposing of all their capital (Acts 4: 34), and it proved a godsend to Paul as a means of holding together the Gentiles of his Churches and the Jewish Christians ('the saints') in Jerusalem (2 Cor. 9: 1–5), as well as being a way of putting pressure on the latter to overcome their reservations about his Gentile mission.

Church government No fixed order of *ministry intended to be permanent is described in the NT any more than an agreed sacramental theology. Developments into Papal, Episcopalian (having bishops presiding over dioceses), and Presbyterian systems came later; they can all claim some continuity with what is in the NT, and so the criterion for assessing them by modern Christians must be theological rather than historical research. What kind of ministry safeguards and expresses the unity of the Christian *community? What kind of ministry gives appropriate freedom to individuals to grow into spiritual maturity and to exercise their personal *gifts?

Eight different kinds of gifts are bestowed on members of the Church, according to Paul (1 Cor. 12: 28) and none is superior, none inferior. There was a 'priesthood of all believers', in that all Christians offered themselves as a *sacrifice according to the pattern laid down by Christ; and all such personal offerings (according to the letter to the Hebrews) are taken up into the one perpetual offering made by the one eternal high priest of the new *covenant—Jesus. All Christian believers therefore comprise a priestly body, called to proclaim the gospel of *reconciliation between God and mankind.

However, Jesus had chosen twelve *apostles and on these Twelve (the defection of *Judas caused a vacancy, filled by the election of *Matthias, Acts 1: 26) rested the leadership of the Church. Paul was added to the group as an extraordinary member (1 Cor. 15: 8–9) with a special responsibility for a mission to the *Gentiles (Gal. 2: 9). Local Churches had local leaders who acted under the general supervision of apostles (1 Thess. 5: 12) and some, like *Timothy, could be dispatched to Churches by an apostle and armed with his authority. At Jerusalem there was a group, or 'college', of *elders (*presbyters) in charge, with *James the Lord's brother as president, and probably other local Churches were similarly organized, following the precedent of the synagogues. One of the leaders would have presided at the *Eucharist. In the *Pastoral Epistles (to Timothy and Titus) there is development: there Paul's delegates themselves delegate authority to successors (2 Tim. 2: 2) and we meet the word *episcopoi* being used of the presbyters (Tit. 1: 7; 1 Tim. 3: 1–7).

Male and female (Rom. 16: 1) deacons have a share in Church government (1 Tim. 3: 8–13) but they are probably not the successors of the Seven appointed to help the apostles (Acts 6: 3), who seem rather to have been the first presbyters, with a duty to 'serve tables', that is, undertake charitable works, although that phrase might also cover presiding at the table of the Eucharist. The *deacons mentioned by Paul have a status as officials, but what they were for is not clear: except that the whole ministry is for the purpose of service.

By the time of *Ignatius, who wrote seven letters while journeying to *Rome, where he was martyred in 107 CE, each local Church is deemed to have a *bishop at its head, and the ministries of charismatic *prophets and *teachers passed into abeyance. By the 3rd cent. the bishop had taken over many of the administrative functions of the diaconate, and presbyters had assumed powers formerly exercised only by bishops. This was caused by the expansion of the Church beyond the towns into the surrounding countryside—where resistance to Christianity had for long been strongest: our word 'pagan' even comes straight from the Latin *paganus* meaning 'villager'.

At the beginning of the 3rd cent., the term 'priest' is used for the first time of a Christian minister, and language reminiscent of the Temple cult was applied to presbyters, so that the OT orders of high priests, priests, and Levites were being used as the model for bishops, presbyters, and deacons.

In 1552 Martin Bucer proposed the use of the titles 'superintendent', 'presbyters of the first order', and 'presbyters of the second order' for the traditional bishop, priests, and deacons, but the Anglican ordinal (the authorized rite for ordination) did not adopt the proposal. So the Church has maintained itself to be a 'royal priesthood', declaring to the world the wonderful deeds of God (1 Pet. 2: 9), through many vicissitudes of pastoral ministry.

Churches, seven Christian communities in the province of Asia. The last book of the Bible, the Revelation to John, also known sometimes as the *Apocalypse, is addressed to seven particular Churches. There were other Churches too, not mentioned, such as those at *Colossae (Col. 1: 2) and *Troas (Acts 20: 5–12), but the author chooses seven because that is the number of completeness (the creation by God in Gen. was completed in *seven* days) and so these seven represent the whole Church. Considerable local knowledge is, however, displayed about the seven selected: *Ephesus, *Smyrna, *Pergamum, *Thyatira, *Sardis, *Philadelphia, and *Laodicea (Rev. 1: 11).

Cilicia A Roman province, in SE Asia Minor. *Tarsus, birthplace of Paul (Acts 22: 3) was one of its towns. Paul twice passed through the area (Acts 15: 40–1 and 18: 23).

circumcision The removal, by cutting, of the foreskin of the male penis. Though widely advocated in modern Western societies on the supposed ground of hygiene, circumcision is particularly associated with practising Jews by whom the rite is immensely valued as the occasion when a male child is initiated into the *covenant of *Abraham, eight days after birth (Lev. 12: 3).

However, as the OT itself (Jer. 9: 25–6) recognizes, circumcision was known amongst other peoples in the Near East, though not all required total amputation of the foreskin. The Hebrews had several versions of the origin of the rite—to *Abraham (Gen. 17: 9–27), or to Moses (Lev. 12: 3) or to *Joshua (Josh. 5: 2–7), and these served as useful justifications from Israel's traditions for the continuing insistence on the rite. In fact the origin may have been a kind of magic: a curious reference to this may survive in the story that Moses' wife saved his (Moses') life by circumcising their son (Exod. 4: 24–6).

At any rate circumcision was always a vital part of Jews' life—valued, exported (according to *Josephus), defended even by *martyrdom (1 Macc. 1: 48, 60) by the fervent, but detested by Jews who welcomed Hellenistic culture and desired assimilation to the Greeks. (Hellenized Jews took painful steps to cover up the embarrassment of circumcision by means of surgery.)

In the NT circumcision was an issue between Paul and the Jewish Christians who expected pagan converts first to become members of the covenant of Abraham by circumcision before being baptized as Christians. For Paul, this demand represented an attempt to add something to Christ, which was an absurdity: indeed it is virtually to reject Christ; it means that one has transferred one's allegiance to Christ alone on to an alternative system, namely the *Law, in which circumcision is basic. The issue was whether the Church had taken over the role of the New Israel, or was a reformed Jewish sect.

There is no mention in the OT of the barbaric surgery of female circumcision traditionally practised in some African countries today and clandestinely in the West by people from those regions.

citizenship Roman citizenship was immensely important in the NT era, and Paul was able to claim it—and more; he was also a citizen of *Tarsus. His Roman entitlement probably derived from his father and grandfather and it provided exemption from degrading punishments. Citizenship of Tarsus conferred merely local privileges. How the tribune at Jerusalem could ascertain that Paul was speaking the truth in making his claim to be a Roman citizen, according to the account in Acts 22: 27–9, is uncertain.

But birth certificates, being copies of official records, have been discovered in *Egypt and perhaps a record was kept at Rome.

One suggestion about how Paul's family secured Roman citizenship is that they were once Jewish slaves taken by Pompey in 63 BCE and deported to Tarsus, and there freed and honoured.

city It was not necessarily the size of its population or its area that made the distinction in Palestine between a city and a village. What mattered was the wall—which was often doubled, with houses built between the outer and the inner wall, while outside might be a cluster of villages which supplied the city with provisions and labour. Palestinian cities were not large and it has been estimated that even 1st-cent. Jerusalem did not occupy more than 200 acres, though it had a population of about 10,000. City walls were provided with gates for entry and exit which were essential but which were also the easiest places for an enemy to breach and therefore were well fortified by the inhabitants. The gate area was often used for meetings and court sessions (Ruth 4: 1–6) and just outside the wall might be a spring which could be tapped for supplying water to the city by means of a tunnel.

The Roman occupation of Palestine brought improved amenities to cities, including Jerusalem. Imposing streets might be flanked by equally impressive dwelling houses equipped with baths and decorated walls. A theatre was built in Jerusalem by *Herod the Great; and streets on the model of a Graeco-Roman city. The magnificently reconstructed Temple was virtually a new building and is sometimes, rather misleadingly, called the Third Temple. Pontius *Pilate (governor from 26 to 36 CE) authorized the construction of an aqueduct for water much needed by the swollen population at festivals—though as he, a Roman, paid for it out of Temple funds, it provoked a riot, which is possibly referred to in Luke 13: 1–2.

With their structure of social interdependence cities have often been part of Christian imagery, from the New Jerusalem of Rev. through Dante to modern writers such as Charles Williams.

Claudius The fourth Roman emperor (41–54 CE), who is reported by the historian Suetonius to have been a weak ruler, though he ordered some major construction works. An edict of 49 CE expelled Jews from Rome on account of rioting (Acts 18: 2), possibly provoked by proselytizing of Jews by Christians.

clean and unclean *Animals and *fish designated 'unclean' were forbidden to be eaten; the 'clean' were permitted. The distinctions are clearly written into the *Pentateuch (e.g. in the time of *Noah, Gen. 7: 2). This suggests that as elaborated after the Exile by the *P source in the tradition, they may have coloured the narratives of earlier periods. Roughly, *birds and animals that devoured their victims were unclean, because they consumed blood. A full list of unclean creatures is given in Lev. 11. People might become unclean—for instance, if they touched a dead body (Num. 19: 11 ff.; cf. Luke 10: 31) or came into contact with *leprosy (Lev. 13: 3) or bodily fluids (Lev. 12: 2). Means of ceremonial *purification from the effects of *defilement were prescribed.

The origin of these distinctions may have been hygienic; for example, the experience of food poisoning. But the effect of the legislation was to emphasize the difference of Israel as a holy people separate from other nations. It was precisely this notion of separateness that Peter was reluctantly persuaded to renounce (Acts 10: 15) as being inconsistent with Christian *fellowship. Much of Paul's energy was given to teaching that for Christians the time of the *Law, including its food regulations, was over (Rom. 14: 15), and Jews who joined the Church should renounce all their ties with the *synagogue and throw in their lot with those converted from paganism. There should be one *community in *faith and *worship and fellowship (Rom. 15: 7) on Pauline principles.

Clement (1) A fellow worker of Paul (Phil. 4: 3). (2) Bishop of Rome, and author of an epistle to *Corinth, 95 CE. It has sometimes been suggested, very improbably, that they were the same person.

Clement, epistles of Writings from the age immediately after the NT period. The first epistle was addressed to *Corinth and is of interest in showing that some of the problems encountered by Paul were reasserting themselves—notably a spirit of faction in which leading *presbyters had been deposed. The epistle was probably written in 95 CE by Clement of Rome and it refers to Paul's own correspondence, mentioning the divisions at that time associated with Paul, *Apollos, and *Cephas. There are also quotations from the OT, cited more for the purpose of edification than as proofs of fulfilment by Jesus. Although it was regarded with respect, the epistle was not in the end included in the NT *canon.

The second epistle is pseudonymous and was probably written half a century later; its origin and its destiny are unclear.

Cleopas One of two *disciples who met the risen Jesus on the road to *Emmaus (Luke 24: 18). Possibly to be identified with Clopas of John 19: 25, since the fourth gospel has various contacts with the Lucan tradition.

clothes According to the biblical view, clothes reveal the essential nature of a person, as with John the Baptist (Matt. 3: 4); but disguise is deceit (Matt. 7: 15). The clothes of Jesus revealed the nature of his mission, when at the *Transfiguration they became dazzling white (Matt. 17: 2). In the Bible the nature of a person and his clothes are so intimately connected that *miracles are recorded through touching clothes (Mark 5: 30; Acts 19: 12). Clothing and unclothing are therefore important both as images (Rom. 13: 12; 1 Pet. 5: 5) and literally in *baptism (Gal. 3: 27). Normal dress for men consisted of a loin cloth, a robe, and a cloak, and appropriate footwear. Women were obliged to wear covering over their heads.

cloud As the giver of necessary *rain, clouds were regarded with awe in ancient Israel, and God is celebrated as dwelling in them (Ps. 68: 4). In the *J tradition the people are said to be led by Yahweh in the form of a pillar of cloud (Exod. 13: 21) which at night became luminous. In an angry address to *Aaron and *Miriam he spoke out of the cloud (Num. 12: 5), and in the Temple at Jerusalem his presence (Hebrew, *shekinah*) is signified by a cloud (1 Kgs. 8: 10–11).

A cloud therefore is established by NT times as a recognizable symbol for transcendence; at the Transfiguration Jesus and the *disciples are overshadowed by a cloud (Mark 9: 7) and at the *Ascension a cloud is described to signify Jesus' exaltation into *heaven (Acts 1: 9).

cock Although not mentioned in the OT, cocks were common in Palestine by the 1st cent. CE and were appreciated for waking up sleepers in the early morning. The third watch of the night, midnight to 3 a.m., was called cockcrow (Mark 13: 35). The cock was a symbol of alertness, and as such often stands today on church steeples.

codex (plural: **codices**) Latin for a 'tablet' and used to denote an ancient manuscript. Ancient writings were on scrolls, but their disadvantage was having to use both hands to unroll them to find and read the particular passage required. An alternative format was developed in the form of a book with pages which could be turned, and this was gradually adopted by Christians as the most convenient arrangement for public reading. These codices also differentiated the Church from the *Synagogues. To make such a book, or codex, sheets of papyrus or parchment were put together.

Two of the earliest codices are from the 4th cent. CE: Codex *Vaticanus and Codex *Sinaiticus.

cohort A Roman military unit of 600 soldiers, a tenth of a legion, commanded by tribunes. Paul was in the charge of a *centurion who belonged to the *Augustan cohort (Acts 27: 1).

collection for Church of Jerusalem In writing to the Christians in *Corinth (1 Cor.

16: 1–2) Paul requests that a collection should be taken every week in support of his fund for the mother Church in Jerusalem, as he had urged in *Galatia. He could also report a good response from *Macedonia (2 Cor. 8: 3–5). Paul spent much time and effort in organizing the collection, and the cost of taking it, with companions, must have eaten into the total to such an extent that evidently suspicions were aroused that he was lining his own pocket with some of it (2 Cor. 12: 14–18). Yet the importance of the gesture was enormous to Paul: it was a necessary act of charity (Rom. 15: 28) to reconcile the two wings (Jewish and Gentile) of the Church. Maybe Paul also had in mind OT predictions of 'many nations' coming to the mountain of the Lord in Jerusalem (Mic. 4: 2). Perhaps he hoped too that such a *gift to the 'so-called pillars' of the mother Church (Gal. 2: 1) would incline them to a more favourable view of his work; but we never hear how they reacted, or even whether they ever received the collection as they had apparently done in 46 CE (Acts 11: 30).

colony Colonies were cities into which Roman soldiers or veterans were accommodated, transplanted from the capital to form Roman outposts amid possibly hostile territory. *Corinth, *Lystra, and *Antioch of Pisidia were colonies; their constitutions were modelled on those of *Rome. In writing to the Church at *Philippi which was a colony (Acts 16: 12) Paul tells the Christians that their true citizenship is of *heaven though they live temporarily in an earthly outpost (Phil. 3: 20).

Colossae A city in Asia Minor not far from *Laodicea, which had come to exceed it in wealth and importance. There was a big Jewish population in the 1st cent., but the greater part was Greek, and Greek was the predominant language. There was a flourishing trade in wool, and possibly Christians enjoyed an affluence similar to their neighbours in greater Laodicea (Rev. 3: 17 ff.). Roman religion did not extirpate the old Phrygian cults and the mother goddess of fertility had her home in neighbouring *Hierapolis. The impression is of a lively and fairly tolerant community in which there were mixed Jewish–Gentile marriages and the dilution of strict Jewish worship with elements of Hellenistic paganism. (The English 'colossal' suggests that our ancestors many generations ago regarded Colossae as larger than life.) Archaeological remains of the city are scanty.

Colossians, Paul's letter to the In the NT, the seventh letter of Paul. Religious teachers had come into the Lycus valley and were disturbing the tranquillity of the Church at Colossae. There has been much scholarly discussion about their identity. Were they Gnostics—of the kind that flourished in the 2nd cent. CE? Or Jewish Christians? Or loyal Jews?

Although *Gnosticism was most fully manifest in the 2nd cent. CE, traces of it are already evident earlier, and some detect it as the false teaching which the epistle to the Colossians seems to reflect. Gnosticism was an eclectic mixture of speculation about the universe which it regarded as divided between good and evil powers who fought in the world. *Salvation was available to the privileged group of initiates who possessed the essential knowledge (*gnosis*) conveyed by a revealed figure. Along with visionary experiences Gnostics promoted an extreme asceticism and distrust of the material world, so that some scholars regard the epistle's discouragement of self-abasement and the worship of *angels (Col. 2: 18) and the concept of **pleroma* (Col. 1: 19) as references to Gnosticism.

On the other hand, the epistle does not seem to be concerned to refute Gnostic *docetism; nor were Gnostics given actually to the worship of angels—indeed these beings were thought to have had a hand in the creation of the evil world. For these reasons, another view is that the teaching which Col. refutes is that of Jews who did have their special days of abstinence and of celebration (Col. 2: 16) and who did venerate angels as messengers of God; they refused devotion to Christ (Col. 1: 15–20) and they held the *Law to be the appointed means of salvation. Moreover, the false *teachers demanded *circumcision. All of this points to *Judaism as the heresy attacked: but they may have been Jewish Christians, since in Col. 4: 11 the author pointedly mentions those few Jewish Christians who still support him. True, the epistle contains nothing of the violent abuse in which Paul denounced Jewish Christians in *Galatia (Gal. 3: 1), but in this case the whole approach is more restrained and tactful and consists more of a positive statement of Christian belief. Possibly this is because later in life and experience, Paul was learning better apologetics. But was Paul the author? It has been doubted. The long sentences are unlike other Pauline letters. Many little words (particles like 'so' or 'but') typical of other letters are absent in Col.; the great theological theme of the headship of Christ over the universal Church and the neglect of a future *eschatology all point to another author.

Yet there remain good reasons for holding on to the traditional view. Paul was quite capable of adapting his language to the occasion (cf. 1 Cor. 9: 19–23) and there remains the expression of a hope for the future (Col. 3: 4; 3: 24) and the theme of the present life in the *Spirit is already anticipated in earlier epistles. But for many readers the obvious connection with the authentic letter of Paul to Philemon (Col. 4: 9) is decisive.

The essence of the epistle's message is that of the unique role of Christ; in him the Christian *community enjoys the certainty of salvation. Paul uses the Jewish concept of *wisdom (Col. 1: 9), which was also attractive to converts from paganism and was an early example of refining concepts in order to express the Christian faith in terms intelligible to new audiences for whom an alien conservative framework was meaningless. It was a very subtle argument. In Christ the wisdom of God was made plain and his purpose executed. Other familiar words were used: 'fulness' (Col. 1: 19; 2: 9) and 'mystery' (Col. 1: 26 f.; 2: 2), by which Paul claimed that the whole being and power of God was present in Christ; there were no secondary intermediaries, such as the Jewish law (Col. 2: 14). Thus Paul, though indeed a man of the 1st cent., firmly rejected its mythologies, whether Jewish or Hellenistic, which interposed various kinds of supernatural barriers between God and humanity. For Paul the transcendence of God was absolute; and God in Christ deals directly with mankind.

If the epistle was written during Paul's imprisonment in Rome, its date would be between 60 and 64 CE.

colours Mention of colours is infrequent in both OT and NT. Hebrews were not interested in aesthetics as were Greeks in *Athens, and the language was deficient for such expression. However, purple, usually a reddish-purple, obtained from shellfish, was appreciated and the word appears in the NT (Acts 16: 14), and Jesus was clothed in purple as part of his maltreatment by soldiers (John 19: 2). Joseph's 'coat of many colours' (Gen. 37: 3, AV) is translated 'a long robe with sleeves' by NRSV, REB ('a decorated tunic', NJB). Other colours mentioned are white (fields ripe for harvesting, John 4: 35), green (of grass, as Mark 6: 39), yellow (of hair, Lev. 13: 30), red (of the sky, Matt. 16: 3), blue (in Exod. of embroidery), black (of a horse, Rev. 6: 5).

comforter Used in AV to translate the Greek paraclete, the term in the gospel of John for the *Holy Spirit, who is sent by the exalted Jesus to be with the *disciples.

coming The translation of the Greek **parousia*, referring to the coming of Christ in *eschatological *judgement (e.g. Matt. 24: 3; 1 Cor. 15: 23; 1 Thess. 3: 13).

commandments The people of Israel are frequently exhorted to keep the Lord's commandments, not however as a burden to be resented, but as a response to his love, with joy (Ps. 119: 47). The word is also used in conjunction with the obligation of subjects towards the sovereign (2 Kgs. 18: 36) and children towards parents (Jer. 35: 14).

Commandments are summarized in two

by Jesus, who puts together Deut. 6: 5 and Lev. 19: 18 (Mark 12: 28–34), and elaborates the meaning in John 13: 34 f.

communion The Greek word *koinonia* may be translated '*fellowship' as well as 'communion' and as such speaks of Christian relationships. But there is also the relationship between believer and Christ through 'holy communion'—corporate Christian sharing in the *body and *blood of Christ (1 Cor. 10: 16) and in the *Holy Spirit (2 Cor. 13: 13).

communion of saints The classic belief affirmed in the creeds that faithful Christians who have died still enjoy a relationship with believers on earth; the germ of the idea is in Col. 1: 12, and the puzzling reference in 1 Cor. 15: 29 to the practice of being baptized on behalf of the dead presupposes some sort of continuing relationship. Requests for the prayers of the saints are first met in the 3rd cent. in the Roman catacombs.

community Much modern social theory and practice has affected biblical studies by showing that a great deal of light is thrown on the NT by investigating the communities out of which the NT writings emerged. Hence the appearance of books on, for example, the Johannine Circle or Community. Studies of 1st-cent. institutions such as *slavery and *marriage explain aspects of community life in the Church and how they either reflected or repudiated standards and values in their environments. The gospel of Matthew, for example, expounds a way of life for its community which distances it from contemporary *Judaism (Matt. 18) while also recognizing how much it still has in common with Jewish communities of the *synagogues. By the time the fourth gospel was written, towards the end of the century, Church and Synagogue are farther apart and by a process of putting a prophecy into the mouth of Jesus (John 16: 2) the author, the evangelist, accepts the inevitability of the separation of the two communities: the Church, not the Jews, is now the faithful people of God.

concubine A secondary wife on whom Hebrew law conferred certain minimal rights (Deut. 21: 15–17). *Sarah and *Rachel gave their handmaidens to their husbands to bear *children when it seemed they were infertile (Gen. 16: 1–3; 30: 3–6) as also did Leah (Gen. 30: 9), who was by no means infertile. Both *Saul (2 Sam. 3: 7) and *David (2 Sam. 5: 13) took concubines, and when *Absalom lay publicly (2 Sam. 16: 22) with his father's concubines it was making a claim to royal status. *Solomon had three hundred concubines (1 Kgs. 11: 3).

confession The word is used in two senses —of an expression of *faith, as Peter's confession at *Caesarea Philippi (Mark 8: 29 etc.), and of the recognition of *sins, as when people came to be baptized by John (Matt. 3: 6). A primitive confession of faith was 'Jesus is Lord' (1 Cor. 12: 3), but, with differing interpretations of this brief confession, it became necessary for the Church to devise more detailed confessions in the form of creeds.

confirmation (1) The Philippians are said to share with Paul in the defence and confirmation of the gospel (Phil. 1: 7), that is, in the establishment of its truth at his forthcoming trial. Possibly he is optimistic about an acquittal which will strengthen the position of the Church of Philippi. (2) Confirmation in the sense of a sacramental rite, though sometimes traced to the teaching of the *apostles, as in the Preface to the service of confirmation in the Anglican Book of Common Prayer, is in fact not known in the NT as a rite supplementing *baptism by *water. When it is recorded that the apostles at *Samaria and *Ephesus used the OT custom of laying on hands (e.g. Moses on Joshua, Deut. 34: 9) this seems to be an indication by the author of Acts that these were turning points in the progress of the gospel from Jerusalem to *Rome. But by the end of the 2nd cent. CE the rite of Christian initiation included baptism in water, anointing by the *bishop, and the imposition of his hand on the head of each candidate. However, by the 4th cent., the latter part of the rite became separated in time from the baptism: the large increase in the number of candidates obliged bishops to delegate baptisms to *presbyters while reserving the second part, 'confirmation', to themselves.

congregation The people of Israel considered as a group for travel (Exod. 16: 1) or worship (Isa. 1: 13, 'convocation', NRSV, 'sacred assemblies', REB, 'assemblies', NJB). The Hebrew word is translated *ecclesia* in the Greek *LXX.

conscience The modern understanding of conscience as a faculty by which past actions or dispositions are assessed or future actions evaluated, leading to experiences of guilt or of innocence, is not that of the Bible—though that is not to say that the Bible is unaware of these feelings (cf. 1 Sam. 24: 5).

In the NT human beings are regarded as morally responsible and 'conscience' is the painful reaction felt when one's ordered human nature is violated. Conscience therefore is something from which a Christian should be free (Rom. 13: 5); it comes into operation after a wrong deed has been initiated. If there is no such reaction ('conscience'), then either the deed is not wrong, or habit or environment has so corrupted the subject that 'conscience' fails to be activated. 'Because of conscience' (Rom. 13: 5) means 'to avoid the pains of conscience'.

This conscience could lead to excessive scruples on the one hand, or be so insensitive on the other that weaker Christians are hurt (1 Cor. 10: 28).

consecration The act of separating persons or things from profane use and dedicating them to the service of God (Exod. 29: 1; 2 Chron. 31: 6).

In modern usage, consecration (e.g. of a church) is the rendering of thanks to God; and a declaration of the function of the person or thing for service, as in the OT.

consolation In the NT, the *redemption to be brought by the Messiah. *Simeon in Jerusalem was looking forward to the consolation of Israel (Luke 2: 25). This *hope was based on Isa. 40: 1–5.

contentment The sense of assurance that God supplies needs (1 Tim. 6: 6). The same word occurs in the Greek of 2 Cor. 9: 8; it means literally 'sufficiency'.

conversation Often used by AV for 'conduct' (as NRSV and REB of 1 Pet. 1: 15).

conversion When someone responds to the call of God, it is remembered as a conversion experience and is authenticated by a changed lifestyle. In the NT conversion is followed by *baptism, as with the Philippian gaoler and his household (Acts 16). Paul's transition from persecuting Jew to evangelizing Christian has been traditionally called his conversion, but it was certainly not a change from unbelief to *faith or from a life of dissipation to one of austerity. He probably did not even see it as a transfer in religious allegiance, but from his inherited *Judaism to true Judaism. Paul was now convinced that he was called to go to the *Gentiles to bring them to faith in Jesus who was after all the Messiah proclaimed by the formerly despised sect of Nazoreans, to whom Paul now attached himself (Acts 24: 5). Those who responded to his preaching were attracted by the strong group identity of the Christian house communities. Paul's converts were often on the margin of society (1 Cor. 1: 26) and women attracted by being valued as equal persons (Gal. 3: 28).

convince Used in AV (as John 8: 46) for *convict* (NRSV, NJB, and REB) and in John 16: 8, where NRSV and REB use 'prove wrong'.

convocation A meeting of people for *worship, as on the *Sabbath (Lev. 23: 1–4) and *feast days (e.g. Tabernacles, Lev. 23: 34–6).

Conzelmann, Hans (1915–89) A German NT scholar who taught at Göttingen; his work on Luke–Acts was influential. He maintained in that work that the structure of salvation-history has three periods: (1) from the *creation to *John the Baptist; (2) the *ministry, *death, *resurrection, and *ascension of Christ; (3) from *Pentecost to the **parousia*.

Coptic The language of ancient *Egypt, into which some Greek words had been introduced; it was spoken in four dialects from the 3rd to the 10th cents. CE. The NT was translated into each of the four. A number of *Gnostic works in Coptic have been discovered, and the language is still used in the liturgy of the Egyptian Coptic Church.

corban Something dedicated to God for a sacrificial purpose and not available for anything else, cf. Neh. 10: 34. In a piece of anti-Pharisaic invective (Mark 7: 5–13) Jesus is said to condemn neglect of duty to parents on the pretext of a vow (*corban*) to offer to God that which was due to them. It seems inconsistent with this principle that Jesus places family responsibilities below that of discipleship (Matt. 10: 37).

Corinth City situated on an isthmus: Corinth was well placed to benefit from commerce between mainland Greece and the Peloponnesus, and boats were hauled across a track from the Adriatic to the Aegean Sea to avoid the danger of storms. (A canal started by *Nero in 76 CE was finally completed in 1892.) The city had been destroyed by the Romans in 146 BCE but was refounded for veterans of his army by Julius Caesar in 44 BCE and Latin was widely used even though most of the citizens in Paul's time would have spoken Greek.

The city had a reputation for sexual orgies and boasted a temple of Aphrodite, the goddess of love. Excavations have also uncovered a Jewish synagogue and some domestic houses of the kind that would have accommodated meetings of Christians. These show that the main room was not large, and if this was occupied by the more leisured and well-to-do members who arrived early, their slaves and employees would have been relegated to smaller rooms, which understandably provoked discontent. An inscription mentions the name Erastus the aedile, a council official, and this may be the man who sends greetings to Rome in Paul's letter to the Romans (Rom. 16: 23), which was sent from Corinth.

Corinthians, Paul's letters to the Placed in the NT immediately after Paul's letter to the Romans. One commentator described the problems surrounding the two extant letters sent by Paul to his turbulent converts as 'excessively dull to all except those who find that literary criticism offers the same kind of interest as a game of chess'. This is surely a very pessimistic view of correspondence which yields fascinating information about the complex social organization of a newly founded, racially mixed, Christian community in the heart of Hellenistic culture. The letters also give unequalled information about Paul himself: his pastoral sensitivity, his human emotions, his guidance on *worship, and his clarification of his teaching in reply to the Corinthians' questions.

Paul came to Corinth from *Athens towards the end of 49 CE at about the same time as *Aquila and *Priscilla arrived; he converted *Crispus, the ruler of the synagogue, and remained in the city for about eighteen months. An accusation against Paul brought by the Jews was dismissed by *Gallio (Acts 18: 16). Still, Paul decided to move on to *Ephesus.

Paul had persuaded the Corinthian Christians that the *End was at hand, before they could expect to die. So he discouraged them from long-term plans and advised an ascetic way of life in the interval that remained; like them, he himself spoke in *tongues. The pagan worship which surrounded them was meaningless, of no account; and the Christians' worship at Corinth was vibrant. But the news that reached Ephesus was disquieting. Paul was unhappy to hear that the Corinthians were proud of their exciting spiritual *gifts and the special knowledge (*gnosis*) which they claimed to enjoy; he therefore wrote to Corinth a letter, sometimes called 'the previous letter' referred to at 1 Cor. 5: 9, warning them to be cautious about contacts with unbelievers, and mentioning *resurrection of the dead, which seemed to imply that some at least would be dead before the End. Evidently this letter came as something of a shock to the Corinthians. They felt that their apostle had changed his views. The Church split into factions, under the banners of Paul, of Peter, and of *Apollos, who had also visited Corinth. (Indeed Apollos may even have written a letter to Corinth which is preserved in the NT as the epistle to the Hebrews.) The divided Church at Corinth sent further information to Paul partly in writing and partly by word of mouth. There were questions requiring his answers, and there were tart comments on the Corinthian situation by those who carried the letters.

Paul replied in 1 Corinthians, though he dispatched *Timothy to Corinth before it was finished. The letter deals with the enquiries about sexual relationships (1 Cor. 7), food sacrificed to idols (8–10), worship (11–14), and the resurrection of the dead (15). Finally he invites them to contribute to his *collection for Jerusalem (16). Paul charges the Corinthians not to be satisfied with merely human *wisdom (2: 5), and expounds a theology of the cross (as in Rom. 5: 6–11). The fundamental conviction of the central section of the epistle is the urgency to build up in *love all who belong to the body of Christ; every gift they have is to be controlled by love.

Timothy rejoined Paul and reported that hostility to Paul had actually increased. Paul himself paid a flying visit, and this too was unsuccessful. Back in Ephesus again he wrote a stern letter of rebuke; many scholars regard 2 Cor. 10–13 as this severe letter, which has somehow got attached to 2 Cor. 1–9 in the wrong place, as could easily have happened. At any rate, the letter, together with a visit paid by *Titus, had better results (2 Cor. 7: 6), and when Paul got this news he sent 2 Cor. 1–9 with Titus (2 Cor. 8: 17), asking an over-zealous minority at Corinth who were unwilling to forgive a delinquent member to be more generous. Again, they were bidden to remember the poor in Jerusalem. Paul paid a third, and shorter, visit (Acts 20: 3), after which there is no information about the Church until forty years later, when fresh quarrels provoked condemnation from the Church of Rome in the first epistle of *Clement.

Cornelius *Gentile *centurion admitted to the Church by Peter (Acts 10), thus marking a turning point in the progress of the Gospel from Jerusalem to *Rome.

corner-stone A large stone placed in the foundation at the main corner of a new building; it might form part of the end wall and part of a side wall, and so it would hold them together. Hence Eph. 2: 20. The occurrences in the OT and NT of the word are mostly metaphorical, as in Job 38: 6. God lays a corner-stone in Zion (Isa. 28: 16)—true security for the man of *faith. In the NT Jesus is referred to as a corner-stone in Mark 12: 10 (where Ps. 118: 22 is quoted) and 1 Pet. 2: 6 (where Isa. 28: 16 is quoted). The psalm explains that a rejected stone has been used as a corner-stone, and this was given a Messianic interpretation by rabbis. For Christians it was evidence from the OT of the divine purpose in Jesus' rejection by the Jews and his exaltation to the Father.

corruption The corruption and decay and death to which the present world is subject. Peter (Acts 2: 27) quotes Ps. 16: 10, which is a thanksgiving for preservation from these evils. At that time it was assumed that the psalm was written by King *David; but David died, so it followed that the psalm must be a prophecy about the Son of David, the *Messiah: he would not suffer bodily corruption. The argument—rabbinic-style—was intended to demonstrate that Jesus' triumph at *Easter over this corruption proved that he was the Messiah.

Cos An island visited by Paul (Acts 21: 1) (not 'Coos', AV). The Romans created Cos a free state within the province of Asia.

councils Porcius *Festus became procurator of *Judaea in 59 CE and confronted Paul, a prisoner at *Caesarea; after consulting his advisory council he confirmed Paul's appeal to *Caesar. Paul's trials follow what was predicted in the apocalyptic chapter of the synoptic gospels (e.g. Luke 21: 12, following Mark 13: 9)—the warning that *disciples will be handed over to councils.

The meeting between leaders of the Church of Jerusalem and delegates from *Antioch led by Paul and *Barnabas described in Acts 15 is often referred to as the Council of Jerusalem. At it a compromise was reached between Jewish Christians and those spearheading the successful mission to the *Gentiles. The former feared for moral standards if Gentiles flooded into the Church without any obligations to keep the Jewish *Law; the latter insisted that the *Holy Spirit was being given to Gentiles without their first becoming Jews (by *circumcision). *James the Lord's brother, presiding, suggested a compromise: no compulsory circumcisions before admission to the Church, but Jewish food laws should be respected in order to ensure a common table-fellowship at meals.

There is doubt whether this harmonious Council took place. Clearly Luke, author of the Acts, was anxious to portray the Church as an institution of *reconciliation and peace and he may have given an imaginative account of a meeting also mentioned by Paul (Gal. 2: 2 ff.) which took place privately at Jerusalem and where the discussion was evidently heated. Moreover, it is odd that Paul never has a word about the decree (Acts 15: 20) which, if it had existed, he had only to quote to settle some of his disputes.

counsellor Used of the coming *Messiah (Isa. 9: 6, NRSV, REB; Isa. 9: 5, NJB) and of the *Holy Spirit in John 14: 16, where RSV uses it to translate the Greek paraclete; but NRSV and REB prefer 'advocate'—literally one who is called to stand beside the defendant in a lawcourt. So it defines the work of the Holy Spirit as a help, as when Christians are on trial before Jewish or pagan authorities.

Among the human counsellors mentioned in the OT are *Ahitophel and Hushai, who took opposite sides when *Absalom rebelled against *David (2 Sam. 15: 30–7).

courage Implied as the human quality opposite to cowardice (Ecclus. [= Sir.] 2: 12 f.) and mentioned positively (Wisd. 8: 7) and exemplified in the Maccabean martyrs (1 Macc. 9: 10). In the NT it is allied with the confidence that comes from trust in God (Heb. 13: 6), who raised Christ from death (John 16: 33; Acts 23: 11).

covenant The Hebrew word *berith* denotes an arrangement between two parties, e.g. *Abraham and *Abimelech (Gen. 21: 25 ff., but the OT is largely the story of a covenant, or agreement, between God and the people of Israel. There was such a covenant with *Noah (Gen. 9: 8) by which God promised there should never again be a universal *flood. Certain conditions were imposed (Gen. 9: 4–6), but in general the OT stresses the element of *grace on the part of God, as in the case of the covenant with Abraham (Gen. 17: 8), which was sealed with the promise of regular *circumcision (Gen. 17: 23). The covenant at *Sinai is more clearly a conditional type of covenant. Thereafter the mutual responsibilities between God and the people is a constant refrain in the history. The account of its establishment in Exod. 19 and the probably more ancient tradition in Exod. 24 may show similarities with arrangements made in treaties between states, e.g. the Hittites in the ancient Near East. Both accounts recognize the saving acts of God for Israel and demand obedience of the people to the *Law in all its ramifications. The agreement or treaty is said to be deposited in the *Ark of the Covenant (Exod. 25: 16), very much as in other nations treaties were deposited before their idols. A ceremonial meal followed (Exod. 24: 9–11) by way of ratifying the covenant.

The details of the covenant made with Moses are contained in 'the book of the covenant' (Exod. 24: 7) spelt out in Exod. 21 and 22.

There was also a promise type of covenant between God and *David, assuring him of the permanence of his dynasty, subject to its obedience to the Law (1 Kgs. 2: 4; Ps. 132: 12), and popular belief in this covenant led to an expectation of the nation's security, which the prophets tried to disabuse. Obedience to the stipulations of the covenant at Sinai was, for them, basic, and when Israel broke the terms of the covenant the agreement could be temporarily rescinded by God. *Jeremiah predicted that in due course there would be a *restoration and a new covenant (Jer. 31: 27–37), when its form would be different, being part of the inward nature of individual Israelites. As such, it would be unbreakable.

The NT sees this new covenant established by Jesus (Mark 14: 24; 1 Cor. 11: 25), and Christians are the human party to this new covenant (2 Cor. 3: 6), a theme especially worked out by the letter to the Hebrews (Heb. 7: 22; 8: 8–13). Paul regards the *covenant promise to Abraham as fulfilled in God's grace in sending Jesus Christ (Gal. 3: 6–18), and the promise to David (Isa. 9: 7) could be used as one OT basis for proclamation of Jesus as Messiah.

covenantal nomism A term coined for the OT belief that God has chosen Israel and given the *Law. God will be faithful to his promise but the nation is required to obey him. The Law provides the means for *atonement which maintains the covenantal relationship.

Coverdale, Miles (d. 1568) A Puritan leader born in Yorkshire whose translation of the Bible into English was the first ever to be printed. His version of the psalms is still used, as it was incorporated in the Anglican Book of Common Prayer.

covetousness Selfish greed, condemned by Jesus (Luke 12: 15) and regarded by Paul as the kind of specific command that makes people aware of their sinfulness (Rom. 7: 7).

creation The biblical *myth of the origin of the universe. There are two accounts in Genesis of the creation by God. Neither deals with the question whether the creation was out of nothing (*creatio ex nihilo*); the first (Gen. 1: 1–2: 3) which was compiled later, the *P source, supposes the pre-existence of an abyss of infinite and formless waters, a chaos out of which God creates order; in the second, and earlier, story, the *J source, God forms *Adam from the soil of a damp, barren plain (Gen. 2: 6) but there is nothing about an existing chaos of waters.

There are parallels to the first account with a *Canaanite and with a *Babylonian creation epic but the biblical author has skilfully refined his picture. There is no longer a story of a god in combat with primordial matter (though there are traces of this elsewhere in the OT, as in Pss. 29 and 104 and Job 38–41): the divine command creates the world, and its inhabitants are made in six days, culminating in the creation of man and woman (Gen. 1: 26–7). The seventh day is assigned for God to rest and contemplate his work.

In the *J account man's creation precedes that of plants, *animals—and woman (Gen. 2: 22). The scene is of a garden, or park, *paradise, in a location named *Eden which enjoys the water of four great rivers, Tigris and Euphrates and two which defy exact identification, Pishon and Gihon. In the garden are trees which are beautiful to look at and generous with their food (Gen. 2: 9) but among them are the tree of life and the tree of the knowledge of *good and *evil, knowledge gained by experience, whose fruits the man is forbidden to taste (Gen. 2: 17) on pain of *death.

In the NT Christ is regarded as God's agent in creation (Col. 1: 15–16), taking up the OT concept of *Wisdom through whom the world was made (Prov. 8: 25–7) and conveying the suggestion that creation reflects an obedience to the laws of God as Christ was obedient to the Father. The coming of Christ is understood as marking the beginning of a new creation of humanity (2 Cor. 5: 17) and of the whole universe (Rom. 8: 19–21). Although there is a revival in *fundamentalist works of 'creationism' which regards the OT creation myths as in some sense historical and to be preferred to scientific theories of the evolution of species over immense eras of time, most modern readers of the OT reject this as incredible. The stories of Adam and Eve and the *Fall are to be regarded as parables expressing insight into the relationship between God and humanity. These chapters of Genesis are to be put into the appropriate literary *genre, which is not that of history but of *myth. They are, however, incompatible with the theories of some modern scientists that the origin and development of the universe are to be explained wholly as a matter of chance.

creed Summaries of Christian belief which are found in the NT at 1 Cor. 15: 13 ff. and 1 Tim. 3: 16, and perhaps also Phil. 2: 6–11; but creeds such as the so-called Apostles' Creed used at *Rome did not come into existence until the 3rd cent. They were found necessary in order to exclude heretical views about Christ, and they were convenient expressions of belief to be learnt by candidates for *baptism.

Crete Island 100 km. (60 miles) south of Greece, with a Jewish settlement (Acts 2: 11). Paul sailed along its south coast on his final journey (to *Rome), but there is no record of how the Gospel reached *Crete, which it must have done before the *Pastoral Epistles were written (Titus 1: 5), since *Titus was to supervise congregations in that area of legendary dissipation (1: 12).

crimes and punishments Broadly, in the OT, crimes are conceived either as religious offences or crimes against society. Both are punishable, even by death in the case of the worship of foreign gods (Exod. 22: 20). *Blasphemy also carries the penalty by stoning (Lev. 24: 13 ff.), as does the practice of black arts (Lev. 20: 27). The punishment for failing to celebrate the Passover was excommunication (Num. 9: 13).

Crimes against the person such as rape if the victim was betrothed were punishable by *death, as were other offences against the family such as incest (Lev. 20: 11 f.) or persistent disobedience to parents (Deut. 21: 18 ff.). Premeditated murder merited death. Crimes against property, such as enlarging one's real estate at the expense of a neighbour (Deut. 19: 14) or using faulty scales (Deut. 25: 15 ff.), were to be made good by compensation.

Imprisonment was unknown until after the Return from the Exile (Ezra 7: 26). Executions took place outside cities, and it is prescribed in Deut. 17: 7 (corroborated by John 8: 7) that a stoning should be initiated by the witnesses in the case.

The OT legislation on criminal justice is contained in four codes of different dates, of which the earliest is the book of the *Covenant in Exod. 20: 22–23: 33 and the latest the *Priestly Code, whose prescriptions are scattered through Exod. (12 and 13; 25–31 and 35–40), Lev. (1–16), and Num. (1–10). In between are the Deuteronomic Code (Deut. 12–18), and the Holiness Code (Lev. 17–26), which is included within the Priestly corpus.

Whether the Jewish authorities had the right under the Romans to inflict capital punishment is debated. It is denied in John 18: 31, supported by a saying in the *Mishnah. On the other hand notices threatening death warned Gentiles against penetrating into the inner part of the Temple, and Stephen was brought before 'the council' and stoned. If Jews did have such a right, then—because Jesus was undoubtedly executed by the Roman method of crucifixion—it is argued that Jesus was accused of inciting to rebellion against Rome: the gospels' accounts of trials before the Sanhedrin are *antisemitic Christian propaganda. Nevertheless a good case can be made for the historicity of the Sanhedrin trial. And the stoning of Stephen could well have been a lynching by the mob during the absence of the governor in Caesarea rather than a judicial execution.

Crispus Leader of the Jewish *synagogue at *Corinth who was converted by Paul and baptized with his household (Acts 18: 8; 1 Cor. 1: 14).

criticism, biblical The examination of the books of the Bible with the resources of historical investigation, archaeology, palaeography, and linguistics. Biblical criticism starts from a conviction that the heterogeneous collection of books which constitute the Bible were written in a variety of *genres for different purposes and readerships by human authors. A biblical critic is a scholar equipped with linguistic skills and literary or historical knowledge who tries to shed light on what the authors of the books were saying, given the thought-world and social and political situations of their own age. The tools for this enterprise have been refined over the past four centuries and are the fruit of the Reformation, the Renaissance, and the 18th-cent. Enlightenment; but, long before, it was recognized that divine inspiration of scripture did not overrule the humanity of the authors. *Origen (185–254 CE), for example, noticed that the letter to the Hebrews could probably not have been written by Paul. The establishment of a *canon of scripture itself witnesses to a process of discrimination and assessment. The first task of biblical criticism is to establish a reliable text; to get as near as possible to what the original authors wrote in Hebrew (or Aramaic) and Greek. This was already being done for classical authors such as Homer when Erasmus (1466–1536 CE) began to apply the same principles to the text of the NT, comparing the variations in the ancient MSS that were available at that time. In the centuries since then many important discoveries have been made and much international research undertaken. The technical investigation is known as textual criticism; sometimes, especially in less recent books of NT introduction, it is referred to as Lower Criticism. For the NT it compares the large number of MSS, together with the many quotations of scripture made by the early Fathers, and also translations of the Greek into *Latin and Syriac; and the verbal differences, mostly only in small details, are very numerous. In a modern Greek NT some are noted in the apparatus criticus at the bottom of each page. It is apparent that scribes copying MSS occasionally wrote twice ('dittography') or omitted a few words, or, in the case of a MS being written from dictation, misheard. It is even possible that scribes sometimes felt at liberty to add something on their own account. The process of tracking down and eliminating such errors is the work of textual criticism. But for all the labour expended, we do not possess the precise words penned by the NT authors. What we can now say with some confidence is that we have the NT text as it was read in the main centres of Christian learning about 200 CE.

Historical criticism, sometimes called Higher Criticism, deals with questions of authorship and date, editorial arrangements of sources, historicity, literary categories (*genres), and doctrinal tendencies. Historical criticism acquired a bad name amongst orthodox churchmen because its pioneers were unbelievers, like Thomas Hobbes (1588–1679 CE) in the *Leviathan*; but early work was also done by Roman Catholics: Jean Astruc, a French professor of medicine (1684–1766 CE), noted that Genesis was a compilation of several earlier documents, following a book (Eng. edn., 1682) of OT history by the Oratorian priest Richard Simon (1638–1712 CE) which pioneered *Pentateuchal criticism and was burnt by Louis XIV. These works were translated into German, which in due course led to the establishment of the view that much of the *Pentateuch was written after the time of the great prophets. The classical 19th-cent. view of the source criticism of the Pentateuch was the work of Karl *Graf and Julius *Wellhausen. The traditional view that these books were written by Moses was shown to be untenable. Archaeological discoveries also suggested that the religion of the Hebrews had marked similarities to that of other peoples. At the same time scientists' work on the origin of species showed that the accounts of *creation in Gen. did not correspond to the fact of biological evolution.

Similar investigation of the gospels followed. Since the three *synoptic gospels often preserve passages where word after word is the same, who copied from whom? Or did some common source which each used exist? The most widely held theory today is that Mark was written first and was used by both Matthew and Luke, but there is no final solution about the non-Marcan material which Matthew and Luke share, though a majority of scholars maintain that Matthew and Luke both had access to a body of teaching material, which has been given the symbol *Q. The scrutiny of the documents in this way is called Source Criticism.

A degree of frustration with Source Criticism, together with interest in recent explorations of folk literature in various cultures, led several German biblical scholars in the 1920s to go behind the written sources and try to show how the traditions embodied in the gospels had come to be shaped during the generation when they were passed on by word of mouth. It was suggested that the stories developed out of a particular life-setting (**Sitz-im-Leben*) a kind of form which followed patterns, or 'laws', which could be inferred. Hence the name '*Form Criticism'. Fundamental to it was the view that the stories and incidents in the synoptic gospels originally existed as independent units and in course of oral transmission were adapted to the needs of the community. These units were then strung together by Mark and his successors without any reliable information relating to chronological or topographical references.

OT scholarship was no less interested in the life-situations in which the literature was developed. Several of the psalms were ascribed to the coronation ceremonies of the kings in the Temple. But on the whole Form Criticism of the OT had a more conservative tone than the corresponding NT discipline.

Subsequent reflection on the gospels has moved toward insisting on the careful coordination and structures of the evangelists —as also in OT scholarship much more regard is had for the final editorial result of the books than in breaking them up into their sources. It is recognized that the gospels are compiled in a sophisticated manner to bring out the theological drives of each author. They are theological interpretations of a piece of history, not simple reports of what actually happened. In Britain, this post-Form Critical outlook was pioneered by R. H. *Lightfoot and A. M. *Farrer, and in Germany where the name *Redaction Criticism became familiar, by Willi Marxsen of Münster and Hans *Conzelmann of Göttingen.

The Bible has a central place in the doctrine and the liturgy of all the Christian Churches but it is not venerated as the Qur'an is by Islam. Muslims hold that the Qur'an was dictated to Muhammad by God and that its inerrant narrative and incomparable, incorruptible Arabic has been properly maintained down the centuries. Any thoughts of 'criticism' would be *blasphemy. Criticism of the OT and NT by Christians is not blasphemous nor does it mean hostile, destructive criticism.

Many in the Churches, however, are content to ignore biblical criticism with its open and unrestricted efforts to reach the truth, and preachers tend to offer an uncritical acceptance of traditional values. There is a fear that once the gospels are assessed for historical reliability and some events judged not to have happened at all, there is no knowing where such free historical enquiry may end. Nevertheless biblical critics have given impetus and encouragement to ecumenical dialogue in as much as scholarship has undermined untenable prejudices and dogmas. And many biblical critics in the areas mentioned, and also those writing on *Canonical Criticism, *narrative criticism and *Reader-Response Criticism, did or do participate in the liturgical life of the Churches.

cross The instrument of torture and execution used by the Romans, and previously by the Persians, was an upright stake to which the condemned man's body was tied or nailed. The Romans usually left the stakes in place for repeated executions, and the convict carried a crossbar which was added to the top of the stake and on which the offence incurring the penalty might be inscribed, or a notice proclaiming it might be hung round his neck.

crown An athlete wears a crown (AV; 'wreath', NRSV; 'garland', REB) as the reward for victory (1 Cor. 9: 25). As a token of kingship, a crown (a different word in

Greek) symbolized power (Rev. 9: 7) and glory (1 Peter 5: 4). Paul looked on his converts as his crown (Phil. 4: 1).

crucifixion Jesus suffered *death by the Roman penalty of crucifixion because the Jewish *Sanhedrin had no power to inflict capital punishment, according to John 18: 31. In the gospels there is certainly a tendency to heighten the Jewish responsibility for the death of Jesus and to mitigate that of the Romans. During a time of conflict between *Judaism and *Rome, when the gospels were compiled, it was in the interests of the Church to stress the differences with Judaism. But a trial before the Sanhedrin is historically credible: it secured a unanimous verdict from the different groups on the ground of *blasphemy. What Jesus was alleged to have uttered and which secured his condemnation (Mark 14: 63–4) is similar to that mentioned in the *Qumran Temple scroll as deserving the death penalty: Jesus was leading Israel astray (Deut. 13: 1–11). Having obtained this convenient unanimity, the high priest could then rephrase the charge for the prefect. Pilate could not ignore the presence in the city of a Messianic pretender (Luke 23: 2).

The death of Jesus by crucifixion was ordered by *Pontius Pilate on a charge of high treason. He was first flogged (Mark 15: 15) and, then being too weak to carry the crossbar, was assisted by *Simon of Cyrene. Jesus refused wine mingled with myrrh, offered to reduce the agonizing pain, but died more quickly than was usual. The body was removed by *Joseph of Arimathaea, who had in mind the injunction of Deut. 21: 23 that a corpse should not remain exposed on a tree overnight.

Such a fearsome penalty was not easy to reconcile with a belief in Jesus' Messiahship, so it inevitably led to intense theological reflection. Paul interpreted the death as signifying the end of the Jewish *Law (Gal. 3: 13) since such a death put the victim under the curse of that Law. It was a sacrifice to deal with sin, and Christians by baptism share in what Christ did on the cross; it is their death to *sin, and the beginning of a new life. In later epistles Paul took up the 'scandal' of the 'weakness' of Christ's death: it represented a judgement against the world's arrogant claims to *wisdom (1 Cor. 1: 18-25), and those who suffer in some measure like Christ (Mark 8: 34) could expect to share in his *resurrection.

cryptogram A phrase which has a secret meaning, as MENE, MENE, TEKEL, PARSIN (Dan. 5: 25) or the number 666 (Rev. 13: 18). The *fish became a Christian cryptogram.

cubits A measure of length which varied at different periods but was approximately half a metre or 18 inches.

Cullmann, Oscar (b. 1902). A NT scholar who has taught both in France and Germany. His historical judgements (e.g. on the primacy of Peter) have been fairly conservative; he maintained against *Bultmann that the Bible is a record of 'saving history'. God's action in Christ was the culmination of the OT. It is the proper object of belief and should not be distorted into subjective terms of human self-understanding.

cultic interpretation The religious history of Israel would be nothing without the cult: *worship, *sacrifice, ritual. Before the first king was anointed by *Samuel, there existed a cult round the *Ark of the Covenant which provided a unifying basis for the loose confederation or association of twelve tribes.

But much of the OT can be understood in relation to the Temple cult at Jerusalem and the festivals celebrated there, such as the blessings and cursings of Deut. 28, and it has been suggested that the book of Job was a drama based on the *myth and ritual of a New Year festival, common in the ancient Near East, when there was a ritual enactment of the *death and *resurrection of a deity. The purpose of the ritual was to assist the well-being of the nation. After a ritual combat in which the god was victorious over his enemies, he was acclaimed as king. The nation's king played a central part in the whole festival by representing the god himself. In Israel before the Exile there was a sense in which the king was identified with the nation; or at any rate its well-being rested on him. This notion of the kingship of the house of David was at the root of the Messianic hope.

It is above all the psalms that reveal how central in the life of the nation the festivals in the Temple were. There was first a New Year festival, though some scholars prefer to regard it as an Enthronement festival, a liturgical reenactment of the *creation myth; among Israel's neighbours the victorious saviour-god was thought to overcome the chaos deity; for every year the lifeless earth when the chaos deity reigned sprang into fertile life as the saviour god returned to rule. Such was the myth taken over in the worship of Yahweh at Jerusalem. Such was the victory celebrated in Pss. 24: 7–10 and 47: 1.

There were also a coronation ceremony (Ps. 110), a covenant renewal festival (Ps. 114), and liturgies of petition when the king recited such psalms as 44 on behalf of the nation.

Thus the psalms were compiled for congregational worship and as such have continued in regular use in the Christian Church, relating the sovereignty of God to people's daily lives.

cultural relativism The view that perception of the world is inevitably determined by the writer's background and environment. It is held by some modern scholars that the thought-world of the writers of the Bible is so different from that of the modern Western world that it is difficult or even impossible for us really to enter into it. All expressions of belief are relative to the particular cultural situation in which they are uttered. It is not that the evangelists or the rabbis were unintelligent—far from it —but that they had a certain perception of the universe from which they could not escape and within which they were obliged to interpret everything else. It is said therefore that in a universe thought to be populated by supernatural powers, possibly dwelling in a space between earth and heaven (1 Thess. 4: 17) human beings are subject to their influence. Hence the belief that people could be possessed by demons—whereas we might describe certain psychic disorders as, for example, Munchausen's syndrome by proxy. Similarly, it is impossible for us to understand sympathetically what the Hebrews intended with their sacrificial rites, or what, therefore, the writers of the NT intended by using the same categories. We do not know what it feels like to live in expectation of the imminent end of the world (cf. Phil. 4: 5). It follows also that orthodox doctrines, such as the *Incarnation, which derive from unbelievable 1st-cent. premisses, have also to be questioned or reinterpreted.

Some scholars, however, hold that there is more continuity between the cultures of different ages than the relativists admit. Moreover, in each generation there are to be found those who repudiate some of the pervading beliefs: the *Sadducees of the 1st cent. rejected the notion of supernatural spirits and *angels (Acts 23: 8). And there are from time to time men of genius who are able to take a quantum leap out of their contemporary thought-world and give it a new direction, as did Galileo in 1613 CE. Those who reject a thoroughgoing relativism hold that communion is possible between diverse cultures. No one has the total truth about God or Christ or human life; dialogue is possible and necessary. The 1st cent. is not our cent.; but what people of that era taught and believed is not to be dismissed as total falsehood or totally meaningless simply because it is not our language.

cum(i) An Aramaic word of command prescribed at Mark 5: 41 and carefully translated for Mark's *Gentile readers 'Get up'.

cummin Seeds of cummin were valued as spice. Some scrupulous *Pharisees gave tithes on things as negligible as cummin, and Jesus is said to have criticized those who did so at the expense of more important duties (Matt. 23: 23).

cuneiform An early kind of writing developed in *Mesopotamia and impressed on tablets of clay. It used stereotyped pictures ('pictographic'), but from representing things and actions, it later represented sounds and concepts. The symbols were wedge-shaped marks.

The earliest texts were *aides-memoire* for scribes working on accounts and were practical, but later there were religious items and narratives which had existed before they were written down. The tablets from Tell el-*Amarna consisting of correspondence between Palestine and *Egypt in the 14th-cent. BCE are written in cuneiform script.

cup Sometimes 'cup' is used as a metaphor, as in Ps. 11: 6, Matt. 26: 39; when Jesus asked *James and *John if they could share his cup (Matt. 20: 22), he was referring to experiences of suffering. In Paul's account of the Last Supper participation in the cup (1 Cor. 11: 26) is explained as a proclamation of the death of the Lord.

curse An utterance intended to harm. Because Israel had been chosen by God and there was a *covenant relationship, the nation could be cursed if it was unfaithful (Deut. 30: 19; Lev. 26: 14–45). Individuals could also be cursed e.g. thieves (Judg. 17: 2) or mutinous soldiers (1 Sam. 14: 24). In the parable of Matt. 25: 41 ff. the cursed are those who have been met by Christ and have refused to welcome him. But in Gal. 3: 13 Paul quotes Deut. 21: 23 and explains that Christ became a curse for us by taking upon himself the plight of those who live under the curse of the *Law, and so redeeming them.

curtains The *tabernacle, or dwelling, where Yahweh met his people in the *wilderness was a portable *sanctuary formed by ten linen curtains (Exod. 26: 1 ff.) sewn into two sets of five and fastened by clasps to enclose the area; it was Yahweh's place on earth, but it symbolized his seat in *heaven, for the linen was adorned with figures of *cherubim.

In *Herod's Temple there was a curtain which divided the Holy of Holies from the Holy Place and the synoptic gospels describe it as being rent immediately after (Mark 15: 38) or just before (Luke 23: 45) Jesus' *death. Such an event (who would have seen it?) is symbolic; a barrier which kept people out of God's presence (everybody, especially *Gentiles) is removed. A representative Gentile at once makes confession of *faith (Mark 15: 39)—or so it is understood by the Church. Perhaps, if he said anything at all, the centurion's words 'This was a son of God' were merely an expression of deep admiration.

Cush The region now known as N. Sudan and called Ethiopia (so NRSV in 2 Kgs. 19: 9) by classical authors. Moses married a wife from Cush (Num. 12: 1), much to the indignation of his sister. Its inhabitants often visited Palestine (Acts 8: 27).

customs Under Roman jurisdiction there was a value-added tax on goods, in *Judaea payable to the Roman governor, or in *Galilee to *Herod Antipas. *Levi was sitting at his tax office or booth (Mark 2: 14, NRSV, NJB; custom-house, REB) when he was called by Jesus to follow him. Levi was employed by Herod at the border with the tetrarchy of *Philip. Tax-collectors were unpopular with their fellow Jews on account of their continual contacts with *Gentiles.

Cyprus Island in the eastern Mediterranean. The name comes from a Greek word meaning 'copper', for which the island was famous. It is known as Elishah in OT (Ezek. 27: 7). Cyprus was conquered by the *Assyrians in the 8th cent. BCE and later by the Greeks and Romans. There were Jewish congregations (Acts 13: 5) who were visited by Paul and *Barnabas (himself a Cypriot), and later by Barnabas and Mark (Acts 15: 39).

Cyrene A city in N. Africa, the homeland of *Simon, who carried Jesus' cross (Mark 15: 21). Jews from Cyrene established a *synagogue at Jerusalem (Acts 2: 10). They were probably freed men, descendants of those who had been captured by Pompey and taken to Rome in 63 BCE and later released.

Cyrus II Ruler of Babylonia, 550–529 BCE, and mentioned by the Greek historians Herodotus and Xenophon, and in the OT books of Daniel, Ezra, 1 and 2 Chron., and the second part of Isaiah. Cyrus' policy was to forestall rebellions among his subject peoples by lenient treatment. In accordance with this, he permitted Jews to return (Isa. 45: 1, 13) from exile and authorized the rebuilding of the Temple (Ezra 1: 1–2). He is even called 'the anointed one' (= 'Messiah') in Isa. 45: 1.

D

D (1) The symbol employed in textual criticism for the *Codex Bezae, a 5th-cent. MS of the gospels and Acts in Greek and Latin, presented to Cambridge University library by the Calvinist theologian Theodore *Beza in 1581. It is the principal representative of the *Western text. (2) D is also a symbol employed in OT source criticism for the *Deuteronomist, and for the Deuteronomic Code (Deut. 12–26), which is sometimes thought to be the book apparently found in the Temple in 621 BCE and presented to King *Josiah (2 Kgs. 22: 10). It requires the suppression of *Canaanite cults (Deut. 12).

Dagon The god of the *Philistines (Judg. 16: 23 ff.; 1 Sam. 5: 2 ff.), and also of various *Mesopotamian countries. According to 1 Sam. 5: 2–7 the effect of the *Ark of Yahweh being captured and placed provocatively in the temple at *Ashdod was first to make the idol fall and then to shatter it in pieces.

Dalman, Gustav (1855–1941) German biblical scholar whose work on the *Aramaic background of the gospels has been influential, especially in the interpretation of Jesus' teaching about the *kingdom of God. He argued that the primary meaning of the term was 'rule' or 'kingship'. He had acquired a great knowledge of Palestinian lifestyle and customs, exemplified in his *Sacred Sites and Ways* (1928).

Dalmanutha A village on the west of the Sea of *Galilee, to which Jesus and *disciples sailed after the feeding of the 4,000 (Mark 8: 10).

Dalmatia In NT times the southern part of the province of *Illyricum (Rom. 15: 19), to which *Titus went (2 Tim. 4: 10).

Damaris One of Paul's few converts at *Athens (Acts 17: 34). His little success there perhaps accounts for the anxiety he felt in coming to *Corinth (1 Cor. 2: 3).

Damascus A very ancient city often mentioned in the Bible, and the capital of modern *Syria. Attacks on Israel were launched by Aramaean kings reigning in Damascus, but the country was conquered by *Assyria in 732 BCE.

The Romans again made Damascus a capital (of Nabataea) in 85 BCE and when Paul came into Damascus after his call to be an apostle (Acts 9: 2–30) the governor in charge was *Aretas IV (2 Cor. 11: 32).

Damascus Document This *Zadokite document accorded the symbol CD was discovered in the Cairo **geniza* by S. Schechter in 1896, and is dated in the 10th cent. CE. It is now known to be related to the *Dead Sea scrolls. In two columns, it regulates *Sabbath observances more strictly than the *Pharisees. On that day it is forbidden even to lend something to a friend. The document appears to emanate from a renewal movement within *Essenism which has parted from the main body.

damnation This word is used in AV for condemnation to everlasting *punishment (Mark 3: 29; 'an eternal sin', NRSV, REB, and NJB.) The idea of an unforgivable *sin is found also in rabbinic writings. Many modern theologians regard the notion as morally incompatible with what God would include in his *providence, though others hold that the *doctrine witnesses to the need to brand some atrocious deeds of inhumanity with 'a curse of supernatural dimensions'.

Dan One of the sons of *Jacob and held to be the ancestor of one of the twelve *tribes of Israel. *Samson belonged to this tribe (Judg. 13: 2, 24). The tribe migrated to the north and rebuilt the city of Laish (Judg. 18: 29), which they had destroyed, and renamed it Dan, which became the most northerly place in the kingdom. It was the site of one of the two national shrines established by King *Jeroboam I of Israel (1 Kgs. 12: 29–30).

dancing Amongst the *Hebrews, and other ancient people, dancing was part of a religious ritual (Exod. 32: 19), but the cultic dance of David wearing only a short linen *ephod as he led the *Ark to Jerusalem (2 Sam. 6: 14) was deemed by his wife *Michal to cheapen the royal dignity. Dancers are mentioned in the NT as welcoming the homecoming of the prodigal son (Luke 15: 25), and as part of children's play (Luke 7: 32), but also, more sinisterly, when the daughter of *Herodias excited the guests of *Herod Antipas (Mark 6: 22).

Daniel, book of Daniel is a legendary figure of *wisdom and *righteousness who features in Ezek. 14: 14 along with the OT characters *Noah and Job. The book of Daniel was published under his name as an appropriate pseudonym, as was usual with *apocalyptic writing, some of whose characteristics are shared by this book.

The book is placed by *Christians, following the *LXX, in the section of the OT after the major *prophets; but in the Hebrew Bible it is included among the *Writings. The first six chapters contain the popular stories of the Burning Fiery Furnace, *Belshazzar's Feast, and the Lions' Den. The second group of six chapters takes the form of visions which purport to be predictions at the time of Babylonian supremacy of what would happen to the four kingdoms of *Babylon, Media, *Persia, and *Greece. Finally (Dan. 12: 1–4) the culmination of all history is proclaimed with God reigning supreme over a kingdom of the saints.

The heathen kingdoms are symbolized by '*beasts' under whom the Jews suffer, but in the face of death Daniel's companions display loyalty to their God (Dan. 3) who miraculously preserves them in the furnace and, later, Daniel in the lions' den (Dan. 6). The message to the readers is: remain faithful to the *covenant and ultimately Israel will be vindicated. It was a book which could be updated to accord with changing historical conditions. Thus *Josephus, at the end of the 1st cent., interpreted the fourth beast as the Romans and assumed that Daniel's reference to 'the *abomination of desolation' (Dan. 11: 31) must have been a prediction of the desecration of the Temple in 70 CE; while Christians—for whom the coming of the *Son of Man (Dan. 7: 13) meant the *death and exaltation of Jesus—were equally aware of the *eschatological significance of the *abomination (Mark 13: 14).

When was the book of Daniel written? It was long assumed, and is still assumed by some conservative students of the OT, that the book is describing the faithfulness of Jews to their *religion under *Babylonian and *Persian overlords in the 6th cent. BCE and was prophesying further horrors in the future. However, modern critical scholarship takes the view that the book comes from the 2nd cent. BCE and that its references to the empires is an impression by a Jew of the past up to the vivid account of the outrages of the 'little horn' who committed the transgression of the sanctuary (Dan. 8: 13). This last event corresponds precisely to the outrage perpetrated by the *Seleucid conqueror *Antiochus Epiphanes when he installed a pagan *altar in the Temple in 167 BCE.

Dating the book in the 2nd cent. is not exclusively a modern theory. The pagan philosopher Porphyry at the beginning of the 4th cent. CE derided the idea that the visions of Daniel were prophecies of the future. Modern critics notice the comparative accuracy of the 'predictions' up to the time of the 'little horn' but then what is recorded does not correspond with historical fact—for example, the account of the death of *Antiochus (Dan. 11: 45)—which implies that the book was composed before his death in 164 BCE, during the *Maccabean period. The literary device of the pseudonymous author (or authors) serves to support his contemporaries in their present sufferings under the guise of lauding the faithfulness of those in the past.

The Seleucid king *Antiochus was determined to create a common *Greek culture throughout his *dominions; and indeed even in Jerusalem some of the younger men welcomed his ideas (Dan. 11: 30; 1 Macc. 1: 11–14). The religious practices of loyal Jews were ferociously repressed, but after the '*abomination' a rebellion was led by *Mattathias; and under his son *Judas Maccabaeus a military success enabled the Temple to be rededicated three years to the day after its desecration in 167 BCE.

Much of the book of Daniel (2: 4–7: 28) is written not in *Hebrew but in *Aramaic. There were also the Greek translations of the *LXX and of Theodotion (2nd cent. CE); these versions included certain additions—the Prayer of Azariah and the Song of the Three Young Men inserted at Dan. 3: 23, and the stories of *Susanna and of Bel and the Dragon after 12. These few items are included within the canonical text of Roman Catholic Bibles (as NJB); but in NRSV and REB they are placed in the inter-testamental *Apocrypha.

Darius (1) Darius the Mede is mentioned in Dan. (e.g. 5: 31) as ruling after *Belshazzar and before Cyrus, but there is no other evidence for such a monarch. (2) Darius I (522–486 BCE) is the Persian ruler who confirmed the edict permitting the rebuilding of the Temple in 520–515 BCE (Ezra 5 and 6). In classical history of *Greece he is famous for his defeat at Marathon in 490 BCE. His palace was at Susa.

darkness A symbol for what is *evil: it encompasses the underworld (Ps. 143: 3), but is not impenetrable by God (Ps. 139: 12). it is the sphere of *punishment (Ps. 35: 6) and will therefore be part of the experience of the *day of the Lord which the *prophets predicted (e.g. Amos 5: 18), thereby overturning popular expectation. In the Manual of Discipline of the *Dead Sea scrolls there is an exposition of two *spirits contending for victory who are called the *Angel of Light and the Angel of Darkness; and a similar contrast is a theme of the gospel of John (John 3: 19).

David Celebrated in the OT as Israel's ideal king and in the NT as an ancestor of the *Messiah, though David's personal life did not seem to portend such a status. He was the great-grandson of a foreigner, Ruth the Moabitess (no comfort here for later narrow nationalism!) and the youngest of eight brothers, sons of *Jesse; and the narrative of 2 Sam. is by no means silent over his misdemeanours and family feuds. At the same time, it is recorded how he bravely confronted the *Philistine champion *Goliath (1 Sam. 17; but was it Goliath? 2 Sam. 21: 19), and was then invited to serve as a musician in *Saul's court and to calm him in his manic phases. David established a deep friendship with Saul's son *Jonathan (1 Sam. 20: 17).

After the death of Saul, David became king of *Judah (*c.*1000 BCE) at the age of 30 and reigned at *Hebron for seven years. But after inter-tribal warfare the supremacy of Judah over the other tribes was consolidated when David captured the *Jebusite city of Jerusalem (hitherto belonging to none of the twelve tribes) and made it his capital, a choice of astute diplomacy. This was reinforced by bringing in the ancient *Ark of the Covenant and so initiating a central *sanctuary for corporate *worship which David's son *Solomon would complete with the building of the Temple (*c.*950 BCE).

David was honoured as a poet (2 Sam. 1) and he may have written some of the earlier psalms; and although this reputation is mostly the product of the nostalgia of later ages, it is possible that David's entourage at court did encourage literary and musical talents which were still flourishing in the time of the 8th cent. prophet *Amos (6: 5).

David was celebrated by Jews as their ideal king, though his orders to ensure the deaths of *Uriah (2 Sam. 11: 15), and *Joab and Shimei (1 Kgs. 2: 6, 9) were not signs of nobility. He did indeed extend the country's frontiers from the border with *Egypt to the upper *Euphrates, and there was great material prosperity. The dynasty founded by David was not to last over the whole nation, but over Judah, at any rate, it survived until the Exile (586 BCE). After the Exile

David's piety and sagacity are recounted anew in the story of 1 Chron.

David, city of The oldest part of Jerusalem, captured by *David (2 Sam. 5: 7–8) and expanded by *Solomon (1 Kgs. 6–7). But in Luke 2: 4 *Bethlehem is so described, on the ground that it was the ancestral home of King David (Mic. 5: 2).

David, son of It is claimed both in the gospels (Matt. 1: 17) and by Paul (Rom. 1: 3) that Jesus was descended from the royal house of David on the human side. But it is a title usually given to Jesus by those who are not members of the inner circle of *disciples e.g. by the two blind men (Matt. 9: 27). For some believers '*Son of David' has to be qualified or complemented by the truth that he is also '*Son of God', even if the affirmation is ironically put on the lips of mockers (Matt. 27: 40 and 43).

It was standard Jewish belief that the *Messiah would be descended from David, and this is cited as evidence of Jesus' Messiahship by Peter in Acts 2: 30–1 and in 2 Tim 2: 8, but the title was rarely used by the Church as it spread out into Gentile areas where Davidic Sonship would have negligible importance. Jesus himself appears to belittle the phrase in the controversy recorded in Mark 12: 35–7, possibly because it contained overtones which he repudiated; yet at the same time he goes on to argue that the Messiah is greater than King David. More probably this is a *pericope which reflects Mark's situation rather than that of Jesus shortly before the *crucifixion: the readers of the gospel are to recognize that Jesus' Messiahship is not only the fulfilment of Jewish hopes; he is the Lord of all. That is the force of quoting Ps. 110: 1 (a favourite Christian proof text) in Mark 12: 36. David, presumed in those days to have written the psalm, himself calls the Messiah 'Lord', so he must be more than David's Son. The verse significantly anticipates the Gentile *centurion's expression at Mark 15: 39.

day Jews in the NT period reckoned each day to start and finish at 6 p.m., and periods of the day were known by such general terms as morning or evening. Days of the week, except *Sabbath, had no special names but were given numbers.

day of the Lord The popular belief in Israel that a day would come when God would dramatically intervene to release the nation from fears and oppressions. Possibly the hope was celebrated (and supposedly its realization accelerated) by an annual festival with imposing sacrificial ritual which was expected to ensure prosperity and victory over enemies. In the middle of the 8th cent. BCE the prophet *Amos declared that the country's welfare was being achieved by exploitation and false *religion, and when the 'day' arrived it would turn out to be a *judgement (Amos 5: 18–27); the capture of Jerusalem in 586 BCE and the previous overthrow of northern Israel were seen as a fulfilment of this prophecy. But beyond the judgement there was still awaited a 'day' when Israel's fortunes would be reversed and Yahweh's rule would be established over all the earth (Isa. 40). In the NT the concept becomes the 'day of the Lord Jesus' (2 Cor. 1: 14), and in Rom. 2: 15–16 the day of final judgement. It will be preceded by warning signs (2 Thess. 2: 1–2) and is not to be regarded as actually present, as if *Christians who possessed knowledge of Christ and enjoyed a sacramental life in the Church had passed already into a state of ultimate *salvation. On the contrary: the 'day' will be the final consummation and with it the judgement of the whole of history and the readjusting of the world.

dayspring Used by AV in Luke 1: 78 for 'dawn' (NRSV, REB) or 'the rising Sun' (NJB). It is a metaphor for the beginning of a new era; and the *birth of Jesus came to be associated with the returning sun after the winter solstice; *Christmas replaced the Roman pagan festival in honour of the Invincible Sun on 25 December early in the 4th cent.

day's journey About 40 km. (25 miles) but on the *Sabbath (Acts 1: 12) it was fixed as not more than 2,000 cubits (about 1,040 metres or 3/4 mile).

deacon In non-Christian *Greek the word is used of officials who act on behalf of rulers, with their authority, such as *ambassadors or *heralds. *Hermes is called a 'deacon', *diaconos*, because he carried out the commissions of the gods. Paul is a *diaconos* of God as being his authoritative mouthpiece. What is common to the various functions is their mediatorial work—Paul is an ambassador to the Gentiles, a reconciler between the two branches of the Church, Gentile and Jewish, holding them in unity through his *collection for Jerusalem and at the same time urging ex-Jews to accept his principles i.e. of separation from the institutions of Judaism.

The deacons in the Church are to embody personal integrity and their role is that of a *scribe, whereas the task of teaching is entrusted to the *presbyter—*bishops (1 Tim. 3: 2, 5). From Rom. 16: 1, where *Phoebe is mentioned, it would seem that in the NT women could be regarded as deacons.

It has often been supposed that the *Seven appointed to assist the *Twelve (Acts 6: 1–6) were the first deacons, since their function was to minister (Greek, *diaconein*), and 'serve tables', but in the event *Stephen and *Philip engage in evangelistic enterprises on their own account, and the Seven were more the antecedents of the presbyterate than the Church's diaconate. The functions of the latter were administrative and liturgical in the early Church, but they were not inferior officers. They acted as the bishop's agents, and they remained deacons for life. Only as the presbyters, under local pressures, gradually assumed some of the bishops' functions did the diaconate in the West become less important. At last, in the Middle Ages, the diaconate became merely a probationary stage on the way to the priesthood (presbyterate).

When Jesus is recorded as saying that he came 'not to be served but to serve' (Greek, *diakonein*, Mark 10: 45), the meaning is that he is not the kind of great person in the world who has agents around to go out and do his bidding; he has his own duties to do himself.

Dead Sea An enclosed area of water, 74 km. (46 miles) long and up to 16 km. (10 miles) broad with high salt content, below sea level, into which the River *Jordan debouches. *Fish cannot survive in it (cf. Ezek. 47: 9, 10).

Dead Sea scrolls Documents first discovered accidentally in 1947 on the west side of the Dead Sea, in what was the country of *Jordan and which has been the Israeli-occupied West Bank, where a Jewish monastic community was established near *Qumran in the 1st cents. BCE and CE. The members were probably *Essenes or an Essene sect separate from the main body (though it has also been suggested that they were *Zealots), and there was a strict discipline of celibacy and *asceticism, with frequent rites of *purification by water. The leader of the community was known as the *Teacher of Righteousness who wrote to the priests at Jerusalem outlining points of disagreement. During the Jewish War of 66–70 CE the sect perished, but not before its library was concealed in neighbouring caves. (A minority of scholars doubt this connection between the community and the *scrolls and believe they were stored in the caves by a group from Jerusalem.) The *Damascus Document, which is related to the *Qumran scrolls, was discovered in Cairo in 1896, but only since 1949 have the caves been systematically searched. Large quantities of the scrolls have been taken to safety and assembled by a team of experts. Many have been published and photostat copies of much material still to be officially published has found its way abroad, for example to the Huntington library in California, USA.

The monastic site has been excavated and features in it correspond to what had already been noticed of the teaching of the sect in the scrolls. The Hebrew scrolls (mostly of leather, some of papyrus, parchment, or wood, one of copper) are 1,000 years older than the oldest of any previously extant MSS and are of great value not only for comparison with the Hebrew *Masoretic text, but also because they illuminate some of the history of the Jews and Jewish thought of the first centuries. There are numerous linguistic and doctrinal parallels with the NT (e.g. the notion of the temple as a community of people), especially with the gospel of John (e.g. the contrast between *light and *darkness). It has also been suggested by a few scholars that *John the Baptist spent his early years with the Qumran community (on the basis partly of Luke 1: 80), but the differences between Qumran beliefs and those of the primitive Church (e.g. Qumran has a belief in two Messiahs) are equally significant.

Controversy is likely to continue in the world of scholarship as to the precise connection between Qumran and Christianity; but it is certain that NT ideas once regarded as Hellenistic can now be shown to have a Jewish origin.

A system of identifying scrolls in bibliographies has been established by numbering them in the first place according to the order of caves plundered. Thus, 1Q indicates manuscripts found in the first cave, and then this code is followed by the first Hebrew letter of each particular text: 1QS is the Rule of the Community. *Pesher* means commentary, so 1QpHab refers to the community's commentary on the OT book Habakkuk and 4QpPs37 to a commentary on Psalm 37. The copper scroll (which some believe to be unrelated to Qumran) is given the *siglum* 3Q15. The leader's letter to Jerusalem which will for a long time be a prominent subject of discussion, is denoted 4QMMT. One surprising discovery, made in the sixth cave to be opened up, was of a MS which contained the same text as that of a medieval MS found in Cairo in 1896, and published in 1910, written in *Damascus and called a *Zadokite work; it is accorded the symbol CD and describes the laws of the community (*Damascus Document). For the most part the scrolls consist of texts of the Bible and of interpretations of OT books, but non-canonical writings (the book of Jubilees and much of Enoch among them) are also included, together with miscellaneous Messianic texts, prophetic pseudepigrapha, calendars, testaments, legal documents, hymns, and magical texts. The largest scroll, the Temple Scroll, rewrites in a revised form the laws of the *Pentateuch.

death The writers of the Bible were not presented with the modern ethical and biological problems about what precisely constitutes human death. How much of a body needs to be dead before it can be said that a human being is dead? Is it a persistent vegetative state with nevertheless some lower-brain activity? Doubtless states of unconsciousness were in the 1st cent. taken to be death which could not be so defined today. Death in the Bible is presumed when all signs of life have disappeared; it is the end of natural life. But it was not part of God's original *creation (Gen. 3). *Sin came in and was the cause of death. There exist in the OT different views about the state of those who have died. It may be just non-existence (2 Sam. 14: 14), or a feeble, twilight existence in **sheol* (Isa. 14: 10; Job 10: 21 f.) without any relationship with God (Ps. 6: 5). But there existed also a belief that departed *spirits could be conjured up from **sheol*, as when the 'witch' of *Endor brought up Samuel (1 Sam. 28). Later the beginning of belief in the *resurrection of the dead appears in Dan. 12: 2, and this becomes clearly expressed in 2 Macc. 7: 9, 11, as also is belief in the immortality of the soul in the book of Wisdom 1–5 under the influence of popular *Alexandrian Platonism.

In the NT dying is regarded as an *evil from which even Jesus himself shrank (Mark 14: 33) but the belief in *resurrection mitigates its horror (1 Thess. 4: 13) and there are comparisons of death with sleep (John 11: 11–13). Sleep is indeed a fairly common metaphor for death in the Bible (e.g. Dan. 12: 2). Jesus rebuked the mourners in the house of *Jairus; 'the child is not dead; she is asleep' (Mark 5: 39). Possibly the girl was in a coma, though Luke (8: 49) interprets Mark as meaning death. Such *miracles as are recorded in the gospels at any rate anticipate Jesus' own resurrection, just as his resurrection constitutes a guarantee of the resurrection of those who believe in him, to the extent that Paul can long for death (2 Cor. 5: 8).

'Death' is also used as a figure for the alienation brought by sin: 'to set the mind on the flesh is death' (Rom. 8: 6); 'I have set before you life and death . . . choose life.' (Deut. 30: 19). 'The second death' (Rev. 2: 11) represents the final state of those who have deliberately separated themselves from God for ever.

On the basis of the perfect tense of the *Greek verb in John 11: 11 an interpretation of the raising of *Lazarus has been proposed: that he was clinically dead, with no signs of life, but not biologically dead, since there was as yet no evidence of deterioration. Lazarus was in a 'near-death' condition and available for resuscitation, which Jesus effected with his command 'come out' (John 11: 43). More probably, however, the story of the raising of Lazarus is an example of the evangelist's creative writing. It is not history but an allegory of the passion, death, and resurrection of Jesus, which are related to human experience. Lazarus represents humanity: we all may be raised by Jesus to a new life. And it was this promise that finally prompted the Jewish leaders to put such a blasphemer to death (John 11: 53).

Deborah A *prophetess and one of the *judges who ruled in Israel. She urged Barak to fight *Sisera; the paean of triumph upon his defeat written in splendid Hebrew *poetry which comes through in the English of the AV (Judg. 5) is often called the Song of Deborah.

debt *Hebrew laws on debt were humanitarian and recognized that falling into debt was a misfortune. All debts were to be cancelled every seven years (Deut. 15: 1 ff.) and only *foreigners were to be charged interest (Deut. 23: 20). Nevertheless there was in NT times much traffic in loans (Matt. 25: 27).

Decalogue *Greek word for the *Ten Commandments (Exod. 34: 28). They are given in two forms (Exod. 20 and Deut. 5), being the earliest laws of the Hebrews and ascribed in the tradition to Moses, though in fact developed and refined over centuries. Moses was given them by God on two tablets (Exod. 31: 18), each tablet bearing all ten commandments, in accordance with the normal practice that there were two copies of every treaty. But Moses smashed them in anger when he saw that the people had given themselves to *idolatry in his absence (Exod. 32: 19) and they had to be reinscribed (Exod. 34: 28) and placed in the *Ark (I Kgs. 8: 9). These basic laws were supplemented with much detail over the years but the Decalogue remained a summary convenient for memorizing and public recitation and was fundamental to the integrity of the nation. The laws require the *worship of Israel's God and no other; they demand respect for human life and human marriage and property; a weekly day of rest is prescribed; there is to be honesty in court.

The Decalogue is divided in two ways: Exod. 20: 3 is the first commandment in many *Christian reckonings, but in the Jewish tradition (followed by Roman Catholics) it is combined with the next—the prohibition of idols—and then the final commandment, against coveting—is split into two, to make ten in all.

Decapolis The territory of ten *Greek cities founded by *Alexander the Great and successors about 323 CE to the east of the Sea of *Galilee. It seems to have been a cosmopolitan area with several languages in use. Jesus went through this region (Mark 7: 31) before presiding over the feeding of 4,000 people (Mark 8: 1–10).

deconstruction A term used in modern secular literary criticism (e.g. by Jacques Derrida) and borrowed by biblical scholars to indicate the awareness of the limitations that language imposes on thought. Texts do not enjoy definitive meanings. Nor can societies and institutions which validate their use go unquestioned. Without denying the legitimacy of traditional historical criticism to determine the author's intentions, it is therefore argued that readers can examine a text and import their own meaning. There are any number of meanings available, all equally valid. It is possible for a Western reader to hold that Paul, as a disturber of the Roman peace, got the punishment he deserved. Others classify him as a Christian martyr. The process of deconstruction also uncovers internal contradictions in a text, as in the relationships attributed to Jesus and the Father in the gospel of John; or in Rom. 1–3 and 6–8. The author of Rev. declines at the end to have conclusively determined its meaning (Rev. 22: 10).

decrees There are in the NT decrees of *Caesar (e.g. Luke 2: 1; Acts 17: 7, NRSV; '*laws', REB, 'edicts', NJB), but where AV used the word modern versions often prefer 'decision' or 'law'.

dedication, feast of An annual celebration in December lasting eight days to commemorate the restoration of the Temple, following the military successes of *Judas Maccabaeus, three years after the *abomination perpetrated in 167 BCE by *Antiochus Epiphanes. The *feast (Hanukkah in *Hebrew) was associated with lights after the relighting of the Temple *menorah, and there grew up the custom of lighting candles in Jewish homes—one additional light each day of the festival. Jesus was present at the feast, according to the time-scale of the gospel of John (John 10: 22–3).

defilement Ceremonial defilement prohibited a Jew from approaching a *sanctuary or offering *sacrifice and forbade human *fellowship. It might be contracted by tending a corpse (which accounts for the behaviour of the priest and the *Levite in the *parable of the Good *Samaritan, Luke 10), by *'leprosy' (Mark 1: 44), or by menstruation. An elaborate system of rules about defilement existed by the time of Jesus, and he taught that moral duties had priority (Mark 7: 2–4).

Deissmann, Gustav Adolf (1866–1937) A German NT scholar of Heidelberg, who made a specific study of papyri discovered in *Egypt which shed much light on the contemporary meaning of many *Greek words used in the NT.

deliverance People and traditions are 'delivered'—Jesus to the Romans by the chief priests (Mark 15: 1), and tradition to Paul in the account of the last events in the life of Jesus (1 Cor. 11: 23; 15: 3).

The purpose of Jesus' *ministry, *death, and *resurrection is the deliverance of his people from distress, *persecutions, and suffering (2 Tim. 1: 10), from the power of *sin (Rom. 7: 18–25), and from the tyranny of death (Heb. 2: 15). Jesus urges *disciples to pray for deliverance from the power of the evil one (Matt. 6: 13); and the idea of rescue (e.g. 2 Cor. 1: 10; Gal. 1: 4) is common.

Demas One of Paul's fellow prisoners who was associated with him in sending greetings to the Church at *Colossae (Col. 4: 14); but according to 2 Tim. 4: 10 Demas later deserted Paul and went (home?) to *Thessalonica.

Demetrius (1) Leader of the silversmiths who instigated a riot against Paul when it seemed that Christianity could threaten his business in connection with the cult of *Artemis (Acts 19: 24–41). (2) A *Christian held in high regard (3 John 12) who was possibly the leader of a group of missionaries and who was entrusted with a letter to *Gaius, the kind of testimonials important to Christian travellers to give them introductions to other Churches.

demons Non-material beings. In the ancient world there was a common belief in spiritual beings, either good or bad, inhabiting the world, independently of any deity, and capable of influencing human beings. In the philosophy of the *Greek Plato the 'demons' are guardian *spirits who determine the fate of individuals, and are thus precursors of guardian *angels. In late *Judaism the demons are wholly bad; they exist as an army under a general (*Satan) and they engage in *battle with God and his angels. These demons were thought to invade human personalities and destroy the power of reason and health. They caused disasters: their proper place was in the *abyss (Mark 5: 13). Jesus, having vanquished Satan in the *wilderness (Matt. 4: 10), is known to be one who casts out demons (Luke 8: 26–39). But whether this was by the power of the prince of demons (Matt. 9: 34) or was a sign corroborating *Satan's defeat (Luke 10: 17) was debated. The genuineness of the healings does not seem to be disputable. Although in John there is reference to the Jewish charge that Jesus is possessed by a demon (John 8: 48–9), there are no accounts of *exorcisms in the fourth gospel, but successful *exorcisms are reported in the early Church (Acts 19: 11–16).

Returning missionaries sometimes describe what they take to be demon-possession as distinct from physical or mental disorder. Several *bishops have appointed official exorcists in order to establish some control over modern *fundamentalist healers with convictions that the NT world-view is still directly applicable today.

demythologization Term adopted by Rudolf *Bultmann to describe the means by which the essential *truth of the gospel could be made acceptable to modern people. The world of the NT is alien to us; we cannot believe in the interventions of God or supernatural beings in the affairs of our lives, and we have long ago discarded the cosmic framework of *heaven, *earth, and *hell which was assumed in the 1st cent., and for long after. Of necessity, the NT writers were bound to use a cultural framework that made sense in their generation: the question is whether the gospel is still intelligible when that world-view is superseded.

Bultmann's work in the historical criticism of the gospels had led him to take a fairly sceptical view of what may be regarded with any confidence as authentic, but as a *Christian apologist he sees this as a positive advantage; *faith should not rest on provable facts. Faith is the decision to choose the new *life in Christ; the choice confronts us when the preacher proclaims Christ crucified. This new life is described by Bultmann in terms of the philosophy of existentialism; the old life of fallenness and alienation is exchanged for the possibility of total integrity and authenticity.

'*Myth' means the description in terms of this world of alleged supernatural events, such as the *virgin birth and the *resurrection. These stories are not history; they are the means by which facets of the meaning of the *Cross can be disclosed.

The gospel may still be a '*scandal', causing offence to modern people, as it did in Corinth (1 Cor. 1: 23), but, according to Bultmann, his demythologized gospel at least puts the 'scandal' in the right place.

Critics of Bultmann maintain that he has been over-zealous in rejecting almost the entire world-view of the NT by relegating it to the mythical, and too enthusiastic in embracing the existentialist philosophy of his one-time colleague Martin Heidegger.

denarius A *silver coin which bore the image of the Roman emperor (Mark 12: 16); it was equivalent to the Greek *drachma as known in the east, which was the cost of a sheep. The *parable of the Labourers in the Vineyard indicates that one denarius was the wages for a day's work by an agricultural labourer, but it was debased by the emperor *Nero.

deputy Used by NRSV in 1 Kgs. 22: 47; 'viceroy', REB. *Edom is being ruled by an official from *Judah's king *Jehoshaphat. In Acts 13: 7 the deputy (AV) at *Paphos is more accurately described by NRSV as a *proconsul; 'governor' in REB.

Derbe A city in the province of *Galatia in central *Asia Minor visited by Paul (Acts 14: 20 and 16: 1). Ethnically, it lay in the district of *Lycaonia, where the inhabitants spoke that language (Acts 14: 11).

descent into Hades The Greek Hades (REB, NRSV, NJB, Luke 16: 23) replaces '*hell' of AV, which is a translation of the Hebrew **sheol* (Ezek. 31: 15, NRSV, REB, NJB). In the OT (e.g. Jonah 2: 2–6) and in the *Dead Sea scrolls there are notions of the idea of a descent into Hades. Such a journey is ascribed to Jesus by the NT in 1 Pet. 3: 19 f., 4: 6. See also Acts 2: 24–31 and Rom. 10: 7. It was known in the Middle Ages as Christ's 'harrowing of hell' and it was held that he was in combat there with Satan to liberate departed *spirits. The belief is possibly found as early as the 5th–6th-cent. apocryphal gospel of Nicodemus. There was a need to explain what the Lord was doing between Good Friday and Easter Day: it was declared that his redemptive love and power extends through the whole of time. A modern explanation is that Christ not only died on the cross but also shared the experience of being dead in solidarity with all sinful humanity and with the extremity of *evil.

desert The desert, or *wilderness, round Mount *Sinai in the south of Palestine where the Israelites wandered for 'forty years' after leaving *Egypt (Deut. 8: 2) was forever in their consciousness, as it was also for the writers of the NT. Jesus' *temptations for forty days in the wilderness (Matt. 4: 2 ff.) and his *Sermon on the Mount (Matt. 5–7) imply a belief in Jesus as the faithful Israelite who fulfils God's purposes. His experiences parallel those of the people in the wilderness and Moses' issue of the *Ten Commandments on Mount Sinai.

Wild animals lived in the *Transjordanian desert (Mark 1: 13) and the area was uninhabited but was not a desert in the sense of the rainless Sahara. There can be such rain that 'the desert shall rejoice and blossom' (Isa. 35: 1).

desire *Ezekiel's wife is called the 'desire' (AV; 'delight', NRSV) of his eyes; the object of his *love, who died (Ezek. 24: 18). The word in *Hebrew expresses strong emotion, and desire could turn to evil *covetousness, as for *money (1 Tim. 6: 9), and to a greed approaching *idolatry (Col. 3: 5).

desolating sacrilege (NRSV; '*abomination of desolation', AV and REB in Mark 13: 14; 'appalling abomination', NJB). The phrase is used in Dan. 9: 27 to denote the pagan *altar set up in the Jerusalem Temple by *Antiochus Epiphanes in 167 BCE; and in Mark 13: 14 it may refer to the failed attempt by the emperor Caligula to install his statue in the Temple (40 CE) or to an event such as the display of army emblems in the Temple in the *war of 66–70 CE. Luke (21: 20) probably has the siege of the city in mind and writes therefore after the sacrilege has occurred. *Mark may be suggesting that Jesus is warning his *disciples about coming trials and urging them not to be taken by surprise.

detachment The renunciation of attractive objects and deep relationships which was urged both by Jesus (Luke 14: 26) and Paul (Col. 3: 9–10) on the ground not that they were intrinsically bad but because they could hinder a total commitment to the *service of God. The practice of detachment became institutionalized in monasticism.

deuterocanonical The term used, especially in Roman Catholic works, for those books in the *LXX but not in the Hebrew OT, and elsewhere designated as the *Apocrypha.

Deutero-Isaiah 'Second Isaiah': the name given to the unknown author of Isa. 40–55 (and by some scholars, of Isa. 40–66) who lived during the period of the Exile in *Babylon (586–538 BCE).

Deuteronomist Either a single preacher or a group of like-minded teachers in *Judah who imposed a theological view with a distinctive oratorical style on the books of the OT especially from Deut. to 2 Kgs which are often called 'the Deuteronomistic History'. The intention was to explain the nation's fate as due to its *apostasy from the true *worship of God. There was a *covenant (Deut. 7: 12) which God for his part would keep, but *peace and prosperity for the people depended on their faithfulness. The prophecies of *Jeremiah (and some would add many other prophetic collections) seem to have been edited to express the Deut. point of view. The collapse of Judah in 586 BCE and the Exile are interpreted as vindicating Deut.'s prophecy of punishment if they were faithless (Jer. 36: 29).

Deuteronomy, book of The fifth book in the *Pentateuch. There is a dramatic account (2 Kgs. 22: 8 ff.) of the discovery of a lawbook during restoration work in the Temple in 621 BCE. It prompted *Josiah the king to expedite his religious and social reforms, which corresponded to the outlook of Deut. It has therefore been supposed that the book allegedly discovered was an early form of the book Deuteronomy, or part of it.

The book is part law and part prophecy. The aim of the book is to bring the whole of Israelite life under a sense of duty to God and thanksgiving for his great acts in the past history of the nation (Deut. 16: 3). It was therefore not inappropriately ascribed to Moses, lawgiver (Exod. 24: 12) and *prophet (Deut. 34: 10), but this traditional view should not be understood literally. Not only does the book describe Moses' death (34: 1–12), but the contents reflect an editorial process of reinterpretation so that old traditions were made relevant for changed historical situations. It was an attempt to maintain the vitality of Hebrew *worship (Deut. 6: 13–15) and the prophets' demands for social *justice (Deut. 15: 1–18). The compilation has been dated in the 7th cent. BCE, but the account in the Deuteronomist book 2 Kgs. 25 of the release of *Jehoiachin from prison suggests a date for its final publication to be soon after 561 BCE.

Deut. is often quoted in the NT, as it was by Jesus during the *temptations (Matt. 4: 1–10) and in his summary of the *Law (Mark 12: 30).

devil The word *diabolos* is used in the *LXX to translate Hebrew *Satan, and 'devil' is an English alternative used in the NT (e.g. in the *temptation narrative, Matt. 4: 1) as an equivalent of 'Satan'. The belief that Satan was once an *angel expelled from *heaven (Rev. 12: 9) is assumed by Jesus (Luke 10: 18). But Jesus has the power to control Satan and expel the *demons who are his underlings and who can take up residence in human bodies (Mark 3: 23–7).

diachronic A term, derived from the Greek, to denote the study of biblical texts as developed over a period of time, and therefore includes the examination of multiple editorial sources, *Form Criticism and *Redaction Criticism. The contrasting method is called *synchronic.

dial According to Isa. 38: 8 the dial of King *Ahaz was approached by steps; or it is possible that the sundial may have been the staircase itself. As the day went on, the shadow would naturally fall, but in the *legend it rises miraculously at the behest of *Isaiah, and *Hezekiah then knows that he will be cured of his illness.

Diana The *Latin equivalent (used in AV, NJB) for the goddess *Artemis (Acts 19: 28, NRSV, REB). But the Artemis worshipped at Ephesus was revered for her fertility powers, whereas Artemis (Diana) of Greek and Roman mythology was a huntress or the moon-goddess.

diaspora The *Greek for *dispersion, and used of Jewish people scattered round the world, outside Palestine. The emigration began with the deportations to *Assyria (8th cent. BCE) and *Babylon (6th cent. BCE). Many comfortable families preferred to remain in Babylon where they were successfully trading within the Persian Empire and with *Egypt rather than take part in a Return to Jerusalem with the agreement of *Cyrus. Their descendants between 500 and 600 CE produced the Babylonian *Talmud. In the Roman Empire Jews were settled in every main city (John 7: 35) often (as at *Alexandria) occupying their own quarter with their own local administration centred on a *synagogue. Their language was *Greek, and the Greek translation of the OT (the *Septuagint, *LXX) was undertaken at Alexandria in the third cent. BCE.

The Jews of the Dispersion maintained links of loyalty with Jerusalem, paying a half-*shekel tax and visiting the Temple when possible (Acts 2: 9–11). In the 1st cent. CE the tax was set at two *drachmas per head. After the destruction of the Temple (70 CE), the

Romans ordered the tax to be transferred to the temple of Jupiter Capitolinus, and the discussion in Matt. 17: 24–7 may reflect the dilemma of the Church of that generation whether or not to pay it.

diatribe The technical term in 1st-cent. *Greek for moral exhortation or dialogue (e.g. Eph. 5–6), sometimes introduced by a rhetorical question (Rom. 6: 1).

Didache 'The Teaching of the *Twelve Apostles': a brief manual in *Greek, of unknown authorship, comprising instruction about conduct and Church order. It contains quotations from the *Sermon on the Mount. It is probably to be dated early in the 2nd cent. CE. Highly valued, a complete Greek MS of the work (dated 1053 CE) was found in 1873, and provoked an intense scholarly discussion.

Dionysius A member of the court of *Areopagus at *Athens converted by Paul (Acts 17: 34). In the Middle Ages Neoplatonic writings of the 3rd to 4th cents. CE were mistakenly attributed to Dionysius and exercised great influence; these writings are now referred to as Pseudo-Dionysius.

Dioscuri *Castor and Pollux, AV, REB; the *Twin Brothers, NRSV; the Twins, NJB (Acts 28: 11); the figurehead of the ship which took Paul to *Rome. There were temples to the Dioscuri in *Athens and in Rome, and they were held to be the protectors of travellers, especially on the seas.

Diotrephes A Church leader, and ambitious aspirant to leadership, who is rebuked by the Elder, the author of 3 John (verses 9–11), for disputing the writer's authority and snubbing his representatives. Diotrephes was a man of strong personality who persuaded a majority of the Church to support him. It would appear that he was a *presbyter who was flouting the claim to apostolic authority of the Elder.

discernment of spirits Paul regards the ability to recognize *gifts of the *Spirit as itself a gift (1 Cor. 12: 10); it enables a person to establish the presence or the absence of God or, a worse case, the activity of an *evil power disguised as good. One criterion for sound discernment is given in the *Sermon on the Mount: 'You will know them by their fruits' (Matt. 7: 16).

disciples Committed followers of a *teacher. Some of the OT *prophets had disciples (Isa. 8: 16) who were pupils or learners, and this was later maintained as a rabbinic custom. *John the Baptist had disciples (Mark 2: 18) and the word is applied to those who accompanied Jesus (Mark 6: 45), among whom were the *Seventy (Luke 10: 1, 23) and the more closely associated *Twelve (Matt. 10: 1). These, in Matt., not only learn from Jesus (16: 21) but are also to teach (28: 20) and even to forgive sins (18: 18).

disease Many kinds of disease are mentioned in the *Bible but it is difficult to identify them by their modern names. The incidence of disease and the means of cure were highly important in biblical times, and its causes were sometimes ascribed to God as *punishment for *sin—as when a plague hit the whole nation because of *David's presumption in counting the population (2 Sam. 24). But such a view of God was modified by the later Chronicler (1 Chron. 21: 1; cf. 2 Sam. 24: 1) who attributes the whole episode indirectly to *Satan. In the NT Satan and his creatures are said to be responsible (Mark 1: 25; Luke 13: 16, 32) for illness. The connection between sin and disease is questioned by Job (Job 2: 10) and by Jesus (John 9: 1–3).

The vast number of disfiguring and painful complaints, the cases of *paralysis and mental illness, malformations of limbs and organs, and the inadequacies of medical care (Mark 5: 26) and diagnosis (1 Sam. 16: 15) all bear witness to the precariousness of existence as an essential part of the context of the biblical narratives.

dispensation The OT system of *law and the ethics of the Old *Covenant are sometimes known as the Mosaic Dispensation;

and the New Covenant of Jesus Christ is called the Dispensation of *Grace.

dispersion The English translation of **diaspora*.

diversity Differences of emphasis and outlook within *scripture. It was for many centuries CE held that the books of the *Bible as bound together formed a coherent and consistent account of God's dealings with humanity. There was a unity between the two testaments and within each testament. The OT anticipated what was fulfilled in the NT: the OT was constructed round the central theme of the covenant between God and Israel, and anyone reading the OT in the 1st cent. would gain a fair idea of what *Judaism was. The NT was held to offer an account of the *salvation brought by Jesus, and the various terms used about him and his work in the gospels and the epistles were not contradictory; they could be interpreted by the standards duly prescribed by doctrinal formulations and were consistent with each other.

In the 18th and 19th cents. this basic unity was questioned. Parts of the OT were diagnosed as different sources with different theological presuppositions. In NT scholarship there was the suggestion that the fourth gospel gave a more advanced *Christology than that of the synoptists who presented a human portrait of Jesus. Paul was sometimes identified as the real founder of Christianity, and within the NT itself a clash could be detected between Peter and Paul.

Although for a time in the 20th cent., during the predominance of *Biblical Theology, there was a revival of support for the idea of the unity of the Bible, especially the unity of the NT, without surrendering the principles of *historical criticism, that movement has spent its force, and it would now be widely agreed that there does exist much diversity of material and interpretation within the Bible. In the OT there is a large quantity of sheer narrative from Genesis to Esther, and within this some of it is rewritten from a new perspective by the books of Chronicles. Into the historical sections large sections devoted to *Law are inserted. Then there are the psalms, a love song, Wisdom contributions, *apocalyptic. It is not easy to assert the unity of such a diverse collection. In the NT there is a similar diversity of *genre—history, parables, letters, prophecy, *myth. There is also a variety of interpretations of Jesus—as *Son of Man, as **Logos*, as *Mediator—and of his work, of *ransom, liberation, and *healing, as the writers struggled to affirm their own experiences of belief in Jesus and life in the Christian *community within changing social circumstances and against a cultural background which changed dramatically as the Church spread out into the Graeco-Roman world and encountered many foes.

Yet despite the diversity there can be discovered a genuine unity within the Bible: belief in the One God which is explained most clearly in the sayings, and the *life, *death, and *resurrection, of Jesus. He is the *Lord whom gospels (John 21: 31) and epistles alike longed for their readers to accept and trust.

Dives Unnamed rich man. Luke 16: 19–31 is the only *parable of Jesus in which a character is given a name (*Lazarus) and it has become usual also to attach the name Dives to the other actor in the story. It is from the Latin *Vulgate (*dives* = wealthy).

divination A supposed method of obtaining information. Divination was widely practised in the ancient world. In 5th–4th cent. BCE Greece the priestess, called Pythia, uttered *oracles at Delphi after inhaling the vapour believed to contain the revelations of the god Apollo. Other divinations were made by observing the flight of *birds or scrutinizing the liver of slaughtered animals. Paul and *Silas encountered a slave girl who had a *spirit of divination ('soothsayer', NJB) at *Philippi (Acts 16: 16 ff.). In *Greek, the girl is said to be possessed by 'a spirit, a *python'; she may have even been a ventriloquist, or possibly she gave utterances while in a state of trance.

Forms of divination were certainly practised in Israel, as when the people 'enquired of the *Lord' after the death of *Joshua (Judg. 1: 1), and **Urim* and *Thummim* were the means of communication of oracles. The disclosure of the divine will by *dreams is common in both OT and NT, but divination is strongly condemned both in the OT (e.g. 1 Sam. 15: 23) and in the NT (Acts 19: 19). The reason is that believers have immediate *access to God and have no need of intermediaries or magical devices to discover what his will is. (Divination pretends to ascertain the future or the deity's will; magic to change it.)

divorce The legal separation of married partners. In the OT *marriage is not regarded as a relationship (such as that of brother and sister) which is in itself by its very nature unbreakable; and divorce and remarriage were permissible under the Mosaic law—though the conditions were weighted in favour of the man (Deut. 24: 1 ff.). In the post-exilic era *Ezra commanded Jewish husbands to divorce wives who were *foreigners (Ezra 10: 11). Although the 'bill of divorcement' presumably required the presence of an official, it seems that there was also the possibility of purely private proceedings, which is what *Joseph contemplated when it was revealed that his fiancée was pregnant (Matt. 1: 19).

The grounds for divorce in NT times were interpreted strictly by *rabbis of the school of *Shammai and liberally by those who followed *Hillel, another well-known exponent of the *Law. The liberal position tolerated divorce for comparatively trivial offences by the wife, whereas Shammai restricted it to proven cases of *adultery. Jesus was invited to say where he stood in this controversy and he apparently sides with Shammai but with the important clause permitting certain exceptions (Matt. 19: 9). He takes the issue back to first principles—God's intentions at the *creation (Gen. 2: 24). Paul advises the Corinthians (1 Cor. 7: 12 ff.) that divorce and remarriage are permissible for a *Christian if the partner who initiated the divorce was a pagan.

docetism The view that Jesus was a divine being who only appeared to be human, explicitly branded as a *heresy by *Ignatius (d. 107 CE), though it may have existed earlier. It derives from *Hellenistic *dualism: the idea that what is created is tainted, 'fleshly', and mortal, whereas the divine realm is good, spiritual, and immortal. So, the docetist argument ran, because Jesus the *Saviour belongs to the divine realm, he cannot have been truly human.

Some scholars believe that there are indications of docetic thought within the gospel of John, especially in Ch. 17. It is claimed that in the fourth gospel Jesus lacks the human characteristics of compassion and *mercy; that he is never properly exposed to the sickness and sufferings of our world; and that he is so manifestly 'always on the side of God' that he has no solidarity with humanity.

Against this it is urged that the hallmark of Johannine *Christology is John 1: 14, 'the Word was made flesh', and that the authenticity of Jesus' humanity is demonstrated in John's gospel by the references to Jesus' mother (2: 1–11) and brothers (7: 3–10); his weariness at the well in *Samaria (4: 6); his thirst (4: 7); his grief (John 11: 33); his *death. On this evidence it is held that the *Christology of John is not docetic: Jesus shares his humanity with all other human persons.

doctrine The word is used in AV (e.g. Deut. 32: 2) for teaching or instruction, referring to the revealed *Law, in the OT. In the NT Jesus is said to rebuke the *Pharisees for 'teaching human precepts as *doctrines' (Mark 7: 7, NRSV); 'they teach as doctrines the *commandments of men' (REB). There is a similar pejorative use of the word in Col. 2: 22 (AV).

Doctrines of the Christian Church were formulated (e.g. in the creeds) in the course of internal controversies and under external pressures from opponents.

Dodd, Charles Harold (1884–1973) English (though born in Wales) NT scholar who was professor in Manchester and Cambridge. He introduced the term 'realized *eschatology', later changed to 'eschatology in process of inauguration', as an interpretation of Jesus' teaching on the *kingdom of God, in contrast to the futurist interpretation of Albert *Schweitzer. Dodd's greatest work was, however, in Johannine studies. Uncharacteristically for one associated with the then current *Biblical Theology movement, Dodd had an immense knowledge of Hellenistic literature and was himself a thinker in the Platonic tradition.

dog Among the Jews, dogs were scavengers; they are mentioned in Jesus' *parable as adding to the distress of *Lazarus the beggar (Luke 16: 21). Because they might feed on corpses, dogs were unclean, and *Gentiles were called dogs as an expression of contempt. However, little dogs were sometimes kept indoors for children to play with (the implication of Mark 7: 28). In Phil. 3: 2 Paul rounds on his Jewish, or possibly Jewish Christian, opponents with the invective that they reserved for *Gentiles; the opponents are like savage dogs who prowl round Christian congregations helping to win over Gentile converts.

dominion The exercise of authority. In the *creation narrative of the *Priestly writer the human species is given dominion over the rest of creation (Gen. 1: 26, 28) but subject to the ultimate dominion of the *Lord (Ps. 22: 28). Dominion which is arbitrary and oppressive is condemned by Jesus (Matt. 20: 25), and persecutors of the Church (Rev. 1: 9) are surely flouting God's dominion (Rev. 1: 6). There was also in some places a belief in angelic powers who existed somehow between God and the world and such beings may be the 'dominions' of Col. 1: 16, where the writer is insisting that such intermediary powers, if they existed, have been defeated by the unique and universal redemptive work of Christ, who is not to be confused with any mere *angel.

Domitian Emperor from 81 to 96 CE (when he was assassinated). Under him the Romans made substantial military advances in Britain. He pressed his claim to divine honours in order to cement the unity under him of the empire, and therefore Christians' refusal to participate was regarded as sedition and rendered them liable to prosecution. This was possibly the occasion of the persecution referred to in Rev. 1: 9.

doublet A second version of a saying or of a narrative. Mark 8: 1–9 is regarded as a doublet of the previous account (6: 35–44) of the feeding of the multitude. But when small units are repeated it is not always easy to know whether these are doublets or deliberate repetitions for stylistic effect.

doubt In the NT doubt is not intellectual agnosticism but hesitancy, a refusal to commit oneself, or even honest scruples which cause timidity (Rom. 14: 23). As such it is deplored by Paul. But in the OT the book Ecclesiastes seems almost a celebration of doubt: the pleasures of the flesh and intellectual enquiry alike are fleeting and futile and there is no purpose in life. Yet now and then the writer recoils from total scepticism (Eccles. 8: 12).

dove A symbol of innocence (Matt. 10: 16) and a messenger of promise since a dove is said to have brought an olive leaf (Gen. 8: 12) to *Noah. *Poor people were allowed to offer doves for *sacrifice in place of a more expensive victim (Lev. 12: 8). The dove was the symbol for the *Spirit at Jesus' *baptism and this passage in the gospels would evoke in believers' minds an association of their own baptism when they received the Spirit as they emerged from the water and were rescued into the Church (1 Peter 3: 20–1). In the *Bible the dove is not a symbol of peace.

drachma A unit of *money in NT times. In the *parable of Luke 15: 8–9 the *silver coins (NRSV, REB), ten drachmas (NJB), refer to coins of *Greek origin, circulating amongst all *Hellenistic peoples, and worth about the same as the Roman *denarius; two of these were sufficient to pay the innkeeper's bill for looking after the victim of a mugging (Luke 10: 35).

dragon An *apocalyptic monster identified with *Satan in Rev. 12: 9.

dreams Supposed supernatural intimations, From *Joseph, son of *Jacob, in the OT (Gen. 37) to Joseph, son of another Jacob, and husband of *Mary, in the NT (Matt. 1: 16, 19) the Bible has many accounts of dreamers who believe that by these means they are in receipt of communications from God. (e.g. Elihu in Job 33: 15–17). *Daniel even told his royal employer (Dan. 2: 28) the content of his dreams and then interpreted them, as well as having dreams and *visions on his own account (Dan. 7: 1–2). However, *Jeremiah suspected that the dreams of some of his contemporaries who prophesied were auto-suggestions rising up from their own subconscious (Jer. 23: 16 ff.) and not to be trusted as divine revelations. *Pilate's wife is said to have complained that she had suffered an alarming dream about Jesus shortly before his *crucifixion (Matt. 27: 19).

dress The clothes of men and women during the biblical periods in Palestine are not described in any detail, but they probably resembled those of neighbouring peoples, of which illustrations exist. They were worn both for warmth and sometimes for their appearance, and out of modesty (Gen. 3: 7), though children before adolescence commonly ran around naked. A man's outer garment was a kind of long robe, which could also serve as a blanket, though it was possible also to change into night attire (perhaps Neh. 4: 23), and not to possess this represented extreme poverty (Job 24: 7). Undergarments supported by the shoulders were also available.

Jesus is described as having a seamless tunic (John 19: 23) which was stripped from him before he was crucified naked, and the unnamed young man who lost his linen cloth (Mark 14: 51–2) during Jesus' arrest must have been wearing a simple garment round his body without a girdle such as *Elijah used (2 Kgs. 1: 8), followed by *John the Baptist (Mark 1: 6). Sandals were generally necessary (Matt. 10: 10), as was a covering over the head. Women wore similar dress to that of men but sometimes also with a veil. There could be some decoration, especially at the time of *marriage (Ezek. 16: 10). Persons of rank were expected to dress accordingly, and sackcloth with ashes was prescribed for times of mourning and penitence.

drink Water and milk were the usual forms of liquid nourishment, but *wine was also greatly valued for occasions of festivity (John 2: 3) and religious rites (Mark 14: 25). Excessive drinking of wine was shameful (Eph. 5: 18). For the *kingdom of God does not consist in eating and drinking but in *righteousness and *peace and *joy in the *Holy Spirit (Rom. 14: 17)—though it is also possible for the symbol of a *banquet to be used of the Kingdom (Matt. 25: 10; Luke 22: 30); and sharing a *meal in *fellowship and *hospitality was for early *Christians a meeting too with their risen *Lord. The Greek word *agape* (love) came to be applied to such meals, for which they could claim the example of Jesus himself (Matt. 11: 19).

drunkenness Although *wine is esteemed as a *gift of God in both OT (Isa. 55: 1–5) and NT (1 Tim. 5: 23) and is used in family *worship (Deut. 14: 26) and in the Christian *Eucharist (1 Cor. 11: 26), over-indulgence is condemned (Gen. 9: 20–7; Gal. 5: 21), for the injustice and irresponsibility it caused (e.g. Luke 12: 45). However, *Jeremiah once compared himself, the prophet, overcome by the holy words of the Lord, to a drunkard overcome with wine (Jer. 23: 9), and the same comparison is even made of God (Ps. 78: 65).

Drusilla The youngest daughter of *Herod Agrippa I; married to Felix after deserting her previous husband. She listened to Paul's defence at *Caesarea, according to Acts 24: 24.

dualism The term for a *belief that there is an absolute distinction between *light and *darkness, *good and *evil, God and *Satan. In *Zoroastrianism the evil spirit is regarded as existing independently of the creator of good, but Hebrew religion always maintained that Satan was subordinate to God. The Christian *doctrine is that God's creation is good (Gen. 1: 31); but human beings in their *freedom have become sinful (Rom. 6: 12). Paul repudiates the heretical doctrine of the *Gnostics that matter is itself evil (Rom. 8: 19 ff.)

Dura (1) A place in *Babylon (Dan. 3: 1) of uncertain location, where a golden image was erected. (2) Dura-Europos, beyond the eastern border of *Syria, on the right bank of the River *Euphrates, where excavations in 1934 revealed the ruins both of a Jewish *synagogue, and of the first known Christian church, constructed out of two rooms of a private house (232 CE). One of these was a rectangular room entered from a central court and adapted for *baptisms. In it was a receptacle shaped like a coffin; it was the font in which the believer was 'buried with Christ' and raised to new life. At the east end there had been placed a raised platform, possibly for the use of the president.

dust Dust is used as a simile for annihilation (2 Kgs. 13: 7). In the NT dust on the head was a sign of *repentance (Rev. 18: 19) but when shaken off the feet it was either a warning of *judgement to come (perhaps Matt. 10: 14) or a gesture of anger or of renunciation of all further responsibility (Acts 13: 51).

dyeing The curtains of the *tabernacle were dyed (Exod. 26: 1) as was the robe in which Jesus was dressed by soldiers (John 19: 2). *Lydia, who was converted by Paul in *Philippi (Acts 16: 14), was engaged in the dyeing business. The dyes were obtained from the fluid of shellfish collected on the coast of *Phoenicia.

dysentery Gastric fever, or some other violent stomach disorder. Jehoram king of *Judah died in agony from chronic dysentery (2 Chr. 21: 18). *Publius' father suffered from the disease, common in *Malta, and was healed by Paul (Acts 28: 8).

E

E The *symbol for that strand in the narratives of the *Pentateuch where **Elohim* is used for the divine name. It is thought to derive in the 9th or 8th cent. BCE from an editor in Israel. But some modern scholars maintain that the different names for God are no longer a compelling basis for the argument which distinguishes two sources (*J and *E).

eagle A *bird recognized by its speed in flight (2 Sam. 1: 23; cf. Rev. 4: 7). Those mentioned by Jesus as gathering to swoop on a corpse (Matt. 24: 28) are probably *vultures (REB, NRSV, NJB). Possibly the readers of the gospel would detect a reference to the eagle carried on the standards of the Roman army; the corpse of Palestinian *Judaism was to be swooped on by the eagles of *Rome.

ear (1) Literally, the human organ (e.g. Luke 22: 50), but also metaphorically of the inward disposition to listen and take heed (e.g. Mark 4: 23). (2) One *sabbath the *disciples picked the grain out of the ears of corn, ('heads of grain', NRSV), 'threshing', as they made a path for Jesus through the fields (Mark 2: 23).

Early Catholicism Term used by some NT scholars to define a development they perceive in Christian life and thought from expectation of an early end and judgement towards sacramental and continuing institutional norms. What happened in the Church after the death of the *apostles? It would seem from the letters of Paul that organization was loose and that members of each Church were credited with different *gifts, which they were expected to exercise (1 Cor. 12: 4–11), such as prophesying and *healing. However, it is argued by some scholars especially in Germany that this early freedom was gradually transformed: teaching and sacramental ministry came to be concentrated in the hands of authorized *ministers who derived their authority from predecessors; *sacraments were increasingly emphasized as the means of sharing the life of Christ; *justification by *faith was replaced by an ethical system. 'Early Catholicism' on this view of the history of the Church therefore represents a stage of decline—when the Church was repudiating the simplicity and the expectations of the first generation and was approaching the legalism and hierarchical, institutionalized religion of the Middle Ages. Traces of this so-called Early Catholicism are said to be discernible in the Acts (14: 23) and in the later epistles, such as the Pastoral Epistles, Ephesians, and 2 Peter. It is not long before *Ignatius of Antioch (d. 107 CE) exhibits a passionate conviction that the good standing of each local Church is validated by its having its *bishop.

It would be widely agreed that there is in the NT a *diversity of forms of *ministry and belief before they became, under the forces of internal controversy and external hostility, established and *orthodox.

earth Sometimes the created *world (Matt. 6: 10) in contrast to *heaven; or else the firm ground in contrast to the sea (Acts 4: 24); or it may designate the whole inhabited area and its population (Rev. 13: 3). In the OT the same word is sometimes to be translated 'earth', sometimes 'world', and it is not always clear which is more appropriate.

earthquakes Frequently experienced in Palestine (Amos 1: 1) as natural phenomena; but also earthquakes are the product of apocalyptic symbolism, such as the earthquake recorded at the time of the *crucifixion (Matt. 27: 51) and at the *Resurrection of Jesus (Matt. 28: 2), perhaps even that during Paul's imprisonment at *Philippi (Acts 16: 26). In the Marcan *Apocalypse (Mark 13: 8) there is a prophecy of earthquakes, which the evangelist may regard as fulfilled at Jesus' crucifixion. Apocalyptic earthquakes are also a sign of the *End (Rev. 6: 12).

*Legends describing earthquakes are those when Dathan and Abiram are swallowed up (Num. 16: 30–4), and the divine theophany at *Sinai (Exod. 19: 18).

east One of the points of the compass, the direction of sunrise, and in combination with the other three points (Ps. 107: 3) there is an expression of totality.

The east is associated with *wisdom (Job 1: 3) and *astrology (Matt. 2: 1).

Easter The word is used by AV for the annual commemoration of Jesus' *resurrection at Acts 12: 4; modern versions prefer '*Passover'. Easter soon became the chief festival of the early Church, and by the 3rd cent. it was preceded by a night vigil. At dawn those who had been prepared were baptized; all received *communion. There was no observance of *Good Friday as a separate memorial day until the 4th cent. For some time Easter, as the *Christian Passover, was observed on 14 Nisan, the date of the Jewish festival, whether it was a weekday or a Sunday. This continued for some time in *Asia Minor, where those who practised this custom were called Quartodecimans.

Its celebration on the Sunday after the Jewish Passover was not ordered at *Rome until 160 CE. It has been suggested that much of 1 Pet. originally formed part of an Easter homily in Rome when there were candidates for *baptism (1 Pet. 3: 21).

Ebenezer The Israelites were defeated and the *Ark captured by the *Philistines at Ebenezer (1 Sam. 4: 1–11 and 5: 1) but Samuel later named a stone Ebenezer when the area was recovered (1 Sam. 7: 2–12). The word means Stone of Help, and as such has often been absorbed into *Christian culture, especially in Wales.

Ebionites A fringe group of *Christians of Jewish descent. Paul's controversy with Jewish Christians expressed vehemently in Gal. did not spell the end of their way of life and thinking. After the Fall of Jerusalem (70 CE) Christians of similar outlook, Ebionites, were found in small numbers in *Syria continuing to observe the *Sabbath and other Jewish customs. The apocryphal gospel of the Hebrews, with a favourable account of James the Lord's brother (cf. Acts 15: 13–21), may have emanated from this circle. They were unpopular with Jews for being Christians and treated by *Gentile Christians as heretics. However, the Christian Apologist Justin Martyr (about 150 CE) defended their way of life as being in line with that of the primitive Church at Jerusalem. *Irenaeus the Church Father (about 190 CE) deplored their rejection of the tradition of the *virgin birth of Jesus. The name, meaning 'the poor' in Hebrew, conveys reminiscences of Paul's references to the Jerusalem Christians (e.g. Rom. 15: 26) but the notion of Tertullian, around 200 CE, that they derived their name from a certain Ebion is improbable. Epiphanius, bishop of *Salamis, about 380 CE attacks the Ebionites for their heretical *Christology. He quotes their *gospel of the Hebrews which omits Jesus' *birth and regards his *baptism as the moment of divine initiative. 'The Ebionites say that the *Spirit, which is Christ, came into him and clothed him who was called Jesus.' They are said to have used only the gospel of Matthew—presumably without its first two chapters.

Ebla Archaeological site in *Syria excavated in 1974–5. Many thousands of *cuneiform texts were discovered written in a language which resembles *Hebrew. They give important information about the Near East in the years 2400–2250 BCE, showing that there was a high degree of urban civilization in *Canaan before *Abraham migrated from *Haran (Gen. 12: 4). But the supposed allusions to people and places in the Hebrew Bible have not withstood closer scrutiny.

Ecclesiastes Much about this OT book is mysterious—who wrote it and when; what its message is. Not once is it quoted in the NT; and in the 1st cent. CE the rabbinic school of *Shammai questioned whether it should be regarded as part of holy *scripture, though the more liberal *Hillel did accept it. (One practical effect of this was that the liberal school felt obliged to wash after touching it, since touching a book that belongs to the *canon renders the hands unclean.)

In Hebrew the title of the book is Qoheleth, which has come to be translated 'the Preacher'; the late form of the book's Hebrew indicates that King *Solomon could not have been the author (cf. Eccles. 1: 12), and its links with Hebrew wisdom literature suggest a date for it of about 200 BCE. Is the pursuit of pleasure or riches or work or womanizing the way of *wisdom (Eccles. 2: 3 ff.)? Not so: all end in death. Time and again the author dismisses what passes for human existence as fleeting, futile, and 'vanity'; and there are strict limits to understanding any guiding purposes. Life's ambiguities had best be accepted and that which is satisfying should be enjoyed while it lasts (Eccles. 8: 14–15) for good things will surely come to an end (Eccles. 12: 1–2). It is not true that God rewards the just; it is all a matter of time and chance (Eccles. 9: 11–12): but it is advisable to keep on good terms with God, and not hurt oneself by useless resentment of what inevitably happens (Eccles. 3: 1–8).

Modern scholarship felt that the combination of scepticism and conventional moral advice was evidence that the book as we have it was the work of more than one author. Possibly a basically sceptical book was worked over by an editor to make it acceptable to more orthodox readers. An alternative hypothesis of the composition of Ecclesiastes was put forward by *Form Critics: a miscellaneous collection of independent proverbs which had been in circulation orally was collected into an anthology without regard to their theological compatibility. These two views about the book are to some extent brought together by *Redaction Critics who suggest that advice to enjoy life while the opportunity lasts (as in 11: 9 ff.) has been put by a redactor into a religious context: we are to enjoy what God gives and to remember that he is the judge at the hour of death (12: 1a). The original pessimistic scepticism of the Preacher is no longer the last word of Ecclesiastes. Certainly that is how generations of readers who have approached it with Christian presuppositions have seen it.

Ecclesiasticus Although about two-thirds of this book in the original *Hebrew have been discovered (e.g. among the *Dead Sea scrolls) it has long been known in the *Greek translation made in *Egypt soon after 132 BCE by the author's grandson, who was anxious in case traditional Jewish learning, *piety, and discipline might be lost amongst the large Jewish immigrant population. The book was first composed in Jerusalem about 190 BCE by *Jesus ben Sirach, a great traveller (34: 12). It is often referred to simply as 'Sirach'. 'Ecclesiasticus' is Latin and is a late title given to it possibly because it was often read in Christian churches, where it formed part of the *LXX; canonical status in the Hebrew Bible was never attained though it was quite widely read.

Ecclesiasticus contains a variety of proverbs and aphorisms and advice for young men on all manner of subjects in daily life with a bias in favour of intellectual as against manual work. It shares the contemporary attitude to women of an androcentric society (25: 16–26). There are also *hymns and *doctrine, as in 33: 7–15, about the coexistence of both *good and *evil, and a call for personal *repentance (17: 25–32). *Wisdom is extolled as in the books of Proverbs and the Wisdom of Solomon, which thus puts Ecclesiasticus in the same literary wisdom category; but it is a wisdom located in the *Torah (Law) and found in the Temple at Jerusalem (Ecclus. 24: 10). Whereas the book of Wisdom has a strong belief in immortality (3: 4), it is clear that Ben Sirach has no hope except for a dismal survival in *sheol beyond the presence of God (17: 27–8). The Greek translator tried to insert a more hopeful expectation (19: 19, NRSV marg.).

ecstasy Ecstasy is not the same experience as a *vision or a *dream; it is an awareness of being taken up outside oneself, and there are only a few such mystical moments recorded in the *Bible. One is by Paul in 2 Cor. 12: 2, when he mentions 'a person in Christ' (presumably himself) being caught up to the third *heaven; but he will not boast of such a *grace; indeed he was given a 'thorn in the flesh' (2 Cor. 12: 7) to forestall any such pride.

Later Christian writers regarded ecstasy as a kind of anticipation on earth of heavenly bliss.

Eden An area in southern *Mesopotamia where *Adam and *Eve were placed according to the *myth of Gen. 2. It contained two fruit trees which were given special significance—one bestowing *everlasting life (Gen. 3: 22) and the other an awareness of *good and *evil. The *garden of Eden came to be thought of as paradise, before the *Fall; and after the Exile in *Babylon the land of the Return was pictured romantically as a new Eden (Isa. 51: 3).

Edomites Inhabitants of land at the SE end of the *Dead Sea. The Edomites lived adjacent to the Israelites, with whom there were frequent conflicts, though their close relationship was indicated by the *legend of their ancestors going back to *Esau, the brother of *Jacob (Gen. 25: 19–26; 36: 1, 8). They finally defeated *Judah in the time of *Ahaz (742–725 BCE, 2 Chr. 28: 17) but were themselves conquered both by the *Assyrians and the *Babylonians. The Edomites had the capacity to survive and even to flourish in commercial enterprise. Some, from whom Herod the Great was descended, emigrated westwards to *Idumaea. The Jews could not forget the hostile actions of Edom after the destruction of Jerusalem in 586 BCE (Obad. 8–14).

education Little is known about methods of education in ancient Israel except that religious, moral, and practical training was in the hands of parents (Deut. 4: 9). So far as is known only boys received education—as is implied by the repeated advice on masculine behaviour in Prov. 1–8, which was probably a textbook against the snares of women. *Punishments were severe. At a later period the teaching *rabbis did not accept fees, and they therefore earned a living by some trade or craft, as did Paul as a leather-worker (1 Cor. 9: 3 ff.) The content of the teaching was the religious traditions of Israel, and much of these had to be learnt by heart (Deut. 11: 19; Isa. 28: 10).

In the NT Jesus educated the crowds (Mark 4: 1–2) and more particularly his usually hard of understanding *disciples. They in turn were to educate their converts (Matt. 28: 20). There is no evidence that the elaborate teaching given to Paul by the famous *Gamaliel (Acts 22: 3) was imitated in the Church. For example, when *Philip encountered the Ethiopian *eunuch, they had a conversation about Isaiah, after which the eunuch proposed *baptism in the adjacent stream, which Philip administered without further ado. However, by the time of the *Pastoral Epistles there existed schemes of instruction (1 Tim. 1: 3; 6: 3) and this was greatly extended with the development of the catechumenate with an elaborate preparation during Lent of candidates for baptism at the ensuing festival of *Easter. Details of the instruction are given by Cyril of Jerusalem (*c.*350 CE).

In the 1st cent. BCE an important element of education for those to enter public life was rhetoric: Quintilian (*c.*40–95 CE) held that the very goal of education was to become a good orator. This regard for rhetoric has influenced parts of the NT, e.g. Paul's letter to the Galatians, which could be analysed as: Paul is the defendant, the Jewish Christians are the accusers; while the Galatians themselves are the jurors. The letter can be shown to incorporate the recognized parts of a contemporary speech. The rhetoric of the gospels (Jesus' preaching) does not conform to such a classical analysis, but is based on Jewish methods. Their intention is to convince the hearers by divine authority rather than by rational argumentation.

The Jewish system of education was based

on the *synagogue (Luke 2: 46). There were 480 synagogues in Jerusalem, with schools attached, according to the Palestinian *Talmud, and the *Qumran community was highly literate; but even before the Exile there was expertise in *writing and reading (Deut. 6: 9; 31: 12–13; 2 Kgs. 22: 3). *Isaiah and *Jeremiah had scribes to record their messages. Much teaching was by rote (Isa. 28: 10), but in the 2nd cent. BCE there were knowledgeable diplomats (Ecclus. [= Sir.] 38: 24–39: 11).

The Babylonian *Talmud claims that schools and teachers existed throughout Israel from the 1st cent. BCE. Although in the NT period the majority of Christians would have been illiterate, there existed opportunities to learn to read and write and before long Christians were to be found in the literary and philosophical schools.

Egypt Egypt and the people of Israel have tangled together throughout the biblical, medieval, and modern periods. As one of the great civilizations of the ancient world, relatively free from foreign invasions until the arrival of the Assyrians (7th cent. BCE), and enjoying a good climate and a prosperous economy, Egypt was at various times a haven for Israelite refugees (1 Kgs. 11: 17; 12: 3), a dominant military power, and a cultural influence on Israel—though never a directly religious influence, even if Moses (an Egyptian name) was instructed in Egyptian *wisdom (Acts 7: 22).

*Abraham went to Egypt for survival. *Joseph was forcibly taken there but rose to eminence. The *Exodus from Egypt was ever afterwards branded on the national memory. Settled in Palestine, the countries of Israel and *Judah were always liable to be squeezed by the warring powers in the NE and Egypt in the SW; captives seized by Egyptian raiding parties during the Israelite monarchy remained there; *Ishmael and his fellow conspirators fled there, dragging *Jeremiah with them after the fall of Jerusalem to the Babylonians. *Joseph and *Mary are recorded (Matt. 2: 13) as taking the infant Jesus to Egypt to escape the wrath of *Herod (*Old Testament in the New Testament).

A military colony of Jews was established in the 5th cent. BCE at *Elephantine on the Egyptian border, and for about two centuries until 73 CE there was a Jewish temple at Leontopolis. There was a considerable Jewish population in *Alexandria; and the historian Eusebius records the tradition that the Church was founded there by Mark the evangelist. Little of the Church in Egypt is known before the great persecution of the 3rd cent., but a *Christian community eventually became the predominant religious group in the country until the Islamic conquest in the 7th cent. CE. Christianity still survives as the small Coptic Church.

eisegesis A comparatively modern term to describe, disapprovingly, a piece of scholarship which appears to find in a given text a significance alien to its context. This might be to provide biblical support for a doctrinal position already held. The term was coined (from the Greek *eis*, in, and *egeisthai*, to guide) as the opposite of *exegesis (Greek *ek*), which means an elucidation *of*.

An example of eisegesis is when theologians holding a theory about *Christian initiation argued for supporting evidence in the gospels. It was suggested that *baptism both effects *forgiveness of *sins and imparts the *gift of the *Spirit, but that this rite became split into two parts, i.e. baptism in water soon after birth and the laying on of hands (episcopal confirmation) at a later date but equally essential to the wholeness of the rite. It was suggested that the two parts of the rite were anticipated at the baptism of Jesus: first the baptism in the river (Mark 1: 9), followed as a second constituent of the rite by the descent of the Spirit (Mark 1: 10).

Elam A country east of the lower *Tigris. Elam is first mentioned in the OT in Gen. 14 where the king is one of a coalition, but it comes frequently into the history of the 8th and 7th cents. BCE when its capital Susa was destroyed by the *Assyrians (646 BCE, Ezek. 32: 24) and later occupied by the Persians (Jer. 49: 34 ff.). The Elamites did their best to prevent the rebuilding of the Temple (Ezra 4: 9).

Some Jewish Elamites were in Jerusalem for the feast of *Pentecost (Acts 2: 9).

elders In Israelite society, heads of families; collectively they form the leadership of the tribe (Josh. 9: 11), or, under the monarchy, of the ruling class (2 Sam. 3: 17). Eldership was never the title of an office to which an individual might be appointed or elected. A son might succeed his father as the elder of a clan. Elders were a body of senior persons with influence varying according to their personal qualities and the importance of the social group to which they belonged, and in succeeding centuries it continued to be the case that 'the elders' were not the holders of any office, though they were accorded respect, and it was from them that individual leaders emerged. Neither in the synagogues nor at *Qumran were elders members of an order or office.

In the gospels elders (Greek, *presbuteroi*) are mentioned five times by Mark. They are associated with high priests and scribes (e.g. 8: 31), and give the impression of being a collective body exercising influence together. In the Acts elders are a constituent part of the Jerusalem authority, and again they appear only as a group, never as holders of an office of eldership or independently of the Jewish rulers and scribes (e.g. Acts 4: 5), and they were not necessarily 'elderly' in the sense of aged.

Surprisingly, perhaps, elders are never mentioned in the early letters of Paul, where the local leaders of Churches are addressed as 'brothers' (NRSV marg.) as distinct from Church members in general (Phil. 4: 21 f.). So also in Acts (15: 40 and 18: 27) it is 'the brothers', the local leaders who are not called 'elders', who commend members of the Church. The apostle himself and his fellow workers were all members of a team working together.

However, as the local Churches expanded, and the apostle moved far away, it became essential for someone in each house-church to exercise leadership and to preside over the assemblies. He was possibly chosen because of ability in teaching. Household churches existed in Jerusalem itself (Acts 2: 46; 4: 31) and if they were looking for a title for their local leaders, they might have found it in the *mebaqqer* of Qumran, *episkopos* in Greek. When these persons met together for corporate consultation, this would be referred to as the elders, *presbuteroi*, of Jerusalem. The situation envisaged by the Pastoral Epistles is similar. Over each local house-church, an overseer (*episkopos*) presided, and when they met as a group they were the elders (*presbuteroi*) of the city's Christians. The writer is urging his readers now to accept the fact that there must be one overseer for all the groups of the place, very much as James was in Jerusalem. The kind of *episkopos* commended would be Onesiphorus (2 Tim. 1: 16–18). Just as there was a local leader for the groups meeting in houses, so there now had to be a leader for the whole Church of the city corporately. Local people were encouraged to serve in this capacity (1 Tim. 3: 1), coming forward from among the *presbuteroi*. By the time of *Ignatius (*c.*107 CE) the single *episkopos* (bishop) is established in Antioch as a safeguard for the unity of the Church. He presides at the Eucharist, together with elders, whose duties are not described, and deacons. Possibly these elders were the *episkopoi* of a number of neighbouring churches, with local prestige which Ignatius feels must now be surrendered in exchange for the ceremonial honour of sitting round the bishop of the city Church.

When the growth of the Church's numbers exceeded the capacity of a building to hold the whole community of a city under its bishops, it became necessary for him to delegate powers to his *presbuteroi* (elders) to preside over local Eucharists—which was to establish the permanent pattern of an ordained ministry of bishops, presbyters, and deacons.

elect lady The addressee of 2 John (AV, NRSV), the 'Lady chosen by God' (REB), 'the chosen one' (NJB), meaning probably the local *Christian *community somewhere in the province of Asia.

election A key idea in the *Bible is that God chose ('elected') people to serve his purposes for the *world. First he chose

*Abraham (Gen. 12: 2) and the nation that sprang from him rather than any other nations, even though they were mightier or more numerous than Israel (Deut. 7: 7). It was a matter of God's free, almost arbitrary, decision—but repeatedly Israel failed, by *idolatry or disobedience, and Isaiah prophesied that only a remnant of the nation would survive to reap the rewards of faithfulness when God's day of *judgement came (Isa. 37: 31 ff.). So the original assumption that the whole nation was elected was gradually scaled down.

The NT represents the climax of this process, in claiming that finally God chose a single individual alone to fulfil his purposes for mankind—Jesus. When that had been accomplished, then the company of the elect could begin to expand again, and 'a chosen race, God's own people' (1 Pet. 2: 9; the Christian Church) replaced the old as the object of God's election and the inheritors of his promises (Rom. 4: 9 ff.). In effect, the Church is the New Israel, though that term is not used in the NT.

As with old Israel, the Church's privileges are complemented by responsibilities; they are to proclaim God's mighty acts (1 Pet. 2: 9) and those who are chosen will respond to the preaching (1 Cor. 1: 24). Paul insists that salvation is entirely the work of God (1 Cor. 1: 30 f.), giving the believers a sense that they need not fear the day of judgement (Rom. 8: 33 ff.), and this conviction is the incentive towards a holy life (Col. 3: 12 f.). The *Gentile response to God's election will be so substantial that, Paul maintains, the Jews will also be won over to *faith in Christ (Rom. 11: 2, 28).

Although Paul praised God for having chosen him and fellow believers in the *Body of Christ, he never suggests that those not so chosen are doomed or destined for *perdition.

elemental spirits The Greek word (*stoicheia*), so translated by NRSV and REB, has puzzled commentators. It occurs at Col. 2: 8 and 20; AV renders 'rudiments of the world', and NJB 'principles of this world'. The same Greek word is used by Paul in Gal. 4: 3 and 9. One view is that the word signifies the basic elements that comprise a whole, such as the letters of the *alphabet; so that Paul's meaning is elementary constituents of a religious system. But more probably the '*elemental spirits' are the '*angels' of Gal. 3: 19. The Jewish belief, shared by Paul, was that the *Law had been given to Israel not directly by God but through the agency of angelic powers.

Hence Christ's death, which destroyed the authority of the Law as the mediator between God and humanity, has at the same time put an end to the dominance of these angelic powers. Christ alone has supreme authority.

Elephantine The site of a Jewish military establishment in the 6th and 5th centuries BCE on the southern border of *Egypt, and subsequently of many civilian Jews who had their own temple in honour of Yahweh, though he was apparently worshipped along with a goddess (Anath-yahu). Aramaic papyri discovered early in the 20th cent. revealed interesting details about Jewish life in the age of *Ezra and *Nehemiah.

Eli A priest in *Shiloh who was also a Judge (1 Sam. 4). He looked after the *Ark of the Covenant and died of a heart attack when he heard that it had been captured by the *Philistines.

Elijah The most notable of the ecstatic prophets of the 9th cent. BCE, coming in time between the orgiastic bands of *prophets who encountered *Saul (1 Sam. 10: 5 ff.) and the classical prophets of the 8th cent. Although the narratives relating Elijah's exploits contain theological ideas (e.g. an idea of a remnant similar to that found in Isaiah, 1 Kgs 19: 18) of a later age or even of post-exilic provisions of the *Law (e.g. Num. 36: 7–9), it is probable that there is a basis of historical fact in the recorded contests which Elijah pursued on behalf of Israel's traditions. Against the introduction of *Baal worship by the Tyrian princess *Jezebel, married to King *Ahab, Elijah triumphantly vindicated his God when fire came down to consume his *sacrifice on *Mount Carmel (1 Kgs. 18: 38). He was told to anoint *Jehu to take over the kingship (1 Kgs. 19: 17), which *Elisha his successor carried out (2 Kgs. 9: 6).

The challenge to Baal was important in that his status as a fertility-god was diminished. When Elijah confronted Ahab because of his appropriation of *Naboth's vineyard (1 Kgs. 21) he was appealing to an ancient Israelite concept of justice which even monarchs were not to override. The two stories show Elijah as champion of Yahweh the only God and the judge who defends the defenceless and exacts vengeance.

Elijah is mentioned by *Malachi (4: 5) in conjunction with the observance of the Law. It became a part of Jewish tradition that Elijah would come before the *day of the Lord (Ecclus. [= Sir.] 48: 10) and in observances of the Passover a place is regularly set for Elijah. He therefore is named in the NT in this role (Matt. 17: 10–13); *John the Baptist is the new Elijah. This identification is, however, denied in the gospel of John (1: 20–1) which is determined to minimize the importance of the Baptist in relation to Jesus. (No baptism of Jesus is recorded in the fourth gospel.) Elijah, representing the prophets, is present along with Moses, giver of the Law, at the *Transfiguration of Jesus (Mark 9: 4). He is mentioned by Paul (Rom. 11: 2) and James (5: 17).

Elisha The designated successor of Elijah (1 Kgs. 19: 19 ff.), who lived in the second half of the 9th cent. BCE. He seemed more at home in an urban environment than Elijah, even though his father was a farmer. He preferred city life (2 Kgs. 6: 13) and had a residence in the city of *Samaria (2 Kgs. 6: 32). He visited religious centres like *Bethel (2 Kgs. 2: 23) and *Gilgal (2 Kgs. 2: 1) but was also attached to groups of ecstatic prophets (2 Kgs. 2: 3–15) who lived in communities.

In the stories handed down in the oral tradition about Elisha and eventually written down, perhaps about 700 BCE, there is a strong element of sympathetic *magic, or the extraordinary combined with prayer to Yahweh, as when the prophet is said to have divided the waters of the *Jordan after receiving the mantle of Elijah (2 Kgs. 2: 13 f.) and when he struck his enemies with blindness (2 Kgs. 6: 18 ff.). There is a similar kind of magic in the story of Elisha recovering a borrowed axe-head from the water and making it float (2 Kgs. 6: 1–7). But it was a demonstration of the power of the God of Israel when he healed *Naaman's leprosy during a temporary lull in the perennial conflicts between Israel and Syria (2 Kgs. 5: 1–19). That Elisha was a prophet of an era before the great classical prophets of the 8th cent. is indicated by the use of music to induce his trance (2 Kgs. 3: 15; cf. 1 Sam. 10: 5–7); and the mockery of his baldness by a group of jeering boys suggests that he had assumed the tonsure, a ritual shaving of the head commonly practised by holy men (2 Kgs. 2: 23). According to the narrative the boys were promptly devoured by bears as a punishment; a coincidence that is thus given a theological, and by no means agreeable, interpretation.

On the political level, Elisha aided the Israelites against Moabites (2 Kgs. 3: 21–7) and Syrians (2 Kgs. 6: 13–7: 23). His motive was to persuade the kings of the absolute sovereignty of the God of Israel, who would tolerate no apostasy.

There are similarities between the story (2 Kgs. 4: 42–4) of Elisha's *miracle in feeding 100 people with twenty loaves, with some still left over, and the account in Mark (6: 30–44) of Jesus' multiplication of loaves.

Elizabeth Wife of the priest *Zechariah (Luke 1: 5–80). After years of infertility, she at last gave birth to *John the Baptist. The story is heavily coloured by reminiscences of the story of the barrenness of *Hannah (1 Sam. 1: 1–20) and the birth of *Samuel. Whatever historical information *Luke the evangelist may possibly have been given has been shaped by recollection of this OT narrative.

At the *Annunciation Mary was told about Elizabeth's pregnancy, and she accordingly paid her a visit. She is greeted by her 'kinswoman' ('relative', NRSV, Luke 1: 36) as 'mother of my Lord', which makes it plain that John is regarded as subordinate to Jesus. But though John and Jesus are therefore cousins, according to Luke, in the fourth gospel John maintains that he did not know Jesus (John 1: 33)—another oblique declaration of his subordination to the Messiah. It was important in the Church to emphasize this in view of the continuing existence of disciples of the Baptist into the Christian era (Acts 19: 3).

Elohim The word is plural in form, literally denoting 'gods' (as in Exod. 20: 3), but normally used for the God of Israel as a plural of majesty, as in other languages. The singular form *El* is also found, especially in the book of Job. The use of *Elohim* in one source of the *Pentateuch from the call of *Abraham (Gen. 22: 1–19) to the revealing of God's name Yahweh (Exod. 3: 13–15) has led to its being called the *E source. It is thought to derive from the northern part of the country in the 9th to 8th cent. BCE period.

Eloi Part of the cry of desolation uttered by Jesus from the *cross (Mark 15: 34), a mixture of *Hebrew and *Aramaic, which are the first words of Ps. 22. Although some commentators point out that this psalm ends in a note of hope and that Jesus would have had in mind the whole psalm, most scholars hold that the words represent Jesus' total identification with humanity and its despair. Jesus feels abandoned even by God his Father (cf. 2 Cor. 5: 21).

El Shaddai The name for God in Ezek. 10: 5 and (without *El*) frequently in Job; but otherwise only in the *P source of the *Pentateuch. The traditional translation is 'God Almighty'.

Elymas Also called Bar-Jesus; a Jewish sorcerer at *Paphos (Acts 13: 6–12). When Elymas tried to dissuade the proconsul of Cyprus from embracing Christianity, Paul caused him to be struck with temporary blindness. The proconsul was 'astonished' (NRSV).

embroidery Needlework skills were used to decorate the sacred vestments (Exod. 39: 3) and the screen of the *tabernacle (Exod. 26: 36) with blue, purple, and scarlet threads. Probably the garments coveted by *Achan (Josh. 7: 20 f.) were trimmed with fine appliquéd needlework.

emerald The third of the precious stones in the high-priest's *breastpiece (Exod. 28: 17), probably of a violet colour, and the fourth in the foundations of the walls of the New Jerusalem (Rev. 21: 19).

Emmanuel The Greek transliteration of the Hebrew Immanuel—'God with us'—and used by Matt. 1: 23 when quoting Isa. 7: 14. The name was a sign given by the prophet to King *Ahaz in 734 BCE and was part of his warning not to put his trust in *Assyria as a defence against the combined forces of *Syria and Israel who were threatening *Judah. They would be destroyed before the child to be called Immanuel shortly to be born to a young woman (the Hebrew word in Isa. 7: 14 does not necessarily denote a virgin) would be old enough to discern the difference between *good and *evil. The sign being offered by *Isaiah was thus of an assurance of the protecting power of God at a dangerous moment.

Emmaus A village possibly 11 km. (about 7 miles) from Jerusalem (Luke 24: 13); but some MSS put it at 32 km. (about 20 miles). The location is therefore in dispute. The risen Jesus appeared to *Cleopas and another disciple as the two were on their way to the village. Their false interpretation of Jesus' Messiahship—of the hopes, now shattered, of political victory—is corrected by Jesus' exposition of OT scripture, showing his death to be a divine necessity (24: 26) by which *Messiah could enter into his glory. On arrival at their destination the disciples are still unaware of the identity of their companion, but when (using language similar to that at the Feeding of the Five Thousand) he breaks and distributes bread, he is at once recognized, and immediately disappears. It is a story of true *disciples coming to know the risen Lord in word and *sacrament, and that is its value for the reader. As history, the double journey and a series of incidents are too many to be accommodated in a single day.

emperor worship The cult of Roman emperors, living and dead, became the State religion throughout the empire, though it originated as a simple act of thankfulness for the peace and stability brought by Rome. Temples were erected in honour of Julius Caesar soon after his death (44 BCE) and to *Augustus in his lifetime, e.g. at *Pergamum. This explains the reference to 'Satan's throne' (Rev. 2: 13). In this city, as elsewhere, the cult developed because local people wanted it not in order to flatter the establishment but out of genuine gratitude for the benefits brought by *Rome, and it was not felt to be a substitute for existing religions. Nevertheless, as the feelings of gratitude faded, the imperial cult became more and more a test of loyalty to the regime. The consequence was that refusal to perform the outward rituals was bound to incur penalties. Both Jews and *Christians were conscientiously unable to burn incense to any human being: Jews, after some initial persecution, got exemption from *Claudius; but Christians suffered when the Church's numbers expanded sufficiently to attract the State's hostile attention (1 Pet. 4: 16). Failure to give divine honours to the emperor or 'to swear by the genius of Caesar' was not the only ground for persecution; but the anti-Christian writer Celsus (about 178 CE) warned Christians of the perils of their lack of civic sense and of their disloyalty to an empire from which they derived many material benefits.

End The Jewish belief in the first centuries BCE and CE was not that this world would come to an end but that the existing world order would be transformed. Mythical language (Isa. 11: 7–9) and cosmic imagery (Isa. 24: 23) were used to prophesy vast political upheavals when the *Gentile oppressors could be overthrown and the righteous nation vindicated. At the conclusion of this present age and the beginning of the new there would be a general resurrection for all Israel; the claim to be the covenant people of the creator God would be exhibited.

The NT suggests that the *life, *death, and *resurrection of Jesus constitutes the beginning of the transformation (Mark 9: 1), although the final culmination is delayed (Matt. 24: 14) and the date uncertain (Mark 13: 32).

Endor A Canaanite city, of uncertain location, where there lived a 'witch' or medium who had survived *Saul's determination to banish witches from the land. However, in desperate need, Saul visited Endor in disguise and cajoled the woman into summoning the prophet *Samuel from the abode of the dead, in order that he could be consulted about the serious threat from the *Philistines (1 Sam. 28: 3–25). Samuel offered no comfort and rebuked Saul for disturbing him; the Philistines were victorious at *Gilboa, and Saul committed suicide (1 Sam. 31: 4).

Endzeit German word (= 'final epoch') used by some commentators as a convenient single term for the whole course of events envisaged in parts of the NT at the end of the world.

Enoch, books of Enoch was a legendary hero; he lived to be 365 (Gen. 5: 23), walked with God, and bypassing death was taken straight to *heaven (verse 24). Such a mysterious figure was an apt character to be the pseudonymous author of sundry *apocalyptic writings, in which he is a preacher of repentance, a *prophet, and one who receives supernatural knowledge of the mysteries of heaven and earth.

The book of 1 Enoch (or the Ethiopic book of Enoch) was originally written in Hebrew during the two centuries BCE. It predicts the punishment of Israel's enemies, and the general resurrection of Israelites. Chs. 37–71, which form a separate section of the book and are absent from copies of the Enoch literature among the *Dead Sea scrolls, are usually attributed to a period later than the gospels. In these chapters the term '*Son of Man' is used for Enoch.

The book of 2 Enoch (or the Secrets of Enoch) has been known in its Slavonic form in the West only since the 19th cent. Originally written in Greek early in the Christian era, it gives a complicated version of the age to come after the *judgement.

The book of 3 Enoch was written in Hebrew and develops the earlier Enochic literature under *Gnostic influence.

envy In the OT there are several examples of the evil operation of envy: *Cain murdered his brother *Abel (Gen. 4: 8), *Joseph was envied and deported by his brothers (Gen. 37: 27), and *Saul resented people's praises heaped on *David (1 Sam. 18: 8). In the NT parable, the welcome given to the Prodigal Son on his return home evoked envy rather than love from the elder brother (Luke 15: 28). 'Jealousy' is sometimes the translation adopted by NRSV, NJB (e.g. at Matt. 27: 18 for the evil dispositions which crucified Jesus).

Epaphras A companion of Paul during his imprisonment, and evidently a leading member of the Church at *Colossae (Col. 4: 12; Philem. 23).

Ephesians, epistle to the In the NT, the fifth of the letters of Paul. But it is more of a doctrinal treatise, dressed up as a letter by having an opening and formal greetings at the end. It was known to *Ignatius (d. 107) and possibly even to *Clement of Rome in 96 CE when he wrote to *Corinth. Although explicitly regarded from the end of the 2nd cent. as a work of the apostle Paul, more modern scholarship hesitates to authenticate the Pauline claim. This is due to the relationship between Eph. and the (Pauline) epistle to the Colossians. Both the reference to the writer's imprisonment (Eph. 6: 20; cf. Col. 4: 18) and to the letter being conveyed by *Tychicus (Eph. 6: 21; cf. Col. 4: 7) argue composition at about the same time, in quick succession. There are remarkable similarities; one-third of Ephesians' 155 verses are parallel to Col. (See e.g. the 'household codes' of Eph. 5: 21–6: 7 and Col. 3: 18–4: 1.) On the other hand, there are striking differences between the two documents in style and content; the word *ekklesia*, Church, refers to the universal Church in Eph., but in Col. both to the universal Church and to the local congregation in Colossae. Key terms are used in different senses: in Eph. 1: 9–10 '*mystery' (which means, not a mysterious secret, but the revelation of a secret), is that the purpose of God demands Gentiles as well as Jews to be equally members of the Body, whereas in Col. the 'mystery' is Christ himself. It is hard to believe that the same author could use so much from a letter he had recently written and yet introduce such differences.

The likelihood therefore is that the author is a disciple of Paul who had access to at least some of Paul's previous epistles and wrote this document towards the end of the century to a group of Churches. (Several early MSS omit the words 'in Ephesus' at 1: 1 and *Marcion referred to it as the epistle to Laodicea. Cf. Col. 4: 16 and Rev. 3: 14.) In accordance with an accepted custom it was dispatched under a pseudonym, a legitimate literary device in those days by which the writer acknowledged his dependence on his master.

The epistle does not deal with the doctrine of *Justification, which so dominates Rom. and Gal., but is an exposition of Christ's supremacy over all cosmic forces (1: 21); he embodies God's plan for the *fullness (Greek, *pleroma*) of time (1: 10) which is the goal of *creation. It is he who draws the whole human race into a unity, Jews and Gentiles alike; and the Church, which is the fullness (*pleroma*) of Christ (1: 23)—completely filled by him—is the means by which God's plan is to be accomplished. Christ's universal work of reconciliation must begin in the Church itself, when unity is achieved by love and mutual service. From the *Body of Christ (5: 30) unity is to spread outwards.

The high doctrine of the Church in the epistle and its references to *ministry (4: 11) and to the headship of Christ in (or over?) the Church (5: 23) have given the document an important place in modern ecumenical discussions.

Ephesus In the 1st cent. CE a port at the mouth of the River Cayster on the west coast of Asia Minor. Today it is Seljuk in Turkey; because of silting, it lies about 5 kms (3 miles) from the sea. In the course of six centuries it passed through a succession of Greek, Persian, and Roman regimes and in the era of the NT (where it is mentioned some twenty times) the site was the fourth largest in the empire—prosperous, with splendid streets lined with colonnades, and a temple dedicated to *Artemis (Diana) which was one of the seven wonders of the ancient world.

Paul took advantage of the constant traffic across to *Corinth. He travelled to and fro (2 Cor. 12: 14; 13: 1) and the Corinthian correspondence was penned in Ephesus. Not that all was calm and uninterrupted; once 'he fought with beasts' at Ephesus (1 Cor. 15: 32), either literally or metaphorically; and a mass meeting of protest was staged in the theatre when *Demetrius, convener of the silversmiths' union, urged that the propagation of Paul's *monotheism was putting in jeopardy their lucrative trade in statues of the great goddess. On this occasion the *asiarchs (appointed to promote emperor worship) were minded to sympathize with Paul for breaking the Artemis monopoly (Acts 19: 31). Nevertheless, Paul discreetly said farewell and sailed for Greece, leaving behind an established Church—*Aquila and Priscilla among the members—with *elders (Acts 20: 17). The Church was destined for a long history: the gospel of John may have been written there by 'John the Elder' (of 2 and 3 John?): even the seer of Revelation (2: 1–7) gives it a favourable review, and in 431 CE the Church Council which condemned the heresy of *Nestorius was held there.

ephod A garment worn by priests in ancient Israel, apparently with numinous quality, for it aided access to the divine (Judg. 17: 5). *David ordered the priest *Abiathar to bring his ephod (1 Sam. 23: 9). It would appear to be the elaborate tunic described in Exod. 28: 39 and used not only in religious ritual but also for divination. The latter use would correspond to the occasions when decorative garments were put on the statues of Egyptian deities. Some references (e.g. 1 Sam. 21: 9) seem more naturally to refer to a piece of furniture than to a garment.

ephphatha Aramaic command, 'Be opened', uttered by Jesus (Mark 7: 34), after he had put his fingers into the ears, and spat and touched the tongue, of a deaf and dumb man, and healed him; a method recorded of other charismatic healers in Hellenistic society, and of the emperor *Vespasian.

Ephraim One of the twelve tribes of Israel which traced its ancestry to the younger son of *Joseph and which occupied an area in the north of the kingdom and west of the *Jordan, including the cities of *Bethel and *Shechem. By the 8th cent. it was regarded by Isaiah and *Hosea as the whole of the northern kingdom. Isaiah advised the king of Judah not to fear Ephraim's alliance with Syria (Isa. 7: 1–17).

Epicureans A philosophical school founded by Epicurus (342–270 BCE), who taught in a garden at *Athens. He propounded an atomic theory of the universe; the atoms which comprise human beings are scattered at *death, so there is no existence after death, and nothing to be feared. In his *Ethics* Epicurus taught that the absence of pain is the greatest good; it consists primarily in a repose of mind. Virtue tends to this repose; therefore virtue is essential for happiness.

Not surprisingly, Paul made little headway when he confronted Epicureans at Athens (Acts 17: 32) in 50 CE. The modern use of Epicureanism to mean self-indulgence does not do justice to the subtlety of the ancient philosophy.

epigraphy The study of inscriptions on ancient buildings and statues, many of which are relevant to the NT. After discovery, they are deciphered and then related to the history of the period. For example, that officials at *Thessalonica were called 'politarchs' (Acts 17: 6–8) is confirmed by an inscription at the western side of the city, and the date of *Gallio's proconsulship at *Corinth is established as 51–2 CE by an inscription found in 1905 at Delphi.

More generally, inscriptions from Palestine in Greek testify to the widespread use of the language and to the influence of Hellenistic culture—previously it had been supposed there was much resistance to it. Light has also been thrown on social life and customs which parallel descriptions in the NT of the *Christian communities, as also on those at *Qumran as revealed in the Manual of Discipline.

epilepsy The account (Mark 9: 14–29) of Jesus' healing an epileptic boy accurately described an attack and its aftermath. The cause of the illness was held to be possession by a *demon, though Matt. (17: 15) uses a Greek verb (*seleniazetai*) which means 'moonstruck'.

Epiphany The word means a manifestation or 'revelation' and is similar to 'theophany' or revelation of the divine. There are several instances of this in the OT, notably God's revelation to Moses at the Burning Bush (Exod. 3: 2). For Christians the supreme manifestation of God was in the *incarnation and two *feasts have commemorated it in the calendars of Western and Eastern Christendom.

From the middle of the 5th cent. CE the Church at *Rome commemorated the revelation of Christ to the *Gentiles as signified by the visit to the infant Jesus of the *Magi (Matt. 2: 11) on 6 January. And from Rome this observance spread in the West, with *Christmas Day on 25 December for the *birth of Jesus, established not later than 336.

It was not always so. From the 3rd cent. in the East the *Epiphany on 6 January had commemorated not only the birth, including the visit of the Magi, but also the *baptism of Jesus and even his first *miracle at *Cana (John 2: 1–11). By the end of the 4th cent. Epiphany concentrated on the baptism of Jesus, while 25 December was borrowed from the West for Jesus' birth.

epistles Traditionally, as in AV, twenty-one of the NT documents have been entitled 'epistles', though modern translations prefer 'letters'. The sense in which they are letters has been much discussed. For generations, they were regarded primarily as doctrinal and ethical treatises, setting out permanent guidelines for the Church down the ages, to which were added opening salutations and closing greetings; though of course it was recognized that they were sent to particular Churches or groups of Churches and read out to the assemblies.

In the 20th cent. the tendency has been to emphasize their status as typical letters not unlike others written in the Graeco-Roman world. Some of the Greek words are those found among the papyri discovered in *Egypt and therefore with meanings of the 1st cent. rather than of classical Greek of 5th-cent. BCE Athens. The papyri were for the most part trivial everyday letters and business communications but they have indicated that the epistles of Paul were genuine letters, though others, such as that to the Hebrews, were of a literary rather than 'occasional' character. However, even the 'occasional' letters of Paul have been shown to conform to some extent to the types and styles of letters of this period, as convention dictated.

epithalamion Greek term for a marriage song—e.g. S. of S. 3: 6–11.

Erastus This name, which was rare in the 1st cent., is mentioned several times in the NT: (1) The city treasurer, or steward, at *Corinth who is included in greetings by Paul (Rom. 16: 23). In the municipal government of the city, which on a much smaller scale was similar to that of republican *Rome, there were four senior magistrates, elected every year, who were assisted by two aediles, responsible for commercial and financial matters. An inscription from the middle of the 1st cent. CE on paving near the theatre names a certain Erastus as an aedile. He could well be the person mentioned by Paul, though he calls him an *oikonomos* which in the local government would have been of lower rank than an aedile. The suggestion therefore is that this Erastus was converted by Paul and became a member of the Church in Corinth, and that subsequently, in spite of his Christian faith, he was promoted to be an aedile. (2) A fellow traveller with *Timothy (Acts 19: 22). (3) A fellow worker with Paul, who spent some time in Corinth (2 Tim. 4: 20).

Because of the rarity of the name, these may all be references to the same individual.

Esarhaddon King of *Assyria, 681–669 BCE. He exacted tribute from Israel and *Judah and rebuilt *Babylon, for which he required materials from subject peoples.

Esau The elder son of *Isaac and *Rebekah (Gen. 25: 26), who forfeited his legal right of inheritance to *Jacob, so that Jacob was regarded as the ancestor of the people of Israel (Gen. 32: 28), while Esau was the ancestor of the *Edomites (Gen. 36: 9). These folk-tales about individuals may reflect the historical relationships between Israelites and Edomites.

eschatology 'The doctrine of the last things' (Greek: *ta eschata*) came in Christian discourse to mean: *death, *judgement, *heaven, and *hell. But in biblical studies the word denotes the basket of ideas in both OT and NT and the inter-testamental literature about the end period of history or existence—'end' meaning both a terminal point and also the events by which everything else is assessed.

In the OT the promise of a 'good land' (Exod. 3: 8) turned sour when the Assyrians and Babylonians subdued it. The *prophets (e.g. Isa. 40 and Jer. 46) predicted restoration when the nation's sufferings had been sufficient to atone for their apostasies, or when the Temple had been rebuilt (*Haggai and *Zechariah). But still there was no sign of peace and security and a form of eschatology was adopted by the apocalyptic writers who maintained that God had revealed the future to his chosen witnesses. There would be cosmic catastrophes to usher in the terrible *day of the Lord, preceded by the return of the prophet *Elijah urging national repentance (Mal. 4: 5–6). The unrepentant wicked would be tormented eternally (Isa. 66: 24). There would then be an era of justice and prosperity and a descendant of *David would reign (Isa. 11: 1).

Israelites who had already died would be raised (Dan. 12: 2)—a more precise definition of the afterlife than had existed before, though it is not the case that the Hebrews had no belief at all in life after *death until the composition of the book of *Daniel in the 2nd cent. BCE.

In the NT, eschatology denotes the complex ideas surrounding the *kingdom of God in the preaching of Jesus, the coming of the *Son of Man, the **Parousia*, and the conditions obtaining in the age to come. By some scholars, classically by Albert *Schweitzer, it has been held that Jesus' expectations were wholly set on an imminent future. This theory is known as 'consistent' or 'thoroughgoing futurist' eschatology. But also in the NT eschatology embraces events which have occurred in history (the life and death of Jesus, for example) as well as those events associated with the *Parousia*, or return of Christ, when departed believers would be raised incorruptible and those still alive would be awarded new bodies suitable for inheriting the Kingdom (1 Cor. 15: 35–53). For example, the fourth gospel recognizes that participating in eternal life can begin here and now (John 5: 25–9; 6: 40). Several NT writers suggest that the *End will come suddenly (1 Pet. 4: 7), but delay did not weaken their Christian *faith since they held that the decisive work and victory of Christ had been achieved by the *Son of Man in the past. In the language of some biblical scholars, it was 'an eschatology inaugurated' but still to be consummated in the future. Modern readers recognize that the expectation of the coming of divine judgement represented an incentive to generosity and goodness (Matt. 13: 30) in the first Christian generation. It is less compelling today. Modern Christians may prefer to give a new interpretation to the sense of urgency which was imparted by NT beliefs about the imminence of judgement. This could be understood as dismay at the continuing evils in the world, and a determination to take political action against them.

Esdraelon The plain west of the *Jordan which divided *Galilee from *Samaria. Esdraelon was the Greek name for Jezreel though strictly it was the marshy area adjacent to the more fertile *Jezreel. Many conflicts were fought in the area; *Deborah against *Sisera (Judg. 4), *Saul against the *Philistines (1 Sam. 29: 1), and *Josiah against the Egyptians (2 Kgs. 23: 30); in modern times (1917) the British against the Turks. The name does not appear in the Hebrew OT but is used in the *LXX at Judith 1: 8 '*Armageddon' (REB) of Rev. 16: 1 is a mountain on the plain of Esdraelon.

Esdras, books of Esdras is the Greek form of the name *Ezra. The first book of Esdras is an alternative Greek version of part of 2 Chron., the whole of Ezra, and part of Nehemiah, and is included in the *LXX together with the primary version of those works. It is therefore a work of history. In the Latin *Vulgate the work is called 3 Esdras and is placed in an appendix in Roman Catholic bibles. (The title 1 and 2 Esdras is accorded in the *Vulgate to the books of Ezra and Nehemiah.)

The book concentrates on the worship of the Temple and opens with an account of *Josiah's reform of 621 BCE and ends with *Ezra's reform. Eastern Orthodox Christians regard 1 Esdras as deuterocanonical scripture.

Chs. 3 to 14 of 2 Esdras, which was written in the 1st cent. CE, are a Jewish apocalypse consisting of revelations given in seven *visions, published under the pseudonym of Ezra; there are *Christian additions at the beginning and end which are sometimes designated respectively 5 and 6 Ezra. The rest of the work is sometimes also called 4 Esdras or 4 Ezra. It is known now only in various translations, including Latin, and is included in the appendix to the Vulgate. There is a remarkably honest discussion about human *sin and how God deals with sinners, with a promise that the righteous will not lose their reward after they have died. Both 1 and 2 Esdras appear in the Apocrypha of Protestant Bibles (e.g. NRSV, REB).

Essenes A group within Palestinian *Judaism, existing from the 2nd cent. BCE until 70 CE and mentioned by *Josephus and *Philo, though not in the NT. Many of the Essenes lived in villages, according to *Philo, but most scholars believe that the community at *Qumran comprised Essenes or a breakaway Essene group and that the Teacher of Righteousness in the scrolls was a priest who led the *community out of Jerusalem when Jonathan Maccabaeus, who was not of the hereditary priestly family, became high priest in 152 BCE.

The Essenes were ascetics and observed the *Law scrupulously; but in *worship they faced the sun rather than the Temple in Jerusalem, for which they felt contempt. Their communities were controlled by priestly leadership and individual freedom was restricted. They held themselves to be 'sons of light' separate from the 'sons of darkness' outside the order.

Esther, book of The seventeenth book of the OT in the traditional English order. Harem politics, *antisemitism, and an audacious Jewish heroine combine to make this story from the later Persian period (early 4th cent. BCE) full of suspense as it unfolds. The beautiful Esther pleads successfully with King *Ahasuerus, her husband, for her people and her adoptive father *Mordecai against the wrath of *Haman, who had suffered a supposed slight. The tables were neatly turned and it was Haman who was hanged (Esther 7: 10).

There is no mention of God in this tale—which is generally thought to be a legend designed to validate the feast of *Purim in March. But in the *LXX there are additions to the book of Esther which give it and the festival associated with it a religious tone previously lacking. The reversal of fortunes which Esther secured, attributed to her beauty and her daring in the book, are described, in the Additions, to her piety. These Additions, written by several hands between the 2nd cent. BCE and the 1st cent. CE, are to be found in the Apocrypha in Protestant Bibles, but Roman Catholics have inserted them in the appropriate places within the text. To make it clear that those sections are not part of the Hebrew text, the Catholic NJB prints them in italics.

eternal life A personal relationship with God unrestricted by time. There is no continuous or coherent belief in the OT in eternal life in the sense of a divinely given quality of life which is uninterrupted by *death or initiated at death. A shadowy, feeble kind of post-mortem existence is assumed by the story of the witch of *Endor bringing up *Samuel to be consulted by *Saul (1 Sam. 28). It is existence without any relationship to God (Ps. 6: 5). But in the 2nd cent. BCE a belief in survival after death, expressed in terms of resurrection, appears in the book of *Daniel, giving comfort to Jews persecuted by *Antiochus Epiphanes. The belief is echoed in 2 Macc. 7: 14. The book of *Wisdom, on the other hand, coming from the Hellenistic environment of *Alexandria, offers the hope of the immortality of the soul (Wisd. 3: 4). However, not all Jews were convinced, and the *Sadducees are portrayed by both *Josephus and the NT as rejecting the doctrine of *resurrection (Mark 12: 18; Acts 23: 8). The motive for right conduct is attributed in the OT to the *covenant with God; the network of human relationships is governed by the covenant between God and Israel rather than by the prospect of judgement and an afterlife.

In the NT, by contrast, admission to the future state will depend on one's actions in this present life (Mark 10: 24–5; 12: 40). The gospel of John mentions that Jesus is God's agent for the eschatological acts of the general *resurrection at the end (5: 19–47) but the emphasis is more on the present experience of eternal life (John 10: 10). *Salvation is the possession of this life both in present existence and eternally. The NT does not offer believers the prospect of eternal life in the future without passing through death (2 Tim. 2: 11–13). While eternal life may be experienced to a degree in Christian existence here and now, its full realization lies in the future on the other side of death. Baptism is the first moment or stage in the process (John 3: 5); death is a further stage; the *resurrection at the last day (John 6: 40; 11: 25) is the ultimate goal.

ethics The principles of right and wrong conduct; the basis for doing what is right, and the discernment of what is right, are fundamental issues throughout the Bible. Indeed many readers of the *Bible would find it hard to separate the religion from the morality: certain actions are forbidden by God (such as murder) because they are wrong in themselves. The *prophets of the OT repeatedly call for a just society where the poor are not oppressed and bribery is outlawed—principles of behaviour which are fundamental and more important than *sacrifices and *burnt offerings (Hos. 6: 6). And yet there are also some injunctions (such as the fourth of the *Ten Commandments, to keep holy the *Sabbath day) which are entirely religious and not part of a universal moral law (sometimes referred to as 'natural law').

It is widely held today that religion and ethics are two separate areas of human awareness, and each has its own autonomy. It is perfectly possible for a non-religious person to have lofty ethical principles, and it is possible for a religious believer to adopt standards of behaviour which are alien to an informed conscience. There are examples in the OT of appalling barbarity performed in the name of God, as when *Samuel hewed *Agag in pieces before the Lord (1 Sam. 15: 33) or the Levite cut up his *concubine (Judg. 19: 29). On the other hand it was a religious insight which forbade human sacrifice, when *Abraham's hand was arrested in the very act of killing his son Isaac (Gen. 22: 12); but it was invoking civilized rules of warfare rather than the voice of God which saved the lives of the trapped Aramaeans in Samaria (2 Kgs. 6: 22).

While, then, religion and ethics even in the Bible are to be regarded as distinct, plainly the relationship is close and there is much overlapping. Perhaps it could be affirmed that moral striving receives from belief in God a measure of hopefulness which supplies it with dynamic; in the OT this hopefulness springs from the covenant which God has made with Israel; it is part of the vision of a way of life; but it is a vision often repugnant to Christians and others today.

Some modern conservative, or *fundamentalist, scholars condone moral standards in the OT which liberal Christians and non-believers find totally unacceptable. For example, the OT enjoins holy *war: the Canaanite inhabitants are to be exterminated by the invading Israelites. So it has been argued that those people were rightly to be destroyed in order that Israel, and therefore the rest of the world, should be protected from the corrupting evil practices of the people of the land. Similarly, *fundamentalist writers have approved of capital punishment for homosexual behaviour (Lev. 20: 13), though the same writers are coy about demanding punishment for breaking Sabbath regulations or the prohibition of usury on which a free-market economy depends.

Most readers of the OT today, however, find it hard to regard the OT as an ethical authority, and they prefer the moral consensus of a modern democratic society which has been influenced by the NT. The command to 'be fruitful, multiply and fill the earth' (Gen. 1: 28) has for them no validity in an over-populated world. They reject the OT attitude establishing both the subjection of women and the system of slavery. It could, however, be argued that moral advances were made in the history of Israel as the primitive and barbaric tribes are instructed by the prophets and influenced by other peoples of the ancient Middle East; it is noticeable how far advanced were some Israelite moral principles—such as the response to the crime of theft, which was not that the perpetrator should be physically punished but that the victim should be compensated (Exod. 22: 1), because for the Hebrews assaults against the person were more serious than crimes against property.

Confidence in its status as the covenant people of God gave Israel inspiration for personal and social behaviour. It became the model for a life of responsibility in family relationships, in business, and in government. In the NT admission to the Kingdom requires a certain kind of behaviour (Mark 10: 24–5), which is succinctly defined as the twofold love of God and neighbour (Mark 12: 29–31). It is spelt out by Jesus: women are disciples no less than men (Luke 8: 1–3), care is for anyone in need (Luke 10: 37), enemies are to be loved (Matt. 5: 44), and legal authorities are to be respected (Mark 12: 13–17). But in so far as the *Kingdom is to be consummated in the future, some of the ethical injunctions cannot be practised in this present life. They represent ideals to which *disciples can only approximate: non-resistance to *evil (Matt. 5: 30), total perfection (Matt. 5: 48), rejection of money (Luke 16: 13–15).

Much of the ethical teaching in the epistles corresponds to that of the Stoics and other philosophers and is part of the 'natural law'. Paul supplies a motivation when he emphasizes the 'law of Christ' (Rom. 13: 8–10), which is to love one another (John 13: 34). *See* Sermon on the Mount.

Ethiopia The area of N. Africa in the *Nile valley, part of the kingdom known in Egyptian as Nubia and in the OT as *Cush (Gen. 10: 6). From 715 to 656 Ethiopian kings reigned in *Egypt and King Tirhakah persuaded *Syria and *Phoenicia to make war on *Assyria. He was also an ally of *Hezekiah of *Judah (Isa. 18: 1–2; 20: 1–6) but the Assyrians conquered Egypt in 670 BCE. Nubia remained independent until absorbed into the Persian Empire (6th cent. BCE) under *Darius. References in the prophets to the Ethiopians (Jer. 13: 23; Amos 9: 7) show that the people were well known in Israel and Judah. The Jewish settlement at *Elephantine was on the border of Ethiopia.

In the NT an Ethiopian eunuch, probably a 'God-fearer', an official in the court of 'the *Candace', the queen mother, is encountered by *Philip (Acts 8: 26–40).

ethnarch NRSV marg. at 2 Cor. 11: 32 for the governor ('commissioner', REB) of *Damascus, a Greek title offered to rulers of subject states in the Roman Empire. It was borne also by *Simon, the last surviving son of *Mattathias, and high priest (142 BCE, 1 Macc. 14: 47, NRSV).

etymology The study of the origin of words. It has influenced *exegesis, especially of the OT. It has sometimes been assumed that the root meaning of a verb in Hebrew, or a related Semitic language, was a clue to its meaning in the *Bible, and that it was even possible to build a doctrinal structure on this basis. It thus tended to ignore the changes made in the course of time by other (e.g. Hellenistic) cultures in the meaning of words and to overlook the contexts in which they were used.

Etymology, in the sense of the study of the formation of words, appears in the NT at Matt. 1: 21, where the name Jesus, the Greek equivalent of Hebrew *Joshua (= Yahweh saves) is explained as 'he will save his people from their sins'.

Eucharist The word used on the basis of 1 Cor. 14: 16 ('your *thanksgiving') from early in the 2nd cent. for the principal rite of Christian *fellowship, or *communion. At the *Lord's Supper Jesus gave thanks (Luke 22: 17) and the Christian rite is essentially an act of thanksgiving for the whole of God's work in *creation and *redemption. The Jewish rite of the *blessing of *bread, as used by Jesus, took the form of a thanksgiving to God. The narratives of the feeding of the multitudes by Jesus (e.g. Matt. 14: 19) may have been influenced by the Church's Eucharistic celebrations.

Eunice Mother of *Timothy, who was converted from *Judaism and gave her son instruction in the Christian faith (2 Tim. 1: 5; 3: 14–15).

eunuchs Persons of the male sex who had been castrated; kings and court officials regarded them as ideally suitable to serve as guards of the royal harems. Whether the defect was by accident or malevolence, eunuchs were excluded from the Israelite assembly (Deut. 23: 1); but they are not treated contemptuously in the OT (Isa. 56: 4–5); indeed the eunuch Ebed-melech the Ethiopian rescued *Jeremiah from his water dungeon (Jer. 38: 10). And in the NT another Ethiopian eunuch was baptized by *Philip (Acts 8: 38).

Jesus said that some might become eunuchs for the sake of the *kingdom of God (Matt. 19: 12), i.e. might be called to celibacy. But Origen (185–254 CE), one of the greatest thinkers of the early Church, took the words literally and mutilated himself —or so it was alleged by the historian Eusebius: but, later in life, Origen, who regarded *allegory as one form of scripture interpretation, rejected a literal meaning of Matt. 19: 12.

Euphrates One of four rivers mentioned in Gen. 2: 14 which seems to have acted as the northernmost boundary of Israelite territory under *David (2 Sam. 8: 3). Its waters irrigated the area towards the River *Tigris and the land became the cradle of civilization. It is often referred to in the OT simply as 'the river'.

Eutychus A young man who dropped asleep during a long sermon by Paul and fell from an upstairs window. Paul was able to assure the company that he was not fatally injured. The readers of Acts may have seen the story as a sign of Christ's power to give *eternal life; the *communion (which then followed the incident) was the means by which believers were drawn from death to life (Acts 20: 7–12).

evangel A Greek word sometimes used in commentaries for *gospel* or *good news*.

evangelist Philip, one of the *Seven (Acts 6: 1–7), is called an evangelist (Acts 21: 8) because he preached the gospel. Before long, evangelists were a distinct order in the Church (Eph. 4: 11), and later still (from the 3rd cent.) the writers of the four gospels are known as evangelists.

Eve *Adam's wife, created by God from one of Adam's ribs (Gen. 2: 22) and later instrumental in persuading him to eat fruit from the forbidden tree. Women's pains in labour were explained by an *aetiological myth as punishment for Eve's *sin. However, masculine sexism in Christianity owes more to Milton's *Paradise Lost* than to Genesis, where men and women are equally made in God's image (Gen. 1: 27).

everlasting life Endless existence uninterrupted by *death. In several parables Jesus taught that this present life leads on to a further phase whose nature depends on our conduct here (Luke 16: 25; Matt. 25: 46). It is argued by some modern theologians that this further phase could not be damnation: such is the *love of God for each human being that it is inconceivable that he could create us with so much potential for responding to that love and then that human *evil could entirely destroy that potential. Cf. 1 John 4: 9.

evil Bad situations, usually involving pain and suffering, attributed to disobedience to God's will, or to natural calamities. Much of the *Bible is about the cause, nature, and consequences of *sin, and God's reaction to it. Evil began according to the *myth in Genesis in an act of calculated rebellion against God and it continued wherever the human race acted in conflict with the will of God. As a result by way of punishment God sent evil disasters (Jer. 26: 19; Amos 3: 6). He is responsible both for *good and for evil (Ps. 78: 49), though fundamentally God's *creation is good (Gen. 1: 32) and his intention to do good (Jer. 18: 8) is only frustrated by human sin (Jer. 18: 10).

However, the power of evil came to be seen as concentrated in particular beings within a system and controlled by a leader, sometimes called *Satan. These rebels against God were bent on defying his will and could even take over the lives of people by 'possession'. Evil is also to be found in the human will (Mark 7: 20–2) and may manifest itself in overwhelming desire for money (1 Tim. 6: 10). The tendency to evil is recognized as fundamental in the human scene and is given various explanations in addition to that of Gen. 3 (and accepted by Ben Sirach and other rabbinic authorities, Ecclus. [= Sir.] 25: 24). By some, it was attributed to the machinations of *Satan; by others among the rabbis to an evil impulse or inclination located within each human being. The evil erupts in people in the form of specific deeds, thoughts, and diseases, and Jesus attacked the whole apparatus by his healings and *exorcisms and by words of *forgiveness. Evils continue within the Church (Rom. 7: 15) and suffering too, but they can be turned to good by God, as, in Christian belief, did the great evil of the *Cross. At the *End there will be a total destruction of all that is evil (Matt. 25: 41; 2 Thess. 2: 8; Rev. 19: 20–1).

excommunication The withdrawing of the *sacraments for grave offences. It is part of Catholic discipline and looks for NT sanction to words of Jesus reported in Matt. 18: 17 and to Paul's injunction in 1 Cor. 5: 2. There was a similar rule for the community at *Qumran. For Paul, though exclusion was a necessity, sin was a matter for grief not complacency. At the time of the composition of the gospel of John, Christians were excommunicated by the authorities of the *synagogue (John 9: 22; 12: 42).

exegesis A Greek word (found in the *LXX but not in NT) meaning 'explanation'. It refers nowadays to commentary on the biblical text to elucidate obscurities and to relate one word or verse or section to others in order to define exact meanings. Modern exegesis therefore makes use of textual criticism and linguistic expertise as well as historical and literary disciplines and can also make use of archaeological discoveries. Mistakes in any of these areas could result in false exegesis, as was the case with exegesis of Jesus' *parables about the *kingdom of God as long as the Kingdom continued to be identified with the *Church.

Exile, the The captivity of the king and people of *Judah in *Babylonia (597–539 BCE). *War between *Egypt and Babylonia was disastrous for little Judah, lying between them. After the Babylonian victory at *Carchemish in 605 BCE and the resumption of hostilities in 604, King *Jehoiakim switched his allegiance from Egypt to Babylonia. But he continued to have half a hope in Egypt and rebelled against Babylonia (2 Kgs. 24: 1). It was a foolish decision and made against the advice of Jeremiah. The Babylonians marched; Jerusalem surrendered (2 Kgs. 24: 10–17), and the leading citizens were deported to Babylon (597 BCE), which represents the beginning of the Exile, or Captivity, of the Jews. About 10,000 people were deported (though a smaller figure is given by Jer. 52: 28), but many still remained in Palestine and a weak puppet king (*Zedekiah) was installed. He too rebelled in 588 and two years later the capital fell to the Babylonians, and there was massive destruction. Many of the inhabitants were taken to Babylonia, though some managed to flee to Egypt.

In 539 BCE Cyrus, king of Persia, captured Babylon and in accordance with his policy towards subject peoples he permitted Jews to return to Palestine, although many preferred to stay put in Babylon where they had become moderately comfortable. The Exile had ended; there was little change in material conditions, but the *Dispersion (*Diaspora) had begun, and with it an era of theological creativity which was to reshape the great bulk of the OT which is pre-exilic material. OT studies therefore give far more attention to the post-exilic period than does the OT itself. This was the era when the editorial work on the *Pentateuch was finalized.

In Babylon the exiles enjoyed a degree of freedom and were able to maintain communication with relatives still surviving in an impoverished Palestine. In exile new importance was attached to sabbath observance and circumcision, and *synagogues, which constituted a revaluation in Jewish worship, may possibly have been established in this period. All this was part of the response to the challenge to faith in Yahweh by the fate of the Jerusalem Temple: it was the footstool of God (Lam. 2: 1), his dwelling place (Ezek. 43: 7), and a visible symbol of the nation's self-consciousness as a chosen race. Now it was a heap of rubble. Moreover, the dynasty of *David, to which so much had been promised, had come to an end which became in fact a new beginning.

Not all the responses to the challenge of exile were equally positive. Some people were deeply nostalgic and tearful (Ps. 137: 5–6) and uttered a savage curse upon their enemies (Ps. 137: 9). Some turned to other gods (Ezek. 20: 32), and some in self-pity blamed calamity on the previous generation (Jer. 31: 29). More optimistically, some expected a new intervention from above (Deutero-Isaiah). Generally, the exiles emphasized, one way or another, their national identity as Jews among *foreigners.

Exodus, the Israel's departure from *Egypt. Tradition had it that the twelve sons of *Jacob and their families, a total of seventy persons (Gen. 46: 27) went down into Egypt to escape the *famine in *Canaan. They grew to a great multitude, to the alarm of the Egyptians, who enslaved them and forced them to build cities; but under the inspired leadership of Moses the Israelites fled from their oppressors, tramped through the desert, and invaded Palestine. By this time there were said to be 600,000 men of military age—which would imply a vast company of more than two million refugees—which is unbelievable.

The historical facts are more complex and by no means certainly established, even with the help of archaeological evidence. Probably the Hebrews originally entered Egypt in several phases, as would seem to be implied by the Genesis narrative itself in that the brothers go down later than Joseph. Some probably never went into Egypt at all. And possibly their departure, or exodus, was also phased over a period, though of the event in some form there can be no doubt; the recollection was deeply engraven in the nation's consciousness (Isa. 51: 9–11) and liturgies (Ps. 114).

A possible historical reconstruction is that the first Hebrew contingent went into Egypt in the late 18th cent. BCE, followed by others in the 14–13th cent., and that some of them left at various times. Possibly the main body left after 430 years in Egypt (Exod. 12: 40).

The tradition emphasizes the reluctance of *Pharaoh to let his Israelite slaves depart, and it was only after the terrifying and miraculous *plagues that consent was won. The most likely date for the final departure is early in the 13th cent. BCE, but it has to be borne in mind that the *Pentateuch, which is our only source of information, was not compiled until the 10th to 6th cents., and details of the geography are unclear. The *miracles are told to express the belief in the overruling providence of Israel's God Yahweh; and the *covenant at *Sinai was regarded as the moment when Israel received the Law. Of the wanderings in the wilderness, it is hard to trace a precise direction: few of the places in the biblical narrative can be definitely identified. First of all, it is improbable that the people ever crossed the tip of the *Red Sea (the Gulf of Suez)—it is too far to the south. (And the Hebrew in any case means Sea of Reeds. The 'Red Sea' derives from the *LXX translation.) It is suggested that the crossing was over shallow water to the east of Avarus near modern El-Qantara on the Suez Canal. The Israelites were trapped between this water and the pursuing Egyptians, but a wind drove the waters back (Exod. 14: 21, 27), while the soldiers of Pharaoh were drowned when the waters returned; this is celebrated in the hymn of Exod. 15. The setback was hardly very momentous for the Egyptians—the successful escape of a party of runaway slaves was not significant, and no Egyptian records of it exist. (There is, however, an Egyptian inscription from the reign of the Pharaoh Merneptah at the end of the 13th cent. BCE which indicates that Israelites were settled in Canaan and no longer in Egypt.) The main group made for Mount *Sinai where their God Yahweh, who had rescued them from Egypt, was to be worshipped, and where he revealed himself. After this renewal of Israel's hopes and the beginning of Israel's election as the people of Yahweh, they made for the oasis of *Kadesh, 80 km. (50 miles) south of *Beersheba. The tradition maintains that they then made a great detour round *Transjordan which included the conquest of the *Amorites of *Heshbon. The invaders skirted the fortified territory of *Edom and *Moab, captured *Jericho (Josh. 6), eventually subdued the whole of Palestine (Josh. 11: 16–23), and distributed it amongst the twelve tribes (Josh. 13–21).

The aim of the biblical narrative is to demonstrate the power of God to punish or to reward his people by defeats or victories in war. It was an attempt by a later generation to make sense of their present situation by giving an idealized account of their past. Historical reconstruction is therefore difficult. It could be that if some Hebrews never went into Egypt, these were joined with those who fled from Pharaoh and so generated a revolt of the underclass against Canaanite overlords.

Exodus, book of The second book of the *Pentateuch. It takes its title from the Greek *LXX and means 'departure'; it is divided by scholars, like other parts of the Pentateuch, into the sources *J, *E, and *P, with the greatest parts in this book being assigned to J and P. The book consists of an account of the birth and calling of Moses (chs. 2 to 6); the contest between Moses and *Pharaoh, and the *plagues, culminating in the death of the first-born (chs. 7 to 12); the march out of *Egypt (chs. 13 to 15); wanderings in the *wilderness (chs. 15: 22 to 18); and the meetings at the mountain (chs. 19 to 40).

In the J source God is the primary agent of the exodus and Moses little more than his mouthpiece. But there is much rebellion and murmuring amongst the people, culminating in the construction of the *golden, or molten, calf (Exod. 32), a story which seems to reflect the apostasy of the northern kingdom under *Jeroboam (about 920 BCE), who set up golden calves for worship in *Dan and *Bethel. It could be that this source in Exodus was written to reassure loyal believers living in the north under Jeroboam—God's promises still hold good, even after backsliding.

The P narrative contains some of the stories of the plagues and the defeat of the magicians, and the inauguration of the *covenant at *Sinai. This way the decisive moment in the nation's realization of itself as the people of Yahweh, as well as all the pain of the wanderings since they abandoned the fleshpots of Egypt (Exod. 16: 3), found a significance. The P narrative concentrates on instructions for *sacrifices and *worship, which was of vital interest to Jews after the Return from Exile when the Temple was being restored: from 520 BCE it became the focus of the nation's very life, replacing in that role the dynasty of *David. What was formerly a Canaanite agricultural festival in spring was taken over and turned into the annual commemoration of the Exodus so that it should never be forgotten. The festival was *Passover and Unleavened Bread.

A reader of the book of Exodus might discern in the J source a hope that in spite of rebellion and discontinuity at Sinai God does not forsake his people; in P a reader might infer that between Sinai and the second Temple there is a divine continuity. A modern reader might express astonishment that descendants of a group of slaves who fled out of Egypt over 3,000 years ago still survive in spite of all the vicissitudes of history, in the land those Hebrews then invaded.

exorcism The expulsion of evil spirits. It was the belief in the 1st cent. CE, and long afterwards, that evil spirits could take possession of a person and cause mental and physical illnesses, especially epilepsy. By exorcism, which could take the form of an incantation, the *demon might be expelled and the sufferer cured. Jesus cast out *spirits (Mark 1: 25; 9: 25) and the *disciples were given the same power (Matt. 10: 1, 8), which they made use of (Acts 19: 13–16), though exorcism is not included by Paul in any of his lists of spiritual *gifts. Some of the *Pharisees, however, were exorcists (Matt. 12: 27).

Indeed belief in demons and demon-possession was common amongst Jews of the 1st cent. CE, and in Hellenistic literature there are accounts of exorcists extracting a demon from the nose of a demoniac with a ring. Jesus' exorcisms were effected by word only. They are not mentioned in the gospel of John, but the undoubted fact of this aspect of Jesus' *ministry witnesses, like his pre-critical views on the OT (e.g. Mark 12: 36), to the distance of his world from ours. There has nevertheless been a revival of the practice of exorcism in some modern churches of a conservative outlook; sometimes there is co-operation with medical and psychiatric opinion.

expiation A means of the removal of *sin. Where AV used 'propitiation' (e.g. Heb. 2: 17), 'expiation' is preferred by REB, NJB, and RSV; NRSV has changed it to 'sacrifice of *atonement'. There is a difference in meaning. The situation is that of God in relation to a sinner and how *reconciliation can be effected. Is it that an angry God must be propitiated? Or is it that a sin needs to be forgiven and its guilt removed—i.e. the offence expiated? In the OT there was a system of prescribed sacrifices to atone for sins; in the NT the work of Christ has expiated human sin.

'Expiation' is a more satisfactory translation than 'propitiation', for it denotes the *love and *mercy of God who has provided means to remove transgressions; whereas 'propitiation' implies a God of wrath who needs to be appeased and his anger assuaged by a transaction of dubious morality: the penalty of human sin being transferred to Christ as our substitute and fully endured and satisfied by him.

eye The eye is found in the Bible in both literal and metaphorical senses (Job 21: 20; Ecclus. [= Sir.] 14: 9). Jesus refers to those without understanding as being blind (Mark 8: 18). An eye in good condition brings the benefit of light to the whole body; so if the spiritual eye is healthy, the whole personality will be sound (Matt. 6: 22–3).

eye-service Used in AV for a service performed only to attract favourable attention, not for the sake of God or at the behest of one's own conscience (Eph. 6: 6; Col. 3: 22). NRSV has 'while being watched'; REB 'to catch their eye'; NJB 'under their eye'.

eyewitness Luke claims (Luke 1: 2) to be relying on the evidence of eyewitnesses; and the author of 2 Pet. (1: 16) implies that he was an eyewitness at the *Transfiguration of Jesus. The latter is part of his apparatus of pseudonymity: it is a way of winning the readers' confidence that the writer is faithfully speaking the mind of Peter.

It has often been held that Luke's statement is a ground for trusting the gospel as a historical record: Luke took the trouble to check on facts recorded in the Acts where he was not himself present, and for his gospel made use of oral and written traditions which in the last resort depended on eyewitnesses. However, the original context of each story had been amended by constant oral repetition before it was set down in the gospels, and all the *evangelists were creative theologians who allowed their interpretations to control the course of their narratives; thus Luke altered the time and place and order of the *temptations to bring out his own understanding of them; he altered his account of the *miracle at *Jericho (18: 35); and the *Passion narratives each have their own distinctive emphases. If there were eyewitnesses at the foot of the *cross, they did not agree about the words of the *centurion (Mark 15: 39; Luke 23: 47); it is more probable that Mark and Luke are imposing their own interpretations upon what had been handed down. Sometimes it has been suggested that the presence of vivid details in a narrative—'on the green grass', Jesus seated on a cushion in the boat, 'Jesus wept'—are the faithful recollections of eyewitnesses. But it has been shown that in the course of repeated oral transmission of a story it tends to grow extra details precisely like that!

The gospel according to Mark was once thought to embody what Mark had learnt from Peter. Certainly this was being asserted as early as *Papias, a 2nd cent. bishop, but it is clear that much of Mark shows signs of its development within the communities of the Church, such as the explanations given about *parables, and the instructions by Jesus to keep quiet about some of his miracles.

There is a claim also in the fourth gospel about the truth of the disciple's testimony (John 21: 24; cf. 1 John 1: 1), and this 'truth' must be understood as the truth about Jesus interpreted by and for the Johannine community. The factual history (John 1: 14) is interpreted by the theology. The theological motifs of the writers do not, however, override the historical facts to the extent of turning the NT documents into mere fiction or legends. On the contrary, many of the chronological, geographical, and political details in the Acts are corroborated from other sources; but in the last resort Luke–Acts is controlled by the author's ideas about Judaism and the Jews, the Mission to the Gentiles, the doctrine of the *Holy Spirit, and the continuing life of the Church. However, while granting these tendencies, Acts is still a serviceable work of history, with some reliable eyewitness testimonies—but even a reliable eyewitness interprets what he sees.

Ezekiel The third of the three major OT *prophets. A younger contemporary of *Jeremiah, and influenced by him, Ezekiel was a priest (Ezek. 1: 3) and a prophet (11: 4) of the Exile. According to 1: 1–3 Ezekiel was among the first group of Jews to be deported to *Babylon, in 597 BCE; but there are reasons for supposing this to be an editorial note and that Ezekiel's message to Judah was delivered on the spot and not from distant Babylon. Ezekiel is remarkably well informed about conditions in Judah and his sermons are addressed to people living there. Some scholars have proposed that Ezekiel did go to Babylonia but returned to Jerusalem to preach doom on the city until its capture in 586 BCE, when, as is suggested, he was again taken captive and carried a second time to Babylon.

Alternatively, it is argued that Ezekiel remained in Jerusalem for the greater part of his prophetic career but perhaps went to Babylon in 586, or—a modification of this theory—that he remained throughout in *Judah but sections which appear to derive from a Babylonian exile can be attributed to a later editor.

Some of Ezekiel's prophetic acts were regarded as bizarre or at least strange by his contemporaries: he ate a scroll (3: 1–3); in front of a rough drawing of Jerusalem under siege, he lay for 390 days on his left side bearing the punishment of Israel and forty days on his right side for the punishment of Judah; he shaved his head and face and weighed the hair, which he then divided into three to symbolize how the population would suffer in three ways. He was afflicted with dumbness for much of the time before 586 BCE. When his wife died suddenly, he refused to mourn as custom dictated, in order to bring home his message of the even more tragic coming destruction of the Temple (Ezek. 24: 15–18).

Ezekiel regarded the nation's history from the Exodus onwards as a story of disobedience (20: 1–18) but, when Jerusalem had been destroyed and his words vindicated, he could turn to hopes for the future. God takes no pleasure in the death of the wicked and wants to restore to life even those in exile (33: 11–20). So he can predict the future restoration of his people in their own country. There would be peace and security, and the Lord would return to the sanctuary (43: 4–7) from which he had once departed (10: 18 and 11: 23). The God of Ezekiel is wholly transcendent, and acts as he thinks fit (Ezek. 36: 22) to produce a people transformed (36: 26). Nevertheless, they are required to respond to God's *grace; individuals are responsible and free. Nobody in the OT more passionately asserts the reality of this responsibility (18: 20).

Ezekiel, book of In its present form, the book was probably edited soon after the prophet's death, some time after 571 BCE, the date of the last prophecy. (Nothing is recorded in the OT of his death, but a later tradition has it that he was murdered and buried near *Babylon.)

The book divides into three sections: chs. 1–24 are messages of *judgement upon *Judah before the fall of the capital. Chs. 25–32 are oracles against foreign nations, and chs. 33–48 are about the future restoration of the people in the promised land. The final section (chs. 40–8) contains detailed plans for the new Temple and its staff of priests of the line of *Zadok (to which Ezekiel himself belonged); thus it was the foundation for later Judaism.

Ezra A priest and scribe in Jerusalem after the Return from Exile. An account of Ezra appears in the books of Ezra (7–10) and Nehemiah (8–9), and it appears that the activities of these two leaders overlapped. One suggestion is that Nehemiah preceded Ezra, whose activity would then be placed in the time of *Artaxerxes II (404–358 BCE). But the usual understanding is that Ezra led the community in restoring the Temple and maintaining the racial purity of the nation (Ezra 9–10). There was a legend that he dictated from memory ninety-four holy books that had been lost in Exile.

Ezra, book of The books of Ezra and Nehemiah were a unity in the Hebrew and the *LXX. But in the Latin *Vulgate the books are separated and called Esdras (the Greek form of Ezra) I and Esdras II. Much of the book of Ezra also appears in a book of the *Apocrypha and is there entitled Esdras I and was part of the Greek *LXX. This Esdras I is known as Esdras III in the Vulgate, or the Greek Ezra, and Esdras II is known as Esdras IV in the Vulgate; neither Esdras III nor IV is recognized as canonical by the Roman Catholic Church but both were included by Jerome as an appendix to his Vulgate. These two Greek works are recognized as deuterocanonical by Eastern Orthodox Christians.

The books of Ezra and Nehemiah were part of the history of the Chronicler, and were probably written in the first part of the 4th cent. BCE. Since Ezra 7: 27–8: 34 and 9: 1–15 is written in the first person (rather like the 'we-sections' of the Acts), it is possible that Ezra was himself the Chronicler.

The books of Ezra and Nehemiah are the climax of 1 and 2 Chron. and stoutly maintain confidence in the God who after the slavery in Egypt and the captivity in Babylon has restored them to their own land of Promise through influencing the mind of the kings of Persia (Ezra 9: 9). It is incumbent on the people that they now observe the law and confess their failures in the past (Ezra 9).

F

fable The description in AV for fictitious 'old wives' tales' (1 Tim. 4: 7); called 'cunningly devised' (2 Pet. 1: 16). NRSV, REB, and NJB prefer the translation '*myths'—but this adds to the difficulty of rescuing the use of 'myth' by modern theologians as a story which conveys a truth other than the historical.

The word does not occur in the OT but *parables about trees (Judg. 9: 8–15 and 2 Kgs. 14: 9 ff.) fall into this category.

face Used both literally of the most highly personal feature of a human being (e.g. Matt. 26: 67) or metaphorically (e.g. Isa. 3: 15). It is used of Moses speaking personally to God (Exod. 33: 11), and in 2 Cor. 4: 6 Paul uses the face of Jesus, meaning the person and work of Jesus considered as a complete whole.

Paul declares (1 Cor. 13: 12) that a *vision of God face to face will replace our present distorted vision, and will be such as Christ himself enjoys now (cf. Heb. 9: 24).

To speak of God hiding his face is a very common way of explaining misfortune in the OT.

faith Trust, especially in the reliability of God. A modern understanding of faith is that it is an inferior kind of knowledge or an acceptance of an opinion or story which cannot be wholly proved. The biblical meaning of faith (the corresponding verb is 'believe') is more in the nature of commitment, though it does also imply that there is a basis in fact which could, by virtue of conclusive historical evidence, render faith unsupportable. OT faith was traced to the *covenant which God had offered to the people of Israel at *Sinai. From then on, the Israelites believed that Yahweh, the creator of the world, had given them the *Law and a promise that their faithfulness would be rewarded. Without such a faith, the people of Israel would not survive, as the prophets warned (Isa. 30: 15–16). Individual Israelites also have faith: a trust that the righteous have in God, come what may (Ps. 73: 1–3); and sometimes an individual's faith is related to the destiny of the nation, as it was with *Abraham (Gen. 15: 6), who believed in God when he was called to leave his country, and when he was promised an heir though he was beyond the normal age of procreation, and yet again when he obeyed the call to sacrifice this heir (Isaac). Through all these trials, Abraham had faith that God would indeed fulfil his promises.

In the NT the classical definition of faith is in Heb. 11: 1, which asserts that the believer in virtue of his faith holds to be true those realities which for the moment are invisible. The realities are, first, as in the OT, God the Creator of the universe (Heb. 11: 3) and secondly God the author of the OT covenant which is now realized in the *salvation brought by Jesus, 'the pioneer and perfecter of our faith' (Heb. 12: 2).

So in the NT the object of faith is God revealed in Jesus Christ. Amongst Jews, faith in God is taken for granted (John 14: 1), but pagans must first believe in God (1 Thess. 1: 9) before they can accept the risen Christ, who reveals this God. It is the gospel of John which especially emphasizes that Jesus came from the Father to reveal him and make him known to humanity (John 1: 18). This is the substance of Peter's preaching after *Pentecost (e.g. Acts 2: 36), and belief is followed up by *baptism.

In the NT the profoundest exposition of faith is in the epistles of Paul, who shows that faith does not remain static; it grows (Phil. 1: 27; 2 Cor. 10: 15) and it issues in love, without which faith is empty. Faith is above all a confidence that the *resurrection of Christ is an anticipation of a general resurrection at the *End (1 Cor. 15: 14, 17; 2 Cor. 4: 14). Faith and baptism confer membership of the *Body of Christ, which is both privileged and painful (Phil. 1: 29).

In Gal. and Rom. Paul links his concept of faith to *justification, and this is sometimes incorrectly understood. Paul taught that believers are justified (put right with God) *not on the ground of* faith but by Christ, in whom we have confidence and whose free grace justifies us. We are not justified by virtue of a subjective experience of conversion.

In the gospels, it is faith on the part either of sufferers or of members of their family (Matt. 8: 5–13) which enables Jesus to heal them. The father of an epileptic boy is assured that 'all things can be done for the one who believes' (Mark 9: 23), and the boy is cured. There are also accounts of healings without mention of faith (Mark 3: 5; 5: 7) because it was present in the human Jesus as Paul declares (Rom. 3: 22, 26, NRSV marg.) and invites us to grow in a like faith.

Fall, the A post-biblical expression for the doctrine of *Adam's transgression and mankind's consequential inheritance of a sinful nature. The story of the disobedience of Adam and *Eve is depicted in the *J version of the *creation (Gen. 2: 4b–3: 24); human *sin is rooted in the desire to be like God and to live lives wholly centred on themselves and their own desires. According to the narrative, the result of this primal sin was expulsion from the garden. God was afraid that Adam, having disobediently eaten from one tree, might be tempted to become immortal by eating next from the tree of life. So the way back to the garden was barred. Paul appears to misunderstand the story as if Adam had the gift of immortality, but lost it by eating the forbidden fruit. He refers to the story in Rom. 5: 21, taking it up from Ecclus. [= Sir.] 25: 24. The doctrine of original sin was much developed by Christian theologians, especially Augustine in the 5th cent. CE and, later, by Protestant Reformers.

falsehood The OT *Decalogue does not contain any prohibition of lying—but there are many terms in which the evils of falsehood are denounced. It is a power of **sheol*, the infernal world (Isa. 28: 15) and it is especially represented by *idolatry (Jer. 10: 14) and by false prophecy (Jer. 29: 9) which leads people to their destruction. Falsehood is illusion.

In the NT falsehood is above all demonstrated in a refusal to accept the claims of Christ, who is the Truth (John 14: 6) whereas the *devil, among whose children the unbelieving Jews are numbered, is a liar and the father of falsehood (John 8: 44).

According to Paul (Col. 3: 9) falsehood is a characteristic of unredeemed persons, and in the gospels it is by engaging in false witness that the Sandhedrin is able to secure a conviction of Jesus (Matt. 26: 59). In the epistles we encounter false apostles (2 Cor. 11: 5) who assert themselves over genuine apostles, possibly teaching a form of *Gnostic doctrine or *ethics. Perhaps even worse, there were those in the time of the early Church who were 'false Messiahs' or 'Christs', that is they imposed on the expectation of their social group by falsely claiming to be such a person (Mark 13: 22; Matt. 24: 24).

family (1) The modern conception of a human family as consisting of father and mother united in lifelong monogamy is alien to the OT. The founder of the nation, *Abraham, had two wives (Gen. 16) but by the time of the monarchy polygamy, common in earlier ages, was limited to the royal household. Each ordinary family was a self-sustaining economic unit; food was stored in the house; animals were quartered there. The father's authority was absolute and children remained under his control at least until *marriage.

Jewish marriage customs presupposed by the NT are that there was first a betrothal, which was binding but could be dissolved by divorce, and subsequently the husband took his bride to his own home, which constituted the completion of the marriage. Only a husband had the right to terminate a marriage by *divorce (Deut. 24: 1–4) but the right was not unrestricted (Deut. 22: 29) and by the time of Malachi (2: 14 ff.) divorce was regarded with disfavour, in that lifelong union in marriage was used as an analogy for the union between God and his people in *covenant.

Jesus taught that the family set-up is not to stand in the way of embracing the *kingdom of God (Matt. 8: 21) and family ties may have to be renounced (Luke 14: 26) and he himself said that whoever does the will of God is part of his family (Mark 3: 35). Indeed, out of the death of the natural family springs the supernatural family which is the Church or household of God (1 Tim. 3: 15), which now becomes the new support to those who have renounced their first family (Mark 10: 30). Paul founds or 'begets' Churches as if he is the father of a family (1 Cor. 4: 15) and baptized pagans are henceforth part of the family of God (Eph. 2: 19).

Not that in Jesus' teaching natural family life is to be discontinued generally. He restored their daughter to *Jairus and his wife (Mark 5: 40) and her son to the widow of *Nain (Luke 7: 11 ff.). The family in NT parlance consisted of parents, children, and servants (e.g. Matt. 10: 25, 35; Rom. 14: 4) and the mutual relationships of all these parties are to reflect the kind of love which Christ has for his Church (Eph. 5 and 6).

The performance and integrating of family life was in the interests of society and to that end the Christian Church frowned on incest and inbreeding and encouraged family life where security and shelter were provided and *prayer and *worship practised. The core of NT and Church teaching on the family can be adapted to changing social conditions, as it already was within the NT itself; for Matt. 5: 32 and 19: 9 already represents a modification in the light of experience. (2) A closely linked group of MSS is known in textual *criticism as a 'family'. All existing MSS are copies of another made either by eyesight or dictation and it is important to ascertain whether a family is descended from an original MS which was reliable.

famine Prolonged starvation, of which the danger and horrors are often mentioned in the OT; it was one of the four acts of God's judgement in Ezekiel (5: 16) and it is one of the curses which God will send on the nation for disobedience (Deut. 28: 48). The best-known famine in the OT is that in *Egypt when *Joseph administered the diminishing food supplies (Gen. 41). In the NT there was a famine which hit Jerusalem during the reign of Claudius Caesar (Acts 11: 28), and Paul and *Barnabas took aid to the *Christians there. The Roman authorities never budgeted on a long-term basis and famines could strike with terrible devastation. There were no contingency plans. It was no unlikely experience for the reader of Rev. 6: 5–8 to read of the third horseman announcing exorbitant prices for food in time of famine. Yet, claims Paul, even famine will not separate us from the love of Christ (Rom. 8: 35).

Farrer, Austin Marsden (1904–68) Primarily a philosophical theologian at Oxford; but Farrer turned his attention to NT studies and enjoyed a considerable influence as lecturer and preacher. He argued that in place of the *Q hypothesis it is more reasonable to hold that Luke used Matthew. Farrer wrote on the gospel of Mark, believing the author to be a creative theologian who interpreted the work of Jesus in the light of the OT by means of a complicated scheme of typology. Later he wrote a commentary on the *Revelation.

fasting Abstention from food which has been, and remains, a widespread religious obligation; it accompanies *prayer, and is a sign of human humility. Fasting among the *Hebrews (Judg. 20: 26) became formalized during and after the Exile, and the Day of *Atonement was established as a national day of fasting. The *prophets (e.g. Jer. 14: 12) protested that the mere act of fasting, without *repentance, did not bring results. Jews fasted in NT times on the Day of Atonement (Lev. 16: 29) and *Pharisees also on two days each week (Luke 18: 12); Jesus fasted (Matt. 4: 2), and the early Church practised fasting (Acts 13: 2). Paul fasted before he was baptized (Acts 9: 9), and this became a common practice in the Church, e.g before baptisms at Easter. It would seem that the interesting passage in Mark 2: 18 ff. reflects the Church's practice of fasting, which was out of step with what was remembered about Jesus, whose disciples were accused of *not* fasting (Mark 2: 18) unlike those of John the Baptist. The reply in Mark (2: 19–20) is that Jesus' disciples *will* fast 'when the bridegroom is taken away'. But bridegrooms are not normally 'taken away'! Nor did the Pharisees have disciples (2: 18). It does look as if an original saying about the joy in the presence of Jesus, like that of a marriage feast, has been expanded by the Church to defend itself in the course of current controversy with Jewish opponents.

father In the OT 'father' can have a wider significance than the modern head of the nuclear *family. He may be head of a tribe (Gen. 32: 9) or a prophet (2 Kgs. 2: 12). In a family, a father arranged *marriages, could sell a son or daughter into slavery, and it was he who gave religious instruction. In the NT, Jesus addresses God as his Father, thus making him an ideal for all human fatherhood (Eph. 3: 14–15). In John, Jesus' activity reflects the love of the Father, and he acts with his authority (John 3: 35). When Jesus forbids his followers to call anyone on earth their 'father' (Matt. 23: 9), it is not a prohibition of children addressing their natural fathers in the obvious way, nor of adults thus addressing their spiritual mentors; it is a warning against ascribing authority to any *prophet or *teacher who is out of step with legitimate tradition.

fear Although there are occasions in the Bible when the word fear represents the human emotion of terror (Exod. 23: 27), there is also a fear which means awe in the presence of the holy (Eccles. 12: 13). This fear induces a hatred of *evil (Prov. 8: 13).

There is also awesome fear in the NT, as when the women were told about the *Resurrection (Mark 16: 8). 'Those who fear God' or '*God-fearers' is a technical term applied to pagans who attended a *synagogue but shrank from *circumcision (Acts 10: 2).

fearful AV uses 'fearful' for both feeling fear and inspiring fear, but modern versions distinguish between the two. 'Fearful sights' (AV, Luke 21: 11) becomes 'terror' (REB), but at Heb. 10: 31 NRSV retains 'It is fearful to fall into the hands of the living God'. (REB has 'a terrifying thing'.)

feasts Occasions of commemoration and religious *thanksgiving. Regular days were set aside for Hebrews when they could relax and enjoy themselves with music and dancing. There were thanksgivings to God for *blessings and for material relief for the poor and oppressed. Apart from the weekly *Sabbath, the release of slaves was prescribed for every seventh year (Exod. 21: 2–6) and debts were to be remitted (Deut. 15: 1–6), while in the year of *jubilee property was to return to the original owner. There were also new moon festivals (Lev. 23: 23–5).

But the three greatest festivals were first, the *Passover and Unleavened Bread, kept for a whole week in the first month (Exod. 12) in commemoration of the *Exodus from *Egypt which has always given hope and self-esteem to the Jewish people through all the vagaries of their history; this festival was established in the Temple under Josiah's reforms (622 BCE) and further changes were made by the priestly laws after the *Exile (Num. 28: 19–24). Secondly, the feast of Weeks took place at the beginning of wheat harvest (Deut. 16: 9–12) seven weeks after Passover, at the end of a busy agricultural season; and thirdly, the feast of Booths, or Tabernacles, was celebrated from the 15th to the 22nd of the seventh month (late September), and was the Israelite Harvest Thanksgiving: the ritual requirement to dwell in booths reflected God's protection of the people during their wanderings in the wilderness (Lev. 23: 39–43). At this feast *Solomon's Temple was dedicated (1 Kgs. 8: 2). After Josiah's reforms crowds of pilgrims visited the Temple. Such were the numbers that it was probably essential for many to erect temporary *booths while waiting their turn. After the Exile this practical necessity was given its theological justification (Neh. 8: 14).

Later feasts were ordered: the feast of Dedication was in thanksgiving for the rededication of the Temple by Judas *Maccabaeus on the 25th of Chislev (December) in 164 BCE, and called Hanukkah (1 Macc. 4: 59;

John 10: 22, but this was not mentioned in the Hebrew OT). The feast of *Purim in Adar (late February), connected with the book of Esther, was perhaps borrowed from the celebrations of New Year by the Persians. It became a minor holiday for Jews.

The gospel of John seems to be written round Jesus' visits to Jerusalem for the feasts, and Paul reports that the weekly celebration of the Lord's death was initiated as a commemoration of the *Last Supper at Passover (1 Cor. 11: 24). Christians in the early Church soon began to hold their regular worship on Sunday as a weekly feast in honour of the *Resurrection, instead of meeting on the *Sabbath.

Felix Governor of *Judaea from 52 CE; he had been appointed on the recommendation of the high priest *Jonathan. Paul was tried by Antonius Felix at *Caesarea (probably in 60 CE).

fellowship The OT does not often mention fellowship. There exists a kind of covenant fellowship between God and Israel, with God as the senior partner. The closeness between them is expressed in Pss. 42 and 63. But the notion of fellowship (the Greek word is *koinonia*) comes into its own in the NT, especially in the epistles of Paul. Believers' fellowship is the unity in love effected first by baptism and then maintained by the community meetings for the Lord's Supper (1 Cor. 11: 17–34). Sacramental eating of the one bread is the appointed means for communicants to enjoy fellowship with the Risen Lord (1 Cor. 10: 16–17) and by its nature it excludes fellowship with *demons (1 Cor. 10: 19–21).

The gospel of John has a principal theme of the *love which believers must have for each other (John 13: 34) and which is illustrated by the *allegory of the True Vine (John 15), of which *disciples are the branches. Christian fellowship is rooted in the relationship of the Father and the Son (John 17: 20–3).

feminism The movement to establish the equality of the sexes. Feminist theology is related to modern feminism in general. It derives from women's experience of suffering and oppression. It argues that traditional Christianity, rooted in the Bible, is too masculine and that the *faith should be reconstructed in order to do justice to the proper needs and expectations of women who, equally with men, have been made in the image of God (Gen. 1: 27). Such a reconstruction would not, however, be primarily in the interests of female liberation but for the sake of the truth of Christianity: it represents a repudiation of patterns of the sexual superiority of men and the subordination of women which have existed at the deepest levels of thought and feeling and are reflected in much of the OT and NT. Readers of the Bible who are alienated by its patriarchal attitudes are inevitably obliged also to reassess traditional views of the authority of the Bible. For example, its bias towards masculine rationalism would be balanced by the rediscovery of female characteristics in God.

Christian feminists point for support to the teaching and attitude of Jesus, who overcame the limitations of his local environment by his welcome to *Gentiles as well as Jews, and to women as well as men. Typical of his attitude is the dialogue with the *Syro-Phoenician woman (Mark 7: 24–30) and the healing of the woman with menorrhagia (Mark 5: 24–34), and he accepted the love of the woman who anointed him (Luke 7: 38). Women were the first of the disciples to be told of his *resurrection (Mark 16: 6). In the gospel of John the mother of Jesus, the two sisters *Mary and *Martha, and *Mary Magdalene are prominent. In the Church, leadership is exercised by such women as *Priscilla (Rom. 16: 3), *Chloe (1 Cor. 1: 11), and *Phoebe (Rom. 16: 1), and *Junia was 'prominent among the apostles' (Rom. 16: 7). The gospel of Luke is often regarded as particularly sympathetic to women—but they do seem to be presented in roles of service and subordination to men (Luke 8: 2–3).

Festus Approved by the emperor *Nero to succeed Felix as governor of *Judaea in 60 CE he died in 62; he had been left the difficult task of suppressing a Jewish Messianic turmoil. Festus was present when Paul defended himself before Herod *Agrippa II (Acts 26) and is said to have agreed that Paul was innocent (Acts 26: 31).

fig The tree is leafless in winter but puts out buds in February; leaves appear in April. Jesus' expression of disappointment at finding no figs (Mark 11: 12, 13, 20 f.) and his subsequent condemnation of the tree must be regarded as symbolic of a disillusionment with Israel. This is apparent from Mark's placing the Cleansing of the Temple and Jesus' clash with the leaders between two parts of the barren fig tree narrative. *Judaism is guilty; it could not tolerate the coming of its *Messiah.

fire *Incense was burnt in the Temple, and animal *sacrifices were roasted with fire. But fire is also a symbol in both OT and NT for an aspect under which God is made known—a 'consuming fire' (Heb. 12: 29). It is associated with God's wrath against *sin, and the *day of the Lord will see the *purification of Israel by fire (Zech. 13: 9). Fire is also a symbol of the presence of God, as in the burning bush (Exod. 3: 2) and the appearance on Mount *Sinai (Exod. 13: 21–2). The *Holy Spirit descended on the *apostles at *Pentecost in the form of fiery tongues (Acts 2: 3).

firmament The division made by God, according to the *P account of *creation, to restrain the cosmic water and form the sky (Gen. 1: 6–8). Hebrew cosmology pictured a flat earth, over which was a dome-shaped firmament, supported above the earth by mountains, and surrounded by waters. Holes or sluices (windows, Gen. 7: 11) allowed the water to fall as rain. The firmament was the heavens in which God set the sun (Ps. 19: 4) and the stars (Gen. 1: 14) on the fourth day of the *creation. There was more water under the earth (Gen. 1: 7) and during the *Flood the two great oceans joined up and covered the earth; **sheol* was at the bottom of the earth (Isa. 14: 9; Num. 16: 30).

first-born The first son and the first fruits of a crop had a special significance for the Hebrews and other ancient peoples in the area, and were dedicated to God to ensure a *blessing on the rest of the family or the crop. The first-born son had special privileges in *inheritance (Deut. 21: 15–17) and authority (2 Kgs. 2: 9). The last of the *plagues inflicted on *Egypt was the death of the first-born, and the ritual of the feast of *Passover owed something to the memory of this (Exod. 13: 11–16). In the NT Jesus is called the first-born of the Father (Rom. 8: 29; Rev. 1: 5).

Paul regards his first converts in *Asia as first-fruits (Rom. 16: 5, NRSV marg.), and Christians in this earthly life have already received the first-fruits of the *Spirit (Rom. 8: 23).

fish Many species of fish lived in the Sea of *Galilee and the *Jordan. They were sometimes caught with a hook (Matt. 17: 27) but more often by letting a net into the water in a semicircle and gradually drawing it ashore (Matt. 13: 47–50). There was an industry of fish-curing at Taricheae. Pickled fish were sent up to Jerusalem.

Early Christians used a fish as a *cryptogram or symbol for Christ, especially in the *Eucharist on account of the fish mentioned at the Feeding of the Multitude (Mark 6: 38). Moreover, additionally, the initial letters in Greek of Jesus Christ, Son of God, Saviour, spelt out the word *ichthus* (fish).

flesh Part of humanity and not of God. Although the word, *sarx* in Greek, is used in the literal sense of the physiology of the human body, or indeed of horses (Isa. 31: 3), meaning that they are mortal, it is also used sometimes of the body itself or of humanity in general. However, 'flesh' tends to have a connotation of weakness (Ps. 78: 39) or even of sin. 'According to the flesh' denotes inadequacy or failure (1 Cor. 1: 26 NRSV marg.). It is humanity viewed apart from God, and is to be contrasted with '*spirit'. Jesus' fleshly descent from *David is less significant than his *resurrection by the Spirit (Rom. 1: 3–4). In the letters of Paul 'flesh' describes the works of the *Law (Gal. 3: 2–3) as well as persons who do not have the Spirit (1 Cor. 2: 12–3: 4). Vices condemned as fleshly are not necessarily carnal or physical (Gal. 5: 20) but all lead equally to '*death' (Rom. 8: 6) because they are egocentric. Christ, however, lived an entirely God-centred life while sharing real human fleshly existence (Rom. 8: 3).

Because of the width of meaning of the word, modern English versions resort sometimes to other translations: 'human weakness' or 'unspiritual nature'.

flood There is no archaeological evidence for the universal flood described in Gen. 6–8, but there are several Mesopotamian stories about how a family marvellously survived such a flood. In Gen. the flood is a *punishment from God who was determined to clear the earth of its gross impurities caused by the misdeeds of its inhabitants. Other flood stories, such as the Gilgamesh epic from 12th-cent. BCE *Babylonia, discovered at *Nineveh in 1872, seem to regard the flood as somehow issuing out of disputes amongst the gods, and the hero who survives is granted immortality. There are so many similarities between the Hebrew and the Babylonian stories that both probably depend upon the same original version.

The fate of *Noah, who survived the biblical flood in his *ark, is different; although he offers *sacrifices he drinks himself to death (Gen. 9: 20–8). The importance of the biblical flood story lies in its narrative of the *covenant between God and Noah: God's promise is that there will never again be such a deluge, while Noah, as representative of humanity, is to observe certain basic laws, especially about the shedding of *blood, and mankind is 'to replenish the earth' (Gen. 9: 1–7). These hints of a covenant, long before the covenant with Moses and the chosen people, were later elaborated by the Jewish rabbis into the 'Noachian precepts', the 'way of all the earth' which they regarded as obligatory on *Gentiles. It has been suggested that the requirements laid on Gentile converts by the Church in Jerusalem (Acts 15: 29) are in fact the Noachian code.

The symbolism of the flood is used in 1 Pet. 3: 20, where Noah's ark is regarded as a prototype of the Church conceived as a lifeboat launched upon the waters of time.

flute-players Employed professionally at weddings and funerals (Matt. 9: 23; 11: 17). Rabbis recommended that funeral processions should always include two flute-players.

follow The invitation to become Jesus' *disciples was given in the imperative 'Follow me!' which demanded total obedience. For Peter, Andrew, James, and John it involved the breakup of the family business and loss of capital inheritance. To follow Jesus would be to share his earthly circumstances (Matt. 8: 19) and suffering (Mark 8: 34). Not surprisingly, Peter was dismayed (Mark 8: 32, followed by 34).

food The Jews were not gluttons, and although they ate mostly products of the soil, they were not vegetarians. *Fish were plentiful, grasshoppers were eaten fried, but other meat, always with a concern for what was kosher, i.e. ritually slaughtered as ordered in the Law (Deut. 12: 15 ff.), was a luxury. Honey was valued, and so were eggs (Deut. 22: 6). Pigs were taboo, and so was the meat of any animal that had died a natural death (Deut. 14: 21).

fool A fool was a person without a proper gift of intelligence (Luke 12: 20; Rom. 1: 21; Eph. 5: 15) but also one who was impious and blasphemous, as usually in the OT (e.g. Ps. 14: 1). Nevertheless in Matt. 5: 22 Jesus condemns the expression of contempt implied by rebuking someone as a fool.

foot Falling at a person's feet indicated homage (Mark 5: 22), walking barefoot indicated awe (Exod. 3: 5) or mourning (2 Sam. 15: 30). One foot placed on the neck of an enemy was an eloquent expression of victory (Josh. 10: 24).

Pupils sat at their master's feet (*Mary with Jesus, Luke 10: 39; Paul with *Gamaliel, Acts 22: 3); Jesus washed the *disciples' feet

(John 13: 5–17) and was himself washed (Luke 7: 38). Shaking *dust off the foot showed scorn or a renunciation of responsibility.

footstool Used in connection with *Solomon's throne (2 Chr. 9: 18), but also metaphorically of the subjection of Israel's enemies to God (Ps. 110: 1; quoted in the NT—Matt. 22: 44; Acts 2: 35; Heb. 10: 13). The earth itself is described as God's footstool by Isa. 66: 1, Matt. 5: 35, and Acts 7: 49.

forehead A *phylactery was to be worn on the forehead of a Hebrew man (Exod. 13: 9, 16; Deut. 6: 8), and in *Ezekiel's vision of the future (9: 4) the righteous are to be branded on the forehead with the letter *Taw* (a cross, X, in the old Canaanite Script). So also in Rev. 7: 3 the seal of God is set on the forehead of his servants. The early Church adopted the taw as an appropriate symbol of the *cross of Jesus, and so a specifically Christian symbol.

'Forehead' can also be used to signify a hard obstinacy of character (Isa. 48: 4; Ezek. 3: 7).

foreigner(s) In the OT a foreigner is a stranger or resident alien living among the Israelites (Exod. 20: 10), protected by *law but obliged to observe the *Sabbath. On the other hand foreigners could be a dangerous influence if they were given to *idolatry (Isa. 2: 6–8; 1 Kgs. 11: 1) and after the Exile there was a prohibition against *marriage to a foreign woman (Neh. 13: 26–7; Ezra 9–10) and an increasing emphasis on the purity of the race.

In the NT believers may be aliens in the world (1 Pet. 2: 11) but are no longer strangers in the Spirit, but citizens with the saints and fellow heirs of God's promises (Eph. 2: 19). The use of foreigner for *Gentile continues (Acts 26: 11) and even for a *Samaritan (Luke 17: 18).

foreknowledge Paul refers to four stages in God's plan of *salvation (Rom. 8: 29). First, God had foreknowledge of all the future and so chose the 'elect' (cf. Acts 4: 28); then he 'called', when those chosen became aware of it; then they were 'justified' by their faith, and finally they would be 'glorified', the *first-born of many brethren.

How God's foreknowledge can be reconciled with genuine human free will has not been easy for traditional Christian theology, and some recent thinkers hold that by his act of creating a temporal universe God was committed to a temporal relationship with it—there were elements in the future unknown to God.

foreordain Although this verb, and the noun 'determination', are fairly rare in the Bible, the idea they stand for dominates both testaments. In a fairly extreme form it is asserted by *Deutero-Isaiah: 'I make well-being, and I create disaster' (NJB, Isa. 45: 7). Everything is in the hand of a sovereign creator, without, however, depriving human beings of their own independence of action (Deut. 30: 15–16). What God ordains is, primarily, the salvation of humanity, and to that end he chooses a particular people (Gen. 18: 18) though not for any merit discernible in that people (Deut. 7: 7). In the NT God's determination is focused on Jesus, who fulfils OT expectations (Matt. 11: 3) and 'is going as it has been determined' (Luke 22: 22, NRSV). In the early Church the items in the apostles' *preaching were said to have been ordained by God in his eternal providence (1 Pet. 1: 20).

forerunner A soldier who ran ahead of the main body of troops as a scout; often *John the Baptist is so regarded, as is implied in Matt. 11: 10, in that he preceded Jesus and prepared his way. But in Heb. 6: 20 Jesus is himself called a forerunner, in that he is a kind of advance guard: he is already seated in *heaven and his presence there guarantees the subsequent entry of all who belong to him.

forgiveness A re-establishment of personal relations after a rupture. Although in the OT God does not forgive automatically, as though that is what he exists for, he is described as always ready to forgive when the right conditions are in place (Neh. 9: 17). To a penitent sinner God's forgiveness is complete (Isa. 43: 25). The OT system of *sacrifices was designed as a concrete expression of repentance enabling God's forgiveness to be bestowed and guilt expiated.

In the NT there is the same emphasis on God's forgiveness, sometimes activated by Jesus himself (Mark 2: 5–6), and in the Church there is to be provision for members forgiving each other (John 20: 23; Jas. 5: 13–16) and the *Lord's Prayer anticipates God's forgiveness of our sins only when we have forgiven the sins of others (Matt. 6: 12). The parable of the Prodigal Son (Luke 15: 11–32) could be more appropriately entitled the Forgiving Father. That of the two debtors (Matt. 18: 21–35) ends with a stern warning: the unforgiving servant will be tortured until his debt should be repaid *in toto*. Possibly there were members of Matthew's Church who were not showing a forgiveness to others expected from those who had been forgiven so much by God. The ending has almost certainly been attached to Jesus' parable by the evangelist himself to accord with common contemporary practice of torturing a debtor by way of putting pressure on his relatives to contribute to his repayment. But it is incongruous. The meaning of the parable is straightforward enough: that God's free forgiveness should be imitated in our own relationships.

Form Criticism The English translation for the German *Formgeschichte*. This technique was developed by a group of German biblical scholars shortly after the First World War. It assumed the widely agreed conclusion of source criticism of the priority of Mark and the view that the Gospel of John was later than the other three but the aim was to penetrate into the period of Church life before even the earliest sources had been written. 'Form Criticism' had been used in Germany since about 1900 to explore some of the OT narratives and Jewish and Hellenistic literature. The literary classifications of prose and poetry were subdivided into (prose) history, legends, and myths; and (poetry) hymns, psalms, and prophetic oracles. It was claimed that in the course of oral transmission these types had developed according to a regular process. Form Criticism, however, is best known in Britain and America as a NT tool since it was introduced by the NT scholars R. H. *Lightfoot and Vincent Taylor in the 1930s.

Form Critics generally accepted the theory of W. *Wrede that Mark's gospel was not a straightforward, though unliterary, reproduction of the reminiscences of Peter but was permeated with a strong apologetic interest: why had Jesus' Messiahship not been recognized by his contemporaries? Because Jesus made efforts to keep it a secret. So, according to Wrede, the gospels, beginning with Mark, were constructed as supports for the early Church. But what was happening to the memories about Jesus in the thirty years or so before anything was committed to writing? There was *preaching and *teaching and controversies with the Jews, and Christian leaders had a fund of stories to draw on as need arose and inevitably their stories developed into characteristic shapes. The audiences needed advice and reassurance, information and answers to problems just as much as the readers for whom Mark was shortly to write his gospel. Indeed Mark's gospel consisted, according to the Form Critics, of precisely all those fragments and isolated units that were used by Christian evangelists to get their message across in a persuasive form. Mark's gospel is not so much a connected, chronological narrative (think of the geography of Jesus' journeys rapidly made from one place to another, from the hills of Galilee to the Lake, from somewhere to Capernaum, from Galilee to the coasts of Tyre and Sidon, and eventually to Jerusalem) as a series of incidents, *miracles, *parables, and injunctions strung together with a minimum of linkage.

These units circulated in isolation during the 'oral' period; Form Critics called them *pericopae, and scrutinized them independently and classified them into 'forms'. Their original historical situation in the ministry of Jesus has been for ever lost: what is discernible is the way the pericopae reveal the influence of the needs and outlook of the Church. A particular example would be the interpretation and explanation of the parable of the Sower (Mark 4: 10–20) which related more to the life of the Church than to the *ministry of Jesus.

After assigning a pericope to the appropriate category or type (e.g. *apophthegm, *chria, *paradigm), the Form Critic next relates it to its social setting or circumstances (in German **Sitz im Leben*, life situation) which explains how a story has been adapted to fit changes in the development of the Church, such as the expanding *Gentile mission. Or there may be additions to a parable (e.g. Luke 19: 27) which reflect growing hostility to the Jews. Several Form Critics eliminated genuine history from some of the pericopae altogether, especially that group of events, such as the *Transfiguration, which have a strong supernatural element, and regarded them as creations of the Church in the light of its belief in the *Resurrection. Rudolf *Bultmann, a leading Form Critic, was held by other scholars to be excessively sceptical and did not allow for factors which controlled any extravagant growth of legendary embroidery. It is for example noticeable that some sayings are accurately preserved (e.g. Matt. 11: 12) which clearly the Church did not understand. But some had remembered!

In due course the detailed concerns of Form Criticism were succeeded by those of *Redaction Criticism, which recognized the theological interests of the evangelists but gave them greater credit as authors. They did more than string together isolated units of tradition. The Form Critics had in a way anticipated Redaction Criticism by recognizing that there is one section—the *Passion Narrative—which is a unity, with reliable topographical and chronological details, but with variations (e.g. in the words of Jesus from the cross) which illustrate the special theological interest of each author.

fornication Various kinds of irregular sexual relationships are covered by this word but the OT takes a fairly relaxed view of most of them. Rahab the harlot is far from being stigmatized (Josh. 2); she comes in the *genealogy of Jesus (Matt. 1: 5). But Lev. 18 represents a harsher attitude. Advice to young men after the Exile was that irregular relationships are inadvisable (Prov. 7: 5–27); but where associated with religious prostitution or heathen rites opposition is total (Lev. 21: 7, 9).

In the NT a metaphorical use is common (e.g. Matt. 12: 39) as well as the literal use (1 Cor. 6: 18). Fornication is condemned as being incompatible with the sovereignty of God, whose intention is to restore all mankind in undivided loyalty to himself. Nevertheless Jesus taught that in the Kingdom outcasts and prostitutes would take precedence over the self-righteous (Matt. 21: 31 ff.).

foundation A carefully selected basis for building (Matt. 7: 25); also a metaphor for such a basis (Heb. 11: 10). *Apostles are the foundation of New Jerusalem (Rev. 21: 14); and in 1 Cor. 3: 10 Paul's preaching of the gospel is said to be the foundation of the Church—though this is corrected somewhat in the following verse: the foundation is Jesus Christ himself.

fountain A natural spring of flowing water, as opposed to a well or a cistern. Fountains are numerous in upper *Galilee. They offer security (Isa. 41: 17–18) and are pictured as among the blessings in the eschatological future (Ezek. 47: 1–12). *Jeremiah (2: 13) describes Yahweh as the fountain of living waters—which is what Jesus promises to be able to give the woman of Samaria (John. 4: 10).

fox Foxes were common in Palestine, and *Ezekiel lambasts false prophets as foxes (AV; 'jackals', NRSV, REB, NJB), that is, scavengers who hope to profit from the ruin of the nation (Ezek. 13: 4). In the NT the fox is not so much a symbol of slyness, or cunning, as in modern parlance, but is despised as being worthless and insignificant, which is what Jesus means by calling Herod *Antipas a fox (Luke 13: 32).

frankincense An aromatic gum found in SW Arabia and Abyssinia, and used in the

Temple (Isa. 43: 23). It was one of the *gifts brought by the *Magi (Matt. 2: 11), part of the stock-in-trade of sorcerers who now laid it in submission at the feet of the infant Jesus. Cf. Acts 19: 13–20.

freedom Release from a state of slavery is both a literal and a metaphorical concept in the Bible. The Israelites secured their freedom from bondage in *Egypt under the leadership of Moses, but if the chosen people were to renounce their side of the *covenant, God would return them to slavery (Deut. 28: 47–8). Israelites who had been sold into slavery by fellow Israelites were to be released every seventh year (Deut. 15: 12 ff.)

In the NT freedom is a prominent notion but it was not freedom from the Roman yoke by means of arms that Jesus advocated but freedom from the domination of the powers of *evil (John 8: 34 ff.). Paul continued this message by teaching that Christians have freedom from the domination of the *law (Gal. 5: 1) and the institutions of *Judaism. Christians are the true *children of *Abraham (Gal. 3: 7).

freedmen In the Roman Empire freemen were those born free, as distinct from freedmen who had been liberated from slavery and whose civil rights were restricted (cf. 1 Cor. 7: 22). Paul declared that such distinctions no longer obtained in the Church of Christ (Gal. 3: 28).

In Jerusalem there was a *synagogue of Freedmen whose members disputed with *Stephen (Acts 6: 9). They were possibly descendants of Jews who had been enslaved by Pompey (63 BCE).

fringes Moses is commanded to instruct the Israelites to make fringes (NRSV; 'tassels' REB, NJB, Num. 15: 37) on the corners of garments, as a reminder of the *Ten Commandments. Jesus is said to have criticized *Pharisees for making their fringes overconspicuous (Matt. 23: 5), but he himself certainly wore them (Matt. 9: 20; 14: 36).

frontlet Used in AV for an ornament or jewel ('emblem', NRSV; 'pendant', REB; 'headband', NJB) worn on the head, probably between the eyes (Exod. 13: 16; Deut. 6: 8) and by implication forbidden in 1 Tim. 2: 9.

fruit In the literal sense, such fruit as *figs (1 Kgs. 4: 25) and pomegranates were familiar in Palestine (Num. 13: 23) as well as mulberries and melons. Children are known as fruit of the womb (Luke 1: 42). Figuratively, human conduct is thought of as the fruit of a tree, either sound or rotten (Matt. 7: 16–20), and Paul asks his converts what fruit (Rom. 6: 21, AV; 'advantage', NRSV; 'gain', REB, NJB) they had in past deeds of which they are now ashamed. The fruit they should produce is that which comes from the *Spirit (Gal. 5: 22).

fulfil Jesus' proclamation (Mark 1: 15) that 'the time is fulfilled' means that the time appointed by God has now arrived, and Paul expresses the same idea: 'when the fulness of time had come' (Gal. 4: 4). The gospel of Matt. frequently observes that an event prophesied has come about at the appointed time by the use of the phrase 'to fulfil what had been spoken by the prophet' (e.g. 1: 22; 2: 17).

fuller A fuller was engaged in the unattractive, evil-smelling trade of cleaning and bleaching clothes in large vats; he trod down articles with his feet. At the *Transfiguration, Mark (9: 3) records that the whiteness of Jesus' garments was beyond what any fuller (AV, NRSV marg.; 'bleacher', REB, NJB) could make.

fullness The translation of the Greek *pleroma*, which can be either active or passive—either 'that which fills' or 'that which is filled up', which is a complication affecting the meaning of passages in the NT. It is used in Mark 6: 43 of the broken pieces of bread which filled twelve baskets, but in John 1: 16, 'we have all received from his fullness', is explained by 1: 14, where the Word is filled by *grace and *truth.

The word *pleroma* is found in the *LXX but in NT times the *Gnostics used it for the totality of all reality and this may be why it is used in Eph. 1: 23 and 3: 19. The fullness of God is in Christ; but these verses are unclear whether the meaning is also that the Church is filled by Christ. Perhaps the meaning is that Christ (head of the Church) and the Church (the *Body of Christ) are a unity; both are filled by God, who fills all (is supreme).

fundamentalism Term applied to strictly conservative religious belief; in Christianity it holds the entire *Bible to be free from historical, theological, and scientific error. Fundamentalists oppose the teaching of evolution and all concessions to modern thought embraced by liberal Christianity. They tend to emphasize the demands of personal morality rather than social issues. The term derives from a series of tracts which began to appear in the USA in 1909 and which affirmed five fundamentals of faith: the verbal inerrancy of scripture, the divinity of Christ, his birth of a virgin, a substitutionary theory of the *atonement, and the physical *resurrection and bodily return of Christ.

funerals After *death, a body was washed (Acts 9: 37) and anointed (Mark 16: 1) and wrapped in cloth (Matt. 27: 59) and was then transported on a bier in a procession (2 Sam. 3: 31; Luke 7: 14), possibly accompanied by professional mourners (Amos 5: 16). Burial as soon as possible was important (Deut. 21: 23). The place of burial would be outside the inhabited area (Luke 7: 12) in a grave (Gen. 35: 8) or cave (Gen. 23: 19), or a tomb cut out of rock in a garden (John 19: 41–2), sometimes whitewashed to be visible at night, as a safeguard against being defiled by accidental contact (Matt. 23: 27). Neither cremation nor embalming was practised by the Hebrews.

furnace Five different Hebrew nouns are translated 'furnace' in AV and refer to ovens, kilns, and domestic fires, as well as the burning, fiery furnace of Dan. 3: 6 ff. A Greek noun is translated 'furnace' by AV and NRSV twice in Matt. (13: 42, 50) and twice in Rev. (1: 15; 9: 2). Modern translations sometimes use 'furnace' (e.g. of *Sodom and *Gomorrah's destruction, Gen. 19: 28), whereas in Exod. 9: 8 it is a 'kiln' (NRSV, REB, NJB). The furnace image was appropriate to apocalyptic passages for the *punishment of the wicked, but the metaphor implies extinction rather than endless torture; so Matt. 13: 42, of *Gehenna.

furniture There was a big difference between the furniture of *Solomon's palace (1 Kgs. 7: 7) and the simple and sparse furniture of a Palestinian household where only the essentials for eating, sleeping, and sitting were provided—perhaps also a lamp (2 Kgs. 4: 10). The Upper Room (Mark 14: 15) probably contained nothing else apart from vessels for the *Passover. The command to the paralytic (Mark 2: 9) was to pack up his pallet and pile it against the wall at home.

The furniture of the Temple is described in 1 Kgs. 7: 48–50.

G

Gabbatha In John 19: 13 this is said to be the Hebrew (though, rather, Aramaic) name for the place of *Pilate's *judgement seat, called in Greek the Pavement, being a street paved with stones. Such a site has been excavated at the *Antonia fortress on high ground north of the Temple in Jerusalem, and there are lines which could have been cut by soldiers for a game. They *could* have been playing it during the mocking of Jesus (John 19: 3). On the other hand, another opinion holds that this street was not laid down until the time of Hadrian (135 CE).

Gabriel One of the chief *archangels, who first appears in the OT as a messenger from God (Dan. 8: 15–26) and in the NT as the herald of the supernatural births of John and Jesus (Luke 1: 19 and 1: 26–36). Gabriel is also prominent in 1 Enoch, and his role in Christian apocalyptic imagery may have been to blow the trumpet to announce the general *resurrection (1 Thess. 4: 16).

Gad A son of *Jacob, regarded as the ancestor of the tribe of Gad, which settled to the east of *Jordan (Num. 32: 20–32). They had enjoyed a reputation as soldiers (1 Chr. 12: 8, 14), but did not escape the *Assyrian captivity in 734 BCE, when the tribe disappears from history (1 Chron. 5: 26). In some translations of Isa. 65: 11 (e.g. NRSV) 'Gad' is regarded as a god of fortune.

Gadara A city of the Decapolis, 9.6 km. (6 miles) SE of the Sea of *Galilee, where Jesus healed two demoniacs in Matt. (8: 28), and one demoniac in Mark (5: 2). (The best Greek MSS of the corresponding passage in Mark (5: 1) gives the place as 'Gerasa' which is 48 km. (30 miles) from the lake, and therefore geographically less plausible.) Gadara was a Gentile city and a centre of philosophical speculation. Whereas Mark (5: 19) records Jesus' instruction to the demoniac to tell his friends about his cure, this is omitted by Matt. This may possibly be because Matt. intends to show that the Church's mission to the Gentiles is yet to come; it will take place, but later.

Gaius The name of several men who have minor roles in the NT. (1) A native of *Derbe who accompanied Paul to Jerusalem (Acts 20: 4 f.). (2) One of the few persons baptized by Paul at *Corinth (1 Cor. 1: 14). (3) The recipient of 3 John. (4) Gaius was also the name of the Roman emperor (37–41 CE), surnamed *Caligula by his soldiers. He behaved generously towards foreign princes, and restored *Agrippa, grandson of *Herod the Great, to his kingdom of *Judaea. But after a serious illness his enemies claimed he had gone mad and he delighted in hideous cruelties.

Galatia The difficulty about precisely identifying the area mentioned in 1 Cor. 16: 1; Gal. 1: 2; 2 Tim. 4: 10; and 1 Pet. 1: 1, is that it may refer to the limited territory in Asia Minor dominated by the warlike Gauls, or to the Roman province of Galatia which also embraced parts of other territories to the south. The cities visited by Paul according to the Acts were in the province but not the territory—Pisidian *Antioch, *Iconium, *Lystra, and *Derbe. So these could have been the Churches to whom Paul's *epistle to the Galatians was addressed.

Galatians, Paul's epistle to the In the NT, the fourth of the letters of Paul. Was this epistle written to the Churches Paul is said to have visited in Acts in the province of *Galatia, or to *other* unmentioned Churches in the ethnically Gallic northern part of the province, the ancient kingdom of Galatia? If the former (southern) destination is accepted, then the epistle will be the earliest of the Pauline epistles, written perhaps in 48 CE to Churches he had evangelized; if the latter, then it is likely to have been written after 1 Thess. and before 1 Cor. Much depends on whether Paul's visit to Jerusalem described in Gal. 2 is that of the 'Council' of Jerusalem of Acts 15 (49 CE). If so, the epistle is late and is a robust response to opponents who alleged that his behaviour at Antioch was inconsistent with the agreement at Jerusalem. If not, then it is understandable that the decision of the council is never referred to, since it had not yet taken place. It is certainly possible that 'the region of Galatia' of Acts 16: 6 could be the northern district, distinct from the urban areas he had visited according to Acts 16: 1–5. There are therefore reasonable grounds for holding both views about the epistle's date and destination: south and early (about 48 CE), or the more usual modern view, north and later. At any rate, Paul's letter is ferocious in its language; clearly he felt his message and his status as an apostle were under threat. A group descended on Galatia (possibly they were Christian *Pharisees) who argued that the *Gentiles who had become Christians should also become members of the covenant people of God (Israel) and observe the *Law, especially by male members being circumcised. It was not the only confrontation between Paul and these opponents: when he had gone to Jerusalem with a Gentile convert, *Titus, certain 'false brethren' had then raised the issue of *circumcision, which Paul resisted (Gal. 2: 3). They claimed too that Paul's apostleship, such as it was, derived from the Jerusalem Church, and he had no authority to deviate from their *teaching. Paul insisted that he was called by Christ himself. It was especially painful to Paul when Peter visited *Antioch (Paul's base) and reneged on his first intention to participate in a meal (possibly including a *Eucharist itself) at which Gentiles were also present.

Paul held that the demand for Gentile converts to submit to the Law undermined his message that we are 'justified' by *faith in Christ, not by the works of the Law. This fundamental theme was famously understood by Martin Luther in terms of his own religious experience and rejection of strenuous efforts to earn salvation by the discipline of a Catholic monk. He thus gave a wrong direction to exegesis ever since. Luther felt an enormous sense of relief when he shook off the burden of a profound sense of guilt: he read Paul's words in Gal. and in Rom. to mean that God reckoned a believer in Christ to be regarded as righteous by reason of his faith alone even though he was a sinner. *Righteousness was *imputed* to him; he was declared righteous by a fiction in virtue of the *mercy of God, though he remained a sinner.

However, Luther's *exegesis was mistaken. Paul was not obsessed with his sinfulness (Phil. 3: 6) but with the transfer from membership of the people of the *covenant who responded to God's *grace by obedience to the law into a *community of faith in Christ. People entered it not by circumcision but by *baptism, and responded with an active life of love involving one's whole being. It was a new dimension (2 Cor. 5: 17) for those made one with Christ (Gal. 3: 28).

Both his opponents and Paul agreed that the OT proclaimed that the God of Israel desired the salvation of all mankind, i.e. Gentiles as well as Jews, but the opponents urged that the Gentiles who accepted Jesus as the *Messiah must first become Jews and obey the laws laid down in scripture, as explained in Gen. 17, for all the descendants of *Abraham. Paul, however, maintained that the Genesis narrative about Abraham shows that what was primarily required of Abraham's descendants was only faith (Gal. 3: 8). And what faith does for the Gentile converts is to make them one in Christ (Gal. 3: 26). It is the Christians who are now the true children of Abraham. If Jewish Christians wished to continue the practice of circumcision, it was up to them (Gal. 6: 15; 1 Cor. 7: 19–20). It was a matter of indifference. But on no account must the demand be imposed categorically on Gentiles (Gal. 2: 21). God's chosen method of *salvation was through Christ—his death and his resurrection—and it is appropriated by faith externalized in baptism. It confers the promises made to Abraham. Christians are freed from slavery to the law and have new life—but not a life of irresponsible abandon. Lev. 19: 18 continues as a guideline: 'You shall love your neighbour as yourself' (Gal. 5: 14).

People, in Paul's view, are not to be judged by their performance but on what they are and have received from Christ.

Galilee The northern part of Palestine. A smallish area—72 km. (45 miles) long—on the west of the Golan Heights and south and east of Lebanon, first occurring in the OT in Josh. 20: 7 and called by Isaiah (9: 1) a land of *foreigners. It saw various fortunes after the settlement of several Israelite tribes in the 12th cent. BCE, which *David consolidated. However, twenty Galilean towns were handed over by *Solomon to *Hiram of Tyre in exchange for timber and gold (1 Kgs. 9: 11). There were invasions by Egypt in 924 BCE and Assyria in 853 BCE and thirteen cities were captured in 732 BCE (2 Kgs. 15: 29). As an *Assyrian province the country became part of *Samaria. It returned to Judah under the *Maccabeans but was later absorbed into the Roman Empire by Pompey in 63 BCE. *Herod the Great was given Galilee as part of his kingdom (40–4 BCE) and on his death it was joined to *Perea as the tetrarchy of Herod *Antipas (to 39 CE). *Nazareth and *Capernaum, where Jesus preached and taught (Matt. 4: 15; Luke 4: 31), are in Galilee and the inhabitants spoke with a northern accent (Matt. 26: 73). Matt. 4: 12–13 refers to it as the land of *Zebulun and *Naphtali but the tribal dispositions had by NT times only antiquarian interest. More important is the description 'Galilee of the Gentiles' (Matt. 4: 15), for many Gentiles were settled in the region: Matthew thus foreshadows the Church's future mission to the Gentiles. During the Jewish war of 66–70 CE *Josephus was the general in charge of Jewish forces in Galilee. He was taken prisoner by the advancing Romans.

Galilee, Sea of In northern Palestine; also called 'the Lake of Gennesaret' (Luke 5: 1) and 'the Sea of Tiberias' (John 6: 1), a freshwater lake separating two parts of the River *Jordan. It spreads out to 12.8 km. (8 miles) across and its length is about 21 km. (13 miles). The situation between hills causes sudden changes of temperature and squalls (Mark 4: 35–41). Fishing has always been an important industry and pickled *fish used to be sent up to Jerusalem. What Mark regards as a 'sea' (Mark 4: 39), Luke from the wider perspective of a world traveller designates a mere 'lake' (Luke 8: 22). It is of great importance in the narratives of Jesus' ministry.

gall Jesus on the *cross was offered a drink of *wine mixed with gall, a herb of intense bitterness which had the effect of dulling the senses. Jesus tasted the drink but did not consume it (Matt. 27: 34). Gall is used as a metaphor for bitter experience in Lam. 3: 19 and Acts 8: 23.

Gallio Brother of Seneca the philosopher; mentioned in Acts 18: 12–17 as proconsul of *Achaia and judge in the case brought by the Jews against Paul. He dismissed the disputants. An inscription discovered at Delphi mentions Gallio and provides a date for his time in *Corinth and so also for Paul's foundation of the Church there, 51–2 CE.

Gamaliel A liberal-minded *Pharisee in the *Sanhedrin who had trained Paul (Acts 22: 3) and who in about 30 CE counselled moderation when the *apostles had been apprehended (Acts 5: 34). The words put into his mouth about the rising of *Theudas are a mistake on Luke's part, since this took place in about 44–6 CE.

games The OT has nothing comparable to the wealth of Greek literature about games, though there is mention of a contest, possibly wrestling, in 2 Sam. 2: 12–17, when twelve young men on each side were slain. However, with the influx into Palestine of Greek culture and in view of Paul's travels round Mediterranean cities, it is not surprising that his epistles often contain metaphors derived from popular games—racing, boxing, and the like (1 Cor. 9: 24–7; Phil. 3: 14).

Boards used for games have been found by archaeologists, and in Jerusalem the soldiers dicing for Jesus' garments (Matt. 27: 35) may have used lines scratched on the Pavement (*Gabbatha).

garden References in the OT (e.g. Jer. 29: 5; S. of S. 4: 12–16) show that gardens were both productive and also much valued. The garden of *Gethsemane (Mark 14: 32; cf. John 18: 1), where Jesus prayed before his arrest, is designated simply 'a garden' by John. In the OT there is the mythical garden of *Eden (Gen. 2: 15) and a garden of eschatological hope (Isa. 51: 3).

gate The defensive walls of a city necessarily had gates for those coming and going; gates were subject to enemy assault and were therefore also specially defended (2 Chr. 26: 9). Even so an enemy might destroy the gate (Judg. 9: 49, 52). Several of the gates of Jerusalem are mentioned in the account of the rebuilding of the city by *Nehemiah. In the NT the *Beautiful Gate of the Temple is mentioned in Acts 3: 2.

Markets were placed outside gates which therefore became the centre of commercial and social life, and also the place for legal decisions (Ruth 4). There the unemployed queued for work (Matt. 20: 3).

The 'gates of hell' which will not prevail against the Church (Matt. 16: 18) is an idiom for *death (Ps. 107: 18; Job 38: 17).

Gath A *Philistine city which was struck by disease when the Israelite *Ark of the Covenant was taken there (1 Sam. 5: 6 ff.). Nevertheless *David was hospitably received in Gath during his flight from *Saul (1 Sam. 27: 3). The claim that David captured Gath made in 2 Chr. 26: 6 is unreliable; a more certain fact is that it was destroyed by the *Assyrians in 712 BCE. 'Tell it not in Gath' (2 Sam. 1: 20; Micah 1: 10) became a proverb signifying disaster.

Gattung German noun meaning a 'species' or 'class'. It has been used by biblical critics to denote types or *genres of literature which can be classified. A *Gattung* follows a certain conventional pattern appropriate to its social circumstances—e.g. in ancient Israel or the early Church—and has acquired its form through oral use such as public *worship or *preaching before being committed to writing. The term may be applied to an *apocalypse or a piece of history, or even to an epistle, and denotes a larger unit than a single *pericope. *Gattungsforschung* is the study of these literary types.

Gauls Several movements of Gauls eastward led to a settlement in Asia Minor and in 189 BCE E. Gaul (called *Galatia by the Greeks) came under Roman domination as a dependent kingdom. It became part of a Roman province in 25 BCE.

Gaza A city in the hands in turn of the tribe of *Judah (Judg. 1: 18), the *Philistines (1 Sam. 6: 17), the *Assyrians, Judah under *Hezekiah (2 Kgs. 18: 8), the Egyptians (609 BCE), and the *Seleucids. It was destroyed by Jonathan the *Hasmonean (145 BCE) and rebuilt by the Romans, after which it was inhabited by both Jews and Christians.

gazelle A beautiful antelope; it was adopted as a girl's name (*Dorcas, *Tabitha, Acts 9: 36).

Gedaliah The governor of *Judah appointed by *Nebuchadnezzar after its capture in 586 BCE (2 Kgs. 25: 22). He enjoyed the support of *Jeremiah and seems to have restored order over what was left of the country. He was murdered by nationalists who regarded him as a collaborator. Later his death (Jer. 41: 2) may have been commemorated by the annual day of mourning on the third day of the seventh month. as mentioned in general terms in Zech. 7: 5.

Gehazi *Elisha's servant (2 Kgs. 4: 12). After Elisha had cured *Naaman the Syrian commander of *'leprosy', Gehazi dishonestly obtained a reward for the cure, and for this crime was afflicted with leprosy (2 Kgs. 5: 27).

Gehenna A valley outside Jerusalem through which the road to *Bethlehem runs. *Children were once sacrificed there to the god *Molech (Jer. 7: 31), and later there was a continuously burning dump. Hence it came to be an image for *hell (*Hades), a place of condemnation and extinction of the wicked (Matt. 18: 8 f. NRSV marg.).

Gemariah There are two people of this name: (1) The son of Shaphan who tried to save *Jeremiah's scroll from being burnt up (Jer. 36: 10 ff.). (2) The son of *Hilkiah who took Jeremiah's letter to *Babylon to encourage the exiles to settle down (Jer. 29: 3). Yahweh could be worshipped there and the destruction of his Temple did not imply the impossibility of Jewish *worship elsewhere.

gematria A Hebrew word for a method of biblical interpretation employed by *rabbis and also adopted occasionally by Christians. It involved the system in Hebrew where numerals were represented by a particular letter, and so a word or name could be assigned a number. Thus in Rev. 13: 18 it could be argued, in reverse, that the number 666 represents a name; the Hebrew letters make the words 'Nero Caesar'.

genealogy A catalogue of names of a dynasty or *family. In the OT and NT there are several lists of ancestors or descendants. In some cases the names (which may not be comprehensive—the Hebrew noun *ben* could mean not only 'son' but also 'grandson', or simply 'descendant') are a means of tracing an individual's history back to an important figure of the past; or the list may be an illustration of human mortality.

The ancestry of the twelve tribes of Israel back to the *patriarch *Jacob is mentioned in Gen. 46 and at the time of the Return from Exile the books of Ezra and Nehemiah record registers of families. Indeed the whole of 1 Chron. 1–9 consists of genealogies, whose existence in the text bears witness to the Hebrews' consciousness of being part of an extended family. The interdependence of individuals on each other, and of different social groups on each other, was a fundamental assumption of Hebrew society. Genealogies were also useful for establishing the validity of claimants to the priesthood.

genealogy of Jesus Both Matthew and Luke, in different places in their gospels, record the ancestry of Jesus. Matt. 1: 1–17 arranges the names into three groups divided by important historical events, from *Abraham to *David; *Solomon to the Exile; and from the Exile to Jesus: two groups of fourteen generations (Matt. 1: 17), by omitting four Davidic kings, and a third group of only thirteen generations, culminating in the birth of the *Messiah. The reduction to thirteen spoils the symmetry and is to be explained either because Matt. thus indicates that by the providence of God the neat scheme is overruled and the nation's destiny is fulfilled, or, more probably, because there is here a typical Matthaean muddle, as at 20: 22 when Jesus addresses James and John, just after Matt. has deliberately introduced their *mother* as their spokesperson.

'Fourteen' is in Hebrew by the method of **gematria* the numerical value of David: D + W + D = 4 + 6 + 4.

Matthew's genealogy includes four women, Tamar, *Rahab, *Ruth, and *Bathsheba, who played important roles in OT history but at the end it is *Joseph, the legal father of Jesus, who is named. The purpose of Matt. here is to show that Jesus is the legitimate heir to the throne of David, who is mentioned five times in 1: 1–17. Two of the four women had disreputable stories; and Ruth becomes great-grandmother of David though she was a *foreigner; Bathsheba is deliberately recalled as 'the wife of *Uriah' the Hittite.

Luke's genealogy comes after the *baptism narrative and before the public *ministry of Jesus (Luke 3: 23–38) and traces the ancestry right back to *Adam 'son of God'; but unlike Matt. the line is taken through *Nathan (2 Sam. 5: 14) instead of his brother *Solomon, and no women are included. It would seem that by going back to the origins of humanity Luke is suggesting that Jesus is the Saviour of the whole world and not only of the people of Israel, as will be repeatedly explained elsewhere in Luke–Acts.

Luke's genealogy is artificially constructed, like Matthew's, but in a different way. He has compiled eleven groups of seven names, and just as the eleven disciples needed to be completed by one more (Acts 1: 26), so the eleven groups in the genealogy imply that there is one era yet to come—the time during which the gospel is to be preached to all nations (Luke 24: 47): eleven is manifestly incomplete.

Genesis The first book in the *Bible (*genesis* is Greek for 'beginning') which describes the origin both of the universe and of the nation of Israel as having their birth outside themselves, in the creative purpose of God. The two accounts (Gen. 1 and 2) of the making of the world are not in the language of science but of theology; humanity exists because it was the intention of God for this to be so, as also does Israel, as his chosen people.

The stories within Gen. were transmitted orally over many generations, and some of them bear close resemblance to the *myths of their Near Eastern neighbours, as is the case with the *Flood (Gen. 6 to 9). Eventually they were committed to writing, and the various narratives were worked over by editors. Much scholarly labour has endeavoured to delineate the different sources which were editorially combined after the Exile to make the final version of Gen. They are given the symbols *J, *E, and *P, and have been assigned probable dates. Nevertheless this theory of composition is not universally accepted, and much interest nowadays is also taken in the narrative as we have it and in the theological stance represented by the work as a whole; there is less interest in breaking up the book into fragments and more concentration on the main themes of the book as it describes God's promises and their fulfilment, though the existence of different literary strands is not denied. Certainly few would wish to return to the pre-19th-cent. view that the entire *Pentateuch was written by Moses and that Gen. 1 to 11 could be regarded as historical narrative. There is on the other hand some support for authentic historical reminiscence in Gen. 12 to 50, in that the *patriarchs seem to be invested with the kind of qualities and activities characteristic of the 2nd millennium BCE.

geniza A place in a *synagogue for storing damaged, outdated, or heretical documents. Valuable MSS were discovered in a *geniza* at Cairo in 1896 which included part of the Hebrew text of Ecclesiasticus and fragments of a Zadokite work, later associated with the Qumran community, and called the *Damascus Document, or Fragments.

Gennesaret, Lake of An alternative used in Luke 5: 1 for the Sea of *Galilee.

genre This French noun (e.g. *le genre comique* = comedy) is used for categories of literature, and by biblical critics to differentiate the various kinds of narrative in OT and NT. It is important for *exegesis not to confuse, say, history with *myth or *poetry, though the form of a particular writing may not be a correct indication of its genre; for example, the fact that Rev. opens with letters to seven Churches belies its true genre, which is a piece of apocalyptic writing. Thus, in biblical studies, the identification of the genre of a major unit of material (such as a book of the *Pentateuch or a NT epistle) will precede the analysis of smaller sections, or *pericopae, within the whole.

Gentiles A word only meaningful in relation to Israel and the Jews for the rest of mankind. Although there was always a good deal of contact between Israelites in Palestine and their Gentile neighbours, attitudes to them were ambivalent. On the one hand a sense of vocation was strong to bring Gentiles to the *worship of the one God (Mic. 4: 2; Zech. 8: 22–3); Israel is a 'light to the nations' (Isa. 42: 6), and there are the books of Ruth and Jonah. But on the other hand there was, for example, the nationalistic apartheid of Ezra and Nehemiah who decreed that non-Jewish wives were to be divorced. Barriers became more rigid in the last two centuries BCE and there were urgent warnings against the *idolatry of the Gentiles (1 Macc. 3: 48) who were generally assumed to enjoy lax behaviour.

It was this last sense of horror that moved some of the Jewish Christians in their opposition to Paul's mission to the Gentiles: if these converts did not keep the Law, would the Church be able to maintain moral standards? Certainly the relaxation of the demand of *circumcision led to an influx into the Church of 'God-fearers' (Acts 13: 26, 48) who were loosely attached to *synagogues. Some Gentiles who had embraced the full rigours of *Judaism (called *proselytes), attracted by the ethics and the strong community life of Jews, also joined the new movement (Acts 6: 5). Probably they hoped to find the same support in the Church (Acts 2: 10; 13: 43), but they would have resented any apparent laxity. Cf. Gal. 4: 21.

Jesus confined his own mission to his immediate home ground (Matt. 10: 5) but there are hints that in the view of the *evangelists he would not have disapproved of Paul's initiatives: he healed the daughter of a Gentile woman (Mark 7: 26) and it would have been impossible for Jesus to have avoided all contact with Gentiles—indeed he might have spoken some Greek. His cleansing of the Temple (Mark 11: 15–17) took place in the Court of the Gentiles, which was carefully separated from the holier parts of the building by a flight of steps and a wall to which were affixed a minatory instruction in Latin and Greek warning Gentiles to proceed no further. Jesus was apparently concerned that there should be no obstacle to Gentiles engaging in *prayer in that limited portion of the Temple which they were allowed to enter. The destiny of the Temple as a sanctuary for 'all peoples' in the Messianic age was prophesied by Isa. 56: 7. The authorities construed Jesus' action as a threatening prediction of its future destruction (Mark 14: 58).

gentleness Paul compares his attitude to his converts to the gentleness of a nurse (1 Thess. 2: 7). So also Titus 3: 2.

Gerasa An impressive city of the *Decapolis in NT times rebuilt by the Romans with theatres and forum. It lies between the *Dead Sea and the Sea of *Galilee and is alluded to in Mark 5: 1. It was incorporated into the province of *Syria in 63 BCE.

Gerizim A mountain in *Samaria where *Jotham spoke a *parable (Judg. 9: 7) and which was a focus for *Samaritan *worship and therefore an alternative to Jerusalem (John 4: 20). Their own temple had been destroyed by John Hyrcanus in 128 BCE.

Gethsemane The name (meaning 'oil-press') of the *garden where Jesus prayed in agony before his arrest (Mark 14: 32) and located by *John (who does not give the name) across the *Kidron valley on the western side of the Mount of Olives. Although Jesus was alone (Luke 22: 41) and the disciples were sleeping (Mark 14: 37, 40), Jesus' *prayer is reported (Mark 14: 36). Presumably the substance of the prayer was inferred and recounted as part of the tradition; it is mentioned in Heb. 5: 7; and the basic authenticity of the event is supported by its inconsistency with the motif running through the rest of Mark. In that gospel Jesus' *passion and *death were decreed by the providence of God (Mark 8: 31—'must').

ghost A disembodied *spirit, supposedly residing in **sheol* (Job 26: 5; Isa. 14: 9) and capable of being brought up by a professional medium (1 Sam. 28: 12–14). The disciples mistake Jesus walking on the sea for a ghost (Mark 6: 49) as they also do when they see him in his risen form after *Easter (Luke 24: 37).

The 'Holy Ghost' (Luke 1: 15, AV) is now generally rendered '*Holy Spirit' in modern versions on account of the former term being obsolete.

giant Before the arrival of the Israelites several races of giants were said to inhabit the land (Deut. 2). *Goliath of Gath was apparently over 2.8 m. (9 feet) tall (1 Sam. 17: 4).

Gibeon A town NW of Jerusalem whose inhabitants tricked the invading Israelites into a treaty (Josh. 9: 4), though later *Saul attacked them savagely (2 Sam. 21: 2). The pool of Gibeon (2 Sam. 2: 13) was discovered during excavations in 1956–62; it provided water in the event of siege.

Gideon One of the Judges (also called Jerubbaal, Judg. 6: 32); he had notable military success against the *Midianites who were expanding westwards. Gideon selected a small band of 300 warriors for their intelligence (Judg. 7: 7) and in a surprise night attack put the Midianites to flight and killed their two princes (Ps. 83: 11–12). His refusal to accept the kingship (Judg. 8: 23; cf. 1 Sam. 8: 7) is part of the anti-monarchical strand in the tradition. *See* king.

gifts In the OT, gifts to God take the form of *sacrifices, but there are also gifts which are bribes (Prov. 18: 16) or wedding presents (Ps. 45: 12). In the NT gifts are mentioned as required at the Temple (Matt. 5: 23) but there is also an emphasis on the gifts of God to mankind (e.g. Rom. 5: 17; cf. Jas. 1: 17).

gifts of the Spirit Moses and others in the OT were seized with an ecstasy ascribed to the, or to a, Spirit (Num. 11: 29), but it is in the epistles of Paul that specific gifts of the *Holy Spirit are listed and said to be the natural endowment of any faithful believer —though differently distributed (Rom. 12: 6). Together they contribute to the health of the whole Body (Rom. 12: 4–5). The lists of gifts in the Pauline and other NT documents divide the gifts into audible manifestations (prophecy, ecstatic speech and its interpretation), practical works of *charity, and the gift of apostleship which comes first at 1 Cor. 12: 28; it derives from the authority of Christ and gives the best witness to his *resurrection.

Paul is embarrassed by the incoherent babel of '*tongues' (1 Cor. 14: 12) and by the time of 1 Pet. (4: 10–11) unintelligible speech is not reckoned amongst the gifts.

Gihon (1) One of the four rivers in *Eden (Gen. 2: 13). (2) A spring in Jerusalem which supplied water to the city and the scene of the anointing of *Solomon (1 Kgs. 1: 33). Hezekiah in 701 BCE had a tunnel constructed to convey water from the spring to the pool of *Siloam (2 Kgs. 20: 20; 2 Chr. 32: 30).

Gilboa A mountainous area where *Saul committed suicide after being defeated by the *Philistines (1 Sam. 31: 4).

Gilead A name for three minor persons (Josh. 17: 1; Judg. 11: 1; 1 Chr. 5: 14), but more importantly of a mountainous area east of the *Jordan (Deut. 3: 16) and famous for its balm (Jer. 8: 22) which was in that reference translated 'triacle' in Miles Coverdale's Great Bible of 1539: hence its popular title 'the Treacle Bible'.

Gilgal The site of the Israelite headquarters as they entered *Canaan (Josh. 4: 19) and later a centre of *syncretistic worship condemned by *Amos (4: 4; 5: 5) and *Hosea (4: 15).

girding up the loins The process of tucking up the long Eastern robes through the girdle to facilitate energetic work or walking (2 Kgs. 4: 29) and therefore also a figurative expression (1 Pet. 1: 13, NRSV marg.) to denote the mental preparation for action (so NRSV, REB, NJB).

girdle Usually made of rough leather as worn round the waist by *Elijah (2 Kgs. 1: 8), and useful for holding up garments when working in the fields. *John the Baptist (Mark 1: 6) wore a girdle similar to that of Elijah, which reinforced his message to repent, like Elijah (Mal. 4: 5 f.) (Mal. 3: 23 f., NJB).

Girgashites A people displaced by the incoming Israelites. The post-exilic editor of Genesis maintains that *Abraham was promised the Gergashite territory in N. Canaan (Gen. 15: 21; cf. Deut. 7: 1; Josh. 3: 10) and *Nehemiah (9: 8) regards the promise as still valid, according to the reported words of *Ezra.

glass Coloured glass has been excavated at many Palestinian sites, but it was expensive and a luxury until NT times when it became as cheap to produce as pottery.

gleaning Humane regulations obliged farmers neither to gather fallen fruit nor to rake the field after reaping, so that the poor and needy could be free to help themselves without penalties (Lev. 19: 9; Ruth 2: 2).

Gloria in Excelsis A liturgical *hymn based on the *angels' song of Luke 2: 14; in early and medieval times of the Church it formed part of the synaxis (readings and prayers), early in the *Eucharist. In the Church of England it was transferred in the Second (1552) Book of Common Prayer to a place preceding the *Blessing at the end. Modern rites have restored it to its ancient place.

glory A biblical concept describing a fundamental experience in the relationship of God and humanity which blends within itself a variety of universal images. The English word translates the Hebrew noun *kabod* denoting 'weight' and hence 'destruction', and was associated with physical phenomena by which the presence of God was made known (e.g. thunder and lightning). Later it is used of God's characteristics, such as sovereignty and *righteousness (Isa. 6: 1–4; 40: 4–5). In *Ezekiel's *vision, the glory of the Lord is said to leave Jerusalem. Later, the *P source of the *Pentateuch regards 'the *cloud' as a kind of covering through which the *glory shines (Exod. 24: 16–17). Thus Yahweh in the later narratives is conceived as both transcendent and having a local dwelling place in the Temple. Later still, in *apocalyptic writings (e.g. 1 Enoch 14: 19–21), 'glory' is ascribed to the *Messiah.

Another Hebrew word, **Shekhinah*, denoting God's accessibility to humanity without impairing his transcendence, was allied to the Hebrew *kabod*, and the *LXX conflated them and used the word *doxa*. The Greek word also had an early meaning of 'distinction'; but the OT concept of the divine brightness, splendour, sovereignty, and righteousness are signified by the word, and in so far as *doxa* refers to this character of God, then Jesus *is* that glory. What is new in the NT is that the glory (*kabod*, weight, destruction) is manifested in the *cross (Heb. 2: 9–10, John 17: 22).

Glory dwells in the Church, the temple of God (1 Pet. 2: 4–5; Eph. 2: 21–2), though it is only a foretaste of the glory to come (1 John 3: 2). It is the Church's work to glorify God, by which is meant not that we can somehow add to God's glory, but we are to recognize it and try to extend its recognition; to submit to it and render praise for it. Just as in Paul's epistles the glory is manifested in Christ's cross and resurrection (1 Cor. 2: 8), so it is also in the apostle's sufferings (2 Cor. 4: 17); and just as the apostle's tribulations are his glory, so too his converts participate in his glory (Eph. 3: 13).

glossolalia Greek for 'speaking in *tongues', one of the *gifts to the Church in Paul's list (1 Cor. 12: 10) but scarcely encouraged by him (1 Cor. 14: 23) and not mentioned at all in another list (Rom. 12: 6–8). It may be the phenomenon at *Pentecost (Acts 2: 4) when those in the upper room are said in a reversal of the confusion at *Babel (Gen. 11: 1–9) to have uttered understandable foreign languages. There are indeed records of similar surprises later in the Church (e.g. among persecuted Huguenots in France) and among Muslims. When, however, linguistic experts have been able to tape-record speakers alleged to be uttering a new language in a state of religious enthusiasm, it has never been possible to identify the language or discern any grammar or structure.

gluttony Not a *sin characteristic of the Jews and not prominently condemned in the NT (though in later tradition it became one of the 'seven deadly sins'). Jesus was denounced as a glutton because of the contrast of his lifestyle with that of *John the Baptist (Matt. 11: 18–19). However, gluttony can lead to unbelief (Phil. 3: 19).

Gnosticism From the Greek *gnosis*, meaning 'knowledge'. The term is used for a kind of religious speculation in vogue in the first two centuries CE and long known through the writings of Christian Fathers, who were hostile to it and feared that it might seduce orthodox believers. But the discoveries in Upper Egypt at *Nag Hammadi of Christian sects who were indeed influenced by Gnosticism has given a much richer impression of the movement. Gnostics had a profound anxiety about *salvation. People, they thought, lived like zombies and needed to be roused to authentic life by receiving knowledge of their true selves and of their destiny with the true God. Equally, they needed esoteric knowledge about their one-time heavenly origin. When this was acquired, the initiate threw off the oppressive bonds of earthly materialism. The knowledge was conveyed by a heavenly revealed figure or Redeemer—the resemblance to Christianity is obvious. Historians have therefore tried to discover whether Christian teachers borrowed elements from Gnosticism or whether 2nd cent. Gnosticism was a deviationist heresy out of the Church or perhaps out of *Judaism. Certainly Paul seems to make use of Gnostic terminology either by way of accommodating his message to the cultural milieu or to refute those who tried to eliminate the essence of the gospel by incorporating Christ into a Gnostic system. For example, the Greek **pleroma* ('*fullness') was a Gnostic concept denoting the range of heavenly powers who were the mediators, or buffers, between the divine and the fallen world. What Paul does is to take this word (e.g. in Col. 1: 23) and hijack it: Christ is the fullness of all being. There is not a plurality of angelic powers (cf. Heb. 1: 4) as in Gnostic speculation. Similarly, the gospel of John contains themes reminiscent of Gnosticism: the descent and ascent of the Redeemer to *heaven (John 3: 13) and the dualism of spirit and flesh (6: 63). But John is no Gnostic: the idea of descent comes from the *Wisdom tradition and ascent from Dan. 7: 13.

goats Hardy animals, well adapted for life in the desert (Lev. 7: 23, Deut. 32: 14; 1 Sam. 19: 13), tended with sheep (but separated for milking and feeding). They provided meat, milk, skins, and hair. Goats were sacrificed, e.g. on the Day of *Atonement, (Lev. 16: 9) even in the time of Jesus (cf. Matt. 25: 32–3), but Christians repudiated these sacrificial rites (Heb. 9: 12–14). *See* scapegoat.

God There are in the Bible *'fools' who do not believe in God (Ps. 14: 1), not on the ground of considered rational argument but by reason of their depraved and selfish lives. The atheism therefore of modern existentialists like Jean-Paul Sartre or positivists who maintain that we have access to knowledge only through the empirical sciences is unknown to the OT and NT alike. God was, is, and is to be (Exod. 3: 14); the Hebrew verb denotes habitual manifestation. This is the first revelation in the *E source of the *Pentateuch of the divine name, and it implies that God is known when he makes himself present to people and they respond.

God in the Bible is conceived as omnipotent, omniscient, and the absolutely good creator of everything that exists. As in other religions it is expected that devotion to God will bring personal and national well-being and avoidance of calamity. This devotion consists in a prescribed way of life, such as eating certain foods, which give the community a sense of solidarity. In the OT the *Torah (Law) contains a code of ritual and ethical behaviour and it is held that these come direct from God. So for Israel belief in God necessarily means also participating in the life of the *community and its rites.

The Hebrews did not arrive at their *monotheism without a long gestation—from a polytheistic environment, via *henotheism (when Israel worshipped one God but admitted the existence of other peoples' gods), to an ethical monotheism. A distinction between Yahweh and other gods which began in the 9th cent. BCE was sharpened by Hosea in the 8th cent. In the 7th cent. (Deut. 13: 6–11) faithful Israelites are to inform on relatives who worship other gods. By the time of the Exile, belief in Yahweh as sole God, author of both good and evil, is accepted by *Deutero-Isaiah (Isa. 45: 7). Yahweh is regarded as Creator of the world but also as the supreme being who has graciously entered into a *covenant with his people. Images which describe God are king, father, and *shepherd: masculine in gender, but later literature in which the personified *wisdom reveals God shows that the feminine is not alien to OT conceptions of God—since the Hebrew for 'wisdom' is feminine; and God's activity is sometimes described in terms of female characteristics (Isa. 66: 13).

God who is transcendent is also a God who condescends to dwell in the midst of his people. It was a Hebrew belief that the divine presence 'tabernacled' among them and in due course dwelt in the Temple. He was regarded as a God who demanded righteousness and justice and his wrath could be poured out on those who disobeyed him.

But it is not true that the God of the OT is primarily a God who exacts *punishment and is terrible to sinners, and that in the NT he is primarily a God of compassion and *forgiveness.

This contrast between OT and NT was overdrawn by the heretic *Marcion (*c.*150 CE), and the Christian Church has always maintained that the God of Israel, the Father of Jesus, is the God whom Christians also worship. There is a continuity of concept. However, the fundamental change brought to the biblical concept of God was that for Christians the main place in which God reveals himself was now in Jesus the Christ (Matt. 1: 23, John 14: 9). God in the NT remains the holy, transcendent God of the OT but it is believed that Jesus has shown in a human life that fundamentally God is *love. Christian experience led further to the doctrine of the *Trinity, of which there are already hints in the NT (e.g. Matt. 28: 19). For Paul's language about believers' participation in the *Spirit (2 Cor. 13: 13) is possibly spelt out by the later formulated doctrine of the Spirit as a distinct 'hypostasis' ('Person') within the very being of the Godhead.

Several names are given to God in the Bible. The Hebrew *El* is often used in the plural **Elohim*. He is contrasted with the heathen *Baal in the *Deuteronomist history and in Hosea 2. Yahweh occurs nearly 6,000 times in the OT, but in the *Masoretic text (6th cent. CE) the vowels of *Adonai* ('my Lord') were inserted into the ineffable name Yahweh, hence the English Jehovah. The phrase Yahweh Sabaoth (= Lord of hosts) is found 279 times in the OT, especially frequently in the prophets; originally the reference was to hosts of war but came to mean the powerful functions of Yahweh in command of nature and the fortunes of men. The NT never denies the holiness of God (much emphasized in the OT) but prefers to use expressions referring to his faithfulness (1 Cor. 1: 9), wisdom (Rom. 16: 27), *truth (John 3: 33). He is the God of *peace (1 Thess. 5: 23), *hope (Rom. 15: 13), and above all *love (2 Cor. 13: 11, and John 3: 16; 1 John 4: 8).

The biblical concept of God as almighty (e.g. Matt. 26: 53), and that his omnipotence can intervene directly in the affairs of this world, must be modified nowadays by the recognition that there is a limit imposed on divine omnipotence in that the universe operates predictably and this is a necessary foundation for human freedom and responsibility. The power of God can be conceived as that of suffering love, of self-emptying which actually gives power to what is created over its creator. God does act in the world (as the Bible maintains) but more in the sense that he has created a world of order, value, and rationality—through which we can conceive of the idea of deity.

God-fearers A class of persons mentioned in the Acts (e.g. 10: 2) as religious, probably adherents of the synagogue but not yet *proselytes who had been admitted to full membership by *circumcision.

Godlessness A kind of thinking which ignores God; or a kind of acting which flouts his commandments, such as pride (Ps. 119: 51) and wickedness (2 Pet. 2: 5–6).

Godliness Not often found in the Bible, where '*faith' or '*righteousness' are preferred for a life which is acceptable to God. Where it is used, it is linked with 'knowledge' (2 Pet. 1: 5–7) and gentleness and steadfastness (1 Tim. 6: 11), and is certainly not to be identified with outward behaviour or observances (2 Tim. 3: 5).

Gog *Ezekiel's name for the leader of a hostile people in the north (Ezek. 38: 2). Rev. 20: 8 uses Gog and Magog as symbols of the enemies of the *kingdom of God.

gold A precious metal found in places near to Palestine (*Egypt, and Ophir—perhaps modern Oman in SE Arabia) and exported to *Judah (Isa. 13: 12). It was much used in *Solomon's Temple, where the entire inner sanctuary was overlaid with it (1 Kgs. 6: 14–31) and in the NT it is recorded that gold was one of the *gifts brought to Jesus by the *Magi (Matt. 2: 11).

golden calf The idol devised by *Aaron during the prolonged absence of Moses as he communed on the mountain with God (Exod. 32: 2–4) and by *Jeroboam I, King of Israel (937–915 BCE) at *Bethel and *Dan (1 Kgs. 12: 28). It is possible that all these calves (Aaron's one and Jeroboam's two) were in fact symbols of Yahweh (Exod. 32: 5).

golden rule The term used since the *Reformation for the words of Jesus recorded in Matt. 7: 12 as being of great value; 'in everything do to others as you would have them do to you' to express the obligation of love to both God and neighbour. There is a similar connection in the *Didache. It is also found occasionally in Jewish writings, but more commonly in a negative form: 'What is hateful to you, do not do to anyone else'; it also occurs in the words of the rabbi *Hillel (60 BCE–20 CE) and in Tobit (4: 15) and also in Buddhist and Islamic literature.

golgotha An Aramaic word translated (Mark 15: 22) 'Place of a Skull', as being the site for executions, outside the city walls but near a main road (Matt. 27: 39). The *Vulgate translation is *calvaria*—hence English 'Calvary'.

Goliath The *Philistine *giant from *Gath, over 2.8 m. (9 feet) tall, who issued a challenge to any individual Israelite to combat, to which *David responded with a sling and five stones. He struck Goliath's forehead and knocked him unconscious, and then seized his sword and beheaded him (1 Sam. 17: 38–51). Some scholars hold that the giant of 1 Sam. 17 was originally anonymous—he is named only at verse 23—and that Goliath was slain by a certain Elhanan (2 Sam. 21: 19).

Gomer The wife of the prophet *Hosea (Hos. 1: 3); her life as a prostitute gave Hosea the impetus to proclaim his message to the nation: by this *marriage Hosea came to see the similarity of the marriage between Israel and God. The infidelity of the woman was paralleled by the nation's attraction to the heathen Baalim. Yet Hosea loved Gomer still; and the parallel with God and the nation is brought out in his prophetic utterances. Some commentators, however, have interpreted Hosea's marital experiences as *parable rather than autobiography.

Gomorrah One of the five cities of the Plain, or Circle of the *Jordan, which was destroyed by fire (Gen. 19: 24), probably some sort of natural cataclysm.

good The view of modern empirical philosophers that calling a person or a thing 'good' is entirely subjective, merely an expression of approval, is not that of the *Bible. God is goodness, so what is good is defined in terms of God (Mark 10: 18). God's goodness is shown in the *glory of *creation and the *blessings of nature. His commandments too are good and obedience to them good. Loving our neighbours is good (Gal. 6: 9 ff.) and acts of *charity are good (Acts 2: 44 f.) because they proceed from the *love of God (1 John 5: 2).

Good Friday The day when *Christians each year commemorate the *crucifixion of Jesus. There was for a long time no special annual observance of the crucifixion of Jesus except that from the 2nd cent. every Friday was a day of fasting . The *death and *resurrection were commemorated in a single *Paschal festival over Saturday and Sunday. But in the 4th cent there was a development of Holy Week at Jerusalem in which the historical events of the *passion were rehearsed, and then Good Friday became a distinct occasion for recalling the Crucifixion, and *Easter Sunday for celebrating the Resurrection. Only the latter allowed the adjective 'good' to be applied to the Friday.

Good Samaritan The usual name given to Jesus' *parable in Luke 10: 29–37. It is the story of a Jew mugged on the lonely road from Jerusalem to *Jericho. Passing by were a *priest and a *Levite, but fearing ritual contamination which would have inhibited them from religious duties, they gave no help. Jesus' listeners were probably surprised to learn the identity of the third traveller—not, as they expected, a lay Israelite (so that the parable would have been a straightforward piece of anticlericalism) but a *Samaritan, hated by the Jews (John 4: 9); he overcame his scruples and gave assistance, and more. So the parable turns out to be anti-racist.

good works A major controversy at the *Reformation was whether a Christian was put in a right relationship with God by reason of his good works, as seemed to be suggested by Jas. 2: 14, or by *faith in Christ, as Luther and others interpreted Rom. 3: 24, 27. The Reformers argued that Catholics insisted on good works—acts of devotion, almsgiving, penitence, pilgrimages, and the like—as the means for earning *salvation and was a relapse into the legalism condemned by Paul. Catholics replied that such good works, which brought 'sanctification', were themselves a response to '*grace'. Paul certainly did not undervalue the evidence of good works (Phil. 2: 12), and his heavy emphasis on 'working for the good of all' (Gal. 5: 22–6: 10) and 'laying aside the works of darkness' (Rom. 13) blunts some of the sharp contrast between *Judaism and Paulinism.

gospel The English word means 'news of joyful events' and is used by Jesus as he proclaims the coming of the *Kingdom (Mark 1: 15) and by Paul of God's work done through Jesus Christ (Rom. 1: 1–2). The NT takes up the theme already preached by the OT *prophet, where it is the good news of the coming end of the Exile (Isa. 40: 9). In writing to the Galatians (1: 6–7) Paul rounds on opponents who were distorting 'his' gospel and teaching 'another' which he regarded as nothing more than the old way of *Judaism slightly improved or modified. It was the old legalism. 'His' gospel was the news that it is by Christ, through faith in him, that we are put right with God. Thus the content has been changed: for Jesus, it was the coming of the *Kingdom: for Paul, it is Christ himself 'the power of God for salvation' (Rom. 1: 16). The content of the gospel has necessarily been changed by reason of the Church's belief about Jesus' death and *resurrection.

In the 2nd cent. 'gospel' came to be put as the title of certain Christian books. This followed the convention that a book should be known by its first word: thus, 'Genesis', the opening book of the OT, acquired its name from the *LXX where the word means 'beginning' (Gen. 1: 1). Almost the first word in Mark is 'gospel'; but before long there were three other accepted accounts of Jesus' *ministry, *death, and *resurrection; therefore to distinguish them their titles became 'The gospel according to Matthew' etc. The word 'gospel' had come to denote a kind of literature about Jesus, but whether it was an entirely new *genre of writing is debated by NT scholars.

Gospels, four The gospels according to Mark, Matthew, Luke, and John were compiled in the second half of the 1st cent. CE, and, in accordance with contemporary custom, were intended to be read aloud. They were all written to satisfy needs felt in the growing Church. At first there was just the oral tradition; Jesus himself did not write a word; but *parables spoken by Jesus and many of his conversations were remembered and in the early days of the Church these reminiscences were constantly passed on. The store of traditions was drawn on by Christian *evangelists who tried to persuade members of the *synagogues, and by controversialists in debates. There were new recruits to the Church who wanted to know how it had come about that a good man (Jesus) had suffered death at the hands of the Romans. Moreover this man was being preached as a Saviour who had been raised from the dead! Not surprisingly, when *Mark was persuaded, perhaps in Rome, to put in writing the reminiscences that were held in the Church, his gospel was a detailed account of the *trials and execution of Jesus, with an introduction containing excerpts of *teaching and examples of about a dozen *miracles. It was left to Mark's successors to fill out his narrative not only with extended extracts from Jesus' teaching but also with narratives about his birth. The gospel of Matthew follows Mark's order in the main, but incorporates it, with much added material, in five great discourses of Jesus, edited by the author of the gospel. He makes improvements to Mark's style by pruning verbosity and clarifies some of the Marcan obscurities. The gospel of *Luke also uses Mark but less slavishly. So these first three gospels have much material in common, as is demonstrated by comparing them word for word, paragraph for paragraph, in three parallel columns. For this reason they are known as the '*synoptic gospels'—i.e. gospels that can be viewed together. The fourth gospel (John) is in another category, was almost certainly written later, and is very different in style, content, and theological emphasis. Much of it spells out what in the synoptics appears in disconnected fragments or veiled allusions.

The four gospels did not represent the end of these literary ventures. They were, however, the works which were in due course pronounced by the Church to be parts of the NT '*canon'; but *Gnostics and others promoted rival versions of the Jesus story, such as the gospel of Thomas (*c.*140 CE), which was discovered at *Nag Hammadi; it is a collection in Coptic of 114 alleged sayings of Jesus but this '*apocryphal Gospel' offers the reward of *salvation (immortality) to those who gain the right knowledge.

The gospels are not *biographies in the modern sense, and the *Form Critics maintained that they were not biographies in any sense. But they could be regarded as biographies such as were written in the ancient world, with the particular differentiation that they describe unique events claiming that the promises of the OT are fulfilled.

government In the Bible God is regarded as the supreme governor of the world and those who execute authority on earth are said to hold it in trust from him (Ps. 72: 1–4; Rom. 13: 1–2; John 19: 11) to ensure that justice is done (1 Kgs. 10: 9). It may be desirable to delegate judicial functions, as Moses did (Exod. 18: 17–23), and it is held that, although Israel was a theocracy rather than a democracy, the people as well as the king shared in the rights and responsibilities of the *covenant with God (2 Kgs. 11: 17).

Even when the Jews were a subject people, it was expected by the *Sadducees that obedience should be given to the government; and in the NT Christians are warned against subversion or defiance (1 Pet. 2: 13–17), though this attitude of submissive acceptance of the government was abandoned under the fire of persecution (Rev. 17 and 18).

grace In the OT, God's kindliness towards the people of Israel (Zech. 12: 10—a 'spirit of grace' in *LXX), which consists in the *forgiveness of *sins (Exod. 33: 19). He gives graces (favours) which are undeserved *gifts of his *love (Isa. 63: 7). This was the basic meaning of the word which is found in the NT when grace is linked to Jesus, who represents in his person God's graciousness towards mankind. Jesus is himself the fullness of grace and *truth (John 1: 14–17). In Paul's thought it is by the grace of God that believers like himself were called, and therefore we are deprived of any grounds for boasting (Rom. 3: 27; 1 Cor. 1: 31); Paul must give God the *glory for the free gift of *salvation which cannot be earned by his own efforts. There now exists a 'rule of grace' (Rom. 5: 21) and to live under grace is to live in the obedience of faith, with holiness as its aim (Rom. 6: 15–22).

What is true of grace, bringing people into the redeemed community, is true equally of the callings to offices ('apostleship', Rom. 12: 3) and the other gifts or graces (Rom. 12: 6) which are bestowed on the Body as necessary for its life (1 Cor. 12: 6–8), though these gifts are not of equal importance. The Greek *charisma*, from *charis* (a grace), emphasizes the free nature of the gifts received; and, by a slight extension, Paul can use the word *charis* to mean an acknowledgement of the gift, which leads to thanksgiving—hence the English 'grace' before meals.

Paul also mentions grace received in suffering (2 Cor. 12: 9; Phil. 1: 19); and the collection for the Church in Jerusalem is a grace (1 Cor. 16: 3), for it is the work of the *spirit who prompts the generosity of believers.

It is indeed difficult to differentiate 'grace' from the 'Holy Spirit' in the NT. Unfortunately in later Christian theology, under the influence of Augustine, grace came to be regarded as a 'thing', a kind of impersonal entity or quasi-physical force or power which lights upon those predestined to absorb it.

gracious In the OT persons of high standing are said to be gracious, or show favour, to lesser mortals, as *Boaz to *Ruth (Ruth 2: 10) or King *Ahasuerus to *Esther (Esther 5: 2). But God too is gracious (Exod. 34: 6) and slow to anger. The adjective is rare in the NT, which prefers the noun, but it occurs in 1 Pet. 2: 3 in AV where modern translations use 'good' (for the Greek has a different word).

Graf, Karl (1815–1869) A teacher of Hebrew in Giessen, Germany, who made the sensational suggestion that parts of the *Pentateuch were compiled during the period of the Exile and therefore reflected the ideas then current.

grapes Grapes were plentiful in biblical times and were eaten either fresh or dried as raisins or made into jelly, though most were crushed for *wine, which, with *bread, formed the staple diet of the time. Wine was regarded as a divine *gift, but excessive drinking was deplored (Isa. 5: 11–13). Some of Jesus' exhortations are illustrated by reference to grapes (e.g. Luke 6: 44).

grass Grass of the kind seen in the fields of northern Europe or on domestic lawns was unknown in Palestine, where the pasture that sprang up after the seasonal rains soon withered. Grass therefore became a symbol for the brevity of life (Isa. 40: 6 f.) especially for those with riches (Jas. 1: 10 f.; 1 Pet. 1: 24).

graven image The OT prohibition of *idolatry (Deut. 4: 15–18) was renewed by *Christians when confronted with orders to worship the statue of the Roman emperor (Rev. 14: 9–11). But the OT (Isa. 46: 6) was too much driven by religious polemic in supposing that Babylonians worshipped the idol itself: it was the god's spirit within the statue that was venerated.

graves Corpses were normally buried in graves, often situated under trees (Gen. 35: 8), as soon as possible after death. The body of Jesus was wrapped in linen (Mark 16: 1) and placed in a tomb hewn out of rock with a rolling stone placed against the entrance (Mark 16: 3–4).

Greece The southern European country between the Adriatic and Aegean Seas, and extending into the Mediterranean. The city-states of classical Greece (*Athens, Sparta, and others) had already lost their independence by the time the Romans conquered the country in 146 BCE, and it is *Christians living in part of the Roman Empire that Paul addresses in the epistles to *Corinth, *Philippi, and *Thessalonica.

Greek The language of the NT writings was a development of classical (Attic) Greek, much as modern English has changed since the publication of AV in 1611. Greek had become a lingua franca through the civilized world in the 1st cent. and it was essential equipment for trade and communication. It was therefore the language used by the four *evangelists, and by *Josephus and *Philo. Even in Rome itself Greek was in general use, Latin being the language of the upper classes engaged in law and administration. Greek of this period is called *koine*, or common, Greek and some of those who spoke it adulterated it with idioms from their first language. In the gospels a few scraps of the original *Aramaic which Jesus spoke have been preserved in a Greek form, and some NT concepts expressed in Greek have to be understood against this OT background (e.g. 'righteousness'), though it is incorrect as a normal principle simply to explain Greek NT concepts in terms of the Hebrew OT.

Greek has seven vowels, since e and o can both be either short (as English 'empty' or long as in 'me'), and this distinction should be represented in public reading of the NT as in Ōmĕga (Rev. 1: 8). Normally a final -e in a proper noun (as in Phoeb-ē, Rom. 16: 1, or Eunic-ē, 2 Tim. 1: 5) is long.

Greeks Although primarily referring to inhabitants of Greece, there was a wider usage; persons who had a knowledge of Greek culture or who lived in sophisticated cities of the Roman Empire were naturally 'Greeks', whereas the local rustics were 'barbarians' (Acts 28: 2, AV, translating the Greek literally). 'Greeks' is also used of *Gentiles in general (2 Macc. 4: 36), so that 'Jews and Greeks' means in effect the whole human race (Rom. 2: 9–10). A deputation of Greeks arrived unexpectedly to see Jesus (John 12: 21); but they were already sympathetic to Jewish *monotheism (John 12: 20).

Griesbach, Johann (1745–1812) A German NT scholar who is best known for the 'Griesbach hypothesis', now widely rejected, that Matthew was the first of the synoptic gospels to be written, followed by Luke, and that Mark was dependent on both.

growth In a series of *parables, Jesus speaks of the growth of the *kingdom of God (e.g. the Mustard Seed, Mark 4: 30–2), where the contrast is made between the tiny almost invisible seed and the final result. Just as growth happened in a mysterious, hidden, but assured process, so the Kingdom will develop from its small beginnings in the *ministry of Jesus. The Kingdom is both present and yet not fully present. For Mark's readers the parable is an encouragement that at the *End the *glory of the Kingdom will stand revealed.

guard The account in Matt. 27: 62–6 that a guard was posted at the sepulchre of Jesus to prevent the theft of the body derives from the controversy current at the time of the writing of the gospel. Jewish propaganda maintained that Jesus was not risen from the dead; the disciples had stolen the body. Guards stationed there had seen it happen. Christians replied that when the guards reported what had really happened, they were bribed to keep quiet and say they were asleep and so did not see the body being removed (Matt. 28: 11–15), and if Pilate were to reprimand them for sleeping while on duty, the priests would look after them (28: 14). That there was any guard (either Jewish or Roman) at all is unlikely; how could the *Pharisees have requested a guard for the purpose of refuting at once any rumour that 'this deceiver's prophecy about rising again' had come true? Were the Pharisees of 27: 62 the same Pharisees who heard the conversation about Jonah (Matt. 12: 40)? Not even the disciples knew about any such prediction.

The story is written up by Matt. from apologetic motives to refute objections to the Christian claim about the empty tomb.

guest-chamber Hospitality was obligatory for Jews and in Jerusalem some rooms were made available at *Passover time for visitors (Mark 14: 14). It was such a guest-chamber that, according to Luke 2, *Mary and *Joseph might have been looking for in *Bethlehem. Guests normally expected water and towel, and perhaps oil and ointment (Luke 7: 44 ff.).

guilt The modern senses of this word are either the legal concept which arises from someone's having infringed a law; or the feeling of remorse and culpability which may not have anything to do with a legal or even social transgression. But in the Bible guilt is the result of a broken relationship between a person and God or one's neighbour, for which there is a personal or corporate responsibility.

There could be responsibility without actually knowing that a wrong action had been committed (Lev. 5: 17–19). In this area therefore it was possible for a great load of guilt to be accumulated, and a means of *atonement and *reconciliation was provided in the OT in the sacrificial system. Because of the Hebrew notion of corporate solidarity, it was possible for a whole group to be guilty of an individual's sin, as with *Achan (Josh. 7)—a view repudiated by *Ezekiel (33: 18–20).

In the NT Paul regards everybody, Jew and Gentile alike, as guilty before God (Rom. 1: 18–3: 20). That is why and how he can maintain that he 'upholds the law' (Rom. 3: 31). *Christians may be deserving of *punishment (1 Cor. 11: 27), but by the *grace of God *forgiveness is available.

guilt offering *Sacrifice of a ram brought to the *altar when some holy thing has been desecrated (Lev. 5: 14–6: 7) and regarded as payment for damage inflicted. The offerer made confession of *sin, but it was done privately. The priest retained the meat.

gulf Used by AV, REB, and NJB in the parable of *Dives and *Lazarus between whom after death a great 'chasm' (Luke 16: 26, NRSV) has been fixed. The saying reflects a contemporary Jewish belief that in the afterlife the *souls of the departed occupied different compartments of the same underworld. Jesus is using a popular story of the time to teach that in the Messianic era there will be a reversal of the states of this present life; he is not giving a geographical description of *Hades.

Gunkel, Hermann (1862–1932) Professor of OT in Berlin and Halle, who was the first to take seriously other Near Eastern religious traditions outside the Hebrew Bible and who pioneered *Form Criticism in OT studies.

H The symbol commonly used by OT scholars to denote the Holiness Code, contained in Lev. 17–26. By observing it, the Israelites made themselves *separate* (the meaning of holy) from other nations. The Holiness Code is part of the Priestly (*P) source, though much of its material is older.

Habakkuk A prophet in *Judah in the final years of the 7th cent. BCE about whom little is known. He was probably a contemporary of *Jeremiah during the reign of *Jehoiakim, son of *Josiah, and was concerned with the duties of public *worship. The OT book which bears his name clearly divides into two sections: the first two chapters (written about 625 BCE) announce the impending onslaught of the irresistible Babylonians ('Chaldeans') who will punish Jehoiakim for his evil tyranny, but in that process some of the less wicked will suffer more. Why should this be? The answer given (Hab. 2: 4) is that 'the upright will live through faithfulness' (NJB): he will live in the highest sense of the word (or, perhaps, simply he will be preserved during the invasion) by maintaining his faithfulness (to God). This verse is cited by Paul in Rom. 1: 17 and Gal. 3: 11 from the *LXX as a scriptural proof for his teaching that a believer is made righteous with God in virtue of his '*faith'.

The third chapter ('one of the noblest in the entire OT') consists of liturgical prayer and *thanksgiving. Come what may (Hab. 3: 17) the prophet refuses to be anything except joyful in the God of his *salvation (Hab. 3: 18–19).

One of the earliest of the *Dead Sea scrolls to be discovered (1947–8) turned out to be a commentary on Hab. 1 and 2, and in the accepted code it is known by the symbol IQpH.

Habiru Sumerian term for refugees of various races scattered round the Fertile Crescent from about 2900 to 1200 BCE. The Tell el-Amarna letters found in 1887 at a town on the River *Nile, were sent to the *Pharaohs in the 15–14th cents. BCE from Egyptian governors in Palestine; they mention hostile incursions of the Habiru, whom some archaeologists identified as the Hebrews of the *Exodus. But the hypothesis has the difficulty that Jerusalem, from which one of the letters was sent, was not taken by the Hebrews until its capture by *David in *c.*1000 BCE.

The term is now thought to be a description of a social rather than an ethnic group.

Hades The Greek translation of the Hebrew '**sheol*', which is the dwelling place of the departed who continue a shadowy existence (Eccles. 9: 10). There was a *legend (Isa. 38: 10; Matt. 16: 18) that entry into Hades was through gates, and Christians supposed that the keys were in the possession of the risen Christ (Rev. 1: 18), Hades is not a place of torment in the OT—except for its appalling boredom (Ps. 88: 12), but in the NT physical pain seems to be envisaged upon their deaths for some (Luke 16: 23), certainly for unrepentant sinners (Mark 9: 48) and for them the word '*Gehenna' in a metaphorical rather than the original geographical ('valley of Hinnom') sense, is used in the Greek (e.g. Mark 9: 43, 48). 'Hades' (Rev. 20: 13, 14) and *Death are regarded as a demonic realm and associated with the *sea (cf. Mark 5: 13). Those who had died at sea could not be buried and therefore could not get through the gates of Hades. However, when the purpose of God is fully achieved, both Death (the Last Enemy, 1 Cor. 15: 26, 54) and Hades, where the already dead repose, will surrender their populations, with the establishment of the eternal reign of God.

Hagar *Abraham's Egyptian *concubine and slave who bore *Ishmael (Gen. 16: 1–4) but was dispatched into the *wilderness by the jealous wife *Sarah. It is used by Paul (Gal. 4: 21–31) to provide confirmation by allegorical interpretation that the religion of the *Law was a form of slavery which *Christians have rejected and from which they have claimed freedom.

haggadah Imaginative interpretation of the OT, excluding sections concerning the *Law. *Legends, *parables, and anecdotes about historical figures (*Abraham, Moses, etc.) proliferate; the purpose is to make the stories come alive for the author's own generation. Something of the method is at work in the books of Chronicles, and it was practised in NT times at *Qumran and within the NT by Matt. (in chs. 1 and 2). Evidence that other NT writers were familiar with haggadic material is in Paul's reference to the Law being given by *angels rather than directly by God (Gal. 3: 19) and to the legend of *Michael the archangel in Jude 9.

Haggai One of the twelve Minor Prophets, living about 520 BCE, and mentioned in Ezra 5: 1; 6: 14. He gives his name to the OT book which puts the blame for drought and disaster on the people's preference to build their own houses after the Return from Exile rather than the Lord's Temple. However, the leaders *Zerubbabel and *Joshua (the high priest) take action (1: 12–14) and there is a promise that eventually there will be a glorious new building (2: 7–9), though there is also a warning not to put a false trust in the security of the rebuilt Temple (2: 14) without true obedience to the Lord (2: 17). The book ends with an expression of confidence that Zerubbabel (2: 23) will continue the royal line of *David.

hair Bald, for the Hebrews, was not beautiful (2 Kgs. 2: 23–4), as the fate of the boys warned. Plucking out hair was a sign of mourning (Amos 8: 10). Although Hebrews admired men's long hair (2 Sam. 14: 26), Paul disapproves of hair worn long by men (1 Cor. 11: 14). Christian women were expected not to have their hair braided (1 Tim. 2: 9) but long (1 Cor. 11: 15).

halakah Commentary or interpretation by the *rabbis on the *Law. The purpose was to bring the written commandments up to date by explaining how they referred to changed circumstances. In their original form they might be inconvenient sometimes, or mutually inconsistent. There had therefore to be constant restatements, accompanied by arguments and debates—e.g. about observance of the *Sabbath. There are examples of *halakic* commentary in the gospels: Matthew's *Sermon on the Mount is reinterpretation of the Law; Jesus demands a more radical righteousness than the rabbis and later he summarizes it in the double commandment of *love for God and for neighbour (Matt. 22: 37–9). Another example of halakic interpretation is at Matt. 12: 5–6, where Jesus reinterprets the law about eating the sacred bread on the Sabbath.

hall In AV, the courtyard of the high priest's house (Luke 22: 55); in the 'common hall' (AV, Matt. 27: 27) of the governor's residence, Jesus was mocked by the soldiers.

hallel The *hymn of *praise sung by Jesus and the eleven *disciples after the *Lord's Supper (Mark 14: 26), as was customary at the end of a Passover meal. It was part of Psalms 115–18, which were praises inspired by the story of the Exodus.

hallelujah A liturgical expression of *praise of Yahweh, prefacing eleven of the psalms (e.g. 146 to 150) and sometimes also appended at the end: sung both in the *synagogues and by the Christian Church. The form '*alleluia', found in AV and NJB at Rev. 19: 1–6 and in Christian hymns, comes from the Greek and Latin spelling of the Hebrew.

hallow Used in the *Lord's Prayer (Matt. 6: 9) with the meaning to regard as holy, but in the OT to make holy (e.g. Lev. 16: 19 of animals for *sacrifice).

Ham One of the sons of *Noah (Gen. 5: 32) and held to be the ancestor of the nations of N. Africa (e.g. Egypt, Ps. 78: 51).

Haman A Persian politician in the court of *Ahasuerus. After taking offence at an alleged insult by the Jew *Mordecai, Haman vowed revenge, but through the intervention of Queen *Esther with Ahasuerus, Haman suffered death on the gallows he had erected for Mordecai (Esther 7: 10). The story is dramatically related in the book of Esther, a literary creation designed to account for the Jewish feast of *Purim.

Hamath A prosperous city on the borders of Israel which eventually fell to the *Assyrians, who ordered a transfer of inhabitants between *Samaria and Hamath (2 Kgs. 17: 24; Isa. 11: 11).

Hammurabi King of Babylon about 1728–1686 BCE. He laid out the capital city which included a *ziggurat which might have inspired the writer of the story of the Tower of *Babel (Gen. 11: 4–9). The Code of Hammurabi is contained on a stele, now in the Louvre in Paris, and there are many resemblances between it and Hebrew laws such as the *Ten Commandments and the *lex talionis* of Exod. 21: 23–5. Unlike the Hebrew laws, the Code prescribes capital *punishment for crimes against property. In the OT the death penalty is reserved for '*abominations' against people (e.g. Lev. 20: 13). *See* ethics.

hand There are a number of uses for 'hand' in both OT and NT. It may refer to God's power (e.g. Exod. 14: 27) or providence (Matt. 4: 6) or to that of a despot (Jer. 22: 3). Direction was indicated by the right hand (i.e. the south) or left (north), and the right-hand side is the seat of honour for a guest (1 Kgs. 2: 19) or for the Son with the Father (Heb. 1: 3).

In Christian iconography, a hand emerging from a *cloud often symbolizes God the Father.

hands, imposition of The gesture is mentioned in both OT and NT as a physical symbol of transfer of authority to a person who is thus commissioned or set apart (Gen. 48: 18; Num. 27: 18; Acts 8: 19; 1 Tim. 4: 14).

hands, washing of The gesture among Hebrews was not an indication that the person would have nothing more to do with the matter, but implied a renunciation of responsibility for what happened (Deut. 21: 6). When *Pilate is said to have washed his hands during the *trial of Jesus (Matt. 27: 24), this is probably a piece of Christian tradition which had developed in order to shift the main responsibility for the *crucifixion from the Romans to the Jews. Such a gesture is not known to have been practised by Romans.

hanging Not a method of execution authorized by the Hebrews, though the corpses of those executed might be hanged provided they were removed before sunset (Deut. 21: 23; Matt. 27: 57). However, the *Bible mentions the custom as prevailing amongst Egyptians (Gen. 40: 19) and Persians (Esther 2: 23).

Hannah Wife of Elkanah and for long childless, but after *prayer in the *sanctuary at *Shiloh she became pregnant and gave birth to *Samuel, who was then dedicated to lifetime service of Yahweh (1 Sam. 1–2).

Haran *Abraham's brother who died in *Ur (Gen. 11: 28) before the migration into *Canaan which Abraham led with *Lot, Haran's son. It is also the name of a thriving city also mentioned in the account of Abraham's journey (Gen. 11: 32) and to which *Jacob fled (Gen. 27: 43). The place was, probably, in Mesopotamia.

hardening In the OT the hardening of the heart is ascribed both to the individual (Pharaoh in Exod. 8: 15) and also to God's action on the individual (Exod. 10: 1). Paul takes up the latter in Rom. 9: 18 as part of the sovereignty of God which is not arbitrary or unjust. In the gospels (e.g. Mark 6: 52) the *disciples' failure to understand the significance of Jesus' *miracles is ascribed to 'hardness' or stubbornness.

hardly Used in the AV for 'harshly' (Gen. 16: 6, NRSV) or 'stubbornly' (Exod. 13: 15, NRSV) or 'with difficulty' (Acts 27: 8, NRSV) or 'scarcely' (Luke 9: 39, NRSV, but REB retains 'hardly'; NJB has 'it is slow', i.e. of the demonic spirit to leave the epileptic boy).

harlot Harlots are referred to in the OT without any note of condemnation (Gen. 38; Josh. 2), but 'harlotry' became synonymous in the prophets for *apostasy (Hosea 4: 12–13). Hosea had himself married a harlot.

In NT times, harlots were outcasts from respectable society (Matt. 21: 31 f.) but were treated gently by Jesus (Luke 7: 37–50; John 8: 1–11). The well-known and accommodating *Rahab (Josh. 2: 1–22) appears in one of the genealogies of Jesus (Matt. 1: 5).

Harnack, Adolf von (1851–1930) Professor at Berlin, and historian of the early Church. He wrote also on the Acts and maintained the traditional view of Lucan authorship. A Liberal Protestant, he believed the essence of Jesus' *teaching to be concerned with the Fatherhood of God and the value of the individual. Harnack was one of the leading intellectuals to sign a manifesto supporting the Kaiser's war in 1914.

harp The Hebrews had a family of stringed instruments under the general category of harp. They included psalteries, lyres, lutes, and dulcimers. The psaltery was triangular in shape; the lyre was U-shaped. There is also a similar instrument translated 'sackbut' (Dan. 3: 5, AV)—trigon in NRSV, triangle in REB, zither in NJB—which had strings plucked either by the fingers or by a plectrum.

harvest In Palestine the harvest was brought in during May, but could begin earlier and finish later, depending upon the land and the weather. Corn was cut by sickles; threshing was by the treading of unmuzzled *oxen (Deut. 25: 4; 1 Cor. 9: 9) and the threshings were tossed into the air so that the wind would scatter the chaff (husks) from the wheat (Ps. 1: 4). The grain was stored in barns (Matt. 6: 26; Luke 12: 18).

Harvest-time is an obvious symbol for the *judgement at the end of the age (Gal. 6: 9).

Hasidim Persons characterized by *hesed*, loyalty, especially to God, in the community of the faithful. They are 'the holy ones in the land' (Ps. 16: 3, NRSV), 'his faithful ones' (Ps. 30: 4; Prov. 2: 8, NRSV). At one time the fact of belonging to the people of Israel was equivalent to being one of the *Hasidim*, but later they were ethically and spiritually 'the righteous' contrasted with the 'wicked', as is implied by the parallelism of Ps. 37: 28–9.

In the 2nd cent. BCE the Hasideans were a group of Jewish zealots for the Law who were opposed to the spread of Hellenistic culture (1 Macc. 2: 42; 2 Macc. 14: 6). As students of scripture they gave intellectual support for a time to Judas *Maccabaeus. A self-portrait of a member of the group may be drawn by Ben Sirach in Ecclus. [= Sir.] 38: 34c–39: 5. They are thought by some to have been forerunners of the *Pharisees.

Hasmoneans The Jewish dynasty deriving from *Mattathias (166 BCE), usually known as the *Maccabees. The sons of Mattathias, *Judas, Jonathan, and *Simon, led the war against the Syrian generals and from Simon's time in 142 BCE the Hasmoneans ruled *Judah until 63 BCE. John *Hyrcanus (134–104 BCE) was the first of the Hasmoneans to call himself king. His successors were *Aristobulus I (104–103), Alexander Jannaeus (103–76), and Alexandra (76–67), wife of the two preceding kings. Because the Hasmonean kings were also high priests, her eldest son Hyrcanus became high priest. Aristobulus II (67–63) was the last of the line before the coming of the Romans; though the younger son of Alexandra, he ejected Hyrcanus from the high priesthood.

hatred Just as Yahweh hates *evil and evildoers (Amos 6: 8; Deut. 32: 41), so must the people of the covenant (Ps. 139: 21 f.). Jesus enjoins love for enemies (Matt. 5: 43 f.); but it would seem that the commandment to love all became a commandment only to 'love' fellow *Christians (1 John 4: 20) or at any rate fellow Christians especially (Gal. 6: 10). Christians must anticipate hatred from the world (John 15: 18 ff.).

Hazael A courtier with King *Benhadad at *Damascus, who became king of *Syria (Aram). When *Elisha the prophet paid a visit to Damascus, King Benhadad sent Hazael to ask if he would recover from his illness. Elisha answered ambiguously (2 Kgs. 8: 10–13), giving Hazael to understand that the king would die and that Hazael would succeed him. He defeated Israel and *Judah in battle (2 Kgs. 13: 3) but was himself defeated by the *Assyrians in 841 and 837 BCE under *Shalmaneser III, and again by Adadnirari III about 805 BCE, according to Assyrian inscriptions.

Hazor An important city north of the Sea of *Galilee which has yielded valuable information in archaeological excavations. According to the books of Joshua (11) and Judges (4) it was taken by the Israelites, fortified by *Solomon (1 Kgs. 9: 15) but captured in the 8th cent. BCE by the Assyrians (2 Kgs. 15: 29).

head Part of the human body which in the OT and NT can be intended both anatomically (Gen. 48: 14; Mark 15: 29) and metaphorically, as of the head of a *family (Josh. 22: 14). In the NT Christ is described as head of the Church (Eph. 4: 15; Col. 1: 18), which could mean either that Christ is its head in the sense of being a king or judge, or protector, over against his Body, the Church, or it could indicate the inseparability of the Head from the Body—Christ and the Church are one organism.

healings In the OT, healings which are apparently miraculous are recorded during the *Exodus of the Israelites from *Egypt (Num. 12: 13) and by *Elijah and *Elisha (1 Kgs. 17: 22; 2 Kgs. 4: 35) and in the NT by Jesus and by the apostles.

It is evident that Jesus was selective in his healings (John 5: 3 ff.) and that the gospels are selective in recording them. There is an element of symbolism in that Mark records twelve *miracles of healing of Jews and one of a *Gentile (Mark 7: 29). The latter could be Mark's way of implying that the Church enjoyed Jesus' authorization for its mission to the Gentiles. Some healings are of the ritually unclean or outcasts (Mark 5: 29; Luke 17: 16). The miracles in the synoptic gospels are to be understood as visible indications of the approach of the *kingdom of God, (Luke 17: 20–1), and in the gospel of John this theological motif is underlined by the use of the word 'signs' for the healings recorded.

Some of the healings in the synoptics are *exorcisms (there are none in the gospel of John) and the symptoms would be differently diagnosed by modern medicine, but Jesus' healings were real enough. This is confirmed by the honest recognition that Jewish exorcists had equal success (Matt. 12: 27). The accounts in the gospels were intended to support *disciples in their beliefs and are explicitly related to the fulfilment of OT prophecy (Luke 4: 21; 7: 21–2). *Exorcisms could also be interpreted sociologically: possession by *demons reflected the sense of economic and political oppression experienced by the Christian community, and the expulsion of the demons was seen as liberation (Mark 5: 15–17).

health Regarded by the Hebrews as the reward for obedience to God (Exod. 15: 26). Disease therefore was directly or indirectly caused by *sin (Num. 12: 10–11), though there were stringent regulations to check infections and water was to be kept pure, food inspected, and periods of quarantine observed (Lev. 13 and 14). In the NT the connection between sin and bodily well-being is not regarded as proven (John 9: 3) and sensible precautions to maintain health are recommended (1 Tim. 5: 23).

heart Frequently used in both OT and NT, but rarely in the physiological sense, since the Hebrews were unaware of its function; but its importance was recognized, and so it acquired a metaphorical sense for what is dearest and best—a 'clean heart' (Ps. 51: 10; Ezek. 36: 26). The heart was also regarded as the seat of emotions (1 Kgs. 8: 38), of memory (1 Kgs. 4: 29), and of *wisdom (NRSV prefers 'mind' instead of 'heart': 1 Kgs. 3: 12), and the new *covenant will be written in the hearts of the people (Jer. 31: 31–4).

Similarly in the NT, Jesus is 'gentle and humble in heart' (Matt. 11: 29). The 'pure in heart' (Matt. 5: 8) are those without an accusing conscience; for 'God searches the heart' (Rom. 8: 27) and being 'cut to the heart' is the prelude to conversion (Acts 2: 37). In Paul's analysis of human nature the heart is neutral: it may be *evil (Rom. 1: 21), but it may also be good (2 Cor. 1: 22).

heathen Those who do not worship God as revealed in the Bible (AV Jer. 49: 15; Ps. 98: 2). After the Exile the 'heathen' were the *Gentiles (non-Jews)—often reproached for wickedness but also sometimes addressed with a yearning to bring them to the knowledge of Israel's God (as in Jonah; Isa. 49: 6). Converts from heathenism were welcomed (Matt. 23: 15) by the *Pharisees.

heave offering Used in AV for the act of dedication, outside sacred precincts, of tithes (Num. 18: 24–9) or votive gifts (Deut. 12: 11). 'To heave' = to select; it was that part of the offering which was retained by the priests.

heaven(s) In the Bible, as in the English language, 'heaven' can refer to the region of the atmosphere or also to a supernatural world. *Birds fly in the heavens (Jer. 4: 25, REB, AV, NJB), and the *clouds are there (Prov. 8: 27–8). 'Heaven' also refers to the firmament or celestial vault (Gen. 1: 6) which divides the waters above from those beneath and is supported on pillars (Job 26: 11). The waters above supply rain which drops to the earth through openings in heaven (Gen. 7: 11). *Stars are attached to or suspended from heaven (Job 22: 12). In visions of the *End, the heavens will be destroyed along with the earth (Isa. 24: 4; Luke 21: 33).

The supernatural 'heaven' is the dwelling place of God, but the word is often written in the plural, as Ps. 148: 4: 'the heaven of heavens'. The Hebrews spoke of the existence of several heavens, and Paul had a mystical experience of being taken up 'into the third heaven' (2 Cor. 12: 1 ff.). Christ ascended above 'all the heavens' (Eph. 4: 10). The book of *Enoch refers to 'seven heavens'.

God's throne is in heaven (Isa. 66: 1), and *angels surround it (1 Kgs. 22: 19), and this *vision is developed in Rev. 4. In the highest heaven there is a '*tabernacle' or 'true tent' (Heb. 8: 2) 'exalted above the heavens' (Heb. 7: 26) where Christ the true High Priest offers spiritual *sacrifices and intercedes for faithful Christians. Similarly, this is where the *Son of Man will sit at the right hand of the Father (Mark 14: 62) and will prepare places there for *disciples (John 14: 2). The names of the redeemed are recorded in heaven (Heb. 12: 23), which is a realm of *joy (Luke 15: 7) and *peace (Luke 19: 38). The descriptions of heaven in Rev. 4, 14, and 21, representing the ultimate triumph of God, have often provided the ground of hope for life after death and have been a fertile source of imagery in Christian *hymns. The essence of the belief is that heaven is the promised fulfilment of a life already begun but incomplete; it is not a continuation in another space and time of the kind of life experienced here and now on earth but a translation into a new dimension.

Hebrew The original language of the OT, apart from a few passages especially in Dan. in *Aramaic. It was the 'language of the Jews' (2 Kgs. 18: 26), and spoken in different dialects (Judg. 12: 5–6), though the word Hebrew is not used of the language until the Hellenistic era. It belongs to a Semitic group of languages of the Canaanite family which includes Phoenician and Moabite. After the Exile Aramaic was adopted for communication with neighbouring states, and increasingly, especially in *Galilee, for daily speech. Hebrew continued to be used for religious matters well into the Christian, or common, era and has been revived as the language of modern Israel. In the Hebrew alphabet there are 22 letters, all originally consonants. The full system, with pointing to represent vowels, dates only from the 5th–10th cents. CE. The script is read from right to left.

Hebrews The usage of the OT does not make it clear whether the Hebrews are regarded as an ethnic group. The description is rare. But they are distinguished from Egyptians (Exod. 2: 11) and identical with 'Israelites' (Exod. 5: 1, 3). The *Philistines who are alarmed by 'Hebrews coming out of their holes' (1 Sam. 14: 11) are referring contemptuously to Israelites. There were rules about releasing slaves who were Hebrews (Israelites), reiterated by *Jeremiah (Jer. 34: 9, 14), some of whom were 'neighbours and friends' (34: 17) of their owners. *Jonah pronounces himself a Hebrew (Jonah 1: 9) who worships Yahweh. As late as the Maccabean period, Jews are distinguished from foreigners as 'Hebrews' (2 Macc. 11: 13) and in the NT those Jews who retained the use of Hebrew and Aramaic rather than speaking Greek are called Hebrews (Acts 6: 1; Phil 3: 5). Generally, however, from and after the Exile, the people of Judah and Judaea are known as 'Jews'.

Hebrews, letter to the About 200 CE *Clement of Alexandria referred to this letter as written by Paul, but there is nothing in the text about its author. There are in fact good reasons for rejecting Pauline authorship; the Greek is too good to be Paul's; favourite theological ideas of Paul ('in Christ', the Holy Spirit, *Justification, 'the *Body of Christ') are all missing; there is more emphasis on the humanity of Jesus. It is wise to regard Clement's attribution as no more than a guess, and fair to speculate on other candidates for authorship of the anonymous epistle. Martin Luther thought it was by *Apollos, and his suggestion has won the support of some modern scholarship: Apollos was from *Alexandria (Acts 18: 24), where the Jewish philosopher *Philo (d. 50 CE) had worked, and there are resemblances between his thought and that of this epistle; Apollos was also active in apologetics to the Jews by quoting the OT (Acts 18: 28) which corresponds to the method of the epistle. A further attractive, but not widely accepted, hypothesis is that Apollos wrote the epistle from *Ephesus to the Church in *Corinth, especially its Jewish Christian members, between 52 and 54 CE. He had visited Corinth (Acts 19: 1) and had made a considerable impression (1 Cor. 3: 6–9). The epistle, in the form more of a homily, was written, it is argued, to demonstrate the supremacy and uniqueness of Christ and the superiority of the Christian faith over the Mosaic *Judaism into which some of the recipients were in danger of returning. Apollos had consolidated his reputation at Corinth by this powerful appeal, and Paul was obliged to defend his own standing at Corinth (1 Cor. 1–4) and in his (first) letter Paul deals with some of the points raised in the letter to the Hebrews, as when he seems to defend his lack of eloquence (1 Cor. 2: 1), which compared unfavourably with the skill of Apollos in speech and writing. As confirmation of this Corinthian destination of the epistle to the Hebrews there are thirteen references to it in the first epistle to the Corinthians written by Clement of Rome in *c.*96 CE. At any rate, whatever the merits of this theory about authorship, the recipients were surely former Jews, perhaps from the Essene group, who were inclined to honour Jesus as more than human but less than divine. They compromised by regarding him as an angel. (Philo wrote of the *Logos* as an angel.) Hence veneration of angels is particularly denounced in Heb. 1: 4–2: 18.

The structure of the epistle is that each statement of doctrine is followed by practical exhortation. The author urges the readers to 'hold fast' (Heb. 3: 6); it is by *faith they must live and die (10: 32–9), but he is worried whether they will in fact endure to the end (Heb. 12: 12), especially as some of them have ceased to attend the assemblies (Heb. 10: 25). They are reminded that Christ is God's final revelation. The old sacrificial system of the Day of *Atonement failed to achieve the redemption it foreshadowed (10: 1–4). Everything Jesus did and the hope he inspires (Heb. 10: 19–25) are proof of the greater 'salvation' (Heb. 2: 3) he offers by his sacrifice which never needs to be repeated.

Hebrews, gospel of A work of the *Ebionites which omitted reference to the conception and *birth of Jesus and appears to have had an *Adoptionist Christology. It was known to Epiphanius (d. 403 CE).

Hebron A town SW of Jerusalem in a mountainous area, captured by *Caleb (Josh. 14: 12 ff.) and the scene of *David's accession to the throne (2 Sam. 2: 4), and where he reigned for seven and a half years (2 Sam. 5: 5) until he took Jerusalem. David's son *Absalom established a rebel stronghold in Hebron (2 Sam. 15: 7 ff.)

hedge Hedges were planted as boundaries (Mic. 7: 4) and to protect vineyards (Mark 12: 1); and could symbolize God's protection (Job 1: 10). The word is translated 'fence' by NRSV and NJB or 'wall' by REB (Mark 12: 1).

heifer The ashes of a heifer (Heb. 9: 13), red and without defect (Num. 19: 2), are ordered to be mixed with water for the *purification of the priest or of a city (Deut. 21: 1 ff.). The epistle to the Hebrews argues that if the old ritual *sacrifices were efficacious, then the sacrifice of Christ must be much more so—indeed of eternal validity.

Heilsgeschichte German, 'salvation history'. A term employed by the *Biblical Theology movement to describe the story of the Bible as that of God's redemptive work in the events of history.

heir(s) Laws of *inheritance were precise in ancient Israel, and gave priority to the eldest son (Deut. 21: 17) of a man's wife, and therefore by rights *Esau rather than *Jacob was entitled to his father's *blessing (Gen. 27: 36). Cf. Luke 15: 31. However, it was possible for a daughter also to inherit (Job 42: 15) provided that she did not marry outside the tribe (Num. 27: 1–8). If there were no children, the brother of the deceased came into the inheritance.

Paul uses the principle of inheritance to describe a person's relationship to God (Rom. 8: 16–17; cf. Heb. 9: 15).

hell The English word is used misleadingly in older versions to translate the Hebrew **sheol*, the Aramaic *Gehenna, and the Greek *Hades. Normally the word is understood to denote a place of eternal torment, especially by fire (Isa. 66: 24), for those who are irredeemably wicked (Rev. 21: 8) and it is confusing that the word also designates the place of rest or waiting, and not of agonizing pain, for the departed, which is the meaning of the word 'Hades' in Rev. 20: 13 and also in the clause of the Church's creed which asserts that Christ 'descended into hell' after the *crucifixion. This is a belief based on 1 Pet. 3: 19, which may well be a Christian appropriation of ancient redemption myths of the descent of deities to the underworld (Orpheus and Eurydice, Persephone, etc.); in the NT context it is an assertion in terms of the belief that Christ's mission is universal—it extends even to those who died before the *incarnation.

Hellenists A group of Jews who read the scriptures in the *LXX translation rather than in Hebrew and who spoke Greek rather than *Aramaic. For nearly two centuries there had been an uneasy relationship between Jews who welcomed Hellenistic culture and those who defended conservative Hebrew customs and distinctions. In NT times Hellenistic Jews in Jerusalem may have had their own *synagogue (Acts 6: 9) and possibly *Stephen was a member who succeeded in converting some of them to Christianity. They were probably already inclined to interpret the law less rigorously than the 'Hebrews' (Acts 6: 1) and were persuaded by Stephen to look beyond Moses and the Temple. Thus such a group of Hellenists could have proved a point of growth for the Church as it spread beyond its Jewish origins into the Greek-speaking *Gentile world.

henotheism It has been held that before Israel had reached belief in one God (Yahweh) exclusively, it conceded that other nations had their own gods whom it was proper for them to worship (Deut. 32: 8–9; 1 Sam. 26: 19), though Yahweh's superiority to other gods was increasingly asserted (Ps.

86: 8). It is, however, possible that *Deutero-Isaiah's unambiguous *monotheism developed from an earlier implicit monotheism rather than from henotheism. Hinduism might be considered a form of modern henotheism.

Henry, Matthew (d. 1714) Presbyterian minister at Chester in England. He was called 'the Prince of Commentators' on account of his great influence as expositor of scripture in the Reformed tradition. His strength was his ability to write clear analyses of large blocks of material.

herald A court official in *Babylon who issued a proclamation from King *Nebuchadnezzar (Dan. 3: 4), but in Isa. 40: 9 one who announces good tidings. Paul is similarly called a herald of good news in 1 Tim. 2: 7, as is *Noah ('a herald of righteousness', NRSV; 'who proclaimed righteousness', REB; 'the preacher', NJB, in 2 Pet. 2: 5).

herbs Herbs were in use by the Hebrews both to give flavour to food and also as medicines. Bitter herbs (lettuce etc.) were eaten during the *Passover meal. Other herbs mentioned in the Bible include balm, for healing (Jer. 46: 11), dill, mint, and *cummin (Matt. 23: 23), and *myrrh from S. Arabia (Matt. 2: 11), a brown resin used in oil and cosmetics, and also for pain relief (Mark 15: 23).

hereafter Used by AV with the meaning 'at some future time' in John 13: 7 where NRSV and NJB have 'later', REB 'one day'. At Matt. 26: 64 NRSV and REB have 'from now on', and NJB 'from this time onward'. The phrase has been added by Matthew to Mark 14: 62 and is a Christian explanation by Matthew for his readers that the glorification of the *Son of Man was imminent, through his *crucifixion and *resurrection.

heresy From the Greek, meaning 'choice' or 'thing chosen', or an opinion. It came to be used (in the Greek) for a sect or a school of philosophy, and of the 'sects' of the *Sadducees and the *Pharisees in Acts 5: 17; 15: 5. It is used by Paul for a protest group in *Corinth (1 Cor. 11: 19) and for a typical kind of divisive action in the *community (Gal. 5: 20), where the word is on its way to its later designation of a deviationist group within Christianity. By the end of the 1st cent. *Ignatius of *Antioch termed theological error a 'heresy' and, in the later Church, heresy, as deliberate adherence to 'false' doctrine, was condemned as sinful.

Hermas (1) A Christian greeted by Paul (Rom. 16: 14). (2) *Origen believed this man (of Rom. 16: 14) to be the author of a highly esteemed work called *The Shepherd*; but it is now dated to the middle of the 2nd cent. For a time it was regarded in the East as part of scripture.

hermeneutics In Greek = 'interpretation': hence, the understanding of biblical texts, including their historical context. The word is also used in a wider sense, to mean the interpretation of a text and its application to the reader after the historical examination has been done; the work of discerning how a NT text written in the cultural context of the 1st cent. CE can be understood in a modern cultural context. It is a movement from 'what it meant' then to 'what it means', taking into account empirical evidence of today. For some NT scholars this has involved detaching the essential message of the gospel from the mythological frame in which it has been enclosed. *Bultmann's hermeneutical vehicle for the reinterpretation has been that of the modern philosophy of existentialism.

Although the term 'hermeneutics' originated in the 17th cent., it is held that it began in the NT itself, when Paul used contemporary Jewish methods of interpreting the OT for preaching the significance of Jesus. In subsequent ages, allegorical and typological methods were practised. At the Reformation the doctrine of *Justification by Faith was adopted as a *canon of interpretation. But only a literal sense of the text was regarded as legitimate by Calvin, who feared that faithful interpretation could otherwise be distorted by merely human values. The history of interpretation is one of the resources at the disposal of modern communicators.

Hermes In Greek mythology, the young attendant of the god Zeus, thought to be particularly eloquent (Acts 14: 12). The Roman equivalent was Mercury (AV). Paul was assumed to be Hermes by the crowd at *Lystra, though he himself is inclined to disparage his oratory (1 Cor. 2: 1) and he seems to have had only limited success at *Athens (Acts 17: 32–4).

hermetic literature A miscellany of Platonic, Stoic, and Oriental teachings in Greek and Latin writings of the Hellenistic age (1st cent. BCE to 2nd cent. CE). The Hermetic tractate *Poimandres* contains words and phrases similar to those of the gospel of John. It suggests that human beings only attain to the vision of God when they become identified with an archetypal man who exists in the perfection of the eternal 'forms', or 'ideas' in the terminology of Plato. Another tractate teaches the possibility of a rebirth with the help of Hermes Trismegistos (the Egyptian god Thoth), and this is reminiscent of John 3: 3–12. But reminiscence does not mean direct borrowing; the gospel may be merely using language which would find a responsive echo in some readers.

The Hermetic collection was preserved in the Christian manuscript tradition and was known in the Middle Ages, but additional texts were among the 1945 discoveries at *Nag Hammadi.

Hermon The highest mountain in Palestine; and being only 22 km. (about 14 miles) north of *Caesarea Philippi, it has often been thought to be the site of Jesus' *transfiguration (Mark 9: 2).

Herod The Greek name of an Idumean dynasty whose members governed Palestine for a century and a half, which included the NT era. Herod the Great (40 to 4 BCE), son of Antipater, was married to Mariamne I, a *Hasmonean. Herod was a friend of the Romans and from 37 BCE, when he seized Jerusalem from Antigonus II, he was their king, though under Roman suzerainty. He was responsible for important building projects—Caesarea on the coast, the fortress-prisons of Masada and Machaerus, and the vast new (strictly, reconstructed) Temple, which was to occupy a quarter of the entire area of Jerusalem, begun in 20 BCE and eventually completed in 62 CE, only to be destroyed in 70 CE. Herod ordered the execution of any potential rivals to his throne, including his wife Mariamne and their two sons. Such cruelty is the plausible background for the story in Matt. 2: 16–17 of the massacre of the infants of *Bethlehem after the *birth of Jesus; but the historicity of the event is doubtful. If he had a reason to suspect that a child in *Bethlehem might become a messianic leader, Herod could have engaged a contract killer. A massacre of such notoriety would certainly have demanded mention by *Josephus, who is silent on the subject.

In 4 BCE Caesar Augustus divided the kingdom (but denied the title of king) to three surviving sons: (1) *Archelaus as ethnarch was given *Judaea, *Idumea, and *Samaria; he died in 6 CE. (2) Herod *Antipas had *Galilee and *Perea. (3) Philip (whose mother was Cleopatra) had Batanea, Trachonitis, and Auranitis; he built *Caesarea Philippi (Mark 8: 27); he married *Salome, daughter of *Herodias and of Philip's half-brother Herod Antipas, and died in 34 CE.

Herod Antipas is described as a 'fox' (Luke 13: 31–2); his marriage to Herodias was criticized by *John the Baptist, who was then executed at Machaerus (Mark 6: 14–29). His capital city was *Tiberias by the Sea of *Galilee. Eventually deprived of his territory by Rome, he went into exile and died in 39 CE. He was succeeded by Herod Agrippa I, his nephew, who ruled from 41 to 44 CE with much approval from the *Pharisees. He died suddenly (Acts 12: 20–3). His son, Herod Agrippa II, too young in 44 CE, was given territory in 50 CE, augmented in 53 CE. He renamed his capital (Caesarea Philippi) Neronia in honour of the emperor. He lived incestuously with his sister *Bernice (Acts

25: 13–26: 32). During the Jewish revolt of 66–70 CE, he took the side of Rome, and died there in 93 CE.

Herodians In Mark 12: 13 and Matt. 22: 16 some Herodians join with *Pharisees in putting to Jesus a difficult question about the payment of tribute money. The Herodians are presumably supporters of Herod *Antipas but their common front with the Pharisees is surprising in that Herod the Great regarded Pharisees as his principal opponents and there is no reason to suppose that his son was any less hostile.

Herodias Granddaughter of *Herod the Great and sister of Herod *Agrippa I. Married (1) to half-uncle Herod, son of Mariamne II, third wife of Herod the Great (Mark 6: 17 calling him Philip is a mistake), by whom she had *Salome (who did marry Philip); (2) to another half-uncle, Antipas, for which she was condemned by *John the Baptist. Hence her call (Mark 6: 17–29) for his execution.

Heshbon A city in the north of *Moab which changed hands several times; captured by the Israelites, it was occupied by the tribe *Reuben (Num. 32: 37) but later by *Levites (Josh. 21: 1–3, 39). It returned to the Moabites in the 9th cent. BCE (Isa. 15: 4); known as Esbus in the NT era.

hexateuch The first six books of the OT, often regarded together as a theological unity in contrast to the more usually discussed *Pentateuch, the first five books, commonly called the Law of Moses. Some NT scholars who have emphasized the typological parallels of the gospel of Matthew with the OT notice that the correspondence is with the OT books Genesis (to which the Matthean birth narrative corresponds) to Joshua (where the entry into the *Promised Land is the 'type' of Jesus' *Resurrection).

The OT narrative itself consists of the accounts of God's gifts to Israel alternating with God's commands—the *Law—to whom Israel was in duty bound to respond under the leadership of God's chosen servants, *Abraham, Moses, and *Joshua.

Hezekiah King of Judah (727–698 BCE) during the time of *Assyrian aggression. *Samaria was taken in 722 and Hezekiah became an Assyrian vassal until 705, when the death of the Assyrian king Sargon prompted a coalition of small states to rebel. In reply, the new king, *Sennacherib, captured some of the outlying cities of *Judah (he claims in an inscription to have besieged forty-six), and reached Jerusalem itself but withdrew. The prophet Isaiah had predicted that the capital would be spared and this was attributed to divine intervention (2 Kgs. 19: 35).

Hezekiah was praised (2 Kgs. 18: 3–6) for his policy of closing the 'high places' in favour of centralized *worship in the Temple.

The dates of Hezekiah's reign are disputed. Alternative to those above, they might be 715–686 BCE. The references in 2 Kgs. 18: 9 and 18: 13 are mutually incompatible.

Hiddekel The name of the third river in *Eden (Gen. 2: 14, AV; also Dan. 10: 4) and given its modern name '*Tigris' by NRSV, REB, and NJB). It is in modern Iraq.

Hierapolis A flourishing city in the Roman province of *Asia. According to Col. 4: 13 Epaphras worked on behalf of the Churches of Hierapolis and *Colossae (19 km. or 12 miles away) and *Laodicea (9.6 km. or 6 miles away). There was a sizeable Jewish population.

high places The conventional translation of a Hebrew word (*bamoth*) whose meaning continues to be disputed. It refers to a place where *sacrifices were offered in the open air. They were hated by *prophets as being part of the fertility religion of the *Canaanites and incompatible with Yahwism (Hos. 2: 2–20 and Ezek. 16: 16), but attracted much popular support. Indeed many local shrines claimed to have been founded by unimpeachably authentic Israelite heroes —*Jacob at *Bethel (Gen. 28: 18–22) and *Joshua at *Gilgal (Josh. 4: 20–4)—which was a means of making originally Canaanite shrines respectable for faithful Israelites. Indeed before *Solomon's Temple was built, they were regarded as acceptable, as when *Samuel officiated at Zuph (1 Sam. 9: 19). The town *Shiloh, north of Bethel, was a religious centre with priests (1 Sam. 1: 9) until destroyed by the *Philistines (Jer. 7: 12–14). But inevitably their existence and attraction led later to Canaanite practices infiltrating into Yahwism. *Josiah is credited with destroying the high places (2 Kgs. 23: 4–20) and no doubt the *Babylonian invaders completed this work. They were never revived when the exiles returned in the late 6th cent. BCE.

high priest Spokesman for the people to God, and for God to the people. *Aaron and his successor *Eleazar were precursors of the institution (Num. 27: 21 ff.) which is mentioned in Lev. 21: 10. He alone was anointed and possessed authority over lay officials. He was clothed with the *ephod in which were set stones bearing the names of the twelve tribes, and on his head he wore a kind of turban or mitre (Exod. 28: 36–7). *Zadok and *Abiathar were chief priests under *David (2 Sam. 20: 25), and Zadok's descendants were high priests in the first and second Temples. *Joshua came back from the Exile as high priest (Ezra 2: 2). Under *Herod the Great high priests no longer held the office for life and *Annas was appointed by *Quirinius governor of *Syria in 6 CE until 15 CE. After several short tenures *Caiaphas, who was Annas' son-in-law, became high priest from 18 CE until 36 CE. The statement in John 11: 49 and 18: 13 that Caiaphas was high priest 'that year' does not mean that the office was an annual one (though it *was* for pagan high priests in Asia Minor and Syria): either John was mistaken or 'that year' means 'that memorable year of the *crucifixion'.

Hilkiah Several persons of this name are mentioned in the OT, of whom the chief is a high priest who announced the discovery of a law book during building work in the Temple (2 Kgs. 22–3). It has been widely assumed that this book, enthusiastically welcomed by King *Josiah, was Deuteronomy, or part of it, and that it precipitated the reforms he instituted in eliminating places of *worship outside Jerusalem, and centralizing the cult in the Temple.

Hillel A *rabbi of great influence who taught in the time of Jesus. He was liberal and lenient in his interpretation of the oral *law; and his followers, who were prominent after the destruction of Jerusalem in 70 CE, were distinguished from the more rigorous school of *Shammai. The controversy within Judaism may be reflected in Mark 2: 16 and Mark 7: 1–8. Hillel is famous for the dictum: 'What you hate, do not do to your fellow; that is the whole *Torah, while all the rest is commentary.'

Hinnom valley A depression just west of Jerusalem where *children were sacrificed in idolatrous rites (Jer. 7: 31), though forbidden by *Josiah (2 Kgs. 23: 10). The Hebrew *Gehenna (= 'valley of Hinnon') came to mean *hell (Mark 9: 43, NRSV and REB; 'Gehenna' in NRSV marg.).

Hiram (1) King of *Tyre who encouraged trade with Israel under both *David and *Solomon (2 Sam. 5: 11 and 1 Kgs. 9: 10–14). (2) A skilled worker in the Temple (1 Kgs. 7: 13).

historical criticism Every writing in the OT and NT emerged from a particular social, political, and cultural environment over the course of many centuries, and historical criticism is a modern and tested method of exploring their various origins and tracing their development and significance. Dates and events, as well as people and places mentioned in the narratives, have all to be scrutinized. The scriptures were certainly studied in detail by the Fathers with abundant use of *typology, as by Justin Martyr (*c.*150 CE) in his *Apology* and *Dialogue with Trypho the Jew*. Inconsistencies between the gospels were noted, and Origen was aware of the necessity of textual *criticism. Medieval theologians also found an application or meaning beyond the literal, through *allegorical, *typological, and mystical interpretations. But with the Renaissance there came a determination to give to the Bible that same rigorous scrutiny of what purported to be history as that which was already being applied to the classical literature of Greece and Rome. The first known critic was a French Roman Catholic priest, Richard Simon (1678–1712 CE). He thought that the uncertainty about scripture brought by criticism undermined Protestant dependence upon it, from which Catholicism was happily free. After him most of the great names are German.

The method involved an examination of the texts to check their authenticity and to establish their probable authorship. Comparison is made with documents from other sources and with external evidence provided e.g. by archaeology. Motives, tendencies, interests, presuppositions will all be taken into account. Vocabulary and style must be scrutinized. A major achievement of the 19th cent. was the recognition by Karl *Graf and Julius *Wellhausen that the Pentateuch was compiled from different sources and reached its final form after the time of the great prophets. Other books were seen to be later than had been supposed: Daniel apparently describes events of the 6th cent. BCE but has been shown to come from the middle of the 2nd cent. in virtue of its accurate account in ch. 11 of Antiochus Epiphanes; the book of Isaiah has been divided to reflect at least two historical periods. Hermann Gunkel (1862–1932) was the founder of OT *Form Criticism after a study of the laws of folk-behaviour in passing on traditions. The OT exhibits characteristics similar to those of early Scandinavia; there exist in the OT recurrent literary categories with a related form, and the social situation in which they were produced can be sought, and the functions which the traditions served can be surmised. Gunkel's OT method was applied to the NT by his pupils, e.g. Rudolf *Bultmann.

The first attempt to map a consistent historical study of the NT was also made in Germany. F. C. *Baur (1792–1860) used the terms thesis, antithesis, and synthesis of the philosopher G. W. F. Hegel (Jewish Christianity met Gentile Christianity and gave birth to *Early Catholicism). J. J. *Griesbach (1745–1812) was a pioneer in synoptic criticism. Research into the life of Jesus took a sceptical turn with D. F. Strauss's *Life of Jesus Critically Examined* (1835), which argued that the apparent historical form of the gospels was but the clothing for legends. Much further work on the quest for the historical Jesus continued with J. E. Renan, Adolf von *Harnack, and William *Wrede; until more recently Rudolf *Bultmann (1884–1976), Gunther *Bornkamm (1905–90), and Hans *Conzelmann (1915–89)—who are a few of the many workers in this field. Scholars in America and Great Britain have been no less active.

It has always been important to determine both the date and authorship of each composition, which is done sometimes by indications within the text itself or, sometimes, by archaeological evidence. The dates of 1 Cor. and 2 Cor. are established by the discovery of an inscription at Delphi proving that *Gallio was proconsul of *Achaia in 51/2 or 52/3 CE. Many hypotheses and guesses have been offered for the places where the books of the NT were compiled: Palestine, *Egypt, *Syria, Asia Minor, *Rome. Sometimes fragments of an author's work have been collected together and edited, and sometimes editorial work has combined several sources into a composite whole, as with the *Pentateuch. Historical criticism is therefore closely related to this kind of analysis, and the effects on historical criticism of *Form Criticism were very marked: it suggested that the gospels consist of collections of small units which have passed through a stage of oral tradition shaped according to the needs of the community. Hard on the heels of Form Criticism came *Redaction Criticism, which emphasized the overriding theological ideas of the *evangelists which governed their selection and placing of the available material, and this method too had important implications for the historicity of the narratives. To what extent did the aim of interpretation (e.g. about *eschatology) affect the narratives of the *teaching of Jesus? Or the necessity to invoke the authority of the OT to buttress the claim for the Messiahship of Jesus? Thus the gospel of Luke is not written to prepare for the imminence of the *Parousia but to teach Christians to live in the continuing future. The gospel of Matthew, especially in chs. 11 and 12, contrasts the new era which Jesus has brought with the old which preceded it, which is part of the doctrine of Messiahship.

Hittites A major power in the 14th cent. BCE from a base in Asia Minor (modern Turkey), conquering north and central *Syria and most of *Lebanon, but they were eventually absorbed into the *Assyrian and *Babylonian empires. There are references to Hittites in the OT (e.g. Gen. 23: 3) and to the kings of the Hittites in 1 Kgs. 10: 29, 2 Kgs. 7: 6. The latter were probably rulers of small states in north Syria. King *Solomon had Hittite wives (1 Kgs. 11: 1) and *David married *Bathsheba, widow of *Uriah the Hittite (2 Sam. 11), who was employed in David's army.

Hivites Occupied part of *Canaan west of the *Jordan at the time of the entry of the Israelites (Exod. 3: 8). They may be known also as Horites; Zibeon is considered a Hivite in Gen. 36: 2 but a Horite in Gen. 36: 20.

Hobab Father-in-law of Moses (Judg. 4: 11, so NRSV and NJB). But Num. 10: 29, in the *J source speaks of Hobab the son of Reuel, Moses' father-in-law, and it is not clear whether this means Hobab or Reuel as the father-in-law. If the latter, then Hobab is Moses' brother-in-law, which is the REB translation at Judg. 4: 11. Moses' father-in-law is also sometimes called Jethro (Exod. 3: 1).

holiness What is separated from ordinary or profane use is 'holy'. Hence, above all, God is holy (Isa. 6: 3) and his holiness extends to human beings who have transactions with him (e.g. priests in the Temple) and equipment which they use, and the festivals they celebrate (Lev. 23). The Holiness Code (Lev. 17–26) is a compendium of ritual and moral precepts based on the holiness of God (Lev. 19: 2) and probably used by *priests and Levites for instruction.

In the NT the holiness which belonged to the Jerusalem Temple is regarded as a quality of Christian people (1 Cor. 3: 16–17), but pre-eminently Jesus is holy (Luke 1: 35), as he is called in the early preaching (Acts 3: 14), and as he addresses his Father (John 17: 11). Because the Church too is holy (Eph. 2: 19–22), indwelt by the Holy Spirit, any behaviour which violates this relationship is reprehensible (Rom. 5: 5; 2 Cor. 6: 16–7: 1).

Holy One of God Jesus is thus addressed by a demoniac (Mark 1: 24) in virtue of supernatural knowledge, and by Peter (John 6: 69). The term is applied to the prophet *Elisha (2 Kgs. 4: 9) and so was not in the OT a Messianic title, but seems to have been used of Jesus by the early Church (Acts 3: 14).

Holy One of Israel A title for the God of Israel often used in all parts of the book of Isaiah (e.g. Isa. 41: 14) affirming the absolute separation of God from humanity, but also his special relationship to the nation of Israel.

holy place Part of *Solomon's Temple, also called the 'nave' (1 Kgs. 7: 50, NRSV) and adjacent to the innermost chamber called the 'Holy of Holies' or 'inner sanctuary' (NRSV, REB). In the holy place were an *altar for *incense, ten golden lampstands, and a table for the bread of the Presence (1 Kgs. 7: 48). The arrangement was reproduced in the second Temple (520–515 BCE) and on a magnificent scale in that of *Herod (sometimes called the third Temple).

Holy Spirit The Hebrew *ruach* and Greek *pneuma* mean 'breath' or 'wind' and are translated by 'spirit', denoting an unseen life-giving force. United with 'holy', the force is said to be divine, though the combination of the two words occurs only three times in the OT: (Isa. 63: 10, 11; Ps. 51: 11).

The spirit in Gen. 1: 2 is the power of

God by which he creates the universe. It is the spirit which enlivens the *community with *hope for the future (Ezek. 11: 14–21) and which breathes life into the dry bones in the *vision of the valley (Ezek. 37: 1–10). It is prophesied that the future messianic era will be marked by the *gift of God's spirit on all people regardless of age or gender (Joel 2: 28).

There is little in the synoptic gospels about the Holy Spirit, except at the *birth of Jesus (Matt. 1: 18; Luke 1: 35) and his baptism (Mark 1: 8), and it is said that the Spirit operates in the *ministry of Jesus against the forces of *evil (Matt. 12: 28). The paucity of references may be due to the belief that the work of the Holy Spirit was only apparent after Jesus' *resurrection (John 7: 39).

After the resurrection the Spirit is said to be the agent of the missionary zeal of the Church (Acts 1: 8) as Jesus had promised (Luke 24: 49); fifty days after *Easter (Acts 2: 1) or on Easter evening (John 20: 22) the Spirit was given, the Spirit of Christ (Rom. 8: 4; 2 Cor. 3: 18). Paul does not identify Christ and the Spirit, even in 2 Cor. 3: 17, but there is an identity of function in the work of *redemption. The Spirit works through the Church, and the Acts of the Apostles is largely the story of the guidance by the Spirit of apostles and evangelists. In the gospel of John the Spirit is five times called the '*Paraclete', meaning a defending counsel. He is to promote the disciples' understanding of the truth concerning Christ.

Just as the Spirit fills the Church (Eph. 4: 4), it is also the power which guides individual believers and endows them with a diversity of gifts for the service of the whole Body (1 Cor. 12: 7), not to be confused with evil spirits (Rom. 8: 15) which foment dissension (1 Tim. 4: 1). 'Spirit' is therefore contrasted with 'flesh', characteristics respectively of the life of the new age and the former age.

In later Christian theology, the Holy Spirit is the third person of the *Trinity—not a doctrine explicitly discernible in the NT, but it is maintained that the NT writers were feeling their way to what the Fathers in due course, operating in a Greek philosophical climate, were to define. They used the terminology of their times to draw out the metaphysical implications of the NT data.

homosexuality The condition of sexual attraction to a person of the same sex. There does not exist a Hebrew or Greek word for this inclination, but fear of sinking to the mores of *Egypt and *Canaan may have led to the condemnation of homosexual genital activity in Lev. 18: 22. The word 'sodomy', derived from the story of Gen. 19: 1–29, is inappropriate for individual homosexual behaviour, for the story is one of gang rape. In the NT Paul cites homosexual relations as an instance of human perversion in the Gentile world. Paul shares contemporary Jewish revulsion at what was not merely usual but socially approved male behaviour in the Graeco-Roman world. The intellectual élite of Athens had long been accustomed to making love to attractive boys; it was part of normal educational life. The same culture prevailed in 1st-cent. Rome. Homosexual relations, however, were far more widely practised—slaves sometimes being forced to play the passive role, according to Seneca. So common and so generally known were homosexual practices in the Gentile world, that Jewish writers were shocked. *Philo condemns homosexual activity as almost as bad as bestiality (*Special Laws* 3: 37–42). It is plain that Paul in Rom. 1: 24–32 shares this Jewish disgust at what, no doubt, many of his Gentile converts had some experience of (1 Cor. 6: 11), and at the least he has in mind the exploitation, as he would have seen it, of minors (child abuse) by older men. His condemnation must also embrace the deviation of heterosexual men and women from their 'natural' relationship. Some scholars believe that this is the extent of his Christian judgement and that he is not necessarily referring to behaviour by consenting adults. Others disagree. At any rate, as with medical knowledge in general, much that is now understood about the psychology and biochemistry of this condition was unavailable in the 1st cent.

honesty In AV 2 Esd. 16: 49 where NRSV, REB prefer 'virtuous'. But 'honesty' has the modern sense in NRSV of Gen. 30: 33.

honey Wild honey was found in Palestinian rocks (Deut. 32: 13) in the carcass of a dead lion (Judg. 14: 8) and in a honeycomb on the ground (1 Sam. 14: 25). It was part of *John the Baptist's fare (Mark 1: 6). Because milk and honey were to be available to the Israelites in the *Promised Land (Exod. 3: 8) it became a custom in the primitive Church to give milk and honey in one of three cups offered to the newly baptized at *Easter.

hope The OT prophets taught that any hope for a satisfactory future depended on the nation's faithfulness to God. Otherwise there would only be chaos and disaster. *Amos warned that they should not imagine that God could intervene for them in the *day of the Lord (Amos 5: 18). In Jeremiah, however, there is expressed an ultimate hope in a new *covenant (Jer. 31: 31–4). In the NT the eschatological hope of Israel is said to be fulfilled in the *resurrection of Christ (Acts 2: 26) and the concept is much developed by Paul. In 2 Cor. 5: 5 Paul assures believers that their faith in what God has already done for them in Christ is a guarantee that their hope of the *salvation of fellow Christians and a re-creation of the whole universe is not misplaced (Rom. 8: 19–21; Rom. 11: 22; 1 Cor. 15: 2). In this present life the Christian has in *baptism embarked upon a road that will end in sharing the glory of God (Rom. 5: 1–3). *Abraham is cited as the great OT exemplar of authentic hope (Rom. 4). 1 Pet. also encourages the persecuted recipients of the letter to be strong in hope.

Hophni and Phinehas Sons of *Eli, priests at *Shiloh, who behaved dishonourably, and were killed by the *Philistines when the *Ark was captured (1 Sam. 4). The news of this disaster caused Eli to have a fatal heart attack.

Horeb The name used by the source *E and in Deut. (e.g. 5: 2) for Mount *Sinai. It is uncertain whether Sinai and Horeb referred to different areas in the same location, were completely different places, or were alternative names for the same place.

horn of salvation A horn is a sign of power (Deut. 33: 17)—literally of an ox (Num. 23: 22) or metaphorically of a people (1 Kgs. 22: 11). In Dan. 7: 8 the horns refer metaphorically to kings. So the 'horn of salvation' (Ps. 18: 2) indicates the saving power of the king. The four horns on the corners of *altars (Exod. 27: 2) afforded sanctuary to a fugitive who clung to them. The horn of salvation in Luke 1: 69 (NRSV marg., AV) denotes royal saving power, now belonging to the *Messiah.

hornets Insects larger than wasps; their descent on the *Canaanites assisted the conquest by the Israelites (Exod. 23: 28, NRSV marg.; *Joshua 24: 12).

horse/horsemen The horse is associated in the OT with warfare, and there are warnings about keeping horses on a large scale (1 Sam. 8: 11–17); this may have been written with Solomon's excesses (1 Kgs. 4: 26) in mind.

There is a description of the horse in battle in Job 39: 19–25. In Zech. 6: 1–8 the prophet describes his *vision of chariots drawn by horses patrolling the earth and reporting peace in the north. In Rev. 6: 1–8 the course of human history is symbolized by four horsemen bringing afflictions to humanity: the first horseman rides a white horse, which stands for war; the second horse is red and stands for civil strife, such as happened in Jerusalem hard on the heels of the Roman assault in Jerusalem in 70 CE; the third horse is black, indicating famine; and the fourth is pale green (NRSV, Rev. 6: 8) and the rider's name is Death, implying pestilence.

Hort, Fenton J. A. (1828–92) Cambridge NT scholar and theologian of wide interests, including science. He edited with B. F. *Westcott an edition of the Greek NT which became the basis for the English Revised Version of 1881.

Hosanna Hebrew for 'Save, we pray' (not 'Hurrah') and taken from Ps. 118: 25 by the crowd as they greeted Jesus when he rode into Jerusalem on a young *ass before his arrest. The whole psalm was recited during the feast of *Tabernacles and had Messianic overtones, which seems to be the meaning of Mark 11: 9–10 and John 12: 13. John's mention of palm trees recalls their use in the triumphal procession in 141 BCE after the recapture of the citadel of Jerusalem by *Simon Maccabaeus (1 Macc. 13: 51). The word was incorporated into early Christian liturgies, and it has been so used ever since.

Hosea A *prophet in the northern kingdom (Israel) from about 745 BCE who was active for about twenty years; he is the only native of the north among the pre-exilic prophets. His messages and judgement upon Israel were built round the sign-names given to his children (Hos. 1: 4, 6, 8), and Hosea's marriage, separations, and reconciliations to Gomer, who was a *harlot, are used to parallel the relationships of God and the nation. The ominous names of the three children are typical prophetic signs to corroborate his verbal message. The message is, however, also illustrated by the vicissitudes of his marriage: judgement is not the last word. As the prophet continued to love Gomer, in spite of infidelities, so God never ceases to love Israel. (Unfortunately we do not have Gomer's version of the relationship!)

Hosea, book of In the past, some commentators have ascribed chs. 1–3 and 4–14 to two different authors on account of the change of content and style. (The Hebrew text of the book is beset with problems, which the ingenuity of translators conceals.) This view may, however, be unnecessary, though it is true that the second part of the book changes gear into denunciations of Israel's religious, social, and political life. Israel's idolatry leads Hosea to nostalgic reflections on the faithfulness of the wilderness experience of the Exodus (Hos. 2: 14–15) and to hope for its repetition. The chaotic situation depicted seems to be that of the northern kingdom soon after the death of *Jeroboam II in 745 BCE. Hosea anticipates divine punishment through conquest by the *Assyrians (9: 3) but there is hope of an end to deportations (11: 11) and a return to the great days of *David (3: 5)—which is a rather unexpected hope for a northerner, suggesting that it may be a later addition to the book.

Hoshea The best-known person of this name in the OT is the last king of Israel, who was defeated by the *Assyrians in 724 BCE after appealing in vain for help from *Egypt (2 Kgs. 17: 4).

Hoskyns, Edwyn Clement (1884–1937) English baronet who studied theology in Berlin under *Harnack and was an influential NT teacher at Cambridge. He translated K. *Barth's commentary on the epistle to the Romans; though an Anglo-Catholic in sacramental doctrine, he was influenced by the Calvinist tone of Barth's dogmatic theology. He was opposed to fashionable liberal theology and once observed that NT *eschatology fortunately protected the doctrine of God from evolutionary theories of history.

hospitality Welcoming a stranger or traveller was a fundamental courtesy in the ancient Near East (Gen. 18: 1–8). Grave consequences followed a violation of this custom (Judg. 8: 4–9). In the NT (Luke 7: 36–50) Simon, a *Pharisee, is rebuked for his lack of appropriate care for his guests. Jesus expected his *disciples to be offered hospitality when he sent them out on a mission (Mark 6: 10). Jesus' own regard for custom is by not upstaging his host: he heals only when asked (Mark 1: 30) and knows the rules of precedence (Luke 14: 8). He was sometimes offered private homes for teaching (Mark 1: 29–34).

The early Church expected its members to receive hospitality if they visited another town (Rom. 12: 13; 1 Pet. 4: 9; Acts 17: 7). Letters of commendation on behalf of travellers were exchanged amongst Church leaders (Rom. 16: 1–2; 3 John 5).

hosts of heaven Used in both bad and good senses in OT: as sun, moon, and stars they were worshipped by Israelites guilty of *apostasy (Deut. 4: 19; 2 Kgs. 23: 4). But stars are also referred to as the host of the Lord (Ps. 33: 6; Isa. 34: 4), who might also be *angels (Isa. 14: 12; Rev. 8: 10–11; 9: 1).

hosts, Lord of The God of Israel as commander of armed forces, both of Israel and in *heaven. God is the general of the armies of Israel and *David is his lieutenant (1 Sam. 17: 45); but God also has his *angels (1 Kgs. 22: 19) and the forces of nature (Amos 4: 13).

hour The Jews divided the day into twelve hours, with the sixth hour being noon. There were four watches of three hours each, after the military pattern. In the gospel of John, Jesus' 'hour' refers to his *passion, *death, and *resurrection. These are considered to be his 'hour' of exaltation and glorification and the whole gospel seems to be looking forward to this 'hour' from the wedding day at *Cana (John 2: 4), through the attempted arrest in the Temple (John 7: 30), and the last visit to Jerusalem (John 12: 27) to the discourse in the Upper Room (John 16: 32).

house For the greater part of Hebrew history, from the settlement in *Canaan (13th and 12th cents. BCE) until the 6th cent., domestic houses were all of much the same size and shape: two rooms up and two rooms down, with roofs of beams filled in with dried mud and brushwood. Animals might occupy the lower portion (1 Sam. 28: 24). A new house was to have a parapet as a safety precaution (Deut. 22: 8). In the 8th cent., as witnessed by the *prophets, the rich enlarged their houses and tended to push the poor into segregated areas. By the NT era houses of the wealthy in Palestine resembled those of Hellenistic cities, with their own water supply. The palaces of Herod were elegant and comfortable, equipped with swimming pools and ventilation. The houses of the poor had roofs easily torn up (Mark 2: 4), evidently still constructed of matting and twigs. But in the sophisticated environment of Luke, the evangelist changes the construction of Mark's simple roof to tiling (Luke 5: 19) to make it comprehensible to his readers.

'House' is also used in the *Bible to refer to a dynasty or tribe (Num. 1: 2) and 2 Sam. 7: 4–11 plays on the double meaning of house as building and as dynasty. A 'house of God' (Mark 2: 26) or 'house of prayer' (Mark 11: 17) was a place for *worship, but in Heb. 3: 3–6 the Christian congregation is a 'house of God'.

household codes Translation of the German *Haustafeln*, used by commentators for a literary type developed for ethical instruction in the Hellenistic world, adopted by Jewish Hellenistic synagogues, and thence by the NT (Col. 3: 18–4: 1, but also Eph. 5: 22–6: 9; 1 Tim. 2: 9–15; Titus 2: 2–10; 1 Pet. 2: 13–3: 7). The codes were an attempt by leaders of the Christian community to establish a pattern of family and social life not unlike that of traditional families among Gentile and Jewish contemporaries in the Graeco-Roman world. The patriarchal style represents a reaction against the egalitarian organization of the earliest Church in Jerusalem (Acts 2: 44–7) and is remarkable for the absence of Paul's radical teaching against sexual discrimination (Gal. 3: 28).

Huldah A prophetess (2 Kgs. 22: 14) who advised King *Josiah that the book of the law discovered in the Temple was genuine and that its provisions ought to be obeyed.

humility Virtue shown in OT by outward demonstration of weeping, *fasting, and rending of garments (1 Kgs. 21: 29; 2 Kgs. 22: 11–20; Prov. 3: 34). In NT humility is to be as powerless as *children (Matt. 18: 4), to decline to defend one's status (Phil. 2: 8–9), or to challenge the dignities of others (Luke 14: 11; 18: 14). This humility is acceptable to God (Eph. 4: 2). But among the Greeks humility was a vice rather than a virtue. Glaucus' father taught him to be 'always the best, distinguished above others' (*Iliad*, 6. 208).

Hymenaeus Denounced in the *Pastoral Epistles (1 Tim. 1: 20; 2 Tim. 2: 17) for the *Gnostic heresy of denying that there will be a future *resurrection.

hymn Broadly, a spiritual theme which could be sung, as were the psalms in the Temple. Jesus and the Eleven sang a hymn, the **hallel* (Mark 14: 26), before he led them to the Mount of Olives, and the early Church used a variety of hymns and songs (1 Cor. 14: 26; Eph. 5: 19). Hymns were composed for use in the Mass in the medieval Church, and at the Reformation became part of popular Protestant piety.

hypocrites In Greek tragedy, actors (who wore masks) were known as 'hypocrites' and so came to mean people who were not what they pretended to be. In Gal. 2: 13 the Greek 'hypocrisy' is translated 'dissimulation' (AV), 'lack of principle' (REB), 'insincerity' (NJB) but 'hypocrisy' by NRSV. This range of meaning suggests that in the NT the word denotes a general self-righteousness, as is clear from its repeated use in Matt. 23 in the denunciations of the *Pharisees.

Hyrcanus A prominent and wealthy citizen of Jerusalem whose money was deposited in the Temple (2 Macc. 3: 11) and only narrowly escaped confiscation at the hands of the Seleucid commander Heliodorus.

hyssop Plants related to marjoram which were tied together in a bunch for sprinkling water in ritual purifications. The reference to hyssop at the *crucifixion (John 19: 29) is puzzling, since the stem of this plant is not tough enough to bear the weight of a sponge, and it has been suggested that there was an early mistake in the Greek MSS and that the evangelist wrote 'on a javelin' (NEB). Yet a reference to 'hyssop' would carry *Passover sacrificial overtones (Exod. 12: 22) which John may have noted for its theological rather than historical truth.

I

Ichabod The name chosen by *Eli's daughter-in-law for her son as she died in childbirth on hearing of the capture of the *Ark of God (1 Sam. 4: 21). *Ichabod* is Hebrew for 'the glory has departed'.

Iconium A Phrygian city, usually under foreign domination, in Asia Minor; it was taken by the Gauls in 278 BCE, and the central part of Asia Minor was then known as Galatia. In 25 BCE the whole of *Galatia became a Roman province. In Acts 13 Iconium and other neighbouring cities are cited as belonging to Galatia, and Paul and *Barnabas are said to have enjoyed some success before being expelled on account of the friction they caused amongst the mixed population of Jews and *Gentiles. They returned to Iconium twice (Acts 14: 21–3; 16: 1–6; *c*.49–51 CE), and Paul's letter to the Galatians may have included the Church of Iconium as one of the recipients. In the 2nd cent. the apocryphal Acts of Paul and Thecla was set at Iconium.

Iddo Eight little-known persons in the OT are called by this name in the English Bible, of whom one mentioned several times is the father (Ezra 5: 1; 6: 14) or grandfather (Zech. 1: 1, 7) of the prophet *Zechariah.

idolatry The cult surrounding a statue of a god or goddess. Such idols were common in Near Eastern religions, but it is not certain whether the worshippers held the statue itself to be a deity or whether the deity was somehow embodied by the image in such a way that the worshipper met the deity through the image. It would seem that at the time of the *Exile, when the worshippers of Yahweh encountered the alien cults of *Babylon, the prophet believed that their neighbours did indeed worship a piece of wood or stone (Isa. 46).

The prohibition against idols is emphatic in the book Exodus and the worship of the God who liberated them from *Egypt is part of the *covenant. It was, however, often tempting to fuse the worship of Yahweh with the fertility cults of the Canaanites: and the dramatic contest arranged by *Elijah on Mount *Carmel (1 Kgs. 18) between Yahweh and *Baal shows the gravity of the challenge (*c*.860 BCE) The efforts earlier (*c*.930 BCE) by *Jeroboam I to dissuade his subjects from travelling south to Jerusalem by installing a sacred calf at *Bethel and another at *Dan (1 Kgs. 12: 29) is depicted as gross *apostasy, though in fact the calves, or bulls, may have been regarded as a throne or seat for the invisible god—whom some of the worshippers may have regarded as Yahweh himself (Exod. 32: 5).

After the Exile the condemnation of idolatry is less shrill (Zech. 10: 2); *monotheism is secure.

Paul warns the Corinthian Christians about a kind of idolatry (1 Cor. 10: 14) which might have been some form of civic ceremony. In the case of meat previously offered in idolatrous sacrifice, the advice is that it is harmless, since the so-called gods are nonentities (1 Cor. 8: 4); but it could be an act of charity towards less instructed Christians to decline such meat (1 Cor. 8: 13). Idolatry is also used metaphorically for evil desires (Col. 3: 5).

Worship of the Roman emperor and obligatory participation in heathen rites was later to prove a fiery test for Christian believers.

Idumaea The Greek name for the land of *Edom. After 586 BCE Edomites began to move into *Judaea from Beth-zur to Beersheba, but the area changed hands several times in the last two centuries BCE. *Herod the Great was an Idumaean and this partly accounts for his unpopularity among the Jews. Some Idumaeans are noted as being in the multitude who followed Jesus (Mark 3: 8).

Ignatius (35–107 CE) Bishop of *Antioch in *Syria; on his journey as a prisoner to *Rome, where he was martyred, Ignatius wrote six letters to Churches in Asia Minor through which he passed and one to Polycarp, bishop of Smyrna. The letters emphasize that each local Church should have a single bishop and, dated about 107 CE, they are important evidence of the development of the Church's *ministry.

ignorance Although ignorance of the *law may not be an acceptable defence in modern society, it was pardonable in Israel (Num. 15: 24) and in the early Church (1 Tim. 1: 13), but the invincible ignorance of *Gentiles (Eph. 4: 18) has led them to depravity.

Illyricum A Roman province which is in the NW of the Balkan peninsula. Paul claims to have preached as far as Illyricum (Rom. 15: 19) at the time of writing. Jerome, who translated the *Bible into *Latin (known as the *Vulgate) was born in Illyricum about 342 CE.

image An object of religious devotion. The English word does not necessarily imply condemnation, as does 'idol', or 'abomination' (1 Kgs. 11: 5), which is commonly applied in English translations to the cultic objects of the nations. In the 8th cent. CE there was a bitter controversy about the propriety of icons, or pictorial representations of Christ and the saints in churches, eventually settled in favour of pictures and images until the Reformation, when Protestants destroyed many statues and shrines.

image of God The term in Gen. 1: 26–27 and 9: 6 for the likeness of human beings to their Creator. It is both male and female persons who possess this image of God and are therefore together 'to be fruitful and multiply' (Gen. 1: 28), reflecting the creative activity of God. In the NT it is Christ who is the perfect image of God (2 Cor. 4: 4; Phil. 2: 6). In Catholic theology it is held that the image was obscured, but not lost, in the *Fall; it is regarded, for example, as an intellectual quality of the *soul. But in much Reformed theology it is argued that the image was destroyed utterly, and moral responsibility with it. Truth about God and his will can therefore only be known by *revelation.

imagination Used by AV for 'thinking' (REB, NRSV; 'arguments', NJB) at Rom. 1: 21. Also at 2 Cor. 10: 4–5, where NRSV has 'arguments', REB 'sophistries', NJB 'presumptive notion'. These translations convey the force of the Greek noun, which is of an evil intention.

Immanuel 'God is with us', the name of Isaiah's child (Isa. 7: 14). *See also* Emmanuel.

immortality Unending life, continuing after *death.

1. *The OT*. There are several beliefs about it in the OT, and the earliest forms of future hope were for the nation's survival rather than for the individual. But such was the belief in the power of their God that Hebrews came to hold that death would not involve extinction even for individuals. Both Elijah and Elisha were reported to have reanimated the dead (1 Kgs. 17: 17–21; 2 Kgs. 4: 18–37), which was proof of God's power beyond death. The expectation, however, was of a miserable, shadowy existence, possibly in a subterranean cavern ('*sheol*'). To be delivered from *sheol* was to be saved from death (Ps. 28: 1; 30: 3; 88: 4). This survival in *sheol* was not a disembodied existence, for the Hebrews did not make a sharp division between body and *soul. But *sheol* was an unsatisfactory concept, and when at last the idea of a future existence in joy and fulfilment took root in *Judaism, it assumed the form of the *resurrection of the body, rather than the escape of an inherently immortal soul from its bodily prison.

Hellenistic Judaism did, however, come to adopt a Greek idea of the immortality of the soul (e.g. Wisd. 3): though some indeed rejected the belief, as does the author of Ecclesiastes (3: 19–21).

Maybe Hebrews were aware of the Canaanite cult of the death and resurrection of a deity (possibly underlying a liturgical theme in Hos. 6: 1 ff.) but *Canaanites did not expect resurrection for themselves. In Israel the belief developed with the growing sense of individual responsibility, and with it came the hope for individuals after death apart from the merely national hope. In the OT the idea of resurrection is first clearly expressed in Isa. 26: 19 and Dan. 12: 2, and it is generally accepted by the time of 2 Macc. 7: 9–11 and in apocalyptic writings including Enoch, 4 Ezra, and 2 Baruch; *Qumran texts also expect a resurrection of the just. It was denied by the *Sadducees (who disappeared after 70 CE) but the belief in resurrection held by the *Pharisees prevailed.

2. *The NT*. Jesus concurred with the Pharisees (Mark 12: 18 ff.; Luke 14: 14), though it will be a different form of existence from that of this present life. The resurrection of believers (but not of this present flesh and blood, from which the Christian awaits, with 'groaning', that is, as though in the pangs of childbirth, to be delivered, Rom. 8: 23) is part of Paul's preaching of Christ as the first fruits of the resurrection of those who are obedient to his will (Rom. 6: 5–6). In John there is a doctrine of resurrection of both the righteous and the wicked to judgement (John 5: 28 f.), and Heb. 6: 2 mentions the doctrine of resurrection of the dead as among the rudiments of Christian faith. As to the fate of the dead pending the resurrection, there apparently existed in early Christian belief an intermediate abode divided at death for the righteous and the evil. This must imply a belief in an immediate judgement at death (Luke 16: 24).

Enoch 22 has the spirit of the sons of the dead distributed into a compartment (which implies a judgement) with an independent existence while waiting to resume a bodily form at the resurrection.

The different beliefs about existence after death found in the Bible are still discussed by modern philosophers of religion. The two main views are: survival as disembodied selves; *or* new life in bodily form. The difficulty about the former is that a human person is a complex entity owing so much to being bodily that it is hard to claim that without a body it could still be a person. The difficulty with the latter view is that it requires the entity after death to be somehow continuous physically with what was once present before death; it implies that a corpse buried or cremated could be empowered by God to live again—if God is a just moral agent who wills to ensure that people survive their deaths.

imprisonment This was not a regular punishment in Israel, but Pharaoh's butler and baker are said to have been imprisoned in *Egypt together with *Joseph (Gen. 40: 3). *Jeremiah was once placed in the stocks (Jer. 20: 2) and later imprisoned (Jer. 37: 15). King *Artaxerxes authorized *Ezra to make use of this punishment (Ezra 7: 26). *John the Baptist was imprisoned (Mark 6: 17), for *Herod the Great had built great prisons to incarcerate any who threatened him. Peter was imprisoned by Herod *Agrippa I (Acts 12: 4). It was a common punishment in the Roman Empire (Acts 16: 24); the epistles to the Philippians, Colossians, Philemon, and Ephesians are all written from prison.

incarnation A term, not found in the Bible, meaning 'becoming flesh', or 'human'. It is claimed that in the NT there are imprecise and sporadic expressions of this doctrine concerning Jesus but there was no precise definition until there had been five centuries of strenuous debate. Councils of the Church eventually agreed that the orthodox faith in Jesus was that united in his one person were the two natures, full divinity and genuine humanity.

There are few indications in the synoptic gospels that Jesus was regarded, or regarded himself, as other than a man with unique *gifts of *teaching with authority and *healing; but soon after the *resurrection there are accounts of apostolic preaching in which

Jesus is proclaimed as Lord and *Messiah (Acts 2: 32–6) and Son (Rom. 1: 3–4). The gospels even mention him as 'beloved Son' of God at his *baptism (Mark 1: 11) and *transfiguration (Mark 9: 7). The two narratives of the virginal conception, in which Joseph is relegated to a powerless background role, push the status of Jesus as Son of God back beyond his birth. There is thus a hint of pre-existence which is confirmed by Paul in Gal. 4: 4. Moreover, Paul identifies Christ as 'the *wisdom of God' (1 Cor. 1: 24), and this echoes the personification of Wisdom (Wisd. 7: 25; 8: 3) in the intertestamental period: for wisdom is said to have pre-existed with God before particular manifestations. In the *hymn extolling Christ's voluntary humiliation in becoming man, Paul appears to accept that he was in the divine arena previously (Phil. 2: 6–11); similarly Col. 1: 15. An incipient doctrine of incarnation in Heb. 1: 1–3 becomes clearer in the gospel of John (1: 14) where Jesus' prior existence with God (John 6: 62) and association with Abraham (John 8: 58) imply a strong notion of pre-existence. But this is not allowed to diminish the clarity of Jesus' humanity; the Word became flesh (John 1: 14); Jesus had human frailties (John 4: 6), he wept (John 11: 35), he died (John 19: 30), and yet without contradicting that humanity he was a revelation of God (John 10: 38; 20: 28). The epistles of John combat the belief that Jesus' humanity was a mere appearance (the belief called *docetism; 1 John 4: 2–3).

The traditional doctrine of incarnation is regarded as unsatisfactory by some modern theologians. For example, an alternative way of interpreting the action of God in Christ has been in terms of the *Spirit. The Spirit works through the whole evolutionary process of creation and evolution, and in this continuous divine activity the decisive and focal point is the person of Christ. In him divine self-sacrificing love achieves its full expression in a human personality.

incense The term is used of those woods and resins which give off a pleasant smell when burnt. *Frankincense, or olibanum, is made from the resin of trees in southern Arabia. It was much used not only in pre-Christian pagan religions in sacrificial rites, but also in the tent of meeting (Exod. 40: 27) and in the Temple, where it symbolized the offering of *prayer (Ps. 141: 2, Luke 1: 10). The Christian Church was reluctant to burn incense in the first three centuries on account of its use in the cult of the Roman emperor; a token offering of it was demanded of Christians as evidence of their patriotism—which for the Church could only be apostasy. Later generations were less inhibited and noticed that incense had been offered to the infant Jesus (Matt. 2: 11). In some Christian Churches persons and objects regarded as representing Christ were and still are censed in the *Eucharist.

inculturation The attempt to make a religious (e.g. Christian) message accessible in and through a local culture. It has been an aspect of evangelism by means of drama, songs, and dance in the S. American base communities of the Roman Catholic Church.

India Although only mentioned in the Bible in the book of Esther (1: 1; 8: 9), there was regular trade between India and the countries of the Mediterranean. The Malabar Christian communities of SW India claim, improbably, to have been founded by the apostle Thomas.

infancy narratives The first two chapters of Matt. and Luke are in agreement that Jesus was conceived in *Mary without the intervention of *Joseph and that he was of the house of *David. Otherwise there are considerable differences: Matt. describes the visit of the *Magi, the journey into *Egypt and temporary residence there, the slaughter by *Herod the Great of the infants—all told from the point of view of Joseph. Luke has accounts of the birth of *John the Baptist, an angelic message to Mary, the visit of the *shepherds, Jesus' *circumcision, the presentation in the Temple, and, finally, the record of Jesus' conversation in the Temple at the age of 12.

There are objections to treating these

narratives as literal history. It is difficult to fit into a satisfactory time-scheme both the flight into Egypt (Matt.) and the return to the house at *Nazareth (Luke): there are improbable features in the story of the travelling star and the journey to Egypt (Matt.) and uncertainty about the universal census under *Augustus (Luke). As the son of an artisan *carpenter it is unlikely that Jesus could have debated in the Temple at the age of 12. No other NT writer reveals any knowledge of the virginal conception. In the face of these difficulties about historicity (it is noted that there is no secular writing to corroborate the crime of Herod's massacre of the infants) many scholars prefer to see the narratives as Matt.'s improvisation on the basis of OT texts, in accordance with accepted rabbinic principles of scriptural interpretation. It is significant that in Matt. the will of God is revealed in dreams, as it was to *Joseph the OT patriarch; and just as the patriarch went to Egypt, so did the NT Joseph. Some scholars would recognize that as history these narratives are fragile, but would also hold that there is a core of fact on which the two evangelists have proceeded to proclaim their Christian *faith about Jesus as *Son of David and *Son of God. This they did by showing that Jesus was the fulfilment of OT prophecy; the *life, *death, and *resurrection were foreordained, not fortuitous. The narratives also answer the objections that Jesus was a Galilean and therefore an unlikely *Messiah and that the circumstances of his birth were irregular.

Later Christian writings had exuberant fancies as they elaborate what they found in Matt. and Luke, much of which has become the stuff of popular piety and children's plays: the birth in a stable, Joseph the elderly widower with children from his former marriage, the three 'kings' and their names, and (which became widely accepted by Catholic theologians) the belief that the birth miraculously took place not only through the conception by the *Holy Spirit but also nine months later without interference with or damage to Mary's physical organs.

Modern theologians who maintain the truth of the virginal conception as literal fact believe it to be congruous with God's action in giving mankind a new beginning; Jesus' miraculous birth constitutes a decisive break with the old order. Other theologians suggest that without the masculine contribution there would be a defect in Jesus' humanity. They are therefore inclined to put the narratives into the *genre of **midrash*, sparked off perhaps by the *LXX translation of Isa. 7: 14 where the Hebrew *ha'almah* ('young woman') is translated, *he parthenos* ('the virgin'). Isaiah was concerned with the troubles of the 8th cent. BCE, not at that moment with a future Messianic age. The prophet declares that within nine months, and the birth of *Emmanuel, the king's enemies would no longer be a threat, and soon there would be peace (Isa. 7: 15–16). *See* Virgin birth.

ink The single reference in the OT (Jer. 36: 18) probably refers to a mixture of soot and gum arabic which was used for writing on papyrus. *Ezekiel (9: 2–3) mentioned an inkhorn which would have contained the ingredients for making ink and could be fastened on to the scribe's belt. Paul (2 Cor. 3: 3) depicts the Corinthian converts as a commendatory letter from Christ, written by Paul, not however with ink but on human hearts.

inner being (or **inner nature**) Paul's term for a person's true self as seen by God, contrasted with the merely outward physical and decaying body (2 Cor. 4: 16). This aspect of one's self desires the right and the good even when one's lower nature indulges in *sin (Rom. 7: 22 f.).

inns Hotels of the modern kind were not known in the NT era, but self-service inns were often available for travellers, possibly with a shop and its proprietor (Luke 10: 35). Private houses might also sometimes offer *hospitality in return for payment.

INRI The initial letters of the *Latin inscription, in John 19: 19, Jesus of Nazareth, King of the Jews.

inscription A writing to provide information at a glance. (1) Jesus was shown a silver coin (Matt. 22: 20). When he asked whose inscription was on it, the reply was 'the emperor's'. The wording will have read: Tiberius Caesar, son of the deified Augustus, Augustus. (2) Another inscription (Matt. 27: 37) was the charge against Jesus at his *crucifixion. It consisted of a placard either hung round his neck or fastened to the *cross, alleging Messianic pretensions.

See also epigraphy.

inspiration The belief in the OT that prophets were so filled with the 'spirit' that their utterances conveyed a divine message (Num. 24: 2), which was sometimes validated by eccentric behaviour (1 Sam. 10: 6). In the NT the *Spirit was also regarded as the cause of prophesying, speaking in *tongues, *healings, and other *gifts (1 Cor. 12: 4–11).

Later the idea of the inspiration of certain writings is mentioned (2 Tim. 3: 16) and this is held by some to be biblical authority for the idea of an inspired text. It is also maintained that 2 Tim. 3: 16 means that written scripture is as much the work of God as author as the oracles of *prophets; both on this theory come straight from God (1 Cor. 6: 16). From these convictions it is a short step to the idea of biblical inerrancy: the Bible is free from error not only in religious matters but also in history, science, and ethics. It is an outlook broadly known as *fundamentalism.

There are, however, serious difficulties about such a view of inspiration: there are the patent inconsistencies and contradictions within scripture itself; there are historical problems; there is an accumulation of knowledge about how books in the ancient world were composed and the clear resemblance of the NT to such books; it is recognized that early Christian Fathers, such as *Irenaeus (*c.*180 CE), did not regard either the OT or the NT (which he often cites) as supernaturally inspired or guaranteed by God; Irenaeus values the gospels simply because they record what Jesus said and did. Following Irenaeus it could be said that the NT provides no more than data on which the Christian faith can be based (or denied).

intercession *Prayer on behalf of another person or group of people; it has a long history in the Bible, from *Abraham pleading for *Sodom (Gen. 18: 22–33), Moses on behalf of the enslaved Israelites (Exod. 5: 22–3), the *prophets (Amos 7: 1–6), and also for the people. In the NT Jesus interceded for Peter (Luke 22: 32) and Paul for the Ephesians (Eph. 1: 17).

The epistle of James urges intercession for the sick (Jas. 5: 13–17). Such intercession has continued down the ages. Its rationale has been questioned (is it an intention to try to change the will of God?) but justified as 'making possible for God to do something which he could not have done without our asking'.

interpretation Unintelligible utterances required the *gift of interpretation (1 Cor. 12: 10); and translation from one language to another was a kind of interpretation (Ezra 4: 7; Acts 9: 36; Heb. 7: 2). The interpretation of the OT by Jesus followed rules prescribed by distinguished rabbis such as *Hillel. Christian scholars in the 2nd and 3rd cents. CE in *Alexandria used the method of *allegory as a means of acceptable biblical interpretation. In Antioch preference was given to the literal meaning of scripture but the method of *typology was employed by John Chrysostom and others. During the Middle Ages the interpretation of the Bible was increasingly brought into the service of doctrine and at the *Reformation Luther and Calvin used the Bible to develop their characteristic theological views, but without allegorical *exegesis. Modern interpreters of the Bible use all the resources of *criticism; the guiding principle is a determination first to elucidate what the text precisely says.

inward parts Used by the AV for 'entrails' (NRSV, Exod. 29: 13) and the seat of truth (Ps. 51: 6) or of wickedness (Ps. 5: 9, which NRSV renders 'hearts'). Jesus used the word (Luke 11: 39) of the invisible evil disposition of the *Pharisees.

iota The small Greek letter *i* used in the Greek NT of Matt. 5: 18 for *yod*, the smallest Hebrew letter. AV translates 'one jot'. It is a saying of Jesus recorded by Matthew to indicate that Jesus was not bent on destroying all that was most precious in the heritage of Israel, though in fact there were few Christians around in Matthew's own day who held that the *Law of Moses was of obligation right down to the last iota.

ipsissima verba Jesu Latin: 'the very words of Jesus himself'; an expression used by commentators making a claim for the authenticity of such words, as opposed to words attributed to him by the *evangelists.

Irenaeus Bishop of Lyons in Gaul from about 180 CE; a champion of orthodoxy against the *Gnostics, emphasizing the continuity of the Christian revelation with the Old Testament. Irenaeus was the first Church leader to set out a *canon of the NT scripture (the four gospels, Acts, and the epistles of Paul) as authoritative alongside the OT, which could be interpreted allegorically. These principles were a foundation for genuine biblical scholarship.

iron The Iron Age in Palestine was from about 1200 to 330 BCE. The *Philistines were makers of iron implements (1 Sam. 13: 19) which David encountered (1 Sam. 17: 7). Because of its hardness, iron is often used as a metaphor: the life of slavery in *Egypt was like being in an iron furnace (Deut. 4: 20, cf. Jer. 28: 13) and the stones of Palestine are like iron (Deut. 8: 9); it is a symbol of strength (Isa. 48: 4); and dissidents can be threatened with a yoke of iron (Deut. 28: 48). Deutero-Isaiah describes the craft of an ironsmith (Isa. 44: 12). The iron gate of Acts 12: 10 is one of the few NT references to iron.

There was probably an iron mine at *Gilead, and the smelting fuel was charcoal (cf. Ezek. 22: 20).

irony A statement in the narrative which bears a meaning in the context which the readers understand differently. An example is in John 18: 14 where *Caiaphas approves the condemnation of Jesus on grounds of political expediency but by the Christian reader the words are taken as an ironical assertion of a greater truth about the *crucifixion.

Other examples in John are at 6: 42; 7: 27; 7: 41 f., 52; and 9: 29. The Jews did not know, as the readers of the gospel do know, that Jesus did come from Bethlehem and had come from God.

Isaac Son of the aged *Sarah and *Abraham; born in accordance with the promise to Abraham of an eternal *covenant (Gen. 17: 15–21). Abraham's *faith was put to the test when he was ordered to sacrifice this instrument of the promise. In surviving the test by his obedience, Abraham recognized that God's promise was a sheer *gift, not something to be possessed by legal entitlement. Isaac was spared at the last moment of the intended *sacrifice by divine intervention, and a ram substituted for the lad. Some expositors have interpreted the narrative as a tale to define Israel's repudiation of pagan child sacrifice or as a stage in the evolution of religion. Paul treats the remarkable birth of Isaac in fulfilment of God's promise as paralleled by the choosing of the Galatians as children of the promise; it is given to those who have faith, without regard to their physical descent (Gal. 4: 28). The epistle to the Hebrews (11: 19) hints at the release of Isaac for the sacrifice as a 'type' of the *resurrection of Christ, but it is not until the so-called epistle of Barnabas 7 at the end of the 1st cent. (and not in the NT) that the sacrifice of Isaac is used to prefigure Jesus' sacrifice on the *cross. It then became a theme for many great artists, as in a mosaic of the 6th cent. in Ravenna. In Judaism the sacrifice of Isaac, called the **aqedah* ('binding'), almost paralleled the significance of the crucifixion of Jesus in Christianity.

Isaiah Son of Amoz, Isaiah was essentially a citizen of Jerusalem; he was married and had two, or possibly three, sons who were given symbolic names (Isa. 7: 3; 8: 3). He advised four kings of *Judah between 740 and 701 BCE at a time of the threat from the invading *Assyrians, urging them to avoid foreign entanglements.

Isaiah, book of Although there is no MS—even the oldest, among the *Dead Sea scrolls—which in any way suggests that the book can be divided among several authors of different dates, it is modern critical opinion that the book cannot be a unity. The reasons for this view are: that the historical situation envisaged in Isa. 1–39 is for the most part that of the second half of the 8th cent. BCE whereas in Isa. 40–66 (and also Isa. 13–14) the references to Cyrus and the end of the Exile in *Babylon clearly derive from the sixth–fifth centuries. This analysis is confirmed by the difference of themes: the 'First Isaiah', the prophet who lived in court circles in Jerusalem during the 8th cent., interprets the *Assyrian threat to *Judah in the light of his belief about God's providence. The 'Second Isaiah', or 'Deutero-Isaiah', the unknown prophet of the exilic period, who is never named, saw the collapse of Babylon and the triumph of the Persian Cyrus in 539 BCE (Isa. 44: 28; 45: 1). So a prominent theme is that of the restoration of the Temple. It is also shown by computer analysis that the Hebrew of 40–66 has many subtle variations from 1–39—even granted certain features common to both parts, such as the expression 'the Holy One of Israel' (Isa. 1: 4; 10: 20 and 41: 14). There is indeed a theological unity about the whole book which makes it understandable that it was compiled as it has been, and in recent years scholars have devoted much attention to the literary and religious features which make Isaiah one book, even though it originates from many different authors.

But conservative Christian scholars who maintain the unity of authorship of the book do so primarily on dogmatic grounds, that this is pre-supposed by the NT in Matt. 3: 3; 12: 17 ff. and Rom. 10: 20 f. They suggest that Isa. 40–66 contains a core of the historical Isaiah's work, but do concede that later disciples added to it and edited it with contemporary allusions.

Isa. 1–39 contains the prophet's denunciations of the sins of Judah and Jerusalem (1 to 12) and invectives against the leaders (22: 15–25). There are oracles against foreign nations (13–23), and prophecies of *hope, denunciation, and *salvation (24–35). Finally, 36–9 is a prose appendix, possibly not written by Isaiah, giving an account of his relations with King *Hezekiah, and parallel to 2 Kgs. 18: 13–20: 19.

First Isaiah is concerned with two major political crises: the Syro-Ephraimite conspiracy (734–733 BCE) when Judah's two northern neighbours in unison tried to force Judah into an alliance against Assyria. The second crisis was that of the Assyrian threat to Jerusalem of 701 BCE. In chs. 1–9 Isaiah tried to persuade King *Ahaz that the two conspirators were doomed to fail. God's promise to *David (2 Sam. 7) was unbreakable (Isa. 9: 1–6; 11: 1–10). However, Ahaz decided to seek Assyrian help (2 Kgs. 16: 7 f.) against the two allies, and Isaiah prophesied that this act of faithlessness, and the social evils of the country, would deserve God's *punishment by *war.

Syria and Ephraim were conquered, and the Assyrians made Judah a vassal state. Assyrian cult objects were introduced by Ahaz into the Temple. Cf. 2 Kgs. 16: 11. However, Hezekiah, who succeeded Ahaz in 715 BCE, threw off the yoke of Assyria, with some encouragement from *Egypt (Isa. 18: 1–2 and 30: 1–7), and at the same time purged the country of foreign *idolatry in a reform which anticipated that of *Josiah (629 BCE). Inevitably *Sennacherib, king of Assyria, marched against Judah, but Hezekiah paid tribute and escaped his wrath (701 BCE).

The situation for Second Isaiah was the Exile of the people of Judah in Babylon (2 Kgs. 24: 12–16). It was a captivity in which the people enjoyed liberty of worship; synagogues may have been established then, and observance of the *sabbath was emphasized. Business prospered. Nevertheless many remembered their homeland with affection (Ps. 137). So the prophet celebrates Cyrus emperor of Persia as God's agent for reversing the exiles' fortunes. In 538 BCE Cyrus issued a proclamation allowing the return to Jerusalem, but the prophecies of 2 Isa. had probably been delivered about 547 BCE when Cyrus was beginning his career of conquest. 2 Isa.'s message is that Yahweh is the only God and he will bring salvation to his people. In chs. 56–66 the style and tone is more critical of the nation, and some scholars regard this section as the work of a third prophet, designated 'Trito-Isaiah'.

An important section of *Deutero-Isaiah is that of the four Servant Songs: 42: 1–4; 49: 1–6; 50: 4–11, and 52: 13–53: 12. Much scholarly discussion has been devoted to the identity of the servant. Was he an individual—e.g. Jeremiah? Or a group? Or the nation as a whole? Or the ideal nation? Because these sections became important for Christian propaganda as seemingly referring to Jesus (Matt. 12: 16–21; 1 Pet. 1: 24–5) they have sometimes been isolated from the rest of the book, although the 'servant' is mentioned in 41: 8; 45: 4, outside the Songs. At any rate their meaning is that it is through failure and humiliation that God's promise is revealed and fulfilled. In addition to the Servant Songs, the book contains the Emmanuel prophecy (7) and the house of David motif (9 and 22), which made Isa., together with the Pss., the most quoted OT books in the NT.

Iscariot Always applied in the gospels to *Judas who betrayed Jesus to distinguish him from another Judas among the *apostles. Iscariot may mean 'man from Kerioth'; or there is a suggestion in one ancient MS which connects it with the Latin *sicarius*, which could imply that Judas was one of the dagger-men, that is the *Zealots, as they were later known, nationalists who favoured armed rebellion against the Romans. *See* Judas (4).

Ishbosheth Son of *Saul (2 Sam. 3: 14); originally called Eshbaal (1 Chr. 8: 33) but changed by editors when '*Baal' became unacceptable. Ishbosheth (Ishbaal in NRSV) tried to rule the tribes at Mahanaim after the death of Saul (2 Sam. 2: 8–10), and to rally them against the succession of his father by *David. But *Judah held out and remained loyal to David. Ishbosheth was murdered and so ended the dynasty of Saul. The two murderers, Rechab and Baanah, brought the head of Ishbosheth in triumph to David. He expressed outrage and had the two murderers executed (2 Sam. 4: 12).

Ishmael Son of *Abraham and *Hagar (Gen. 16: 15), brought into the *covenant (Gen. 17: 25) but in a lesser role than *Isaac (Gen. 17: 19). He is to be the ancestor of a great nation (Gen. 21: 18) but life for him will be harsh (Gen. 16: 12). His descendants were related to the *Edomites (Gen. 28: 9), since *Esau was said to be the ancestor of the people of Edom (Gen. 36: 1).

The traders who bought the patriarch *Joseph from his brothers (Gen. 37: 25–8) were called Ishmaelites in one source; another (possibly the *E source) calls them *Midianites (Gen. 37: 28).

It is part of Muslim belief that adherents of Islam are descended from Ishmael.

island Many islands are mentioned in the Bible. Among them are *Crete (Caphtor), from which the *Philistines came (Amos 9: 7); *Titus was sent to pastor that Church (Tit. 1: 5–14). Other islands are: *Cyprus (Jer. 2: 10; Acts 4: 36); *Malta (Acts 27: 39 ff.), on which Paul was shipwrecked; *Patmos (Rev. 1: 9), off the Ionian coast, on which the seer of Revelation was imprisoned; and *Tyre (Ezek. 26–8), famous for wealth, and connected to the mainland by a causeway built by *Alexander the Great.

Israel Used for the twelve tribes, who trace their descent from *Jacob, who was also called Israel (Gen. 32: 28). After the one kingdom of the twelve tribes was separated into north and south, the northern kingdom from 924 to 721 BCE is called Israel, as distinct from *Judah. But the OT text is not entirely consistent. By a kind of nostalgia for the united kingdom of *David and *Solomon, even after the disruption we can read of the 'people of Israel who were living in the towns of Judah' (1 Kgs. 12: 17). After the northern kingdom has disappeared and many inhabitants have been deported to

*Assyria, the term 'Israel' can be applied to Judah (Ezek. 2: 3).

The people of Israel were *foreigners in the land of *Egypt according to the earliest traditions, and made their escape into Palestine. Whether this 'exodus' included the whole people or whether there were incursions by a few tribes at a time, with piecemeal settlement, is uncertain. But the biblical writers claim that the occupants of Canaan were eventually dispossessed and the land distributed among the twelve tribes of Israel who became a nation around the end of the first millennium. When David, who was at first king only of Judah (2 Sam. 2: 4), united his kingdom with Israel in the north (2 Sam. 5: 1–3), his dominion stretched from '*Dan to *Beersheba'.

The people of Israel believed themselves to be specially chosen by God to occupy this *promised land. They also believed that they had been given the privilege of the *Law (*Torah) at *Sinai. They were called to be a *blessing to all the rest of the human race and the Torah was the compendium of instructions by which they would work out this vocation.

The divided monarchy was an era of rivalry and hostility among kinsfolk. Both kingdoms were weak, though under *Omri and his three successors the north had a period (885–843 BCE) of prosperity, tempered by social injustices and conflict between Yahwism (championed by *Elijah) and Baalism (supported by Queen *Jezebel). Threats from Assyria ended with the conquest of the north by King *Shalmaneser V in 721. Judah survived until the Exile and the deportations to *Babylon in 597 and 586 BCE. In Babylon the Jews began to see their *faith in the One God as a faith which should be propagated among other nations (Ezek. 39: 21). The fate of Israel (meaning here the exiles from Judah) will demonstrate the power and justice of the true God (Ezek. 37: 28; 38: 23). Other nations will be drawn to Israel (Isa. 45: 14).

Sadly, as the biblical historian passes his judgement, after the Return from Exile and the rebuilding of the Temple (520 BCE), the people of Israel failed to live according to the requirements of the *Torah. The belief much later was that when at length they did so the '*kingdom of God' would have arrived. Various groups within *Judaism held themselves to be the true Israel who might put the divine plan into effect. Such a group was the *Qumran monastic community by the *Dead Sea, who were organized into twelve tribes headed by twelve overseers—the very number being indicative of their claim to be the true Israel. The early Church, too, regarded itself as the true Israel, the legitimate descendants of *Abraham, and inheritors of the *covenant promises, with a vocation to be a blessing to mankind (Gal. 3: 8). The Jews were no longer in the true succession (Rom. 4: 13). The Church had become as it were the twelve tribes (Jas. 1: 1; Gal. 6: 16; 1 Pet. 2: 9) of the authentic Israel—a claim that was to afford additional ammunition for Christian antisemitism.

Israel of God Paul asserts that the promises made by God to the Jews have now been inherited by Christians, who are the faithful remnant (Rom. 9: 6). Jesus' choice of the twelve indicates some sort of continuing relationship to the twelve tribes of Israel, even if the form of the saying at Matt. 19: 28 reflects the eschatological understanding of the Christian generation after Jesus. James 1: 1 addresses the Church as 'the twelve tribes in the Dispersion' (cf. 1 Pet. 1: 1) and Paul in Gal. 6: 16 calls the Church comprising both Jews and *Gentiles 'the Israel of God'.

Issachar The ninth son of *Jacob, born to *Leah, and ancestor of the tribe occupying part of northern Canaan west of the Sea of *Galilee (Josh. 17: 10) which was overrun by the *Assyrians in 732 BCE. His brother was *Zebulun (Gen. 30: 18–20).

Italian band A cohort of the Roman army probably composed of Roman citizens, of which *Cornelius was a *centurion (Acts 10: 1).

Italy By NT times the name Italy denoted the area of modern Italy, and *Rome, its capital, controlled an empire with good roads radiating in every direction. These encouraged both merchants and immigrants (including many Jews). According to Acts 18: 2 *Aquila and Priscilla came from Italy to *Corinth after an edict extraditing Jews. In Heb. 13: 24 'those from Italy' send greetings; but it is uncertain whether these are Italians resident in another country who send their good wishes to friends at home, or whether the greetings are sent from Italy to readers elsewhere.

Ituraea A wild region NE of the Sea of *Galilee ruled by *Herod the Great; on his death it passed to *Philip (Luke 3: 1) along with *Trachonitis.

ivory Caches of luxurious ivory jewellery and knives have been discovered in excavations at *Ugarit and *Megiddo, and may have been made from the tusks of Syrian elephants. *Solomon's ships carried ivory amongst the varied cargo (1 Kgs. 10: 22) and the royal throne was inlaid with ivory (1 Kgs. 10: 18). Amos denounces the excesses of the rich who flaunt their wealth with 'beds of ivory' (Amos 6: 4). Ivory was sold by merchants in Rome in NT times, and is mentioned in Rev. 18. 12.

J

J The symbol (from the German Jahvist; Yahwist in English) used by German OT scholars and followed internationally to denote one of the sources of the *Pentateuch, which uses Yahweh for the name of God. It was probably written in the south of the country in the 10th (or 9th) cent. BCE, though a few recent OT scholars think it might even be post-exilic. J is characterized by *anthropomorphisms (e.g. Gen. 8: 21) but above all records the faithfulness of God to the promises he made to the *patriarchs.

Jabbok A tributary of the *Jordan; a ford across this stream was the scene of *Jacob's wrestling with God (Gen. 32: 22).

Jabesh-Gilead A town on the east of the River *Jordan. *Saul came to the rescue of its people (1 Sam. 11) when they were threatened by the *Ammonites; their gratitude was later expressed when they removed the dead bodies of Saul and his three sons from desecration by the *Philistines (1 Sam. 31: 12). *David praised the heroes of Jabesh-Gilead for this, at the same time pointing out that the kingship had now been assumed by himself (2 Sam. 2: 4–7).

Jachin A bronze pillar on the right (the south side) of the porch of *Solomon's Temple (1 Kgs. 7: 15–22), hollow and free-standing, about 8 m. (26 feet) high and 5 m. (17 feet) in circumference, surmounted by a bowl-shaped capital decorated with pomegranates. There was a corresponding pillar on the north side. The independent position of the two pillars (Jachin and Boaz) contrasts with Doric pillars in ancient Greece as in the Parthenon at *Athens, which were an integral part of the building, with a logical and aesthetic place in the whole structure. Differently arranged, they would look bizarre, and indeed the building (it was a treasury) might not even stand. But Jachin and Boaz had no functional or architectural *raison d'être*. They were symbolical. They were the gateposts for the Temple, which was Yahweh's dwelling place and into the inmost sanctuary the public were forbidden to enter. The pillars were an impressive reminder to Israelites and visitors alike that Yahweh was within and that Solomon who built the Temple was his appointed king.

Archaeological excavations have shown that similar free-standing pillars were a feature also of Canaanite temples.

Jacinth Not the modern (reddish-orange) jacinth, but a blue stone; it is mentioned at Exod. 28: 19 as being in the high priest's breastpiece. In Rev. 9: 17 the colour of the *breastpieces of jacinth (AV) worn by the horsemen in the seer's *vision is likened to 'sapphire' (NRSV), 'turquoise' (REB), 'hyacinth-blue' (NJB). Jacinth is also a stone in the wall of the New Jerusalem (Rev. 21: 20).

Jacob Son of *Isaac and *Rebekah, and brother of *Esau whom he twice outwitted (Gen. 25: 29 ff.; 27). The traditions about Jacob were compiled over several centuries but in so far as it can be assumed that Jacob was a historical person he is likely to have lived in the period between about 1750 and 1570 BCE. Some authorities believe there are indications that he lived about 1200 BCE at the beginning of the early Iron Age, but this date would assume a surprisingly late date for the *Exodus. The combination of the Jacob stories from the two sources *J and *E and then with the addition of *P material (in Gen. 35) is in a remarkably sophisticated arrangement reflecting, by the genealogical relationship of Jacob to his sons, something of the actual social realities at the time of compilation.

Jacob is portrayed unattractively: he steals from his father-in-law *Laban; he deceives his brother Esau and so obtains his birthright; and he tricks his father into giving him the blessing of the *first-born which was Esau's privilege. In his old age, Jacob comes into the *Joseph narrative and dies near Joseph, his favourite son (Gen. 49: 33).

In the NT Jacob is mentioned in the speech of *Stephen (Acts 7: 8) in his recital of the promises of God to his people; but more importantly the preference of Jacob over Esau is used by Paul (Rom. 9: 13) to illustrate the sovereign *grace of God, and by Heb. 11: 9 as one of the examplars of true *faith. In the conversation between Jesus and a Samaritan woman, she asks whether he is greater than the ancestor Jacob (John 4: 12). It is an example of John's irony: she thinks it is absurd—but the readers of the gospel know otherwise.

Jacob's well The scene of the conversation described in John 4 of Jesus with a woman of *Samaria. He promises her 'living water' instead of the dull, still water she is to draw from the well. The well was probably near *Shechem and may have been in fact no more than a tank for rain-water. Theologically, however, John is using the symbolism of a gushing, lively spring; this is an evocative image to express the joy and excitement of the new life which Jesus offers, beginning in *baptism.

Jael Wife of Heber the Kenite (Judg. 4: 17); she invited *Sisera, commander of the *Canaanite army of King Jabin of *Hazor, into her tent as he fled from the battle in defeat. While he was sleeping, she drove a tent-peg into his temple. The exploit gave Israel protection from the Canaanites and is memorably celebrated in the song of *Deborah, a prophetess who had lent her support to the victorious campaign (Judg. 5: 24–31).

Jahveh Used by German scholars for the name for God in the *J source of the *Pentateuch which English scholars normally write as *Yahweh.

Jairus The leader of a *synagogue (Mark 5: 22) in *Galilee (*Capernaum?) who asked Jesus to heal his daughter. Jesus did more; he raised her from what onlookers took to be death. The account of this miracle is interrupted by the narrative of the healing of the woman suffering the embarrassment of continuous uterine bleeding, who had spent all her resources on medical treatment to no avail (Luke 8: 43). Both *miracles have in common the *gift of *faith in Jesus and the explicit mention of '*twelve'—the girl was 12 years old; the woman had suffered for twelve years. Readers of the gospels may have picked up the signals that the Lord of the Church, the New Israel, (old Israel comprised twelve tribes) offers the hope of their own future new risen life.

James A common name in the 1st cent. CE ('Jakobos' in Greek) and in the NT several men bear it: (1) A son of *Zebedee and brother of *John (Mark 3: 17) and one of the *twelve; with Simon Peter and John he was also one of the more intimate group of three with Jesus, being present at the *Transfiguration and in the garden of *Gethsemane. He was beheaded by Herod *Agrippa I (Acts 12: 2). (2) Son of Alphaeus, and also one of the Twelve (Mark 3: 18); he may be the 'James the Less' (AV; 'James the Younger', NRSV, REB, NJB, Mark 15: 40) whose mother *Mary was present at the *crucifixion and became a witness of Jesus' *resurrection (Mark 15: 40; 16: 1). (3) A brother of Jesus (Gal. 1: 19). The Greek does not necessarily demand that he should be another son of Mary, younger than Jesus, and as the belief in Mary's perpetual virginity grew in the Church, it was held either that James was a son of Joseph by a former marriage—Joseph's first wife having died—or that the word could denote 'cousin'. Jerome (*c.*400 CE) took the view that James (and the other brothers, Mark 6: 3) were the sons of another Mary, a sister of the mother of Jesus (John 19: 25—the wife of *Cleopas).

James, the Lord's brother, saw the risen Jesus (1 Cor. 15: 7) in spite of earlier hesitations about him (Mark 3: 31–3) and became leader of the Church in Jerusalem (Acts 15:

13). He sent a delegation to investigate the affairs of the Church in *Antioch (Gal. 2: 12), where Paul had converted *Gentiles. He had a reputation in Jerusalem for great piety and became the Church's second known martyr, Stephen being the first. About 61 CE, after the sudden death of the *procurator *Festus, he was stoned by order of the high priest, Ananus, according both to the Jewish historian *Josephus and the Christian Hegesippus (quoted by *Eusebius).

James, epistle of Addressed to a group in the Church, probably of Jewish background, and suffering poverty. It is a plea for good works, such as compassion (2: 14–26) and honesty (4: 11–12). It deprecates worldliness (Jas. 4: 4) and obsequiousness to the well-off (Jas 2: 3). The epistle was little esteemed by Martin Luther, who reckoned its exhortation to good works to be contradicting Paul's doctrine of *justification by *faith and not works. But as the good works which James urges are not the works of the *Law which Paul does not wish to impose upon his *Gentile converts, such as *circumcision, there is therefore no real conflict between James and Paul. James is worried that the community is becoming obsessed with money (Jas. 1: 11) and gossip (4: 11).

It is uncertain who the 'James' of 1: 1 might be. Traditionally, he has been regarded as the Lord's brother, in Jerusalem, the only James who could write with such authority. But could he write such Greek? It is true that Greek was more widely known and spoken in Palestine than used to be asserted by scholars, so it is not impossible that he wrote the epistle. More depends on the apparent knowledge of Paul's epistles (2: 19–20); such knowledge would require a date later than James, who is reported by the historian Eusebius and by *Josephus to have been martyred before 70 CE. If the epistle is pseudonymous, it may have been formed from a miscellaneous compendium of advice from Jewish Christian sources. An editor gave a unified structure to the homilies and used the name of James as author.

Jamnia A city (also spelt Jabneh or Javneh) south of *Joppa where after 70 CE a body of Jewish teachers met by way of succeeding to the role of the *Sanhedrin but without any formal constitution. It has been suggested that the *canon of the Hebrew OT was settled here, but this is unlikely. It was at any rate in the Jamnian period, up to 135 CE—when it was important to come to terms with the destruction of Jerusalem and the Temple and to reconstruct a new and authentic *Judaism—that the status of books like Ecclesiastes would have been discussed.

Jannes and Jambres Names given by *rabbis in the first centuries BCE and CE to the Egyptian magicians who opposed Moses (Exod. 7 and 8), and who were regarded as sons of *Balaam. They were known also by the *Qumran community, and in the NT are mentioned by 2 Tim. 3: 8 as typical opponents of the truth. There was a tradition that they possessed the power of flight; and in the 6th cent. CE a Christian work, no longer extant, tells of their ultimate conversion.

Jashar A Book of Jashar is quoted in Josh. 10: 13 and 2 Sam. 1: 18 and apparently consisted of songs and poems.

Jason In Greek mythology, Jason was leader of the Argonauts, who fetched the golden fleece. It became a common Hellenistic name and was even used by Greek-speaking Jews as the equivalent of Hebrew Joshua. A Jewish high priest called Jason is mentioned in 2 Macc. 4: 7 ff., and in Acts 17: 5–9 a Jason was Paul's host in *Thessalonica, and perhaps was with him at *Corinth (Rom. 16: 21).

jasper A green stone in the fourth row of the high priest's breastpiece (Exod. 28: 20) and in the wall of the New Jerusalem (Rev. 21: 18).

Javan Hebrew for Ionia and hence the Greek area of Asia Minor which included *Ephesus and *Miletus (Acts 20: 17; Rev. 1:

11) and Greece in general, so that the king of Javan (Hebrew of Dan. 8: 21) is a reference to *Alexander the Great.

jealousy Resentfulness towards another or suspicion of someone else is rebuked by Paul (Rom. 13: 13); but there is a godly jealousy, meaning an exclusive zeal for the Lord (2 Cor. 11: 2), and God is said to be jealous (Exod. 20: 5) in the sense of having a passion for total loyalty (Num. 25: 11).

Jebusites Dwellers in an area in and around Jerusalem. Apart from the city itself, the principal stronghold, the land was occupied by *Joshua (Josh. 11: 3). *David (2 Sam. 5: 9) and *Solomon (1 Kgs. 9: 20–1) completed the conquest and Jerusalem became the capital of the united Israelite kingdom under both these rulers. David bought some land from one of the Jebusite inhabitants, *Araunah, on which to build an *altar (2 Sam. 24: 18–24), and Solomon built the Temple (2 Chron. 3: 1) on the same site.

Jehoiachin Also called Coniah (Jer. 22: 24); he began to reign over *Judah at the age of 18 in 597 BCE and soon afterwards surrendered to the *Babylonians (2 Kgs. 24: 12) but in 561 BCE had his status in captivity raised (2 Kgs. 25: 27–30; Jer. 52: 31–4) and his fortunes reversed.

Jehoiakim Son of *Josiah; he was imposed upon the nation as king of *Judah in 609 BCE by the Egyptian Pharaoh *Neco, who changed his name from Eliakim; but when the Babylonians defeated the Egyptians at *Carchemish in 605 (Jer. 46: 2), Jehoiakim found himself a vassal of *Babylon. Unwisely, he rebelled (2 Kgs. 24: 1)—contrary to the advice of Jeremiah—and *Nebuchadnezzar besieged Jerusalem shortly after Jehoiakim died.

Jehoshaphat Son of *Asa; he became king of *Judah about 873 BCE and reigned for twenty-five years (1 Kgs. 15: 24; 2 Chron. 20: 31) as the contemporary of *Ahab of Israel (1 Kgs. 22), with whom he formed an alliance against *Syria. The prophet *Micaiah tried to dissuade the two kings from fighting, but they ignored his advice and Ahab (though he disguised himself in Jehoshaphat's robes) was slain. The account is rewritten by the Chronicler (2 Chr. 18) in the light of his theology; Jehoshaphat is praised for his piety and rewarded with a substantial victory over *Ammonites and *Moabites (2 Chr. 20: 1–30).

Jehovah The divine name for the Hebrews was YHWH; but so sacred a word was not pronounced by Jews—and they substituted *Adonai* (Lord). But when vowels were added to the Hebrew script, the vowel points inserted in YHWH were those of Adonai, producing the form Jehovah as in AV and frequently in English usage.

Jehu King of Israel, 842–815 BCE. After the death of *Ahab, he was encouraged by *Elisha to stage a coup in which Ahab's widow *Jezebel was killed. She had supported the *Baal cult and incurred the wrath of *Elijah. Jehu also slaughtered the worshippers of Baal (2 Kgs. 10: 25), but in spite of this zeal for Yahweh he continued to maintain the *sanctuaries of *Bethel and *Dan which were rival attractions to Jerusalem. Jehu's atrocities are condemned by the prophet *Hosea (1: 4).

Jephthah An Israelite 'Judge' for six years, probably in the century before the establishment of the monarchy of *Saul (*c.*1050 BCE). After a period of exile, Jephthah was recalled to *Gilead by the elders and persuaded to lead an expedition against the marauding *Ammonites. He accepted the responsibility but made a vow that if successful he would sacrifice to Yahweh whatever or whomever emerged first from his house to meet him (Judg. 11: 30–1). This proved to be his daughter. The vow took priority over his humanity and the girl was sacrificed after a delay of two months for lamentation for her virginity. The story is related as a rather shocking incident, which perhaps implies that human *sacrifice practised amongst Israel's neighbours was normally unacceptable in Israel.

During an inter-tribal conflict with the Ephraimites (Judg. 12: 1–6) Jephthah instructed his men to use the word *shibboleth*, which their enemies could not pronounce, as a password (cf. Heb. 11: 32).

Jeremiah A major prophet who lived in *Judah *c.*640–587 BCE and who spent his last years in old age as an enforced refugee in *Egypt. Jeremiah came from a priestly family and began his prophetic *ministry reluctantly (Jer. 15: 10—'Why was I ever born?') in his village of Anathoth (a short distance from Jerusalem) much to the indignation of neighbours and family (Jer. 11: 21; 12: 6). There is a remarkable account of his prophetic inspiration in Jer. 23: 9. Jeremiah had easy access to the king and he used his knowledge of international affairs to offer advice fearlessly. In his early years he opposed the official policy of an alliance with *Egypt against *Assyria (2: 14–19, 36–7), but when Egypt marched to help Assyria against the rising power of *Babylonia, King *Josiah attempted to block the path of Pharaoh *Neco at *Megiddo and was killed (609 BCE). Soon afterwards Judah became a Babylonian province. Jeremiah was convinced that it was essential for Judah to maintain a good relationship with the Babylonians. But King Jehoiakim foolishly rebelled; almost immediately after his death, the city was besieged, but the young king *Jehoiachin submitted (597 BCE) and the puppet king *Zedekiah, Jehoiachin's uncle, was installed. In the reign of Zedekiah (597–587 BCE) Jeremiah frequently urged obedience as the safest course; the existence of the Temple was no guarantee of security. For this 'blasphemy' he was arrested (Jer. 26: 8). But in 589 Zedekiah withheld tribute; the Babylonians marched (Jer. 37–9). When the Babylonians were outside the gates of Jerusalem, some of the leading politicians denounced Jeremiah as a traitor and he was imprisoned. King Zedekiah transferred Jeremiah to a more humane prison, where he remained until the Babylonians released him when the city was captured in 586 BCE. The victorious Babylonians appointed Gedaliah to be governor of Judah, and Jeremiah attempted to muster support for him (Jer. 39: 14). But a group of conspirators opposed to the Babylonians murdered Gedaliah, and after seizing Jeremiah fled in a group to Egypt.

Jeremiah, book of The second of the three great prophetical books following Isaiah and preceding Ezekiel. Many of Jeremiah's prophecies were preserved by his secretary *Baruch (Jer. 36), but scholarly attempts to determine which parts of the book were written on that scroll have not been generally agreed. Substantial editorial work was done on all parts of the book. The influence of the *Deuteronomist circle has left its stamp on the theology: it was the aim of this school to update the words of Jeremiah in such a way that they addressed the situation of a later generation. This is clearly the case with Jeremiah's famous Temple sermon (ch. 7) and the following units, which contain Deuteronomist words and phrases. Authentic words of the prophet have been edited, and it is likely that the material took its final shape among the exiles in *Babylon. These are the people addressed in Jer. 29. Moreover, there can be discerned in the book traces of the contempt felt by the exiles for those Judaeans who remained in their native land. The emphasis on the *Law, which took the place of the Temple as the focus of devotion, and the development of the Jews in exile as the People of the Book, were significant features in Babylon which led to an antagonism against those not deported. This feeling was intensified after the Return (Ezra 4: 1–5) but claims for the moral superiority of the exiles ('the remnant') are made in Isa. 46: 3, contrasted with the desolate land of Judah (Isa. 43: 28), and especially in Ezek. 33: 23–9. In Jer. 29–33 words of the prophet Jeremiah himself are adapted to suit the community in exile who are the true remnant of Israel. After the account of the destruction of Jerusalem in 586 BCE (Jer. 37–9) comes the indication that the future of the *covenant people lies with those in Babylon, whereas those left in Judah are strongly condemned. It is therefore argued that this material in the book of Jer. reached its shape in the

Babylonian Exile and was intended to support and encourage the Jews who were there. The people left in Judah after 586 BCE are dismissed by Jer. 40: 7–44 as offering no hope for the nation's future. That is to be found in the Babylonian exile (2 Kgs. 27–30).

Jeremiah, epistle of A book in the *Apocrypha. It is printed in AV, RV, and REB, following the *Vulgate, as ch. 6 of Baruch. Under the guise of a letter from Jeremiah to the Babylonian exiles, it is a warning to Jews by a pseudonymous author in the Hellenistic era against apostasy.

Jeremias, Joachim (1900–79) German NT scholar who taught at Göttingen and wrote an influential work on the *parables based on extensive research on the Jewish environment of Jesus and on knowledge of Aramaic.

Jericho On the west bank of the *Jordan, a city 10 km. (6 miles) north of the *Dead Sea, which comes into OT history when *Joshua captured and destroyed it (Josh. 6: 20–1) at the beginning of the Israelite invasion of *Canaan. Jericho has been thoroughly explored by archaeologists, and John Garstang in the 1930s believed that the ruin of a mudbrick wall was evidence of the city's fall about 1400 BCE. However, more recent excavation suggests that Jericho can tell us nothing about Israelite settlement as it was unoccupied for most of the 2nd millennium. It gradually recovered (1 Kgs. 16: 34) and by the 6th cent. BCE it was a centre of administration. On the banks of the Wadi Qelt a palace was built by *Herod the Great, who died there in 4 BCE.

Visits by Jesus to Jericho and its neighbourhood are recorded: he healed blind *Bartimaeus as he left the city (Mark 10: 46), or as he approached it (Luke 18: 35); here he met *Zacchaeus the tax collector (Luke 19: 1–11) and spoke the parable of the pounds (Luke 19: 12–28). Travellers from *Galilee to Jerusalem passed through Jericho in order not to be harassed by *Samaritans.

Jeroboam (1) The first king of Israel after it had broken from *Judah under *Rehoboam; he reigned from 922 to 901 BCE. Jeroboam instituted religious change in order to consolidate his position in the north; the sanctuaries at *Bethel and *Dan were established with bull images and people were discouraged from worshipping at Jerusalem. This was regarded by the southern editors as *apostasy, and he is condemned by 1 Kgs. 15: 26. *See* idolatry. (2) King of Israel 786–746 BCE, who used the end of intervention from *Assyria to expand the borders of Israel (2 Kgs. 14: 25, 28) and increase his personal wealth. Distribution of land to Jeroboam's supporters created a class of the poor and dispossessed, and this social injustice, together with worship of *Baal, was bitterly denounced by *Amos (2: 6–8).

Jerusalem In the central hill country, the key city in the historical events of the OT and NT; yet it was unattractive in its geographical endowments with a poor water supply, uncomfortable climate, and unproductive soil. But in the hands of the *Jebusites it lay conveniently in the very north of the territory of *Judah on the boundary with the northern tribes. *David's genius was to recognize this neutral city as the ideal capital from which he might rule both north and south in a religious and political unity. Under his commander *Joab, the city was stormed and taken (2 Sam. 5: 8); at the northern end the royal palace rose, and on Mount *Moriah a site was selected for the Temple, built by David's son, *Solomon. From now on Jerusalem was to be a religious and political centre, often prosperous, repeatedly assaulted, and the destination of millions of pilgrims. Sadly, the ideal of Jerusalem as the centre of religious unity for both parts of the kingdom did not last. Even in Solomon's time various pagan cults were permitted on the outskirts, and after the disruption, when Israel in the north detached itself from Judah, sanctuaries in the north were built to dissuade worshippers from travelling to Jerusalem. However, Jerusalem continued to expand; walls were strengthened by *Hezekiah, and the water supply augmented by the pool of *Siloam (John 8: 7). The city's defences were never seriously tested when *Sennacherib brought the Assyrian forces to its gate, but the prophets' warnings (e.g. Mic. 3: 12) that this immunity would not last for ever were fulfilled when the Babylonians plundered the city in 597 BCE, deported many of the leading citizens, and returned to raze Jerusalem to the ground in 586 BCE.

*Cyrus king of Persia permitted the exiles in Babylon to return and rebuild in 538 BCE. Some did: but the rebuilding, including that of the Temple, was on a small scale, and the walls constructed under the direction of *Nehemiah in 445 BCE surrounded only a small area, so that the city no longer enjoyed its pre-exilic prestige; it was nevertheless responsible for maintaining Jewish religious ideals. This earthly poverty inspired apocalyptic dreams of a new Jerusalem where God would be hailed as the victorious king (Zech. 14: 16) and to which the nations would be attracted and *Gentiles would make pilgrimages (Zech. 8: 21).

In the time of *Alexander the Great, Jerusalem's political importance recovered, but Hellenistic ideas and customs intruded under the *Seleucids of Syria in the 2nd cent. BCE (1 Macc. 1: 14), to the dismay of traditionalist Jews who re-established and fortified the city between 160 and 134 BCE, the *Hasmonean era. After the victory of the Romans in 63 BCE the city grew greater and buildings of grandeur were erected by Herod the Great, who was appointed king in 37 BCE. Walls and fortresses and city gates appeared, and the Temple mount was dominated by the Antonia fortress where Paul was to be imprisoned (Acts 21: 34). Houses to accommodate a population of some 25,000 inhabitants, trebled at the time of the *Passover by visiting pilgrims, were needed, and a reservoir called the Sheep Pool (John 5: 2) was dug. Above all Herod began the building of the huge Temple, destined to occupy a quarter of the area of the city, though always regarded by strict Jews with an uneasy conscience in view of Herod's mixed (Idumaean) race.

When tension between Jews and Romans flared into open rebellion in 66 CE, it was only a question of time before the city was again destroyed (70 CE), as Jesus had warned if the people rejected the way of peace which he proclaimed (Luke 19: 41–4; 21: 20–4). But on the site of the Temple another generation of Romans determined to build a shrine in honour of Jupiter. Jewish nationalists could be restrained no longer, and rose to revolt under *Bar-Kochba in 130 CE. The rebellion was crushed; a pagan temple was built (cf. Mark 13: 14); Jews were banished; Roman soldiers were installed in the city now renamed Aelia Capitolina—after the family name of the emperor Hadrian (Capitolina referred to the god Jupiter). It was not until the emperor Constantine (*c.*330 CE) established the city as a Christian centre with churches that any further major changes took place. In 451 the bishop was elevated to the status of patriarch. The city was more than once taken by Muslim invaders and was in their hands from the 13th to the 20th cents., and the site of the Temple, the Dome of the Rock, became a shrine, venerated in the belief that it was from here that Muhammad ascended into heaven. Jews believe that the rock was used by *Abraham for his intended sacrifice of Isaac. Jerusalem is therefore sacred to three great religions.

Jesse Father of *David, grandson of *Boaz and *Ruth the Moabitess (Ruth 4: 17–22), who farmed successfully (1 Sam. 16: 20). He is named in the *infancy narratives of Matt. (1: 5–6) and Luke (3: 22) as one of the ancestors of Jesus.

Jesus The focus of Christian devotion; born a Palestinian Jew *c.*4 BCE.

1. *The name*. A common personal name for Jewish males in the centuries BCE; it is the Latin form of the Greek translation of the Hebrew *Joshua or Jehoshua, and means 'He whose *salvation is Yahweh'. Joshua was the name of Moses' successor who commanded the Israelite forces when they occupied *Canaan, and it was the name of the author of Ecclus. [= Sir.]. (When Joshua is mentioned in NT at Acts 7: 45 and Heb. 4: 8, the name appears in AV, misleadingly for modern readers, as 'Jesus'.) *Josephus mentions four high priests called Jesus, and one

of Paul's fellow workers was called Jesus Justus (Col. 4: 11). Later the name was dropped for Christians out of piety and by Jews out of revulsion, though in modern times 'Jesus' has been used at *baptisms for children in Spain and the Philippines.

The name Jesus occupies many columns in NT concordances, often with 'of *Nazareth' added, but 'Jesus Christ' appears only twice in the synoptics, though it is common in the epistles, and Paul uses 'Jesus' (by itself) only eighteen times.

2. *The life of Jesus: the evidence in ancient historians*. Outside the NT, there is little recorded about 'Jesus of Nazareth': a reference in *Josephus' *Antiquities* (xviii. 63) describes Jesus as a wise man who performed astonishing deeds, and a teacher; Tacitus (*c*.110 CE) mentions his death, and Pliny (*c*.110 CE) refers to Christians who worship 'Christ as a God'; Suetonius (*c*.120) mentions the expulsion of Jews from Rome in 49 CE when they had rioted about a certain 'Chrestos' (probably a mistake for 'Christos'). There are a few references to Jesus in rabbinic writings but they yield very little historical evidence; they reject his claim to be *Son of Man, and there are several references to the *crucifixion, and suggestions in connection with *miracles and the *resurrection that Jesus was a magician. However, the discoveries of the *Dead Sea scrolls and the *Nag Hammadi codices have made possible a new evaluation of Christian origins and a more reliable assessment of the Jewishness of Jesus and his *teaching.

3. *The life of Jesus: the NT evidence*. It is necessarily from the gospels that evidence for Jesus' life and teaching must be drawn. And because they were written from within the Christian *community in response to its own internal needs, they must be examined (as they have been) with all the resources at the command of scholarship, and their claims tested by appropriate criteria of coherence and methodology, as well as in the light of what is known of social, political, and intellectual conditions of the 1st cent. CE.

The study of the historical Jesus has had many repercussions and consequences, even in politics and in international relations. It was held in the gospel of Barnabas (in the 15th cent. CE) that Jesus prophesied the coming of Muhammad and that it was Judas who was crucified—though this work was probably a medieval forgery by a convert to Islam. In modern times Jesus has been portrayed as an economic reformer (an early Marxist), as an anti-Roman Black Messiah, as a Liberator of the poor; and as a pacifist; a Socialist, and a Freemason. It has been alleged that Jesus was not crucified by the Romans but stoned to death by the Jews. In these mutually incompatible assessments of the evidence, the Jesus of history is blurred. Some have said he cannot be recovered at all. But there are grounds for greater confidence.

In this study it has long been recognized that the three synoptic gospels must have pre-eminence for extracting the historical core from the interpretation rather than the gospel of John, which represents theological interpretation, primarily, while preserving the outline of the story. There are other accounts of the lives of teachers in the ancient world which are not unlike the synoptic gospels, but as biographies the gospels omit much that a modern writer would expect to use and to psychologize about. Still, Matt. and Luke record the birth of Jesus and both agree that it was when *Herod the Great reigned. Luke also mentions that Caesar *Augustus (27 BCE–14 CE) was the ruling Roman emperor, and that a *census was held when *Quirinius was governor of Syria which necessitated the journey of Joseph and Mary to *Bethlehem. These assertions raise difficult historical questions—as also when these two gospels say that Jesus was miraculously conceived 'by the Holy Spirit'; and also that he was descended from King *David: both assertions have been important for NT *Christology. As Herod died in 4 BCE, the date of Jesus' birth was in that year or a little before. (The reckoning as the year 1 *AD was a 6th-cent. miscalculation.)

Jesus' family lived at Nazareth, where *Joseph worked as a carpenter. This was a lowly occupation and implies that Jesus would not have had the intellectual education so admired by Ben Sirach (Ecclus. [= Sir.] 38: 24), but he did have a gift for vivid and popular speaking.

Much modern scholarship stresses that Jesus must be understood within contemporary *Judaism. As a *teacher and healer he was a charismatic itinerant leader or a Galilean holy man. The evidence is against *docetism of any kind and clear about Jesus as a human being within the social and religious order of 1st-cent. Palestine.

Jesus' public *ministry began when he identified himself with the work of *John the Baptist. He then called *twelve men to be immediate *disciples—not because they were men rather than women (he did not distance himself from women (Luke 8: 2–3), unlike the generality of Jewish teachers) but because they were to correspond to the twelve *patriarchs who were the ancestors of the twelve tribes of the old Israel and therefore the foundation of a new community, a 'new Israel'. As an itinerant *teacher, Jesus was a charismatic personality whose authority created a strong impression, and bitter opposition, reinforced by his association with outcasts and pariahs (Luke 15: 2). He did not set out to undermine the *Law or ridicule the principles of *Judaism: on the contrary, there is no record of his repudiating any of the legislation in the *Torah which was the framework for civilized life; he spoke in *synagogues (Mark 6: 2) and wore 'fringes' (Mark 6: 56), the four blue tassels worn by a Jew in accordance with the Law (Num. 15: 38); but he said that the relief of human need or suffering might override the observance of the *Sabbath (Mark. 3: 4; Matt. 12: 3–14). The essence of the Law could be briefly summed up (Matt. 22: 37–9) in the duty to love God and our neighbours and the '*Golden Rule' to do to others as you would wish others to do to you (Matt. 7; 12). But he demanded more than his contemporaries: not only neighbours but enemies must be loved (Matt. 5: 44); the rich young ruler must sell everything and follow him (Luke 18: 22). It is an ethic in view of the Kingdom. For the message that Jesus proclaimed was that the *kingdom of God is coming. Therefore some traditional duties would have to be abandoned (Matt. 8: 21–2). Jesus at no point describes this Kingdom, but it was assumed to refer to the time of deliverance, when the wrongs suffered by the people would be reversed, and a period of justice and peace would be ushered in. The coming of the Kingdom is pictured as in the future—the near future (Mark 1: 15)—and it will be a time of judgement (Matt. 20: 1–16; Luke 12: 16–21; 18: 1–6). Hence the call for repentance, which means not 'being sorry' but a change of outlook. Jesus insists also that this coming judgement has already begun (Luke 11: 20): Jesus' own ministry is included within God's plan of *salvation. His mission is to show that the eternal Creator and Father is a near and approachable God, and can be known as compassionate and merciful. Those people who acknowledge their need of his *grace are welcome into God's *covenant. Jesus is God's instrument for bringing the new age in which ritual and ethnic conditions required by *Sadducees or *Pharisees or *Essenes are ignored.

The method of Jesus' teaching included extended *parables (there are thirty-nine in the synoptic gospels) and brief proverbial sayings, metaphors and similes, and prophecies. He makes no explicit Messianic claim for himself but his regular use of '*Son of Man' is not a denial of such a claim, and unorthodox references to the Temple ensured the hostility of vested interests (Mark 14: 58). The cleansing of the Temple (Mark 11: 15) followed by the withering of the fig tree (symbolizing the nation of Israel) could be interpreted as a Messianic action.

Like many Hellenistic teachers of the period, Jesus is portrayed as a healer and *exorcist, but unlike some contemporaries he speaks plainly in his native Aramaic and does not use drugs or charms. The healings are done in a context of *prayer (Mark 1: 23) and against an OT background. As an eschatological prophet, Jesus was influenced especially by Isaiah 40–66 (e.g. Luke 4: 18–19). *Faith is either presupposed by a healing or it is required, and the onlookers are amazed (Mark 5: 1–20). People who were affected by diseases which rendered them 'unclean' were welcomed into *fellowship with Jesus: it was part of Jesus' revision of current attitudes.

4. *The last week*. Not surprisingly, then, Jesus excited opposition from the moment he went public (Mark 2: 6–7) and a final fatal clash became inevitable. At a farewell supper with his disciples, sharing *bread and *wine, Jesus gave his interpretation of his coming death in terms of a new covenant. Yet he had not journeyed to Jerusalem in order deliberately to die; he went in the hope of giving his contemporaries a last chance for a change of mind, for the long-promised Reign of God had drawn near. The Temple, the special place where God was present amidst his people, would be destroyed (Mark 13: 1–4) and a new faithful people established by the new covenant (Mark 12: 1–9). So if death *was* to be the end of his journey, then that death could be the means of salvation, 'a ransom for many' (Mark 10: 45). The authorities feared his popularity, disbelieved his eschatological message about the Kingdom, and were terrified by his outrageous disturbance in the Temple when the city was crowded with pilgrims for the Passover (Mark 11: 15). The Jewish leaders, angered too by the challenge of some of his implicit claims, resolved to lay a charge against him of inciting to insurrection, which the *prefect could not possibly ignore. Thus the death of Jesus by the Roman method of *crucifixion was engineered (29/30 CE) on the basis of the very claim which he himself rejected, of aspiring to be a political 'king'.

The gospels were not the earliest of the writings of the NT. The epistles of Paul are to be dated ten years at least before the gospel of Mark; and 1 Cor. 15: 3–7 is therefore important testimony to the belief of the first Christians that Jesus was raised from the dead and appeared to many of his followers. Many of the details in the accounts of the *resurrection are difficult to reconcile, but what happened at *Easter that year created a new faith, a new community, a new future.

Jethro The father-in-law of Moses (Exod. 3: 1) and a Midianite. He gave Moses advice on the structure of legal administration (Exod. 18). Elsewhere (e.g. Exod. 2: 18) Moses' father-in-law is called Reuel and Hobab.

jewellery The high priest's breastpiece (Exod. 28: 17–20) had *twelve jewels, or precious stones, three in each row. The foundations of the New Jerusalem are also said to be adorned with twelve jewels (Rev. 21: 19–20). Other types of jewellery were bracelets and ear-rings (Gen. 24: 22) and necklaces (Gen. 41: 42). Some gold and copper pendant items discovered in excavations suggest that before the Exile jewellery which had polytheistic significance was worn. This would explain the denunciations of the *prophets (Isa. 3: 18–22; Ezek. 16: 17).

Jewish Christians The first Christians were practising Jews. It took time before the revolution, which was their belief in Jesus, established a new relationship with *Judaism and there was a bitter internal conflict in the Church. Judaism was not a monolithic system—*Pharisees, *Sadducees, *Essenes, and perhaps *Samaritans being among the diversities. According to Acts 2: 46 the apostles frequented the Temple and a number of the Pharisees (Acts 15: 5) had joined the Church. The common element that bound these early Christians together was their belief that the *Messiah was Jesus and that the scriptures were fulfilled in him. There was a major crisis when Paul, the former Pharisee, engaged in large-scale evangelization of *Gentiles. It was his conviction that Jesus was not just the Jewish *Messiah, and Christianity was not a kind of reformed Judaism. *Salvation was offered to all mankind by Jesus, crucified and risen, and the Jewish *Law was not an obligation for converted Gentiles. Some of Paul's Christian opponents in Jerusalem feared that this liberalism would lead to the abandonment of ethical standards of behaviour altogether. They continued to observe the whole Law and were Christian Jews rather than Jewish Christians. Others just wanted to go on observing the dietary habits of a lifetime for themselves and were prepared to accept a compromise; indeed Paul's own practice was a compromise—he observed at least part of the Law himself (e.g. in entering the Temple, after purifying himself, Acts 21: 26) while fighting to ensure the Gentiles' exemption from it. After the destruction of Jerusalem in 70 CE, the future for the Church was in Paul's Gentile Christianity, but the controversies in the gospels of Matt., Luke, and John suggest that Jewish influence in the Church was still a force up to 100 CE. It is also the period when the Jewish-Christian epistle of James was written. But from the beginning of the 2nd cent., survivors of Jewish Christianity were marginalized into the *Ebionites and Nazoreans, whose works are quoted by *Origen and Jerome. The former group had an *adoptionist Christology. Within the NT the epistle of James is the most obviously Jewish document; there are only two references to Christ (1: 1 and 2: 1), both of which could have been later Christian additions to a Jewish writing. James's examples of patience under suffering are from OT prophets and Job, not Jesus (5: 10 f.). The Revelation to John is certainly indebted to Jewish Christianity; it owes its imagery and style to Ezekiel and Daniel, and its acceptance into the Church was slow and reluctant on account of its alien apocalyptic character. The Church did not deny its own Jewish roots and Rom. 9–11 was there to remind the winning side that what God had promised in his covenant would embrace both Jews and Christians alike: but there were extremists who were at the opposite pole to the Ebionites; they were *Marcion and his followers, who rejected the continuing validity of the OT and even much of the NT. Paul's epistles and a truncated gospel of Luke were their minuscule scriptures.

There are also NT writings with a Jewish background which occupy a mediating position in the Church. Thus the gospel of Matt. is heavily indebted to the OT and emphasizes the validity of the *Law, but Matt.'s *Christology in which Jesus is identified with *Wisdom (Matt. 11: 19; 23: 34–6) goes beyond the *adoptionism of the Ebionites. The epistle to the Hebrews is thoroughly Jewish in tone, but is hostile to the Temple cult, as were the *Ebionites (the Temple did not exist after 70 CE) but the epistle dismisses the (Ebionite) view that Christ could be accorded the status of an *angel (Heb. 1: 4–2: 18).

Jews The inhabitants of the state or province round Jerusalem after the return from the Exile in *Babylon; the name also for the ethnic-religious group who remained there or lived elsewhere in the *Diaspora, distinguishing them from neighbours who were *Gentiles.

Jezebel Wife of King *Ahab of Israel and a worshipper of *Baal (1 Kgs. 18: 19). She was a princess of *Tyre and overruled Israelite law and customs in the interests of the royal estates by ruthlessly appropriating the vineyard of the obstinate *Naboth (1 Kgs. 21). *Elijah prophesied her downfall, which was executed when *Jehu seized power (2 Kgs. 9: 30–7).

Jezreel A town by Mount *Gilboa where *Ahab had property adjacent to *Naboth's vineyard, which *Jezebel obtained, and the town where she met her death. The valley of Jezreel separates *Galilee from *Samaria. *Hosea prophesied (Hos. 1: 5) that the nation of Israel would be finally destroyed in the valley of Jezreel.

Joab Commander of *David's army (2 Sam. 8: 16) and pre-eminent in securing Jerusalem and in campaigns against the *Ammonites. His loyalty to David extended to ordering the execution of *Absalom, David's son, after Absalom had rebelled against his father, this in spite of David's express instruction that his son's life was to be spared (2 Sam. 18: 5). In the contest for the succession after David's death Joab took the losing side (that of *Adonijah) and the successful *Solomon had him put to death.

Joanna The wife of *Herod's steward (a manager, with some kind of political status; Luke 8: 3), who helped maintain Jesus and the disciples on their travels. Joanna was in Jerusalem for the *crucifixion and was one of the women to be told of the resurrection (Luke 24: 5–10).

Joash (1) The son of Ahaziah of Judah who reigned for forty years from about 836 BCE. Under the watchful care of Jehoiada, the high priest, Joash renewed the *covenant with Yahweh (2 Kgs. 11: 17) and extirpated Baalism. He was forced to surrender much treasure in the Temple to the Syrians, and eventually was murdered in a household conspiracy. According to the Chronicler (2 Chr. 24: 20–2) this was because Joash had ordered the outspoken *Zechariah to be stoned, a crime referred to in Luke 11: 51. (2) King of Israel, 801–785 BCE, who defeated Benhadad of Aram and recovered several towns, as *Elisha had predicted (2 Kgs. 13: 14–19). He also defeated *Amaziah of Judah (2 Kgs. 14: 8–14), which the Chronicler in a late narrative embroiders: Amaziah had wickedly toyed with the gods of Edom (2 Chr. 25: 20).

Job, book of A major book of the *wisdom literature of the OT, of unknown authorship, probably written during or soon after the Exile. The book questions the justice of a God who was expected to offer protection and prosperity in return for loyalty and obedience. How then explain the suffering and pains of people who were pious, 'blameless and upright' (Job 1: 1)? Job was such a one in this dramatic poem. In a short time he lost children, property, and health. If God was just, why did he let this happen? The book of Job has a special poignancy in the age of the Holocaust—so many prayers answered by total silence.

The structure of the book is that of a prose prologue and epilogue, while chs. 3 to 42: 6 are the poem in which there is a dialogue between Job and three friends; there are three cycles of six speeches, with a reply to each by Job. This is followed by a further intervention (chs. 32–7) from a fourth friend, the young Elihu. Job's 'comforters' (16: 1–2) explain why it is not unreasonable that he should be suffering such distress and they exhort him to endure it with fortitude and without complaint—much to Job's indignation. Job rails against God, blasphemously, cursing with his mouth and proclaiming his innocence. The day of his birth should be deleted from the calendar (3: 3–7). He issues a challenge to God, even if he is killed by doing so. There is only silence. There is no one to act as umpire between himself and God. Then at last God *does* reveal himself and Job's obstinate pride is brought low. He can no longer assert his righteousness in the face of God speaking out of the whirlwind (38: 2–4). What God says is very much what the friends had argued, but it reduces Job to repentance and God's generosity responds—though it is rather a crude response in that it is simply that Job's former prosperity is not only restored but doubled (42: 10), and his three new daughters are the most beautiful in the land (42: 15).

Job's three friends, Eliphaz, Bildad, and Zophar, would vindicate God by claiming that there is a correspondence between sin and suffering. In the epilogue God is shown to rebuke the friends for this naïve orthodoxy (42: 7) which in effect he has himself affirmed: he is inscrutable (38: 4) and the laws of his creation are just, regular, immutable, and magnificent, by no means to be questioned (40: 8).

The orthodox doctrine was that the good and the wicked receive their just deserts in life; the author of the book of Job felt that this was irreconcilable with the facts as he knew them. He has left a superb poem, but as the eighteen illustrations by William Blake (1825) suggest, the book is ultimately mysterious.

Joel, book of One of the Minor Prophets, of uncertain date, but since the Second Temple was standing (Joel 1: 14; 2: 17) it may have been written in the 5/4th cent. BCE. The country had been ravaged by a plague of locusts which devoured everything in sight, even though they themselves were edible (Lev. 11: 21–2; Mark 1: 6). The prophet regards the disaster as foreshadowing the *day of the Lord (Joel 2: 2) of which *Amos (5: 18) and *Isaiah (13: 6) had spoken. Joel summons the people to lament and suggests that God will then respond with an outpouring of his prophetic spirit 'on all mankind' (2: 28–29), for which Moses had prayed (Num. 11: 29) and which the Church later believed to be in principle fulfilled on the day of *Pentecost (Acts 2: 17). Israel's enemies will be destroyed in the valley of Jehoshaphat (Joel 3: 1–14).

John the Apostle John was a common Jewish name in the 1st cent. CE—a shortened form of Johanan. It was borne by the grandfather and brother of *Judas Maccabaeus (1 Macc. 2: 1; 2: 2). John son of *Zebedee (Mark 10: 35) became with *Peter and his brother *James one of the group especially close to Jesus.

Traditionally he has been identified with the *Beloved Disciple of the fourth gospel (John 13: 23; 19: 26; 20: 2 and 21: 7, 20. He could also be the unnamed disciple of 1: 35 and perhaps 18: 15 f. and 19: 35).

John the Baptist ('Baptizer', NRSV, Mark 1: 4); his preaching preceded that of Jesus, who aligned himself with John's *ministry. The gospels betray an increasing reluctance to say without qualification that John baptized Jesus (the fourth gospel does not mention such an act at all). They do not wish their readers to be in doubt that John was merely a forerunner, probably because for some time there existed disciples of John (Acts 18: 24–8; 19: 1–7) and it was important to establish a distance between the Messiah of Christian belief and the prophet who preceded him, as scripture had anticipated (Mal. 4: 5). But it is clear that John's effect on the population was electric (Mark 1: 15). He castigated the evils of his time: *baptisms in the river, unprecedented in *Judaism, offered the hope of *salvation. John's rite was once for all, unlike the daily ritual washings at *Qumran. Other differences existed between John and the Qumran community: he was essentially a public figure addressing all who came; they were a closed and austere sect. Nevertheless it is possible that John had been for a time a member of the community—he emerged out of the desert (Luke 1: 80) to preach the coming of the kingdom. John's clothing, however, made of camel's hair (Matt. 3: 4) and his leather belt were more in the prophetic style than that of the white garments of Qumran. His audacity in rebuking Herod *Antipas (Mark 6: 18) for a marriage illegal according to Jewish *Law led to his imprisonment and execution in the fortress of Machaerus by the *Dead Sea. *Josephus, however, gives a different reason for John's execution: Herod feared his popularity and regarded him as a threat to established law and order. John's activity by the River *Jordan could have been regarded as a rallying point for *apocalyptic plans of rebellion.

John the divine, Revelation to Also called the *Apocalypse; the last book in the Bible, traditionally ascribed to John the Apostle, but the language and content make it unlikely that this work could have come from the same hand as the fourth gospel, and no more likely that this John (Rev. 1: 4) is the John the Apostle or John the Elder (of 2 and 3 John). Irenaeus (*c.*190 CE) held that John's vision was put into writing towards the end of the reign of the emperor *Domitian, which would give the date of composition as *c.*95 CE when Christians experienced another round of repression.

Rev. was written to give encouragement to Christians at the end of the 1st cent., not to predict events centuries hence—though the book has been an inspiration to enthusiasts of all ages. The images, the visions, the symbols have been a rich field for bizarre interpretations. The work's first three chapters take the form of a letter to seven Churches in Asia Minor whose failings are well known to John and are sternly rebuked.

The book continues in a series of *sevens: seven seals (4: 1–8: 5), seven trumpets (8: 2–11: 19), seven visions of beasts (12: 1–15: 4), seven bowls (15: 5–19: 10), seven further visions, followed by a description of the New Jerusalem (19: 11–22: 5), and an epilogue (22: 6–21). Each set of seven represents a miniature apocalypse in itself, and seems to answer to the six workdays of God in the Gen. narrative of creation, plus the *Sabbath. John represents the most extreme form in the NT of a Christianity which is against its surrounding culture. There is no question (as in Rom. 13) of accommodation towards the emperor—*Rome is '*Babylon' (17: 5) with its seven hills (17: 9) and destroyed Jerusalem in 70 CE as the ancient Babylonians did in 586 BCE.

Rev. is heavily indebted to OT *apocalyptic writings—to Ezekiel, Zechariah, and Daniel—and with these books in mind the first readers of Rev. would have grasped the seer's message of no compromise with pagan religion. It may have been possible for the Egyptian cult of Isis to be assimilated to the worship of *Caesar, but this was unacceptable to Christians, even though the purpose of Caesar-worship was to fuse together a hundred different races into a single state. No exemption from the imperial cult was granted to Christians (as it was to Jews). For they could not recognize the State as the ultimate good, nor that force and wealth were to be acknowledged as the realities before which all must bow. So under cryptograms the Roman emperors are pilloried: 666 (Rev. 13: 18) may stand for Nero by transliterating the letters into Hebrew, where their numerical values add up to 666. *Nero, the worst of all the emperors, stabbed himself in 68 CE. The beast of Rev. 13: 3 may be a reference to Caligula, who had a mortal wound, because he was murdered.

Rev. is a mosaic of a book, exceptionally difficult for modern readers to disentangle, but with a clear message: the Christian life is one of conflict and struggle, in which death has its part, but the *resurrection of Jesus enables us to begin to come to terms with suffering and death.

John, epistles of Three epistles traditionally, but improbably, attributed to John the Apostle. It is unlikely that an apostle would identify himself simply as an '*elder' or *presbyter (2 John 1: 3 John 1). The connection with the fourth gospel is, however, close; many of the characteristic terms of the gospel reappear in the epistles—'light', 'love', 'world', 'truth', for example. The letters are written in the light of problems: some members, and former members (1 John 2: 19) are hating rather than loving, even though they claim to be without sin (1 John 1: 8), and this seems to be associated with their heretical Christology; they deny the total humanity of Jesus (1 John 4: 2), who was baptized and crucified (1 John 5: 6).

The third epistle rebukes a certain Diotrephes, who appears to be usurping the elder's leadership role. He has been inhospitable (3 John 5 ff.) in defiance of Christian custom (Rom. 12: 13; 1 Pet. 4: 9), though 2 John 10 comes very near to the discourtesy of the upstart Diotrephes by urging its readers ('the children' loved in the truth, 2 John 1) to exclude any who deviate from strict orthodoxy.

*Ephesus would seem to be the probable place of composition. It was a centre where *Judaism and Hellenism met, both of which may have attracted some members of the Church; Ephesus was associated by *Irenaeus with John.

John, gospel of Often called 'the fourth gospel' to mark its distinction from the three *synoptics. The gospel is certainly a narrative about Jesus and in this resembles the others and is quite different from the epistles in the NT. The gospel of John in its first chapter describes Jesus' encounter with *John (the Baptist); then *disciples are called; there is a public teaching *ministry of Jesus and opposition to it; after a triumphal entry into Jerusalem (12: 12–19), Jesus and the *Twelve gather in the upper room; there follow the arrest, *trial, *crucifixion, and *resurrection appearances of Jesus—as in the synoptics. But the differences between John and the synoptics are substantial: whereas Matt. and Luke have their infancy narratives recording Jesus' birth, John makes a great Christological affirmation at the outset; that the *Word became flesh (1: 14). Next, the *miracles recorded in John are never of exorcisms, or the healing of lepers, as in Mark (1: 21–8, 40–4), and John calls those miracles he records, *seven in all, such as the changing of water into wine at *Cana, 'signs'. But whereas in the synoptists the miracles are indications of the coming of the *kingdom of God (e.g. Luke 11: 20), in John the Kingdom is not at all central in Jesus' *teaching and the 'signs' are to validate Jesus' claim to be the *Son of God. The scene of Jesus' teaching, usually in *Galilee in the synoptics, takes place in John in the capital and revolves round the celebration there of the Jewish *feasts on the themes of 'light', 'life', and 'glory'. Both the synoptists and John record Jesus' visit to Jerusalem at the feast of *Passover at the end of his life but they disagree about the precise date. They also disagree about the date of the Cleansing of the Temple which John (2: 13–22) places early in the ministry, the synoptists at the end (Mark 11: 15–18). When John records the Lord's teaching, it lacks the epigrams and the *parables of the synoptics and instead consists of long discourses, often in the form of a controversy with the Jews and sometimes in the form of allegories (e.g. that of the Good Shepherd; 10: 11–18); its function is to expound the nature of Jesus' own person rather than, as often in the synoptics, to give ethical teaching for the disciples. There is no account in John of the *Transfiguration.

Because of the great differences between the synoptics and John, it has sometimes been said that, whereas they give us history, he gives us theology. (Clement of *Alexandria towards the end of the 2nd cent. CE remarked that after the first three gospels had been written John composed a 'spiritual gospel'); but this exaggerates the differences, for each of the synoptists is giving his theological interpretation of the history, and John's theology is certainly rooted in history: it is claimed that there are thirteen accurate pieces of topographical information in John, and that this is evidence of his reliability about facts (e.g. the pool of *Siloam, 9: 7).

The evangelist states his aim in writing (20: 31). He hopes his readers will be confirmed in their *faith that Jesus, the *Messiah, is the Word of God incarnate. He is not relating 'bare facts' but offers an interpretation. He can assume that his readers know the synoptic narratives. His work was to pull together the isolated fragments of the synoptics into a coherent system, which proved enormously influential. Without the fourth gospel it is difficult to see how the Church's thinking about Jesus could have maintained its belief in his distinctiveness. John supplied the necessary basis for later Christology. Although some scholars have been attracted to the view that John was influenced by *Gnosticism, seeing Jesus as a divine being from *heaven striding across the earth to save the elect, this ignores the gospel's emphasis on the humanity of Jesus (4: 6), which is very different from Gnosticism.

Authorship and date of the gospel of John. From about 170 CE, that is, after Irenaeus, it came to be generally accepted that the fourth gospel was written by John the Apostle; before that date there is some uncertainty, for there is a confusion about whether a certain John who lived in Ephesus was the apostle (as Irenaeus thought) or 'the elder' who wrote the epistles of John. There is, however, good evidence that the gospel was circulating in the Christian world well before the time of Irenaeus. A papyrus fragment of John 18, dated about 140 CE, now at Manchester, is the oldest MS of the NT in existence; it was written in *Egypt and, allowing time for the book to travel in the ancient world, the gospel must have been written by 120 CE. Another papyrus MS of the gospel, now in Geneva, is almost complete, and has been dated by several experts as about 150 CE; but so early a date is not widely accepted and can hardly be used as evidence for pushing the gospel's composition back to 110, still less to 100 CE.

That John the Apostle was the author is still held by a few scholars but rejected by the majority. The former emphasize the gospel's familiarity with the geography of southern Palestine, with Jewish rites and ideas, and the similarity of thought and language, e.g. about '*truth' and '*faith' with certain of the *Dead Sea scrolls from *Qumran. The 'disciple whom Jesus loved' would seem to claim authorship (21: 20, 24)—at least of that chapter (sometimes regarded as an appendix to the preceding twenty chapters)—and this disciple, who was at the *Last Supper, and at the foot of the cross, and at the tomb on *Easter Day, could be John. It has also been held that the 'beloved disciple' was *Lazarus, on the basis of John 11: 5. It probably has to be said in the end that the identity of the author is simply unknown. The difficulty about accepting the view that John the Apostle wrote it is that it clearly comes from the last years of the 1st cent., when the conflict between Church and *Synagogue was fierce (cf. John 9: 22; 12: 42). The style, language, structure, and theology are not such as would come from John the Apostle. They are not the reminiscences of a very old man of what he himself saw and heard, but a piece of creative theological interpretation by the evangelist at the end of a process of tradition. The name of the a postle John became attached to the gospel in order to give it the necessary authority, possibly because John was in fact the originator half a century before of the traditions which eventually took shape in the fourth gospel. And in this case the claim in 21: 24 is, in an indirect way, true; the Beloved Disciple is both John and the evangelist.

Jonah, book of One of the Minor Prophets in the OT; the 'hero' of the book lived in the 8th cent. BCE (2 Kgs. 14: 25) but the book is an extended *parable probably written in the 6th cent. BCE. It describes how Jonah was sent to warn *Nineveh, capital of *Assyria, of its great evil and of the consequences. Jonah embarked on a ship to avoid this uncongenial mission, but in a storm he was suspected by the crew of causing divine anger and was cast overboard. The sea at once calmed; but Jonah was saved by a great fish (Jonah 2), after which Jonah accomplished his mission with success: but the prophet deplored the success; he thought the people of Nineveh deserved to suffer. For this curmudgeonly attitude, he was soundly rebuked. As a parable the book therefore seems to imply God's universal mercy on all mankind, *Gentiles equally with Jews, and is a tract repudiating the narrow nationalism associated with *Ezra after the Exile.

In Matt. 12: 38–42 the Pharisees who demanded a sign from Jesus are told that they are to have only the sign of Jonah—the necessity of repentance. However an editorial addition, which looks like a piece of early Christian reflection, regards Jonah as a *type in the OT of Jesus' resurrection (12: 40).

Jonathan Son of King *Saul, and devoted friend of *David, who was to succeed Saul and whom Jonathan protected from his father's jealous wrath (1 Sam. 19: 1–7; 20: 30). Indeed Saul was even jealous of Jonathan's military success against the *Philistines (1 Sam. 14: 44). Jonathan and Saul were both killed and David is credited with a poignant lamentation for Jonathan (2 Sam. 1: 25–7).

Joppa A town with a natural harbour north of Israel's border with *Egypt occupied by the tribe *Dan (Josh. 19: 46). It was useful for importing materials during the building of the Temple (2 Chr. 2: 16), but the town repeatedly changed hands; Simon Maccabaeus (142–134 BCE; 1 Macc. 13: 11) captured it, but it was later taken by the Romans, though added to the territory of *Herod the Great. It remained Jewish in character, and *Peter stayed in the city (Acts 9: 36–10: 23); but *Vespasian destroyed it in 68 CE during the Jewish War.

Jordan The river which runs from north of the Sea of *Galilee to the *Dead Sea is central to the history of Israel, ancient and modern. Towns and villages near it are mentioned in the NT—*Capernaum, *Bethsaida, and Gergasa. It flows below sea level and is about 320 km. (200 miles) long. The Israelites crossed the Jordan (Josh. 4: 10) when the flow of the river temporarily ceased, as has happened at other times as a result of earthquakes or landslides, and in modern times in 1927. It was the chosen scene for the *ministry of *John the Baptist; because it was the means by which the Israelites had entered the *Promised Land, it acquired symbolic importance for apocalyptic enthusiasts in the 1st cent. CE.

Joseph (1) The favourite son of the patriarch *Jacob (Gen. 37–47) who had dreams of future domination. He became the victim of his brothers' jealousy, and was sold by them to merchants on their way to *Egypt. There, however, Joseph's administrative ability brought him great power. When the brothers came to buy corn during a *famine, a dramatic reconciliation followed, and the whole family removed to Egypt. In the eyes of the narrator the story probably typifies the constant internecine conflict amongst the twelve tribes but also the hope of an ultimate reconciliation. There may also be a dim recollection of a historical fact that parties of *Hebrews of different tribes had temporarily settled in Egypt at different times. (2) The husband of *Mary the mother of Jesus and a descendant of *David (Matt. 1: 20) who like the preceding Joseph also had dreams and in the gospel is commended (1: 19) for his refusal to divorce his betrothed Mary when she was found to be pregnant. Joseph disappears from the gospel narratives apart from a sneering reference in John 6: 42—'Is not this Jesus, the Son of Joseph?' (3) Joseph of *Arimathaea. (4) Joseph *Barsabbas.

Josephus, Flavius Jewish historian who lived from about 37 to 100 CE; during the rebellion against *Rome he was commander of the forces in *Galilee but was taken prisoner. He became the interpreter for *Vespasian during the siege of Jerusalem in 70 CE and afterwards resided in Rome, well-regarded by the emperors, Vespasian and *Titus. Josephus wrote the history of the Jewish War, laying blame for the revolt on a minority of fanatics. He tried to explain the Jewish faith to *Gentiles in his *Antiquities of the Jewish People* but a reference in it to Jesus was expanded and made more favourable at an early date by a Christian scribe. The four extant works of Josephus are fundamental for our knowledge of Jewish life and history for the last two centuries BCE and the first CE.

Joses (1) A brother of Jesus (Mark 6: 3); and also AV of Matt. 13: 55, where modern versions read 'Joseph'. (2) The first name of *Barnabas (Acts 4: 36, AV), where modern versions read 'Joseph'.

Joshua An *Ephraimite (Num. 13: 16) designated as Moses' successor (Num. 27: 18–23), who is therefore the leader of the invading tribes as they enter *Canaan, and commissioned by God himself (Deut. 31: 14–23). Joshua is written up in the narratives as one who through his obedience to God was highly successful in his undertakings. 'Joshua' becomes 'Jesus' in Greek—hence the misleading Heb. 4: 8 in AV.

Joshua, book of The sixth book in the OT. It describes the occupation of the *Promised Land under Joshua, Moses' successor. Although in Jewish tradition the book belongs to the Former Prophets, the historical and theological links are with *Deuteronomy. It is a collection of diverse material probably edited by the Deuteronomic school during the Exile in *Babylon and cannot be considered a reliable historical narrative. For example, the reported capture of Jericho (Josh. 6) has no support from the archaeological investigations of Kathleen Kenyon in 1952–8. The views of this school are impressed upon the narrative: the land was a gift from Yahweh and, so long as the people were faithful to the *covenant, they would retain it. But there must be no religious syncretism with Canaanite cults; the existing inhabitants were to be annihilated (Josh. 11: 20), though in fact they were not totally destroyed (Judg. 3: 1). Much of the book details the arrangement for the distribution of land amongst the tribes. It emphasizes the importance of the tribal structure and of the family within the tribe.

Josiah Became king of Judah at the age of 8 in 639 BCE at a time when the power of *Assyria was contracting, which enabled him both to expand the territory and to institute important religious reforms, said to be prompted by the discovery of a law book in the Temple (2 Kgs. 23) usually thought to be part of the book of Deuteronomy in some form. Religious practice was centralized in Jerusalem and this was inaugurated with a great *Passover celebration.

Josiah was killed at the battle of *Megiddo in 609 BCE when he intervened to prevent the Egyptians from assisting the weakened *Assyrians at a time when they were being threatened by the rising power of Babylonia (2 Kgs. 23: 29).

jot Used in AV at Matt. 5: 18 to translate the Greek iota, the smallest letter of the alphabet, which was itself a translation of *yod*, the smallest Hebrew letter. Jesus' saying asserts that the law was given to Israel by God and retains its validity for him.

Jotham (1) The youngest son of Jerubbaal (*Gideon) who protested when the men of *Shechem chose *Abimelech to be their ruler (Judg. 9: 7–21). (2) Son of *Uzziah of *Judah, king 750–734 BCE and therefore a contemporary of *Hosea, *Isaiah, and *Micah. He failed to remove the high places where there were *sacrifices to *Baal (2 Kgs. 15: 35) but built the upper gate of the Temple—an ambiguous record which gets only moderate acclaim (2 Kgs. 15: 34).

joy In the Bible, more than an emotion. It combines a sense of happiness with a state of blessedness. But in the OT it is marked by public excitement at times of festival (Deut. 12: 6 f.) and by relief when an individual had a grievance which he could bring to the Temple for settlement (Ps. 43: 4). In the NT the note of joy is prominent in Luke's gospel (2: 10; 19: 37) and in the Acts (13: 52), where it is a characteristic *gift of the *spirit (Acts 8: 39; Gal. 5: 22).

jubilee The celebration of every *seventh year, and especially every fiftieth year, which was after 7 × 7. Jewish slaves were to be released and mortgaged land returned (Lev. 25: 8–13, cf. Luke 4: 18–19). This was more of an ideal than actual practice, though it may have at least discouraged the slave business. The word 'jubilee' is from the Hebrew for a 'trumpet' which was blown to inaugurate the year of jubilee.

Jubilees, book of A Jewish work containing a solar calendar differing from the official; composed in the late 2nd cent. BCE and used by the *Qumran community. So called because its chronology is worked out in periods of 49 years.

Judaea Greek and Latin form of *Judah, and in the Maccabean era an independent state (1 Macc. 5: 45); it sometimes refers to the whole of Palestine, e.g. the territory of *Herod the Great (37–4 BCE). In NT times, Herod's son *Archelaus ruled as tetrarch over Judaea (4 BCE–6 CE), which did not include *Galilee and *Perea; but after Archelaus Judaea formed part of the Roman province of *Syria, ruled by a Roman prefect or *procurator residing in *Antioch. Sometimes the NT uses 'Judaea' in the broader political sense (e.g. Matt. 19: 1) to include Perea, east of the *Jordan, bordering on Samaria, and not part of 'Judaea' administered by the Romans. (Perea, with Galilee and Samaria, was ruled by Herod *Antipas.)

Judah Son of *Jacob and *Leah (Gen. 29: 35) who gave his name to the tribe which settled in the south of Palestine, and thence to the country itself. *David was anointed king of Judah (2 Sam. 2: 4), but after capturing Jerusalem he became king also of the northern area (Israel) and thus the *twelve tribes were united until the death of *Solomon. After the division or disruption (922 BCE) there was civil war between Judah and Israel but Judah's geographical position proved a bonus for its survival, and when Israel and other states were swallowed up by the *Assyrians Judah survived, even when *Sennacherib besieged Jerusalem in 701 BCE. The religious history of the country was an oscillation between collusion with the non-Israelite population by legalizing Canaanite cults, followed by Yahwist reforms as in the reigns of *Hezekiah (727–698 BCE; 2 Kgs. 18: 2) and *Josiah (639–609 BCE; 2 Kgs. 22: 3–20). Josiah was killed at *Megiddo when attempting to repel an Egyptian army which was hastening to the help of the ailing Assyrians being assaulted by the rising *Babylonians (2 Kgs. 23: 29). After the Babylonians defeated the Egyptians at *Carchemish in 605 BCE, Judah came under their domination, but

King *Jehoiakim rebelled and Jerusalem fell to the Babylonians in March 597 BCE. The deportation of leading citizens to Babylon followed. More seriously, a further rebellion by *Zedekiah led to *Nebuchadnezzar's siege of Jerusalem for two years and the destruction of the city in 586 BCE, followed by the Exile (2 Kgs. 25: 11) of most of the population, although the prophet *Jeremiah, who had favoured surrender to the Babylonians, was captured by a group of nationalists and removed with them to Egypt. As an independent kingdom, Judah was finished, until the *Maccabees threw off the Seleucid yoke. There was then an independent *Hasmonean Kingdom of Judah from 164 BCE until Pompey's victories in 63 BCE.

Judaism The term used of the religion and culture of the Jewish people from the time of the Return from Exile to the present day, but especially from the Maccabean age. 'Early Judaism' is often dated from 167 BCE when the Temple was desecrated (2 Macc. 6: 4). In the three centuries BCE/CE Judaism was pluralist, embracing nationalists like the *Hasmoneans, *apocalyptists who wrote the book of Daniel, quietists like the *Essenes, as well as the *Pharisees and *Sadducees. There were also Herodians (Mark 3: 6) who entered into friendship with the royal family, and *Zealots who 'came out' during the Jewish Revolt of 66–73 CE and fought the Romans, and the heterodox *Samaritans. There were many disputes, even within the Pharisee group, about how the *Law was to be applied, but later rabbinic literature tended to generalize disputes into the rival schools of *Hillel and *Shammai. Jesus had many disputes with Pharisees though he was also on visiting terms with some of them (Luke 7: 36; 14: 1). But the bitterness between Church and *Synagogue at the time of the composition of the gospels may be responsible for the harsh condemnations of Matt. 23 etc. and the portrayal of Jews as murderers whose children will suffer the consequences of their sins (Matt. 27: 25).

At *Jamnia the foundations of later Judaism were laid. Amid all the diversity of Judaism, the common elements were the rite of *circumcision, observance of the *Sabbath, veneration of the *Torah, an obligation to the Temple while it remained, and worship of the one God and rejection of all images.

Judaizers Usually understood to refer to the Jewish Christian opponents of Paul's mission to the *Gentiles. Judaizers demanded that converts should submit to the full rigour of the Jewish *Law as a precondition of Christian *baptism (Gal. 2: 14).

It has been thought that Paul also had opponents in *Corinth and *Philippi, who argued for the same policy, but it is equally possible that the opponents were Gentiles attracted to Judaism.

Judas (1) Judas Maccabaeus, a son of *Mattathias, who succeeded his father in fighting the Syrians and after victories purified the Temple (1 Macc. 4: 36–61). He was killed in 160 BCE. (2) One of the *Twelve, son of James (Luke 6: 16), probably the '*Thaddaeus' of Mark 3: 18. (3) A brother of Jesus (Mark 6: 3). (4) Judas Iscariot, one of the Twelve; the name may derive from the Greek *sikarios* = assassin; or it could mean 'man of Karioth', making Judas the only apostle from *Judah. He betrayed Jesus to the authorities. *Schweitzer held that he betrayed the secret of Jesus' *Messiahship, but more probably he betrayed information to enable Jesus to be arrested without publicity. Afterwards Judas suffered remorse and committed suicide (Matt. 27: 5). However, Christian tradition began a process of rehabilitating Judas by showing that even his great sin might be forgiven; so in Acts 1: 18 his death is attributed to an accident.

Jude, epistle of Placed in the NT between the epistles of John and Rev. Although traditionally ascribed to Judas, one of the brothers of Jesus (Mark 6: 3), the epistle is more likely to have come from the next generation, who are exhorted to remember the words of the apostles (Jude 17). The letter has none of the usual greetings to particular individuals and belongs to the group sometimes known as the 'catholic' or 'general' epistles. The heretics who are condemned may have been *Gnostics, who regarded themselves as free from moral constraints (verse 4). Warnings to them are taken from the OT and from 1 Enoch and the *Assumption of Moses (as also 2 Pet. 2: 1–18). Jude was probably written in Palestine by a Christian of Jewish background.

Judgement In the OT God is depicted as a Judge who will reward people according to their deserts—with wrath on the evil and blessings on the righteous (Deut. 10: 18; 32: 41). In the NT, while people's attitude to Jesus is itself a self-judgement (John 3: 18), there are also suggestions of a future judgement (e.g. Matt. 25: 3 ff.; Rom. 2: 15 f.) in which the evil and the righteous will be for ever separated. The criterion of judgement is spelt out in the *parable of the sheep and the goats (Matt. 25: 31 ff.).

Judgement, Day of *Amos (5: 18–20) and *Joel (1: 15) prophesied that there would be a *day of the Lord when Israel, or *Judah, or all the nations, would be punished—unless they repented (Mal. 4: 5–6). At the end of the OT era there is expectation that the righteous and the wicked will get their deserts in the life to come after death (Wisd. 3: 1–9; 2 Esd. 14: 34–5).

A day of judgement is mentioned in the NT. In 'that day' (Luke 17: 30) the *Son of Man will be revealed, and although John suggests that judgement is already taking place among Jesus' hearers (John 9: 39) there is also to be a future judgement (John 5: 28–9). Paul speaks often of the 'day of Jesus Christ' (e.g. Phil. 1: 6) when Christ in concert with faithful Christians would judge the world (1 Cor. 6: 2). Some of his converts, in their enthusiasm, believed that the day of the Lord was already present (2 Thess. 2: 1–12).

Judgement, Last Not a term found in the Bible, but one which came into Christian usage, sometimes as the 'General Judgement' (possibly from Matt. 25: 31–46) to mean the final judgement by God on the whole human race after the *Resurrection of the Dead, in contrast with the Particular Judgement of individuals upon their deaths.

judgement seat A tribunal, or judicial bench. Jesus was brought before *Pilate who was seated on his judgement seat (Matt. 27: 19), at a place called, in Aramaic, *Gabbatha (John 19: 13), probably in Herod's palace on the western hill of the city. A judgement seat has been excavated at *Corinth, and this may have been used by *Gallio (Acts 18: 12).

judges The archetypal judge was Moses, who delegated some judicial responsibilities to local officials (Deut. 16: 18–20; 17: 2–13). Before the monarchy there was no standing army in Israel and charismatic leaders arose who acted with courage and earned respect, and were called judges. The best-known were *Deborah, *Gideon, and *Samson. There were also others, often called 'minor judges' (Judg. 10: 1–5; 12: 8–15). Their relation to the charismatic judges is not clear.

Judges, book of OT book setting out some of Israel's traditions from Joshua to Samuel, a period of about 300 years. The material was collected early in the monarchy, but the final edition was probably made after the Fall of Jerusalem in 587 BCE. The Deuteronomic editor has shaped the material to conform to his theology: after *idolatry came the divine punishment; so there is a cry for divine assistance and a hero emerges who gives the land a time of rest under the judge. The book includes the story of the prophetess *Deborah (Judg. 4) and the Song of Deborah and Barak (Judg. 5) and a war for the plain of *Esdraelon (*c.*1200 BCE).

The first chapter of Judg. goes over ground covered by Josh. 14–15. The middle chapters describe an era of anarchy. The book ends with a cry for the monarchy, such was the chaos when 'every man did what was right in his own eyes' (Judg. 21: 25): the tribe of *Dan engaged in theft and massacre, and *Benjamin was almost exterminated by other tribes.

Judith A Jewish heroine whose legendary feats are told in the apocryphal book bearing her name. She assassinates the Assyrian general Holofernes, but the book contains a mixture of allusions to *Assyrian, *Babylonian, and *Persian eras, and may be a 2nd-cent. BCE comment on Maccabean political ambitions. In the story, Judith returns home after her victory (Judith 16: 21), which could be a hint or warning to the *Maccabeans. There is a powerful painting of Judith holding aloft the head of Holofernes by Gustave Doré (1832–83) in Swansea, S. Wales, but many painters of the Renaissance were attracted to the gruesome scene, including Michelangelo, in the Sistine Chapel.

Julia A Christian woman greeted by Paul (Rom. 16: 15). It was a common name in the 1st cent.

Jülicher, Adolf (1857–1938) German NT scholar and professor at Marburg who wrote a seminal work on Jesus' *parables. He rejected the historical authenticity of explanations ascribed to Jesus (e.g. Mark 4: 13–20) that are in terms of *allegory. Parables were stories making one point only: the explanations are, he judged, the work of the next-generation Church.

Junia(s) A Christian woman convert greeted by Paul (Rom. 16: 7); NJB renders 'Junias', which is masculine—a name not found elsewhere. If 'Junia', with most recent translations, is preferred, she may have been the wife of *Andronicus. Paul was at one time imprisoned with them, and observes that they are prominent among the apostles; here 'apostles' must be used in its wider sense, beyond that of the *Twelve and Paul, to mean a further group with a commission and having an authority. It is significant that a woman is put into that category.

Jupiter The chief god in the Roman pantheon, equivalent to the Greek Zeus, to whom many temples were dedicated in the Roman Empire. At *Lystra, after the healing of a cripple, a crowd acclaimed *Barnabas as Jupiter (Zeus) and Paul as Mercury (Hermes); the apostles deprecated this (Acts 14: 15).

Jupiter was associated with storms and thunder, but was especially venerated as the protector of the Roman people.

justice An attribute of God (Ps. 89: 14) which is to be pursued by his creatures (Mic. 6: 8). It is justice which involves punishment (Hab. 1: 12) but also promotes God's work of *salvation (1 John 1: 9). So God's justice also means his faithfulness to the people of the *covenant (Rom. 1: 17).

justification The establishment by God of a new relationship with mankind. The Protestant *Reformation put this term at the heart of theological dispute centred on the epistles of Paul to the Galatians and to the Romans. When God 'justified' people, did he *make* them righteous (the natural meaning of the Greek verb), as Catholics maintained, or did he *declare* them to be righteous, i.e. impute righteousness to them, by an *act of fiction*?

However, it has now been established that the verb 'to justify' is concerned with the restoration of a relationship rather than making, or pretending to make, a new character. It is not so far from an act of *forgiveness (Rom. 4: 6–8), and it derives from God's *righteousness expressed above all in Christ. Paul's earliest exposition of justification is in Gal. where he was anxious to preserve the unity in the community of both Jewish Christians and *Gentile Christians; he was repudiating the view being foisted on some of his converts that Gentiles should keep the Jewish *Law. In other words, that they should first be *circumcised before they could be *baptized—otherwise the Jewish Christians would not share a common meal with Gentiles. For Paul, however, there could be no way to Christ except by *faith. It is irrelevant whether or not the convert has been circumcised (Gal. 6: 15). Every single person has sinned and falls short of the *glory of God (Rom. 3: 23) and as such both Jews and Gentiles are under the condemnation of God (Rom. 1: 18) and so they would remain unless God had taken action by the sending of Christ.

The importance for the individual is that he is accepted by God by receiving righteousness as a *gift, not by earning it by good works. It is a matter of either/or, in the sense that to accept the latter is to reject the former.

The meaning of justification for the individual is the theme of the epistle to the Romans (1: 16 to 3: 20). If *salvation was by means of the Law, there would have been no need for Christ to come; but he did come, and die, and rise, and that proves that the universal day of the Law is over. By faith, it is possible to 'escape from the old creation by sharing Christ's death'; being justified by faith means becoming one with Christ, a member of his body, being taken out of a group without a future into one which will be saved. The consequence of this change is the fruit of the *Spirit—just as fruit is normally produced from a sound tree, so good deeds proceed naturally from becoming one person in Christ.

Justus (1) The surname given to Joseph Barsabbas (Acts 1: 23). (2) Titius at *Corinth (Acts 18: 7). (3) A Jew called Jesus (Col. 4: 11).

K

Kadesh The site of the Israelites' encampment on the road from *Egypt to Judah (Num. 13: 26). Spies were sent out from Kadesh to explore the country of *Canaan and returned with such an ominous report that the people hesitated. A request to pass through *Edom was rejected (Num. 20: 18).

Kähler, Martin (1835–1912) Professor of Theology at Halle who maintained that the quest for the historical Jesus was futile; it is in the preaching of the Church that Christ is truly met. He is famous for observing that the gospel of Mark is really a passion narrative with an extended introduction consisting of isolated units. The study of these individual units was to be the chief concern of *Form Criticism.

Käsemann, Ernst (b. 1906) German NT scholar and a professor at Tübingen who is best known as a pupil of Rudolf *Bultmann; he came to regard Bultmann's views on the historical worth of the synoptic gospels as too minimal. On the fourth gospel Käsemann, making special reference to ch. 17, held that its *Christology is docetic, in that it denies the genuine humanity of Jesus. As a theologian of the Church, he thus diminished the authority of that gospel. At the same time he appeals to mythological elements in epistles of Paul in support of his opposition to social and political creeds of his time: Paul preached a doctrine of the Cross and Resurrection and Justification, which Paul saw as God asserting and extending his rule over history; on this NT basis Käsemann as a Church theologian participated in the struggle for social justice and world peace as being part of the will of God.

Kedesh Several cities have this name in the OT: (1) The town from which *Deborah summoned *Barak (Judg. 4: 6). (2) A town in *Galilee established for refugees from the private vengeance of members of a family seeking retribution for a supposed injury (Josh. 20: 7). (3) A city in the south of *Judah (Josh. 15: 23).

The name, like that of *Kadesh, means 'holy', and thus it is likely that each of these places had a sanctuary.

Kenites One of the groups in *Canaan before the Israelite occupation (Gen. 15: 19), to which Moses' wife's father belonged (Judg. 1: 16). It has been suggested that the *worship of Yahweh came to Israel from the Kenites. An association of Kenites with Israel is suggested by *Saul's friendly dealings with them (1 Sam 15: 6) and the presents sent to them by David (1 Sam. 30: 29). This relationship continued after the Exile (Neh. 3: 14), when a *Rechabite repaired a gate; the Rechabites (Jer. 35) were Kenites (1 Chron. 2: 55).

kenosis Greek for 'self-emptying' and used by Paul in Phil. 2: 7 for Jesus' renunciation of the state of *glory with the Father in order to share human life and death. The 'kenotic theory' of the *incarnation proposed by several German Lutheran and English theologians (Charles Gore, P. T. Forsyth) was an attempt to combine an orthodox Christology with a recognition that Jesus' humanity shared the limitations of his age: divine omniscience was renounced; Jesus of course had no knowledge of biblical criticism, and for him Ps. 110 was assumed to be written by David (Mark 12: 36) even though it is now thought to have been composed at a much later date.

Kerioth A town in the south of *Judah which was the birthplace of *Judas Iscariot (according to one ancient MS at John 6: 71, NRSV marg.).

kerygma Greek, 'proclamation': and so those parts of the NT identified by some modern scholars as preaching material rather than teaching material or descriptions of deeds. It includes the proclamation of Jesus' redemptive work in the *passion and *resurrection (e.g. Luke 11: 32; 1 Cor. 1: 21) usually with references to the fulfilment of OT prophecy. And there is a summons to repentance.

ketib Hebrew for 'what is written'; it refers to the correct reading of a section of the OT Hebrew text, as preserved with both consonants and vowels, and differentiates it from 'what is said' (*qere*). *See* Yahweh.

key A symbol of authority as in Isa. 22: 22 and as when *Peter was given the keys of the kingdom of *heaven (Matt. 16: 18) as the reward for his stumbling confession of the *Messiahship of Jesus. The privileges would seem to be that of admitting persons into the Kingdom; certainly at *Pentecost (Acts 2: 41) Peter admitted many into the Church in response to *faith. Peter's authority is further explained as that of *binding and loosing, which was a rabbinic term for forbidding and allowing, i.e. interpreting questions of right and wrong; a privilege extended to others, such as Paul.

kid A young he-goat, having tender flesh which made it an appetizing dish for guests and so provided as part of the *feast by his father on the return of the Prodigal Son (Luke 15: 29).

kidneys The best portion of an animal *sacrifice (Exod. 29: 13). They were thought to be the seat of human emotions, as in Jer. 11: 20 where modern translations prefer the metaphorical use of '*heart' ('reins' in AV).

Kidron A valley which runs from the eastern boundary of Jerusalem to the *Dead Sea, across which *David fled when his son *Absalom tried to seize the kingdom from him (2 Sam. 15: 23). Pagan objects of worship were destroyed in the valley by kings loyal to Yahweh—*Asa, *Hezekiah, and *Josiah. Jesus and the eleven disciples crossed the Kidron valley after the *Lord's Supper (John 18: 1).

kindness God's kindness is shown both in his ordering of *creation and the natural processes of rain and sunshine and of *harvests (Acts 14: 17), but also in the abundance of his *grace in Christ Jesus (Eph. 2: 7; Tit. 3: 4–7). It is an attitude that is to be imitated by human beings (Matt. 5: 48; 2 Pet. 1: 7).

king Kingship and deity were closely associated in the ancient world—Egyptian *Pharaohs and *Assyrian emperors, for example; and at a later date Roman *Caesars. In Israel before the monarchy there were no hereditary rulers and when the twelve tribes were settled in *Canaan they were ruled by local elders who invited the '*judges' to gather an army for action when necessary.

*Samuel's anointing of *Saul is told (1 Sam. 8) as an act of *apostasy in one of the sources of the tradition; Samuel acted reluctantly under popular pressure. There was also the threat from the *Philistines (1 Sam. 13: 19–21). On the other hand, a second source (1 Sam. 9: 1–10: 16; 1 Sam. 11) is favourable to the monarchy; Samuel, sought out by Saul to help in the recovery of lost donkeys, anoints Saul, who is then possessed by the 'Spirit of God', like the ecstatic prophets. The monarchy in Israel was never totally despotic, and it was subject to restraints and to the judgement of the *prophets (as when *David, reckoned to be the ideal king, was rebuked by *Nathan). Saul did not found a dynasty. When he was regarded as a failure, the prophet anointed David, and his house lasted for 400 years. From 104 to 37 BCE some of the high priests were called king.

Kings of Israel and Judah have some religious leadership, and the coronation ritual in the Temple was important. But there was never a suggestion in these kingdoms that their king was divine, unless a disputed text (Ps. 45: 6) be understood in that sense. God was the great king above all gods (Ps. 95: 3).

kingdom of God The central theme of Jesus' preaching according to the synoptic gospels, and a major subject of scholarly investigation for more than a century.

The term does not occur in the OT; it is mentioned in the book of Wisdom (10: 10) about 50 BCE and in the *targum* of Isaiah (*c*.100 BCE) and was current, though not common, in the time of Jesus. Matt. usually prefers the term 'Kingdom of Heaven', which is not to be understood as the realm of the departed hereafter; 'heaven' simply reflects Jewish reluctance to utter the divine name.

The primary meaning of 'kingdom' is 'rule' or 'sovereignty' or 'kingship', and Jews could not believe that the existing state of the nation, subject to Roman rule, was compatible with the justice of God and the *covenant with his chosen people. God their king was bound to intervene. The proclamation of Jesus was that this kingship of God *was* indeed to break in on the world. Albert *Schweitzer wrote the classical exposition of the view that for Jesus the Kingdom lay in the near future. In the Beatitudes, the Kingdom is promised as a future reward. In the *Lord's Prayer, the disciples are to pray that the Kingdom will come. Schweitzer maintained that Jesus regarded himself as the *Messiah to come and that he went up to Jerusalem to take upon himself the 'Messianic woes', the period of suffering sometimes expected by Jews before the coming of the Kingdom, and thus force the hand of God. He was willing to die because God would be obliged to vindicate him. Jesus did not tell the public about his role, and imposed a seal of secrecy on the lips of the disciples, though Judas betrayed this secret to the leaders.

An alternative reading of the synoptic evidence is that Jesus preached that the Kingdom was actually present in his own *ministry, as demonstrated by the *exorcisms (Luke 11: 20). The main thrust of the *parables is that of the mysterious arrival of the Kingdom—e.g. the Hidden Treasure, the Costly Pearl. A greater than *Solomon was there! It is these sayings, of the presence of the Kingdom, that make it difficult to accept Schweitzer's theory that Jesus regarded the Kingdom only as God's future intervention. For the Kingdom will not come with apocalyptic signs to be observed but could be discerned already—it is 'among them' or 'within their grasp' in their own society (Luke 17: 20–1)—if only they would recognize it. The presentness of the Kingdom is obscure and expressed in the parables in which the seed is hidden in the ground or so small that it is almost invisible. Jesus did not encourage expectations that there would soon be a dramatic manifestation of God's rule. He did not foretell an eschatological battle or the intervention of a host of *angels, as in the OT book of Daniel, or in the War Scroll at *Qumran, or in the *Assumption of Moses (a Jewish work probably written during the lifetime of Jesus). Evil, in his view, was to be eliminated by suffering and refusing to retaliate (Matt. 5: 38–48).

The Kingdom is still future in the sense that the Rule of God is not yet fully operative in the world. Like the mustard seed, the rule of God will continue to grow and this is how it will be to the end of time (Mark 4: 26–9). So disciples are to act 'as if' they were already members of the Kingdom, 'as if' the new Age was already here. Absolute obedience in our human conditions may not be possible, but Jesus laid down the guidelines which should be our aim.

Modern scholarship has made it plain that in the gospels the Kingdom cannot be identified with the Church, as it has often been since the time of St Augustine, nor can it be envisaged in terms of human virtue or social *righteousness, 'building the Kingdom'. Nevertheless both these interpretations have relevance: the rule of God implies a realm in which rule can be exercised, and the Church is the society which aims to keep alive the incentive and the attraction of the Kingdom. And although the kingdom of God is not to be equated with a human Utopia, there *are* important ethical and social consequences of embracing or entering the Kingdom, the coming of which is to be sought (Matt. 6: 10). Social hierarchies and class discriminations are irrelevant (Matt. 22: 9–10) and evidently Jesus himself lived out these principles (Luke 7: 33–4). The Rich Young Ruler was asked to give away everything; there must be unquestioning trust in

God and selfless love of others; *Peter was told to forgive seventy times seven; the *Samaritan of the parable did help a wounded Jew.

Kings, books of The narrative of four centuries of Israelite history. In the Hebrew Bible the books are included in the Prophets section. In the *LXX they are grouped with 1 and 2 Samuel and are then together called the 1st, 2nd, 3rd, and 4th Books of the Kingdoms (followed by the *Vulgate, and some Roman Catholic translations). The whole work is essentially a religious compilation from an editor of the *Deuteronomist school who probably inspired the reforms of *Josiah in 621 BCE. A supplement was added to bring the story down to the Exile (586 BCE) and to mention *Jehoiachin's fate (2 Kgs. 25: 27–30). The complete work might have been published around 550 BCE. It is an attempt to explain the harsh fortunes of the people as God's punishment for the sins of *apostasy and injustice. The editor has used a number of historical records for the work, which he names—e.g. the chronicles (or annals) of the kings of Israel (1 Kings 14: 19), and the whole forms a clear structure, though slightly more detail is included for the kings of *Judah than for those of the north. The theology assesses every king by the criterion of his faithfulness to the God of Israel. So the northern kings are all condemned; the southern kings at least include two, *Hezekiah and Josiah, who are praised for their religious reforms; they got rid of local shrines and they centralized worship at the Temple in Jerusalem. The kings of the north all followed the way of *Jeroboam I with his *golden calves at *Bethel and *Dan. But the fact that Jeroboam was elected king by the people of the north who rejected Rehoboam is testimony to a kind of democratic principle in the community.

The history begins with the death of *David and the coronation of *Solomon (1 Kgs. 1: 1–2: 46) and continues with the stories of the divided kingdom; the story of Israel is taken to its destruction (2 Kgs. 17: 1–41) by *Assyria, which is explained by the Deuteronomist editor, in accordance with his principles (Deut. 12), as a just punishment from God. The histories of Israel and Judah are related in tandem up to this point (722 BCE), after which the narrative continues with Judah's fortunes up to the Babylonian Exile (2 Kgs. 24: 8–25: 21) and the appointment of *Gedaliah as governor over the surviving remnant in Palestine (586 BCE).

Included in the history are the somewhat different stories of the prophets *Elijah and *Elisha (1 Kgs. 17: 1 to 2 Kgs. 9 and 13: 14–21). *Elijah opposes the kings of Israel for an autocratic style of life alien to the traditions of Israel. *Elisha is said to perform miracles which gave great practical assistance to two kings, but he also cured *Naaman, the commander of the army of Aram. These prophetic narratives may have originally been a document of the 8th cent. BCE from the north which the editor of the book incorporated.

kinsman Family relationships were important to the *Hebrews. Members of the same tribe felt their kinship and derived themselves from a common ancestor. Wives were taken from among their own kindred (Gen. 24: 38 ff.). There were obligations on kinsmen; for example, the brother of a man dying without issue was to marry the widow and raise up a son for him (Deut. 25: 5 ff.; Matt. 22: 25 ff.) and a kinsman was bound to redeem the property of a bankrupt relative (Lev. 25: 25 ff.).

Kiriath-jearim An important city of the *Gibeonites where the *Ark of the Covenant was kept for twenty years after its return from *Philistia (1 Sam. 7: 1; 2 Sam. 6: 2) until *David removed it to Jerusalem.

Kishon A river which flows from northern *Samaria into the Mediterranean. Normally sluggish, it overflows during the rains and creates a surrounding area of swamp, which caused disaster to Sisera's chariots (Judg. 5: 21), as celebrated in Ps. 83: 9. It was also the scene of *Elijah's massacre of the prophets of *Baal after the contest on Mount *Carmel (1 Kgs. 18: 40).

kiss It was, and remains, a common greeting in the east, and the holy kiss (Rom. 16: 16, 1 Peter 5: 14) was practised in the Christian fellowships. *Judas gave Jesus a kiss as a signal to the soldiers to arrest Jesus (Mark 14: 45). A kiss was the conventional greeting of a *disciple to a *rabbi.

Kittel, Gerhard (1888–1948) German biblical scholar; professor of NT at Tübingen. His work on the NT led him to strong views about the supersession of the Jews by the Christians and in the Third Reich he was sympathetic to the Nazis. Kittel's greatest achievement was as first editor of the *Theological Dictionary of the NT*.

Kittim A son of Javan (Gen. 10: 4), whose supposed descendants settled in *Cyprus and became sea traders (Num. 24: 24). In the *Dead Sea scrolls the Chaldeans (Kittim) are named as the last of the oppressing *Gentile powers against whom the faithful 'Sons of Light' wage war. The Kittim of the scrolls are probably to be identified as the Romans.

knees, kneel Knees are said to be particularly affected by emotions—fear (Dan. 5: 6) or weakness (Is. 35: 3; Ps. 109: 24). Kneeling was an attitude of *worship (1 Kgs. 19: 18) or reverence (Matt. 17: 14). Paul and the elders of *Ephesus knelt to pray (Acts 20: 36).

knowledge In the Bible knowledge is not merely intellectual apprehension. It includes the emotions and personal relationships. Israel has a knowledge of God denied to other nations (Jer. 10: 25; Isa. 37: 20), and Moses is said even to know God face to face (Deut. 34: 10). Among human beings sexual relations are spoken of as 'knowing' each other (Gen. 4: 1; Luke 1: 34, AV, NJB) and *Hosea uses this language of intimacy to describe God's feelings for Israel (Hos. 4: 1–2).

In the NT knowledge of God is the knowledge of his work fulfilled in Jesus (1 Cor. 2: 2; Phil. 3: 10). There are indeed certain facts which have to be known (2 Tim. 2: 25) as well as the OT scriptures. The knowledge of God's mighty acts in *creation and *redemption is more than a theoretical knowledge, for it issues in love (1 Cor. 13: 2) and is a *gift of the *Spirit (1 Cor. 12: 8).

The later books of the NT are aware of some of the ideas that later issued into the religious movement of *Gnosticism (*gnosis* = 'knowledge')—a theology, cult, and ethics which regarded *salvation as acquired by an esoterical knowledge, often conveyed by a heavenly revealer. Some scholars believe that the Johannine writings have been penetrated by this type of belief. The writings discovered at *Nag Hammadi provide good evidence of the beliefs of Gnosticism.

Kohathites Claimed to be descended from a son of *Levi (Gen. 46: 11) and therefore to have an important role in *worship. They were responsible for the Holy of Holies of the *Tabernacle (Num. 4: 4–20) and prepared the *shewbread (1 Chr. 9: 32).

Koheleth The Hebrew title of the book Ecclesiastes, meaning 'the Preacher'.

koine The form of Greek (*koine* = common) which was the international language after the death of *Alexander the Great (323 BCE) both in cities of Greece and throughout the Hellenistic world. It is a simplified form of the Attic Greek in which the dialogues of Plato and the classical tragedies were written. It was the natural vehicle for all the writers of the NT and even for the Church in Rome until it was superseded by Latin in the middle of the 2nd cent. CE.

Korah The leader of a revolt against Moses and *Aaron (Num. 16), who complained that the leaders were keeping the priesthood to themselves. The narrative may conceal a historical probability that there was a dispute about priestly prerogatives—who had the right to offer incense before the Lord? According to the narrative, those who objected to the Levitical privileges were swallowed up alive (Dathan and Abiram, Deut. 11: 6), though the sons of Korah were spared when he died (Num. 26: 11).

kurios Greek, 'Sir', 'master', 'owner' (e.g. Matt. 25: 11), but also used in *LXX to translate Hebrew *Yahweh. In Christian speech it was applied to Jesus as 'Lord', being raised above the human level (e.g. Rom. 14: 8–9) and Paul designates the brothers of Jesus as 'the brothers of the Lord' (1 Cor. 9: 5; Gal. 1: 19), which seems to point to the Jesus of the earthly ministry as nevertheless a more than human being. According to *Josephus the Jews refused to call the Roman emperor 'Lord' (*kurios*), because this was the title reserved for God. Paul, however, transfers to Jesus passages in the OT (e.g. Isa. 45: 23) which refer to God: it is to the Lord Jesus Christ that every knee shall now bow and whom every tongue shall confess (Phil. 2: 10 f.).

The liturgical form *Kyrie eleison* ('Lord, have mercy') became a litany in Christian worship from the 4th cent. CE.

L

L A symbol frequently used to designate matter contained only in the gospel according to Luke, such as the *parables (e.g. 10: 29–37) and accounts of the appearances of the Risen Jesus in and around Jerusalem (24). The *infancy narratives (Luke 1–2), though peculiar to this gospel, are not always included as L material on the ground that they seem derived by the evangelist from another source.

Laban Father of *Rachel who required *Jacob to do fourteen years of agricultural work as the price for Jacob to marry her (Gen. 29: 1–30). His rapacious greed received a set-back when Jacob and Rachel fled secretly to *Canaan taking with them not only the **teraphim*—household images that conveyed rights of *inheritance—but also Rachel's elder sister *Leah, whom Laban had tricked Jacob into marrying first.

labour Human labour is regarded in the OT as sharing in the divine purpose (Exod. 20: 8; 34: 21) because God's work was described in Gen. 1 as creative. Jesus 'worked' (John 5: 17) by healing on a *Sabbath, which he justifies on the ground that God continued to maintain the universe even during the prescribed Sabbath day of rest: the obligation to rest can be set aside to enable Jesus to do God's work. Paul worked for his living (Acts 18: 3), and exhorted his converts not to be idle (1 Thess. 4: 11).

On the other hand, work is part of the penalty for *sin (Gen. 3: 17), and much labour described in the *Bible was arduous and unrewarding. For some of it the Hebrews used slaves (Exod. 21: 2–6) and captured *foreigners (Josh. 9: 3–27); and slaves are frequently mentioned in the Roman period by the NT, e.g. the epistle to Philemon. Women had heavy manual work such as fetching water (Gen. 24: 15; John 4: 7) but could also aspire to be entrepreneurs (Acts 16: 14).

Lachish A large city 48 km. (30 miles) SW of Jerusalem, much fought over, which has yielded important discoveries to archaeologists. It was captured by *Joshua (Josh. 10: 31 f.), rebuilt by *Rehoboam according to the Chronicler (2 Chr. 11: 5 ff.), captured and destroyed by the *Assyrians in 701 BCE (2 Kgs. 18: 14; 19: 8), and after having been rebuilt destroyed again by the *Babylonians (588 BCE; Jer. 34: 7). After the Exile some of those returning settled in Lachish (Neh. 11: 30).

Among the archaeological finds at Lachish are the 'Lachish Letters', fragments of correspondence to the military commander at Lachish, written in black ink on potsherds in elegant Hebrew. They can be very accurately dated and they contain information about conditions in this fortified city and Judah during the time when the Babylonians were attacking both it and Jerusalem (Jer. 34: 7).

lacuna Latin, 'hole'. The term used by textual critics and others where there is a gap in a sentence in a MS as a result of damage.

laity In the present-day Church the laity are those members who are not ordained *ministers. In the NT (1 Pet. 2: 9–10; 1 Cor. 12: 4–7) the distinction is that God's *gifts, of equal value, are diverse but necessary to the well-being of the whole body; the Greek *laos* refers to the whole Church or people of God, not to a subsidiary group within the whole.

Lamb of God Used to describe Jesus by *John the Baptist (John 1: 29, 35) and in Rev. as a symbol of Christ slain as a ransom for *sin (Rev. 5: 6–14). Behind the expression in

John 1 may be the influence of the Suffering Servant of Isa. 53, and in some apocalyptic writings the lamb was also a figure destined to overcome *evil and sin which are symbolized by beasts. In the gospel of John (but not in the synoptic gospels) the death of Jesus is placed at the time of the slaughter of the *Passover lambs in the Temple (cf. 1 Cor. 5: 7).

Lamentations In Hebrew, a poetic work in acrostic formation, and in the *LXX attached to the end of the prophecies of *Jeremiah, with whom the book has been traditionally associated. He was in fact a composer of dirges (Jer. 7: 29; 2 Chron. 35: 25). Modern scholarship does not accept the traditional authorship. Regret that hope placed on foreign alliances had come to nothing (Lam. 4: 17) is out of step with Jeremiah's thought, whereas the practices he condemned (Jer. 5: 7–8; 9: 1–5) are not castigated in Lam.

Although there is no historical reference in Lam., it is agreed that the book derives from the time of the fall of Jerusalem (586 BCE) and the desolation of the city (chs. 1, 2, 4, and 5). Ch. 3 describes the plight of a single individual, abandoned by God, who recognizes the justice of his suffering and believes that after genuine repentance he will be helped (Lam. 3: 26–30).

Both Christian and Jewish commentators have regarded the book as prophetic. Christians in the early and the medieval Church held that it referred to Christ's sufferings and as such it was read at services called Tenebrae before Easter. Jewish commentators have seen Lam. as a prophecy of the capture of Jerusalem by the Romans in 70 CE, and the lamentations are the sorrows of God at having inflicted such sufferings on the people of the *covenant.

lamps and lampstands A great variety of lamps and lampstands have been excavated in Palestine, revealing interesting differences as techniques developed. The *oil was obtained from olives and the wicks were usually made from flax, and would stay alight for up to four hours. Lamps were placed either in niches or on shelves, and they could have important symbolic or metaphorical significance. In the OT the word of the Lord is likened to a lamp which illuminates one's way (Ps. 119: 105). In *Solomon's Temple there were ten golden lampstands (1 Kgs. 7: 49), though in the Temple after the Exile there was but a single lampstand (Zech. 4: 1–6, 11–14). In Herod's Temple the lampstand was fashioned with seven branches with a light on each, and this became a characteristic Jewish symbol. In the NT the lamp is used to illustrate the vigilance of a faithful disciple (Luke 12: 35) and his witness in the world (Matt. 5: 14–16). In Rev. 1: 12 the Churches which, in varying degrees, shine in the world are symbolized by lampstands. *See* menorah.

landmarks Boundary stones or double furrows which were recognized by mutual agreement. Their removal was strictly forbidden (Deut. 19: 14) for Israel had a strong respect for hereditary property rights.

languages The OT was written in Hebrew, apart from a few passages in the books of Ezra and Daniel which are in *Aramaic, the language which predominated in Palestine from the 4th cent. BCE. (Aramaic is another Semitic language closely akin to Hebrew.) With a decline in spoken Hebrew, translations into Aramaic were necessary, and when these had developed into fixed forms, they were then written down and known as *targums*.

The NT was written in Greek, which was an international language in the Mediterranean basin until it was gradually superseded by Latin by the 6th cent. CE. It is known as *koine ('common') Greek; a common tongue superseded regional variations, and classical or Attic Greek of the 5th–6th cents. BCE had absorbed words and idioms from the Persian Empire conquered by *Alexander the Great. Many *papyrus documents discovered in *Egypt have thrown light on the meaning of koine Greek words in the NT and the considerable use of the language in 1st-cent. CE Palestine suggests that Jesus himself may have sometimes spoken it.

According to John 19: 20 Pilate put a notice on Jesus' cross in Hebrew, Latin, and Greek.

Latin was spoken by educated people in the western part of the Roman Empire from the 2nd cent. BCE but hardly at all in Palestine; it was the official language in the army and for administration in NT times, but Greek was the main language for communication, and was even used by the Church in Rome for corporate worship until Latin became more widely used by the 3rd cent. CE, and a greater proportion of educated (Latin-speaking) people had joined the Church. The last Western theologian to write in Greek was Hippolytus (d. 236 CE).

The OT was translated into Greek in *Alexandria from the 3rd cent. BCE and so became the scriptures of the early Christians. The NT was translated into Syriac by *Tatian in Rome in the middle of the 2nd cent., and into Latin, first in the Roman province of Africa, in the 3rd cent.

Laodicea A city in the Roman province of *Asia which enjoyed great prosperity until laid waste by an earthquake in 60 CE. Its uncertain water supply reached it through an aqueduct in a lukewarm condition—which is how the author of Revelation (3: 16) castigates the condition of the Church there. Although there is no record of a visit of Paul to Laodicea, he apparently wrote a letter to the Church since it was to exchange their letter with one sent to *Colossae (Col. 4: 16). Such a letter is not extant, though some have argued that it is the epistle to the Ephesians; but the Pauline authorship of both these epistles is widely disputed. There does exist a letter to the Laodiceans dating from the 2nd or 3rd cent. CE. It is ascribed to Paul but the pseudonymous author's purpose in writing the work is unknown. The only MSS of this letter are in Latin.

Last Supper *See* Lord's Supper.

Latin The language originally of the people of Latium, the Latini, who occupied territory with several towns in Italy in the neighbourhood of *Rome. The Romans gradually extended their hold over the whole area and in 340 BCE defeated the Latins at the battle of Mount Vesuvius. The language spoken by these peoples, many of whom moved into the capital, was adopted by the educated, by administrators, and by soldiers, although Greek remained the principal language amongst people generally in the first centuries BCE and CE.

Latinisms A few words in the Greek NT are clearly borrowed from the Latin, such as *denarius* (Mark 6: 37) and *praetorium* (Mark 15: 16).

laughter There is not much in the Bible in praise of laughter as an expression of pleasure and paradox; Ps. 126: 2 is one of the rare instances. However, there are those who laugh at others' misfortune (Job 30: 1), as well as those who laugh their way self-confidently through disasters (Job 5: 22), and the adversaries laugh at Jerusalem in her downfall (Lam. 1: 7). But at least there is a promise of laughter in the *kingdom of God (Luke 6: 21).

laver(s) In *Solomon's Temple there were ten lavers (AV, RV, REB in 1 Kgs. 7: 30), which were 'basins' (NRSV, NJB, REB in Exod. 30: 18) of copper or bronze designed for priestly ablutions, made by order of Hiram of *Tyre.

Law Translation of the Hebrew **torah*, though this word has a meaning much wider than the legal: 'interpretation' might be preferable. It is used in the Hebrew Bible for the *Pentateuch, in which law as a system of commands occupies only a part of the five books.

Law in the narrowest sense was the basis for the administration of justice which was done by local elders at the city gate, though difficult cases were referred to the Temple authorities at Jerusalem. The king exercised judicial functions and his judgement created precedents which, with his entourage of prophets, led to the consolidation of the several codes in the *Pentateuch. The historical narratives record events which contributed to this process: there was *Samuel placing a book of royal privileges in the *sanctuary (1 Sam. 8), followed by *Jeremiah (Jer. 36), and the discovery of the book of the *covenant in the reign of *Josiah (2 Kgs. 23), and Ezra's proclamation (Ezra 7: 10).

The oldest of the codes is known as the *Covenant Code (Exod. 21–3) with rules governing slavery, murder, and theft, together with various humanitarian provisions. The Priestly Code, so called because of its kinship with the source *P of the *Pentateuch, contains some early material but is mainly concerned with religion and ritual, and embodied in it is the Holiness Code of Lev. 17–26.

The Deuteronomic Code (Deut. 12–26) is widely identified with the book found in the Temple in the time of Jeremiah. It takes the form of a speech by Moses to the people of Israel before their entry into the *Promised Land. The observance of the Law is made a condition of the covenant, for Moses is regarded merely as the mouthpiece of God, whose commandments Moses transmits. To break the law was to infringe the will of God.

After the Exile the law was increasingly restated, elaborated, and brought up to date by various groups in *Judaism, as Deut. itself had anticipated (Deut. 18: 15–22). The oral law of the *Pharisees which eventually was collected into the *Mishnah, about 200 CE, was only one form of supplement. The community at *Qumran elaborated a code of purity in separation from Jerusalem and the Temple; Philo interpreted the Law by means of *allegory; Jesus did not 'come to destroy the law' but in the *Sermon on the Mount gave it a radical reinterpretation; Paul did not so much oppose the Jewish Law as such but did reject the view that its observance was the means of *salvation for Christians (Gal. 5: 4). The Law was basic to membership of the Jewish community, and Paul's aim was to detach his Churches from Judaism. He saw that there was no future for the *Gentile Churches if converts from paganism were required to be circumcised and committed to observing the precepts of Judaism before being accepted for *baptism.

lawgiver The supreme lawgiver in Israel was Moses, who played a major role at the time when the nation was born. The *Pentateuchal narrative describes Moses as receiving the Law from God, engraved on tablets. Later, Moses became a judge and arbitrated in disputes until, overwhelmed by the number of cases, he delegated lesser matters to assistants (Exod. 18: 24–6).

lawyers Luke's word for the scribes (Luke 11: 45), who gave themselves to meticulous and dedicated study of the law, written and oral. Many were *Pharisees (Matt. 23: 13), but membership of that group was not essential to their professional legal qualifications. Those of Mark (2: 16) *were* probably Pharisees and had a sincere devotion to the *Torah which led them to condemn those, including Jesus, who apparently fell short of their own standards.

laying on of hands A ritual gesture in the OT; it is the means by which a man could enable a sacrificed animal to be his representative (Lev. 1: 3–4); and the means by which God's authority is transmitted, as by Moses to *Joshua (Num. 27: 18). In the NT Jesus laid his hands on the sick to heal them (Mark 6: 5). In the Church it was a ritual act within the rites of initiation (Acts 8: 14–17) conferring the *gift of the *Holy Spirit, or marking a new stage in the progress of the Gospel (Acts 13: 3; 19: 6).

Lazarus (1) The only character to be given a name in Jesus' *parables (Luke 16: 19–31) where he is the archetypal beggar in contrast with *Dives, a billionnaire, who is given no name—though often referred to as *Dives. (2) The brother of *Mary and *Martha who lived at *Bethany; he was raised from the dead by Jesus although he (Jesus) deliberately delayed his response to the sisters' call for help (John 11). Because Lazarus is said to be 'loved' by Jesus, he has been identified by a few commentators as the 'beloved disciple' who wrote the fourth gospel (John 21: 24), but the view has not attracted much support.

lead A metal found in the *Sinai peninsula and (Ezek. 27: 12) in *Tarshish (a town in the west, Jonah 1: 3) and mentioned in Job (19: 24), where he pleads that his words could be chiselled out of a rock and filled with lead to ensure that they would never perish.

Leah The elder daughter of *Laban, who contrived by a ruse to make her *Jacob's first wife; she bore six sons and a daughter (Gen. 29: 21 ff.) and was buried at *Machpelah (Gen. 49: 31).

leather Tanners used skins of unclean animals and so were not highly esteemed by Jews, though the trade was recognized as important (Acts 9: 43). However, tanned rams' skins (Exod. 25: 5, NRSV) were used for the *tabernacle in the *wilderness. Skins were used for containing *wine (Matt. 9: 17), and the skin and hair of goats were used for making tents. *Tarsus, where Paul was born, was a centre of the industry and Paul supported himself by this trade (Acts 18: 3).

leaven Yeast, added to dough. In the *Bible leaven had two incompatible significances: as fermented dough left over unbaked from a previous batch it was added to cause further fermentation. It could therefore be regarded *either* as a living substance which should not be burnt in *sacrifices on the *altar *or* as a corrupting influence which caused a kind of decay. The latter was the understanding of the *rabbis: leaven symbolized the corruption of human nature. Hence Jesus' condemnation of the leaven ('yeast', NRSV) of *Pharisees and others (Mark 8: 15) and Paul's denunciation in 1 Cor. 5: 6–8 of the Corinthians' arrogance and in Gal. 5: 9 of the wiles of his opponents. On the other hand leaven as life-giving is used in the parables of Matt. 13: 33 and Luke 13: 20–1.

During the feast of Unleavened Bread, which became combined with the *Passover, Jews ate only unleavened bread (Exod. 13: 7–8).

Lebanon A mountain range in *Syria which runs for about 280 km. (180 miles) parallel to the eastern Mediterranean coast, divided by the River Jordan from the Anti-Lebanon range. Great cedar trees grew on the mountain slopes, and their wood was used in the construction of *Solomon's Temple (1 Kgs. 7: 2–3). The mountains were beautiful when the peaks were covered in snow (Jer. 18: 14).

lectionary An orderly compilation for reading scripture. Jewish *synagogues read the scriptures according to a prescribed arrangement; a lectionary designated the passages to be read in public *worship on the sabbaths in a continuous sequence, but for feasts the sequence was interrupted for appropriate readings, e.g. Exod. 12 for *Passover. Christians took over the lectionary principle and it has been suggested that the gospels may have been written with a lectionary use in mind: thus, when the synagogue offered the account of *creation in Genesis, Matthew had a complementary narrative to expound the Christian doctrine of re-creation beginning with the *birth (genesis) of Jesus (Matt. 1–2). Learned attempts to explain the construction of the synoptic gospels as compiled to fit a cycle of public readings through the year have not been judged successful.

lees What is left at the bottom of a jar of *wine after fermentation (Isa. 25: 6, AV), preferred to new wine, hence NRSV translation 'well-aged wines' and REB 'well matured'. *Jeremiah compares *Moab to wine 'on his lees' (REB, AV, NJB in Jer. 48: 11; NRSV, 'wine on its dregs') meaning that Moab is rich and self-satisfied and due to be 'decanted' (48: 12, NRSV).

legend OT scholars use the term 'legend' to designate stories as a literary type without regard for historicity. Since most such stories, especially those in the *Pentateuch, have acquired their form through a long process of oral tradition, they are often non-historical as they come down to us, but they are not imaginative fairy tales. In some cases 'legends' are the accounts of individuals which reflect the migrations of tribes bearing the names of the alleged ancestors. Other legends are intended to explain the origin of a custom or institution (cf. Exod. 12: 26). In NT studies the German *Form Critics used the term for narratives about Jesus which were related in order to provide the communities with information they desired about Jesus and his followers, and although these legends may have had a historical basis, the process of repetition within the Church led to non-historical embroiderings.

legion The Roman army was organized in legions of between 4,000 and 6,000 soldiers; each legion was divided into ten cohorts. When Jesus asked the name of the *Gerasene demoniac whom he had cured, he replied 'Legion', indicating that he believed a large number of *demons possessed him (Mark 5: 9).

leprosy The disease known today as 'leprosy' with its ulceration and paralysis was not the 'leprosy' of the Bible, which was rather a skin complaint (as with *Naaman, 2 Kgs. 5) possibly connected with an emotional disturbance. Lepers in the Bible were tormented with greenish or reddish spots (Lev. 13: 49); the disease was held to be contagious, and the sufferer excluded from society. On recovery, purificatory rites had to be performed by a priest (Lev. 14: 2–20; Mark 1: 44). Jesus healed several lepers and visited Simon, a leper (Mark 14: 3) at *Bethany, which may have been a leper colony.

Levi A son of *Jacob (Gen. 29: 34) whose name was taken by one of the six tribes who claimed to trace their ancestry back to Jacob's wife *Leah. The Levites did not participate in the country's share-out by *Joshua (Josh. 13–19) but were regarded as chosen for the cult (Josh. 18: 7) in various minor capacities (1 Kgs. 8: 4; 1 Chron. 9: 26). They did not enjoy the priestly status of those descended from *Aaron (Ezek. 44: 10–14) through *Zadok (Ezek. 40: 46). This distinction is traced to Moses himself by the post-exilic *P (Num. 18: 2).

leviathan A mythological sea monster who was said to have been defeated by the god *Baal. The myth surfaces in the OT where Yahweh has replaced Baal (Ps. 74: 13–14); and the destruction of Leviathan symbolized for Isaiah (27: 1) the final death of the unredeemed—who did not include Israel (Isa. 26: 2 f.; 27: 12–13).

Levirate law An injunction that if a married man died without children, it was the duty of a brother or other near relative to marry the widow, and the son of the union would be reckoned to be the son of the first husband (Deut. 25: 5–10). The law did not forbid a man to be married twice (Deut. 21: 15–17), but it was possible for a brother or kinsman to relinquish his right to marry a widow by taking off his shoe and giving it to a neighbour (Ruth 4: 7). Levirate law seems to be presupposed in the dialogue of Matthew 22: 23–30 between Jesus and the *Sadducees—religious conservatives, who did not believe the comparatively recent doctrine of *resurrection but did acknowledge the authority of the *Pentateuch. Jesus argues that life after death is of a different order from that of the present, and the Levirate law does not apply to the case cited by the Sadducees. Jesus quotes Exod. 3: 6.

Levites According to Deuteronomy the Levites were priests who offered *sacrifices and passed on the *teachings of the law (Deut. 17: 18; 33: 10). They are scattered round the towns, but in view of the centralization of *worship in Jerusalem Levites were invited to join the Zadokite priests descended from *Aaron through *Zadok, priest to *David (2 Sam. 15: 24), but always in a subordinate role. They were suspected of dabbling in rural Canaanite rites (Ezek. 44: 10–14) and became Temple vergers, musicians, and slaughterers. This was the situation after the Exile and the *P tradition in the *Pentateuch assigns the distinction even to the time of Moses (Num. 18: 2–6). Levites retained their office, supported by tithes, as long as the Temple lasted (until 70 CE). The NT mentions them as accompanying priests to see *John the Baptist (John 1: 19), as a character in the parable of the Good Samaritan (Luke 10: 32), and Paul's companion *Barnabas is said to be a Levite from *Cyprus (Acts 4: 36).

Leviticus, book of The third book in the OT; the name (from the Latin *Vulgate) is due to its principal theme: the staff who served the Temple, descendants of *Levi. It is a post-exilic compilation of about 500 BCE which incorporates the Holiness Code (chs. 17–26), which comes from an earlier date, and legal practices from the monarchical centuries (before the Exile). Lev. is a priestly work and the intention of the compilation was to provide a kind of blueprint for the new age of the Return from Exile. Laws were represented as the legacy of Moses (Lev. 16: 34), and the nation's sufferings were given a theological explanation as being punishment from God for failing to heed the *covenant of *Sinai (Lev. 26: 44–5), as prophets had warned. The important task at the time of the editors was to go on repeating the ancient laws to ensure the Jews' survival in the future; the *Sabbath (Lev. 23: 3) and ceremonial purity (Lev. 11–15) are prescribed as giving distinctive customs to mark off Jews from the rest. *Sacrifices were to be renewed as in the past (Lev. 4: 1–6: 7); these the Christian Church was to regard as superseded by the one sacrifice of Christ (Heb. 7: 11). Although the book Lev. is dominated by ritual prescriptions, these are combined with moral insights (Lev. 19: 17, 34) which express a message of equality that marks out Israelite society.

lewd Used in AV for 'wicked' (e.g. in Acts 17: 5). The reference is to a gang of ruffians, not rapists.

lex talionis Latin, 'law of retaliation'. It is expressed in Exod. 21: 23–5, and the intention is to terminate an ongoing process of acts of revenge once and for all, by ensuring that punishment did not exceed the original injury. Even this limited response is prohibited in Jesus' Sermon on the Mount (Matt. 5: 38 f.).

liberation theology Several Roman Catholic theologians working in impoverished communities in Latin America have emphasized that both in the OT and in the NT injustice and oppression have been condemned in the name of God. The books of Exodus and 1 and 2 Macc. encourage the oppressed to refuse to submit to their sufferings. Jesus was on the side of the poor and against the exploitations by the rich (Luke 16: 19–31); the first Christians liberated themselves from inequalities of wealth (Acts 4: 32). In some countries (e.g. Peru, Brazil, Central America, South Africa) the *Bible has been used as the basis for, and justification of, political action on behalf of those discriminated against.

Liberation theology has not been favourably received by those theologians who detect in it a Marxist analysis of society. They fear that Jesus is interpreted in predetermined categories and made to fit into them, and also that much liberation theology is indifferent to established critical scholarship. But the influence of liberation theology (and movements such as *feminist theology which are akin to it) is enormous. There is steady progress in teaching oppressed groups to understand their situation and to change it—a process called 'conscientization'.

One or two theologians in the West have adapted liberation theology to apply to what they call the 'culture of contempt' which erodes standards of moral and intellectual achievements. To reverse the ever-declining value set on high standards of competence and integrity, they welcome liberation theology as offering a challenge to the ideals of a decent, caring society to look beyond its own self-interest.

Libertines Used in AV for 'freed men' (NRSV, REB, NJB) in Acts 6: 9 for members of a *synagogue who opposed *Stephen. They were possibly Jews who had been taken prisoner, or Jewish slaves living in *Rome; they had purchased their freedom and moved to Jerusalem.

liberty Freedom for nations and individuals. The Jews lost their political liberty from 586 BCE when the Babylonians overthrew the state. Independence was recovered during the *Maccabean period and then lost again until the emergence of the modern state of Israel in 1948 CE. For some, the hope of a Messianic kingdom was a hope for political independence. Individuals were sometimes released from slavery as in the year of *Jubilee (Lev. 25: 8–17) and in NT times (Acts 12: 7–11; Heb. 13: 23) imprisoned Christians sometimes had their liberty restored.

Liberty is an important concept in the NT as a description of one aspect of *salvation in Christ (John 8: 32). 'Enemies' from which Christ liberates believers are listed as *sin (alienation from God), *law (a system of salvation by works), *Satan (the power of darkness, Col. 1: 13), and *death (a power of extinction and corruption). Thus liberated, believers can respond in *joy and give themselves in *love and service to God and neighbour (Gal. 5: 13–14, 22); the new freedom is so complete that Paul can paradoxically describe this state as a new bondage (to *righteousness; Rom. 6: 18–22) leading to abundant life, in the *Spirit (2 Cor. 3: 17).

Libya A country on the coast of N. Africa which participated in several assaults on *Judah in collaboration with its neighbour, *Egypt (2 Chr. 12: 3; 16: 8; Dan. 11: 43; Nahum 3: 9). People from the city of *Cyrene in Libya, where there was a large Jewish colony, were present in Jerusalem at *Pentecost (Acts 2: 10).

lie False statements and accusations are sternly forbidden in the OT (Lev. 6: 2 f.; Jer. 14: 14; Hos. 12: 1), though *David is recorded as lying to *Saul (1 Sam. 20: 6). In the NT *Satan is called the 'father of lies' (John 8: 44) and Christians are forbidden to tell *falsehoods (Col. 3: 9). Paul held that paganism was rooted in falsehood (Rom. 1: 18), as is denial of the Messiahship of Jesus (1 John 2: 22).

life In the OT the normal reference is to present existence on earth, ended by *death, but given by God (Deut. 30: 19). In the NT the word often denotes *eternal life (Rom. 6: 13), especially in the gospel of John (e.g. John 6: 48).

Life, book of A list of names of people who are destined to enter the New Jerusalem (Rev. 21: 27).

light A powerful symbol mentioned at the beginning (Gen. 1: 3) and the end (Rev. 22: 5) of the Bible for goodness and *truth. Hence the *Law is described as 'a light for my path' (Ps. 119: 105); 'the *day of the Lord' was expected to be light (Amos 5: 18), and people were disconcerted when the prophet predicted darkness. Light and darkness are commonly encountered as antitheses in the *Dead Sea scrolls, and this is a feature which they share with the gospel of John; Christ is described as the light of the world (John 8: 12), sent by God who himself is light (1 John 1: 5), and those who are sent by Christ become in turn the light of the world (Matt. 5: 14).

Lightfoot, Joseph Barber (1828–89) English NT scholar. After a long academic life at Cambridge, he became bishop of Durham in 1879. His main work was on the epistles of Paul and the Apostolic Fathers, and he defended the trustworthiness of Acts against the speculations of the *Tübingen school. On the origin of the Christian *ministry he maintained that government of a diocese by single bishops emerged from a more primitive system of a college of *presbyters.

Lightfoot, Robert Henry (1883–1953) English NT scholar. In Oxford he introduced NT scholarship to the works of German *Form Critics and was widely misunderstood as being extremely sceptical. Before his death his work on the gospel of Mark partly foreshadowed in Britain the new *Redaction Criticism.

lightning Awe-inspiring phenomenon regarded by ancient peoples as a manifestation of the power of deities. By the Hebrews it was also associated with God's revelation at *Sinai (Exod. 19: 16) and it was a feature of *Ezekiel's inaugural *vision (Ezek. 1: 13–14). Lightning is mentioned in connection with the *angel who announced the *resurrection (Matt. 28: 3), which is a kind of *theophany; and in *apocalyptic passages (e.g. Matt. 24: 27) the lightning image of the OT reasserts itself.

lilies Symbols of beauty, in S. of S. (2: 1–2; 5: 13), where the flower referred to is probably a blue hyacinth. In one of the rare references to natural beauty in the NT, Jesus speaks of the glory of lilies (Matt. 6: 28–9), which probably denotes a variety of spring flowers of different colours, perhaps of crown anemones.

Lilith A female *demon (Isa. 34: 14) borrowed from *Babylonian speculation. There was a late tradition, mentioned in the early medieval work *Alphabet of Ben Sira*, that she was the first wife of *Adam and far more assertive than the domesticated *Eve.

lime There is plenty of limestone in Palestine, and lime was made by heating this in a kiln (Isa. 33: 12). It could be used to make plaster to cover large stones (Deut. 27: 2) and create a smooth surface. Amos (Amos 2: 1) condemned the *Moabites for burning the bones of the king of *Edom to lime—i.e. his total destruction.

linen Cloth made from flax which grew well in the low, hot areas of the *Jordan valley. Fine woven cloth was a luxury (Judg. 14: 12–13), and was much sought after by the wealthy (Luke 16: 19). Because of its quality, linen was used for the *tabernacle curtains (Exod. 26: 31) and priestly garments (Exod. 28: 6). It was used in the royal tombs of *Egypt, and the custom spread among the Jews of wrapping corpses in linen (John 11: 44), as was the body of Jesus (Matt. 27: 59). A linen shroud exhibiting the outline features of a body, preserved in Turin Cathedral, was for centuries supposed to be that used by *Joseph of Arimathaea for Jesus, but the carbon-14 test exposed it as a medieval forgery. The same form of test, carried out at the University of Chicago, disclosed that linen cloth in which some *Dead Sea scrolls were wrapped yielded a date between 187 BCE and 233 CE.

Linus A Christian who sent greetings to Timothy (2 Tim. 4: 21) and whom Roman tradition regarded as successor of *Peter as bishop of *Rome.

lion Lions were not extinct in Palestine until the 14th cent. CE, and hunting them was a popular sport for the aristocracy (2 Sam. 23: 20) and an essential part of a shepherd's duty (1 Sam. 17: 34–6; Amos 3: 12). The lion answered his human enemies with special ferocity and therefore became a symbol for *Satan (1 Pet. 5: 8). Parts of the lion's physique are ascribed to the apocalyptic beasts in Revelation (teeth, 9: 8; head, 9: 17; mouth, 13: 2; roaring, 10: 3).

Lion of Judah A Messianic title for Christ (Rev. 5: 5) recalling *Jacob's blessing (Gen. 49: 9), and the lion is here a symbol not of ferocity but of pre-eminence and strength. The same title was assumed by the former emperor of Ethiopia on the ground of alleged descent from *Solomon and the queen of Sheba (1 Kgs. 10: 13).

lips Just as the human *eye is sometimes described as though it acted on its own without the direction of its owner (as in Isa. 5: 15; Deut. 7: 16, AV, RV), so the lips can shout for *joy (Ps. 71: 23) or guard knowledge (Prov. 5: 2) or lie (Prov. 12: 19), and *Hosea (14: 2) calls for words of repentance, 'the fruit of our lips' in place of *sacrifices.

literary criticism In biblical studies, this includes the investigation of sources and problems of authorship. More recently, the term is used as in the study of *poetry, drama, and novels, as an attempt to understand the biblical writings as literature. This involves, for example, appreciating the wealth of symbolism, metaphor, paradox, paronomasia, irony, and characterization and plot in the text.

liturgy An act of *worship; from a Greek word meaning an act of public service (e.g. donating a trireme for the Athenian navy). Paul uses the word (Phil 2: 30) for acts of service provided for him by converts. It came to be applied especially of service to God, and specifically of the *Eucharist (e.g. by Justin Martyr, *c.*150 CE). About 215 CE parts of the NT were being used in the liturgy of Hippolytus at *Rome, and Luke–Acts was fundamental in establishing the basis of a Christian calendar.

liver Regarded as the very seat of life itself (Prov. 7: 23), it was the part of a sacrificial animal that was burned (Exod. 29: 13). It was used in *Babylon for divination (Ezek. 21: 21).

lizard Different types of this reptile are mentioned in the OT and the English translations are uncertain. A list of six varieties is given in Lev. 11: 30 as unclean.

loans Money or goods given on condition of future repayment. In the simpler and more closely knit tribal society of Israel there was no system of loans, as in the modern world, whereby a person could launch a new business or buy a farm. Loans were for the relief of need and those who supplied credit were supposed to be acting out of generosity with few legally acceptable economic advantages. Interest could be charged from non-Israelites (Deut. 23: 20) and the existence of impoverished debtors (Jer. 15: 10) argues for wider extension of such charges. The *parable of the talents implies that in the time of Jesus the payment of interest on sums lent was widespread (Matt. 25: 27). Debts were supposed to be cancelled every *seventh (Sabbatical) year (Deut. 15: 1–11) and there were numerous safeguards to protect those who were obliged to borrow. Bailiffs could not enter a house to seize a pledge (Deut. 24: 10–11) or take the clothing of a widow (Deut. 24: 17), and a pledged garment of a poor person was supposed to be returned for the night (Exod. 22: 25–6); but evidently it was not always so (Amos 2: 8).

locust Insect eaten in Palestine (Lev. 11: 21; Mark 1: 6) and considered very nutritious. A plague of locusts could utterly destroy a harvest, and as such they were sent against *Egypt as *punishment for refusing to release the Israelites from slavery (Exod. 10: 12–20), and against the Israelites themselves for disobedience (Deut. 28: 42), and are regarded as heralding the *day of the Lord (Joel 1: 1–20; 2: 1–17). The prophet challenges the nation to repent.

logia Greek noun in the neuter plural used in the *LXX for words coming from God and likewise in the NT (Rom. 3: 2 'oracles', NRSV, REB; 'message', NJB) of the words of God in the OT. *Logia* comes to be used in the early Church of the sayings of Jesus, and about 110 CE *Papias of Hierapolis compiled a work, not now extant, entitled 'Expositions of the *Logia* of the Lord'. A generation of NT scholars considered the possibility that this work might have been the hypothetical document *Q which was taken to be a collection of Jesus' sayings embodied in different ways by Matt. and Luke to supplement the material they took from Mark.

logos Greek noun usually translated 'word' but also 'reason' or 'meaning' and familiar in Greek philosophy from Heraclitus (6th cent. BCE) to the Jewish theologian *Philo of Alexandria (1st cent. CE) for the principle of coherence undergirding the universe. In the *LXX *logos* translates the Hebrew *dabar*, the creative 'word' of God, which is parallel to *sophia* (wisdom) as a mediator who acted for God in relation to his *creation (Wisd. 9: 1–2). In the gospel of John (1: 14) and the Revelation (19: 13) Jesus is called the *Word of God. This is an important development for Christology: it is an assertion that the Word who was God's agent in creation was to be identified with the human figure of Jesus of Nazareth (John 1: 46).

Lois Grandmother of *Timothy (2 Tim. 1: 5). She is described as a woman of *faith, which suggests both that Timothy is a third-generation Christian and that the Church is well established.

Loisy, Alfred (1857–1940) French biblical scholar. He was a Catholic priest who became one of the leaders of the Catholic Modernist movement which attempted to reconcile Catholic theology with scientific thought and biblical criticism. Modernism was condemned by Pope Pius X in 1907 and Loisy was excommunicated. He maintained that the coming of the *kingdom of God was the essence of Jesus' *teaching, but that it was the Church which, in fact, arrived.

long-suffering The quality of being 'slow to anger' (Joel 2: 13) of God; in human beings it is commended as 'patience' (NRSV, REB, NJB, Gal. 5: 22) towards others.

Lord Used in both OT and NT, as in addressing Moses as leader (Num. 32: 25) and masters by slaves (Matt. 18: 25); also by Jesus in reference to the authority of the *Son of Man over the *Sabbath (Mark 2: 28) and by others in speaking to Jesus as a person with authority (Matt. 7: 21).

In the OT 'Lord' is also used to translate Hebrew words for God, and in the NT it is used tentatively of Jesus, as in 'come, our Lord' (1 Cor. 16: 22). The formula 'God has made him both Lord and Christ' (Acts 2: 36) may well be an authentic phrase from the Aramaic-speaking Church of Jerusalem, for there is similar use of 'Lord' and 'the Lord' in the *Dead Sea scrolls, without the pronoun 'my' or 'our'.

The term was a convenient bridge when the Gospel spread to *Gentile lands, for there were many pagan 'lords', and it was later used of the Roman emperor. For Paul 'Lord' and 'Christ' are almost synonymous (1 Cor. 8: 6; 1 Thess. 4: 17; 5: 12).

Lord's day Sunday; the day of *resurrection. Its sole occurrence in the NT is at Rev. 1: 10, when John the Seer was 'in the spirit'. It is used in the *Didache near the end of the 1st cent. when reporting *Eucharistic gatherings. Weekly celebrations of the Eucharist were possibly the normal practice in Pauline congregations (1 Cor. 16: 2) and *Sabbath (Saturday) observance was not encouraged (Gal. 4: 10; Col. 2: 16).

Lord's Prayer Name of the prayer given by Jesus to his *disciples at their request (Matt. 6: 9–13, and in a slightly different and shorter version, Luke 11: 2–4). It is an essentially Jewish prayer, without specific Christian content, and each clause has parallels in Jewish prayers recited in the *synagogues. It is primarily a prayer that God will fulfil his purposes in the world, as they already are achieved in *heaven. The prayer is thus set in an *eschatological context and the petition that we may be given our 'daily' *bread should probably be translated 'our bread for tomorrow'; it is a prayer for today's necessities in order to be ready for the morrow, whenever that may be, of the coming of the Kingdom.

In some later MSS there is a doxology—'For yours is the kingdom . . .' (Matt. 6: 13), an acknowledgement of the privilege of knowing God as Father.

The prayer invites Jesus' disciples to share in the intimacy with the Father which he himself enjoys but at the same time does not diminish the otherness or holiness of God.

Lord's Supper The last meal before his *trials and *crucifixion which Jesus held with his *disciples in the Upper Room; the name derives from Paul's reference to it (1 Cor. 11: 20).

It has been much disputed whether this meal was a *Passover, as apparently assumed by the synoptists, or some sort of solemn fellowship meal (Paul's word, *'koinonia'* in Cor. 10: 16 can be translated 'sharing' or 'fellowship' or 'communion') which was being celebrated before the Passover. The gospel of John puts the *death of Jesus at the time when the Passover lambs were being slaughtered in preparation for the meal (John 19: 14, 31).

Unfortunately there is a problem about the text in the account given by Luke: 22: 19b–20, which records Jesus' instruction to continue the rite, with the cup as the symbol of the new *covenant: the verses are absent in certain Greek MSS. They are therefore omitted by REB; it is thought they were mistakenly introduced by an early copyist who recollected 1 Cor. 11: 25. On the other hand, these words are included by NRSV and NJB, probably on the assumption that it is out of character for those MSS which have the shorter version, since they are normally given to expansion rather than omissions. Luke regards the meal as a Passover on either text. If the longer text is correct, then we have a tradition in the synoptists and Paul (there is no mention of these words in John's account of the Supper) of the meal being a Passover, now given new significance by Jesus to explain the meaning of his crucifixion on the Friday. The facts that the meal was held in Jerusalem, at night, with a group of 'at least ten' as the requirement went, and Luke's mention of a second cup, all give support to the view that the meal was indeed a Passover. In that case, John has made a theological point by transferring the date in order that Jesus could be shown to be the true Passover *lamb by dying at the hour of slaughter in the Temple. One tentative suggestion to reconcile the two dates is that the Last Supper was held on the Tuesday in accordance with the calendar of the sectarian book of *Jubilees, whereas according to the official calendar the Passover meal was held on the Friday evening, shortly after the crucifixion.

The celebration of the Lord's Supper, or *Eucharist ('thanksgiving') was at first attached to a common meal, at which there were some irregularities (1 Cor. 11). Social differences were exaggerated rather than diminished. So Paul argues that insensitivity to other members of the Body was a violation of the new covenant effected through Christ's death. The Lord's Supper must be a real bonding not an occasion of division (1 Cor. 10: 17).

Lot *Abraham's youngest brother's son; the narrative describes how when the city of *Sodom was to be destroyed Lot and his family were rescued by divine intervention (Gen. 19: 1–19), but Lot's wife looked back at the mayhem and was transformed into a pillar of salt (Luke 17: 28 ff.). The *Moabites and the *Ammonites were believed to have originated in incestuous unions between Lot and his two daughters (Gen. 19: 30–8).

lots A common means of making decisions among the Hebrews which continued into NT times, although the procedure of casting lots is never described. It was believed that God indicated his preferences by means of the lot, as when *Saul was invited to be king (1 Sam. 10: 19–26). In the NT the soldiers divided Jesus' coat at the *crucifixion by lot (Mark 15: 24), and *Matthias was chosen by lot to replace *Judas as one of the *Twelve apostles (Acts 1: 26).

love The whole range of meanings of this English word, physical, emotional, intellectual, are found in the Bible. There is the love of God for Israel (Hos. 3: 1) where it denotes his fidelity to his own *covenants. There is also a command that his people shall love God, and this is not the deep affectionate love for a personal God but rather an injunction to loyalty. The word in Hebrew is found in ancient treaties where a subject king is enjoined to 'love' (be loyal to) his overlord. But 'love' is also used in quite a modern sense as when *Jacob served seven years for *Rachel and they seemed but a few days because of the love he had for her (Gen. 29: 20). *Hosea used the images of love, marriage, and faithlessness to describe the relationships of Israel to God (Hos. 6: 4–6). And *Deutero-Isaiah, also to proclaim the love of God for Israel, compares it to the love of a mother for her child (Isa. 49: 15).

In the NT the Greek noun *eros* (sexual love) is not found for the love of God, or love for God, or for love within the Christian community. Jesus brings together from Deut. 6: 5 and Lev. 19: 8 the commands to love God and neighbour but goes farther and emphasizes the duty to love one's enemies (Matt. 5: 43–6). An episode in Luke 7: 36–50 indicates that the woman who anointed the feet of Jesus was so liberated by *forgiveness that she gave out a wealth of love.

In Paul's epistles love is united with *faith and *hope (1 Cor. 13: 13) as a *gift of the *Holy Spirit; it is not a human achievement and not therefore a ground for boasting (1 Cor. 13: 4). The *sacrifice of Christ on the *cross was the supreme sign of the love of God (Rom. 8: 39), and therefore Christian lifestyle was to be in imitation of it (Luke 9: 23; 1 Cor. 11: 1). This is the outworking of faith and involves special care for the weaker members of the community (1 Cor. 8: 11–12). Love is a prominent theme in the Johannine writings and the mutual love of the Father and the Son (John 16: 28) is to be reflected in the *disciples (John 17: 26). At the end of the gospel, Peter reverses his threefold denial of Jesus by a threefold affirmation of his love. The epistles of John maintain the same teaching on the primacy of love ('no new commandment', 1 John 2: 7).

love feast The meal which preceded the *Eucharist, called in Greek an *_agape_, which means literally 'self-giving love'. These meals, mentioned in Jude 12, and possibly 1 Cor. 11: 17 ff., were continued in the early Church (cf. Acts 20: 7 ff.) in the evening of the Lord's Day, and were intended, as the name suggests, to be a time for God-centred community love. However, it is clear that this admirable intention was gradually eroded by the acids of unrestrained paganism and by the 4th cent. CE it was deemed advisable by the Western Church that the meal, in spite of being an occasion for acts of charity, should be disconnected from the Eucharistic commemoration of Christ's living presence. There then grew up the contrasting custom of *fasting before communion.

loving-kindness One translation of the Hebrew *hesed*, an attribute of God, associated with his *mercy, and, also of human beings (Ps. 141: 5), sometimes translated 'great kindness' (Gen. 19: 19, NRSV; 'unfailing care', REB). It is used by Hosea of the care he promises for his wife (Hos. 2: 19). The NT 'grace' is an approximate equivalent.

Lucifer A sarcastic reference (Isa. 14: 12) to the king of *Babylon calls him the Light-Bearer, or the Day Star (NRSV), or Venus, 'fallen from heaven', i.e. from his lofty power, and is translated Lucifer by the *Vulgate and AV. Because the verse is partly quoted by Jesus (Luke 10: 18) with reference to *Satan, Lucifer has since the time of Jerome been identified with Satan.

Lucius The name of two Jewish Christians in the early Church: (1) A prophet and teacher from *Cyrene; (Acts 13: 1). (2) A companion of Paul in *Corinth (Rom. 16: 21).

Luke A physician and companion of Paul, according to information provided in the epistles (Col. 4: 14, Philem. 24 and 2 Tim. 4: 11). Since about 190 CE there has been a tradition, unbroken until modern times, that this person was the author of the third gospel and of the Acts (two volumes of a single work). The traditional argument is supported by an examination of the two books (whose vocabulary and style are similar, and both enjoy the same dedication, to *Theophilus) which confirms that the author was well educated—though nothing can be proved about a knowledge of medicine, since no technical medical terminology existed in the 1st cent. It is also probable that where the author falls into using the first person plural for parts of the Acts ('we' and 'us' in Acts 16: 10–17, 20: 5–15; 21: 1–18; 27: 1–28: 16) he is using his own personal reminiscences, although the use of the first person could be a mere editorial device to add interest to the narrative. On the whole Luke of the epistles seems to fit the data for establishing the authorship of the NT work Luke–Acts.

Luke, gospel of The third gospel, which comes from the same hand as the Acts of the Apostles, according to a long tradition; the internal evidence of their common dedication to *Theophilus and the similarity of their literary style confirms that view. From the time of Irenaeus (*c.*190 CE) the author of the two-volume work has been regarded as Luke, the companion of Paul (Col. 4: 14; Philem. 24), but doubt has been thrown on this by scholars who find it hard to explain how so close a companion of Paul should be so ignorant of his fundamental beliefs (e.g. about *justification). And how odd that he gives an account in Acts of Paul's visits to Jerusalem which does not conform with Paul's own travelogue in the epistle to the Galatians. Nevertheless, objections to the case against Luke can be met; it has to be borne in mind that Paul wrote in the white heat of controversy and naturally made out the best case for his independence and apostolic status, equal to that of the *Twelve in Jerusalem, whereas Luke, writing a generation later, about 85 CE, was anxious to paint a picture of the calm and unity of the early Church.

In the prologue of the gospel Luke warns that he was not an *eyewitness of the events described, and he relies on various existing sources. One of them is undoubtedly the gospel of Mark, which Luke freely takes over. A theory formerly held by several English scholars was that Luke composed the gospel in two stages: first he combined the source *Q with material to which he had access (*'L'); this resulted in 'proto-Luke'. Later the evangelist came into possession of the gospel of Mark, and conflated the two documents, adding the *infancy narratives (chs. 1 and 2) to make the existing third gospel. However, the proto-Luke hypothesis has not won favour, and it is more usually supposed that Luke combined Mark with Q, though those who have doubts about the Q hypothesis suggest that Luke used the gospel of Matthew to supplement Mark.

Luke is a *Gentile, possibly a native of *Antioch, and that his intended readers were presumably Gentiles is suggested by the dedication (to *Theophilus) and his omission of matters of particular Jewish interest which Mark (e.g. 7: 1–23) had included. The second volume is the story of the progress of the Christian message from Jerusalem to the centre of the Gentile world (Rome), of which, in the gospel, he can only give hints—such as the narratives which show Jesus as breaking with the traditional Jewish hatred of the *Samaritans and as having numerous contacts with non-Jews; all people are to share in *salvation, not merely the descendants of *Abraham (Luke 4: 25–7). There is a discernible tendency in Luke as compared with Matt. and Mark to release Christianity from any impression that it is merely Jewish: even the census of 2: 1 is held to be (improbably) applicable to the whole Roman world, just as in 10: 1 Jesus sends out *seventy disciples, corresponding in number to the nations listed in Gen. 10: 1–32.

However, the whole story is firmly anchored not only in the mainstream of Roman history, by relating Jesus' *birth to a decree of *Caesar about a universal census, but also to the revelation given to Israel. Luke frequently (though not with the obsessiveness of Matt.) intends his readers to understand that all that he records has taken place in accordance with a definite divine plan which fulfils the redemptive history of the OT. Many traditional titles of the *Messiah are applied to Jesus, and he is called 'Saviour' (2: 11) and a suffering Messiah (24: 26). Luke portrays the Church as the New Israel—hence his emphasis on the *Twelve 'Apostles' as successors of the Twelve OT *patriarchs who gave their names to the Twelve tribes of Israel (Luke 22: 30). The Church is the heir to the promises to Israel. Jesus' connection with Jerusalem from an early age (2: 22, 41) is therefore stressed, and the Church is shown to have originated in Jerusalem (Acts 2: 1). Luke answers the question surely asked in the Church—'What about the Jews and their literature now?' It was thus important to establish that the crucial events concerning Jesus took place in or around Jerusalem, the place of revelation and eschatological expectation. The *eschatology of Luke has received a shift of emphasis compared with Matt. and Mark: the sense of the *parousia* and all the events associated with the *End is much reduced, and instead Luke invites his readers to share in the joy of living in the Spirit-filled *community. Luke wants believers to be ready for the End to come suddenly at some incalculable time (12: 39–40) but not necessarily soon. The urgent summons to *repentance is replaced by the exhortation to be imitators 'daily' (Luke 9: 23) of Jesus.

There are other distinctive themes in the third gospel. It presents Jesus as faithful to Jewish customs, and not the sort of innovator who might justify Roman suspicion of Christianity. Jesus is declared innocent by *Pilate (23: 4, 13, 22) and Paul by *Festus (Acts 25: 25). More than the other gospels, Luke stresses Jesus' concern for the disadvantaged and the despised *Samaritans. There is a sympathetic regard for women, although Luke portrays them in traditional roles of *prayer and almsgiving and in supporting the missionary labours of men (8: 3). But Luke's message of *hope for the oppressed certainly does apply to women, who in every age have been the poor in the population.

lust In the AV this word is used for any strong desire (e.g. Gal. 5: 17), but also in the more limited sense of sexual passion (e.g. 1 John 2: 16), where modern English translations prefer 'desire' or 'bodily desire'.

LXX The symbol employed for the *Septuagint, a Greek translation of the OT, on account of the legend that it was the work of *seventy (or seventy-two) scholars in as many days, in *Alexandria. More probably, it was done less rapidly at some time between 285 and 150 BCE, and was frequently revised. There are numerous differences between the LXX text and the Hebrew *Masoretic text. The LXX, containing the books of the *Apocrypha, became the scriptures of the early Christians.

Lycaonia An area in Asia Minor north of Pamphylia and Cilicia which contained towns (*Derbe, *Lystra, *Iconium) visited by Paul (Acts 13: 51–14: 21 and 16: 1–3). It was incorporated into the Roman province of *Galatia in 25 BCE, but under the emperor *Trajan (98–117 CE) was transferred to the province of *Cappadocia.

Lycia At the port of *Patara on the Lycian coast, Paul and his companions embarked on a different ship *en route* for Jerusalem (Acts 21: 2), and also at *Myra in Lycia *en route* to Rome (Acts 27: 5). Lycia had become a Roman province in 43 CE, but in 74 CE was attached to Pamphylia.

Lydda An important commercial centre, 18 km. (11 miles) SE of *Joppa. Some of the returning exiles settled at Lydda (Lod; Ezra 2: 33). After being outside *Judaea, it was returned to *Jonathan and the Jews by King Demetrius in 145 BCE (1 Macc. 11: 34) and was visited by Peter (Acts 9: 32–5).

Lydia Paul's first convert in Europe; she was a *Gentile attracted to *Judaism; possibly a freedwoman and a widow; she sold purple-dyed cloth, and was sufficiently well-off to give Paul and *Silas *hospitality at *Philippi (Acts 16: 15, 40).

Lysanias The reference to Lysanias in Luke 3: 1 as the tetrarch of *Abilene NW of *Damascus may be a mistake, since the only Lysanias certainly known died in 36 BCE. There was, however, possibly another, unknown, Lysanias in the fifteenth year of the reign of Tiberius (28 or 29 CE) to which Luke is referring, since an inscription bearing the name Lysanias was found at Abila, the city in Abilene, and dated between 14 and 29 CE.

Lysias The general in charge of the Seleucid army against the Jews (1 Macc. 3: 32 ff.) who was defeated at Bethsura (1 Macc. 4: 34 ff.) and who was killed by order of King *Demetrius (1 Macc. 7: 2–4).

Lysias, Claudius The Roman commander who rescued Paul from a mob in Jerusalem (Acts 21: 31; 23: 26). He claimed to have bought his citizenship.

Lystra A Roman colony 40 km. (25 miles) SW of *Iconium; it boasted a temple to Zeus (Acts 14: 11–13) and when Paul and *Barnabas visited the place they were taken to be the gods Zeus and Hermes after enabling a cripple to walk (Acts 14: 8–12).

M The symbol employed to designate material in the gospel according to Matthew which does not appear in either of the other synoptic gospels, such as the *parables of the hidden treasure and the pearl of great price (13: 44–6) and the charge to Peter (16: 18). It is thought by some NT scholars to derive from traditions preserved in Jerusalem. The *infancy narratives (Matt. 1–2), though peculiar to Matt., seem to come from another source of information.

Maccabees, books of Four books recording the struggle of loyalist Jews against the forces of Hellenist absorption in the 2nd cent. BCE. Of these 1 and 2 Macc. (there is considerable overlap in telling the history) exist only in the Greek of the *LXX, but were incorporated into the *Vulgate and are reckoned by the Roman Catholic Church to be *deuterocanonical and having the authority of scripture. They are placed in the *Apocrypha by Reformed Churches.

The Greek OT includes 3 Macc., which is inappropriately named, since it is not concerned with the Maccabees but is a fictitious work compiled by a Jew in *Alexandria at about the turn of the eras concerning the triumphs of Jews over their enemies about 220 BCE.

Also written about the time of Jesus, 4 Macc. is attached to the *LXX as an appendix and enlarges on the sufferings of the Jews in the Maccabean period.

Access to historical records in Jerusalem about its political fortunes seems to be shown in 2 Macc., but only 1 Macc. contains reliable history of the Jewish opposition to the attempt by the kings of *Syria to bring about religious unity in their empire by imposing Hellenistic religion and culture. The story begins in 167 BCE when *Mattathias and his sons, who were members of a priestly family in the village of Modein, raised the flag of insurrection. The military leadership soon passed to the son *Judas, who had the surname '*Maccabaeus', meaning a 'hammer', who launched a guerrilla campaign, recaptured Jerusalem, and rededicated the Temple (164 BCE), which the Syrians had defiled. After Judas was killed, the war was continued by his brothers *Jonathan and *Simon. The latter obtained some recognition by the Syrians but was murdered in 134 BCE, to be succeeded by his son, John Hyrcanus, who survived for twenty years and was both ruler and high priest. Under him the *Hasmonean state (Hasmonean was the name of the Maccabean dynasty) flourished and expanded.

The purpose of 1 Macc. seems to be to justify the military adventures on the ground that political freedom was essential for Jewish religious freedom. The purpose of 2 Macc. is to assert the importance of the Temple and the *Law. Advice is given in 3 Macc. to Jews in the *Dispersion when they are persecuted (3: 3), but 4 Macc. extols the sufferings of Jewish martyrs as being atoning *sacrifices for the whole nation (6: 28–9). Cf. Heb. 12: 1–2.

Macedonia Territory in NE of *Greece. Its rulers Philip II and *Alexander III ('the Great') created an empire from *Egypt to India, and Greek culture spread throughout the world. Alexander died in 323 BCE and the empire fell apart. In 148 BCE the Romans established *Macedonia as a province with a capital at *Thessalonica. They built a great road, the Via Egnatia, to Byzantium in Asia Minor, and this must have been invaluable to Paul (Acts 16: 11–17: 1). Paul's visits to Macedonia are mentioned in his correspondence (1 Thess. 2: 2; 2: 17–3: 2; Phil. 4: 15–16; and also 2 Cor. 2: 13; 7: 5).

Machpelah The place near Mamre where the *patriarchs *Abraham, *Isaac, and *Jacob and their wives were buried (Gen. 49: 29; 50: 12–13). The site was in or near *Hebron (Gen. 23: 19), but in *Shechem according to Acts 7: 15–16. Walls possibly built in the time of *Herod the Great surround the Hebron site, which is venerated by both Jews and Muslims.

madness Attributed to a demonic power seizing possession of a personality, notably of *Saul (1 Sam. 16: 14–17), whose 'evil spirit from the Lord' may have been a form of manic-depression. Young *David was his therapist, and music the instrument of calm. In the synoptic gospels there are records of Jesus' *exorcisms, by which *demons were cast out of madmen—evidently a different sort of affliction from *Saul's. Jesus was himself alleged to be 'possessed' on the grounds of transgressing contemporary social or religious customs (Mark 3: 21–2).

Magdala *Mary Magdalen (Luke 8: 2; 24: 10) came from this place, but the reading 'Magdala' of Matt. 15: 39 in AV should follow the better MSS and read Magadan, as in NRSV, REB, NJB; and although a few MSS also read Magdala at Mark 8: 10 in place of Dalmanutha, it is probable that Magdala does not appear in a reliable text of the NT.

magi The 'Wise Men' from the East (Matt. 2: 1–18), probably astrologers. They are portrayed as *Gentiles, ignorant of the OT (Matt. 2: 2–6), and presumably three in number (2: 11). Later tradition supposed, on the basis of Isa. 60: 3, that they were kings, and even accorded them the names Caspar, Melchior, and Balthazar. The story contains impossible features—a moving star (could the magi have only travelled by night?) which stopped over exactly the right location (2: 9), though in spite of this celestial guide the magi were obliged to ask directions from *Herod (2: 8–9). It is better to regard the story as **midrash* rather than history and Matthew's way of showing Christ's universal kingship (Gentiles acknowledge it) and of the submission of magical arts and superstition to the Lord—the point made by Luke in his account of the destruction of magical texts at *Ephesus after listening to Paul (Acts 19: 19).

magic and sorcery Magic is a means by which human beings manipulate deity to obtain a benefit; the appropriate formula and substance (metal or animal, etc.) is essential, otherwise disaster may ensue. There are stories in both OT and NT which approximate to magic, where physical contact with a man of God acted as the channel for power from God to people—e.g. *Elijah healed the son of the widow of *Zarephath by lying on the child three times (1 Kgs. 17: 17–24). Or the contact may be quite indirect—Paul's handkerchiefs (Acts 19: 11–12) or Peter's shadow (Acts 5: 15).

Related to magic is *divination—a non-rational process for learning about the future.

It was held among Hebrews that God revealed the future by dreams or by **Urim* and **Thummim* (Exod. 28: 30) or through calling up the dead (1 Sam. 28). King Joash expected a victory in the light of the flight of an arrow (2 Kgs. 13: 18). Nevertheless magic and divination were condemned (e.g. by Isa. 8: 19) for we do not need such intermediaries between us and God (Jer. 27: 9). The condemnation is reaffirmed in the NT (Acts 8: 9–24).

magistrates Officials in the Roman Empire, holding various titles, had duties on the bench—Asiarchs in *Ephesus looked after patriotic events, and were friendly towards Paul (Acts 19: 31), politarchs in *Thessalonica after a riot released Jason on bail (Acts 17: 9), the pro-consul of a senatorial province, *Gallio at *Corinth, dismissed a charge against Paul (Acts 18: 17), and *strategoi* at *Philippi apologized to Paul for false imprisonment (Acts 16: 39).

Magnificat Latin for 'magnifies', the first word of the canticle or psalm of Mary (Luke 1: 46–55) expressing her joy at being chosen to be the mother of Jesus. There is, however, a suggestion, based on references in the Fathers, and a few Latin MSS evidence, that some in the early Church regarded the poem as that of *Elizabeth, the mother of

*John the Baptist. Moreover, the transition in verse 56 is strange: 'with her' would be the more natural expression if Elizabeth was the speaker. This is not generally accepted by scholars, but it is widely recognized that the poem is modelled on the song of Hannah rejoicing over the birth of *Samuel (1 Sam. 2: 1–10). This is Mary's song of praise as she realizes her destiny and she speaks words typical of the pious poor of certain of the psalms (e.g. Ps. 37). Luke through Mary gives an advance interpretation of the birth of Jesus whose message will continue in this gospel to be a ministry to the poor (4: 18).

Magog Son of Japheth, grandson of *Noah (Gen. 10: 2). *P, the Priestly source, lists a table of nations under the name of Japheth, probably a seafaring people (Scythians?), including Magog. In Ezekiel (38: 2, 14–22; 39: 6) Magog is an apocalyptic figure for a northern heathen nation led by Gog which invades Israel, and in Rev. 20: 8 Magog is one of the nations assembled by *Satan for an assault on the saints.

Magus Commonly used as the surname for *Simon (Acts 8: 9–24) and meaning 'magician'. He was rebuked by Peter for wanting to buy spiritual power—hence the English word 'simony'. This Simon seems to have been regarded as a *prophet by certain *Samaritans and may have been taken on as patron by a Gnostic sect in the 2nd–3rd cents., according to Justin Martyr (*c.*150 CE).

Maher-shalal-hash-baz The symbolic name of a son of Isaiah (Isa. 8: 1–4), meaning 'the spoil speeds, the prey hastens', to indicate the prophet's conviction that the alliance between *Damascus and *Samaria was not a threat to *Judah, because *Assyria would subdue both of them.

Malachi In Hebrew, 'my messenger'; the title given to the last book of the OT, taken from the words in 3: 1. The book itself is anonymous. Its date of composition is estimated at about 450 BCE, after the restoration of the Temple (515 BCE) but before the reforms of *Nehemiah (445 BCE) and *Ezra. The author was closely connected with the Temple and its rituals and he deplores a lack of reverence (1: 14) in the performance of *sacrifices (1: 8) and payment of tithes (3: 8). The priests have so signally failed that God has withdrawn his favour from the people of the *covenant (2: 2). The people have been equally faithless and given to offering in sacrifice animals that were feeble or defective (1: 13), and those who succeeded in riding out the economic depression were oppressing the helpless (3: 5). There was a remedy: return to the Lord (3: 7). Then there will be abundant *blessings (3: 10) for God-fearers, but evildoers will be destroyed (3: 13–4: 3). The day of judgement will be preceded by the advent of the prophet *Elijah (4: 5; cf. Matt. 17: 10–13).

Malchus The high priest's slave who bore this name (John 18: 10), common among Nabataean Arabs, meaning 'king', had his right ear cut off by a *disciple during Jesus' arrest. Only John mentions the name, and that it was Peter who struck—so adding to the offence of Peter's denials (18: 25–7). It is characteristic of Luke, who several times notices the compassion of Jesus, that he records that Jesus healed (restored?) the severed member (22: 51).

Malta 'Melita' (Acts 28: 1, AV); the island was captured by the Romans from the Carthaginians in 218 BCE and the 'barbarous people' (AV, giving a literal translation of the Greek; 'nations' in NRSV, REB; 'inhabitants', NJB) retained their Punic culture. Paul spent three months among them after his ship was wrecked on the journey to *Rome.

mammon Not a Hebrew word, but an Aramaic noun used in the Greek of Matt. 6: 24, translated 'wealth', NRSV, 'money', REB, NJB; familiar in modern English in a pejorative sense; ill-gotten gains, or self-centred covetousness, which takes possession of a person in place of the service of God.

man In English usage, a generic term for human beings in general but also for a male representative of the human species. Because of this ambiguity, modern translations tend to prefer 'human being' or 'mortal' where a male person is not specifically intended. This avoids the danger of linguistic sexism which has been discerned in a bias towards the masculine gender. Hence 'Let us make humankind in our image' (Gen. 1: 26, NRSV; 'human beings', REB; but 'man', NJB).

man of lawlessness An *eschatological figure expected by Paul to be the culmination of evil to herald the final events (2 Thess. 2: 3). But because he was not yet come, those who believe that the *End is already here must be mistaken.

Manasseh (1) Joseph's son; elder brother of Ephraim (Gen. 48: 1), but in the national history the weaker of the two tribes (Gen. 48: 19). (2) Son of *Hezekiah and king of *Judah (698–642 BCE) and regarded by 2 Kgs. as overturning all the religious reforms of his father. He even sacrificed his own son (2 Kgs. 21: 6). He was loyal to his *Assyrian overlords and once helped Ashurbanipal in a campaign against *Egypt, but once also he was suspected of treason and imprisoned (2 Chr. 33: 11), perhaps during 652–648 BCE. It was inevitable that Judah's subjection to Assyria would involve some infiltration of the imperial cult, and equally that this should be condemned by strict Yahwists. According to 2 Chron. 33 he experienced a religious conversion, but there is no mention of this in Kgs. and it may be the Chronicler's way of explaining the length of his reign.

Manasseh, Prayer of A non-canonical work inserted towards the end of the Apocrypha. Manasseh reigned as king of *Judah from 698 to 642 BCE and was totally condemned by the Deuteronomic author of 2 Kgs. 24: 3 and held responsible for God's punishing the people by the Exile in *Babylon. In course of time his reputation was rehabilitated somewhat (see preceding article). This prayer was composed in Greek by an Alexandrian Jew in the 1st cent. BCE as an example of what his prayer of penitence might have been. It was not included in the *LXX and was never in the Hebrew OT, but in some Greek MSS it was added to the Pss. and it appears in the *Apocrypha in Protestant Bibles. It was not, however, recognized as one of the *deuterocanonical books by the Roman Catholic Church, but may be printed as an appendix.

manger A trough used for feeding animals, often carved out of the stone in a side of a cave. One was used as a bed for the infant Jesus at Bethlehem (Luke 2: 7).

manna Hebrew for, perhaps, 'the what is it?'; the food which sustained the Israelites during their trek through the *wilderness. It was lying on the ground (Exod. 16: 13 ff.), white in colour and sweet in taste. On Fridays a double portion fell (Exod. 16: 22–6) to avoid breaking the *Sabbath by working, to accord with the *Priestly editor's insistence on Sabbath observance (Exod. 20: 8–11). An explanation of this food is that it was sap expelled by insects which turned into a kind of dew with a taste resembling honey or the cocoon of the parasitic beetle *Trehala manna*. Its presence is described as a *miracle (Deut. 8: 3). Jesus refers to manna in his discourse about the *bread of life (John 6: 31–2) and it is mentioned as food for believers in the new age in Rev. (2: 17).

mansions Used by AV (John 14: 2) for 'dwelling places' (NRSV, REB; 'places', NJB) in the Father's house, in which there will be room for everyone.

maranatha Aramaic expression used by Paul (1 Cor. 16: 22), meaning 'Come, our Lord'; a prayer again used in Rev. 22: 20. It is an interesting link between Greek-speaking Christians and the mother Aramaic-speaking Church in Jerusalem. It is used in the *Didache (10: 6), where it probably means 'our Lord has come'.

Marcion (85–160 CE) Influential heretic who came to *Rome in 140 CE and established schismatic congregations. He held that the God of the OT was a God of Law and wrath, and not to be identified with the NT concept of the God of Love, revealed by Jesus.

Paul was, he said, the apostle who understood this, and accordingly ten of his epistles, suitably edited, formed the major part of Marcion's canonical scriptures. Of the gospels, he accepted only an abbreviated version of Luke. Such radicalism may have encouraged Marcion's 'orthodox' opponents to consider which Christian writings were authoritative for the life and faith of the Church.

Mari Ancient city on River Euphrates. Between 1933 and 1955 20,000 tablets in Akkadian language were found. They date from the 2nd millennium BCE and give information about the surrounding culture in the time of the OT *patriarchs of Gen.

Mark, gospel of The shortest of the gospels and by nearly universal agreement the first to be written, and later used as a principal source by Matt. and Luke. Traditionally it has been ascribed to *John Mark, companion of Peter (1 Pet. 5: 13) in Rome. An early bishop, *Papias of Hierapolis, in 130 CE even mentions having been told that Mark wrote down 'accurately but not in order' Peter's recollections of Jesus' sayings and deeds. So close a connection with Peter's recollections is, however, unlikely, since before Mark put pen to parchment individual stories about and sayings of Jesus ('*pericopae') had been circulating in Church groups by word of mouth, and what Mark did was to select, arrange, adapt, and interpret what he had learnt. The Greek is hardly eloquent, but Mark's pioneering work is evidence of a sophisticated and intelligent author, determined to answer some of the problems which beset the Roman Christians about 65–7 CE. If the gospel was indeed written in Rome (which cannot be regarded as definitely established), the kind of questions that might have been asked were: 'How was it that Jesus, regarded as a good man, died the death of a criminal and a disturber of the Roman peace?' 'What did Paul mean when he had written to the Church some years previously (Rom. 14) that "none of us are living to ourselves or dying to ourselves: if we live, we live to the Lord, and if we die we die to the Lord"?' 'Why does Paul explain Christian fellowship in terms of *death and *resurrection, of humiliation and glorification?'

Mark's theme therefore is of the suffering *Messiah who aroused anger by reason of apparent claims, and the shadow of the *cross lies over the whole gospel. Jesus is the mysterious Son of God—ultimately triumphant, but only because he presses on towards a tragic culmination in Jerusalem through which alone that triumph could be won. Mark explains the quarrel between Jesus and his enemies by relating only the essential facts, beginning at 2: 7 after the baptism and temptation. There is no infancy narrative, such as Matt. and Luke were to provide, but the final days are anticipated—as when the prediction of the *Passion in Mark 8: 31 is followed by the *Transfiguration 'after six days' (Mark 9: 2). Each of the three Passion predictions is followed by a summons to the *disciples to share his 'baptism' (of suffering; Mark 10: 35–45). Mark's gospel is surely intended to strengthen the community in the face of persecution. The pastoral concern of the evangelist is demonstrated by the accounts of Jesus' healings: he is the Lord who cares.

Mark also has an important secondary theme, of the Church as the rightful successor to Israel in the providence of God, and this may account for his repeated use of '*twelve' (following the *twelve* tribes of Israel): twelve disciples are chosen, the woman with menorrhagia had suffered for twelve years, a girl aged twelve was raised to life, twelve baskets of broken bread were taken up after the Feeding of the 5,000.

Mark's themes are clearer than his travelogue: his description of Jesus' itineraries has been likened to 'the meanderings of an intoxicated fly'. At the conclusion there is puzzling and sinister depth with the cry of dereliction (15: 34) and the unexpected darkness (15: 33) and the astonishing, abrupt ending (16: 8) with the fear of the women, and their disobedient silence. The existence of the Church and this gospel implies that they were not silent for ever, and Mark's sentence that the tomb was empty (16: 6) does show his belief in Jesus' vindication.

The reason for Mark's ending at 16: 8 has fascinated readers, as do the unfinished Eighth Symphony of Schubert or the Borodin quartet, and early attempts to supply what was thought to be an appropriate conclusion are represented by the 'shorter ending' of a few late Greek MSS and by 16: 9–20, the 'longer ending', probably written early in the 2nd cent. and known to *Tatian and *Irenaeus. A possible explanation of the abrupt ending at 16: 8 lies in this gospel's unfavourable depiction of the disciples who constantly fail to understand Jesus (e.g. 8: 17) and are sternly rebuked in the person of the spokesman Peter (8: 33). If Mark was influenced by Paul, then he would discern in the disciples the leaders of the Jerusalem Church who were Paul's adversaries. The gospel understands the significance of the death of Jesus as the disciples did not. So it is of a piece that Mark offers no appearance of the risen Lord to any of them. Just three women see the empty tomb, and flee.

Mark, John A common Latin name. In the Acts a certain John Mark lived in his mother's house in Jerusalem (Acts 12: 12) and was the companion of Paul and *Barnabas on their mission of aid for Jerusalem (Acts 12: 25). However, Mark deserted them during a later journey (Acts 13: 5, 13), and when Barnabas suggested him as a companion once more, Paul angrily severed the partnership with Barnabas, leaving the latter to go off to his native *Cyprus with Mark (Acts 15: 39). It is possible that a deeper reason for the separation was Paul's determination that *Gentile converts should not be subject to the Jewish law—a quarrel between Paul and Barnabas which had blown up at *Antioch (Gal. 2: 13).

A Mark (a 'cousin of Barnabas') is mentioned in Col. 4: 10; and in 2 Tim. 4: 11, Mark is praised as a useful assistant to the writer. He is referred to affectionately in 1 Pet. 5: 13. It is possible that these, and the Mark who sends greetings to Philemon (24), all refer to the same man, and that therefore a reconciliation took place between Paul and Mark. That is the traditional view, and it is certainly more plausible than the legend that Mark founded the Church in Alexandria and that his body was eventually buried in Venice in 829 CE. Mark's emblem of a winged lion is ascribed to him on account of the supposed portrayal in his gospel of regal characteristics of Jesus. The emblem is familiar in Mark's patronal city of Venice, and there is a sequence of mosaics of him dating from the 13th cent. in the Capella Zeno.

market (place) In Palestine of the gospels the market was the place of merchandise (Mark 7: 4), but in the Graeco-Roman world of Paul it is the city centre or 'main square' of *Philippi (REB, Acts 16: 19) and of *Athens (Acts 17: 17).

marks Paul carries the marks of Jesus branded on his body (Gal. 6: 17) as a result of scourgings, and this was a means of sharing Christ's sufferings (Phil. 3: 10). In the *vision of the end of the age in Rev. 19: 20 those who have the mark of the beast (the symbol of the Satanic Roman Empire and the priests of its imperial cult) on their foreheads or hands will perish but faithful Christians with no such marks will be admitted to reign with Christ (Rev. 20: 4).

marriage In the OT marriage is assumed to be a normal relationship, ordained by the Creator (Gen. 1: 26 f.) though the initial period of innocence and *joy gave way to '*hardness of heart' (Matt. 19: 8). Then rules became necessary, though there are many references in the OT to happiness in marriage (e.g. Prov. 5: 18 f.; the S. of S.; Tobit 7: 8–10: 13), and the means of ensuring posterity—even sometimes a marriage within what were later to be forbidden degrees of kinship (Gen. 24: cf. Lev. 18: 6–18). Marriage with *foreigners became forbidden after the Return from Exile (Ezra 9–10), but numerous wives and *concubines had been available for kings in the period of the monarchy (1 Kgs. 11: 3). Husbands were allowed to divorce a wife (Deut. 24: 1–4) and were in general allowed more licence in relationships than women, but there was also a strong strain of objection to the idea of divorce (Mal. 2: 14 ff.), based on the analogous relationship of the *covenant between God and Israel—human marriage being a reflection of that bond.

In the NT there is a similar theme of the covenant relationship, but the parallel is now that of Christ and his Church (Eph. 5: 22–33). That marriage is the normal form of adult life is endorsed by Jesus' presence as a guest at the wedding in Cana (John 2: 1–12), and some if not all of the twelve apostles whom he chose were married (1 Cor. 9: 5). The sayings of Jesus about marriage that are recorded do not imply that a marriage is indissoluble in the sense that a relationship has been created by God which cannot be terminated by human action, though he warns couples of the great human responsibility they undertake under God's general provision for marriage (Mark 10: 6–9). It is characteristic of Jesus' ethical teaching not to lay down precise rules but to indicate qualities of behaviour: anger is as bad as murder; absolute non-resistance is preferable to retaliation. Lifelong marriage is better than *divorce—but divorce is not inherently impossible, and perhaps sometimes (as in the case of a Christian married to an unbeliever) it could be desirable (1 Cor. 7: 15), and the divorcee is no longer 'bound', that is, he or she is free to remarry.

In the time of Jesus a marriage was initiated by two families who negotiated a betrothal between their offspring which was a more solemn and binding commitment than our engagements, even though the bride and bridegroom were much younger than is usual in modern western countries. The marriage was finalized, sooner or later after betrothal, when the man led his wife from her parents' home to his own. Joseph was disconcerted according to Matt. 1: 18–19 when Mary was found to be pregnant after betrothal but before they were cohabiting and, in accordance with Jewish law (Deut. 24: 1), he resolved to terminate the contract. But being of a generous disposition his intention was to act without attracting any publicity.

Mars' Hill Used in AV at Acts 17: 22 for the Areopagus in *Athens, where Paul addressed the council.

Martha Sister of *Mary; they entertained Jesus at home (Luke 10: 38) but Martha resented Mary's conversing with Jesus while she was left with the chores. In John 11: 1 their village is named as *Bethany and Martha is presented as a woman of *faith (John 11: 21) whose brother *Lazarus is raised to life.

martyr In Greek, 'witness' and so one who bore witness to *faith in Christ by his or her death, as did *Stephen (Acts 7: 60; 22: 20). Witnesses to Jesus by death are praised too in Rev. 17: 6, and from the end of the 2nd cent. martyrs were commemorated on the anniversary of their deaths.

Mary In Hebrew 'Miriam', the name of Moses' sister (Exod. 2: 4–8) and in the NT the name of seven women—the mother of Jesus (see below); the sister of *Lazarus of *Bethany (John 11: 1); Mary Magdalen (see below); the mother of *James and *Joses (Mark 15: 40); the wife of *Cleopas (John 19: 25); the mother of John Mark of Jerusalem (Acts 12: 11–17); a woman greeted by Paul (Rom. 16: 6).

Mary, mother of Jesus Both Matt. and Luke have accounts of the *birth of Jesus to Mary in *Bethlehem, and both assert that Mary's pregnancy was 'by the *Holy Spirit' and not by her husband *Joseph—hence she is known as 'the Virgin'. In Matt. the child Jesus receives a visit from the *Magi, and Mary and Joseph then temporarily retreat to *Egypt to escape the wrath of *Herod. Luke's narrative is told from Mary's point of view; it includes the *Magnificat, the visit of the shepherds, the postnatal purification of the parents in the Temple, and greetings from *Simeon and *Anna.

There is no birth narrative in Mark, and the references to Mary in that gospel are not very flattering. In Mark 3: 21 Jesus' family thought he was mad and when Mary and his brothers arrive and remain outside the house (3: 31), Jesus said that it was those who do the will of God, not those of his family standing without, who are his true family. In Mark 6: 3 Jesus is said to be 'the son of Mary and brother of James, Joses, Judas, and Simon' by the shocked and unbelieving congregation in the *synagogue. Mary the mother of James and Joses is also mentioned at Mark 15: 40 as 'looking on from a distance' at the *crucifixion, and this would appear to identify her as the mother also of Jesus. In Mark 16: 8 the same group of women fled from the empty tomb (which they alone are recorded to have seen), disobeying the command to tell the disciples of the resurrection, because they were afraid (16: 8). In Mark's view those nearest to Jesus did not demonstrate faith in him any more than did Peter and the other eleven disciples in the earlier part of the gospel or the members of the synagogue in his home town (Mark 6: 6; assumed by Luke 4: 16 to be Nazareth).

Such disparagement of Jesus' family (his brother James became a leader of the Church in Jerusalem, Acts 15: 13; Gal. 1: 19) was unwelcome to Mark's successors, who modified his references to Mary to enhance her reputation, not only by including the beautiful *infancy narratives but also by omitting Mark's statement (Mark 3: 21) that the family intended to arrest Jesus; and Matt. (28: 8) adds the account, at the end, of the two Marys running with joy to the disciples. They believe. Luke makes similar editorial alterations, and at Acts 1: 14 he has Mary and the brothers in the Upper Room along with the eleven disciples.

John's gospel does not mention Mary by name, and there is something of a return to Mark's theology in the apparent discourtesy of Jesus to his mother in John 2: 4 at the wedding in *Cana. In 19: 25–7 Mary stands at the foot of the cross and is taken away by the *Beloved Disciple. In this incident she seems to be portrayed as a symbol of *Judaism the mother of Christianity, which is represented by the believing Beloved Disciple.

Mark's theology was that God would save; though when this would be even the Son did not know (Mark 13: 32). We can only trust, without any support from those nearest to Jesus. Unsurprisingly, the later evangelists complemented this stark message, and Mary, like her infant son, grew in stature. By the time of the apocryphal book of James in the 2nd cent. Mary's virginity was regarded as 'perpetual' (the 'brothers' of Jesus could be regarded as 'cousins' in the Greek), and at the Council of Ephesus (431 CE) Mary is accorded the title Theotokos, 'Mother of God'; which safeguarded the two-natures doctrine of Christ, God and Man.

Mary Magdalen The leader of a group of women who supported Jesus in his itinerant *ministry (Luke 8: 1–3). Her home according to some MSS was in *Magdala, probably to be identified with Tarichaea, where there was a Hellenistic fishing industry, and she had been healed of a serious disease ('seven demons'). She was apprised of the *resurrection at the empty tomb and gave the news to the disciples (Matt. 28: 5–9), though at first she was disbelieved (Luke 24: 11, 22–48). An appearance to her alone is recorded by John 20: 11–18. There has been much modern speculation about the relation between Jesus and Mary Magdalen, but this says more about the authors' imagination than about the biblical text.

Masada A vast fortress on the west bank of the *Dead Sea, extended and strengthened by *Herod the Great during his reign (37–4 BCE) and used to accommodate his relatives in safety during his absences. It was taken over by the Jewish rebels early in the war against the Romans in 66 CE and was the last stronghold of resistance. It was said by *Josephus to be large enough to grow produce to supply the refugees inside. However, in 73 CE all except seven women and children committed suicide after a ferocious assault by the Romans.

Masoretic text The 'official' version of the Hebrew Bible, often referred to by the symbol MT. It was settled by *rabbis from the end of the 1st cent. CE and notes were appended to ensure accuracy of transmission and pronunciation of the text, which was written only with consonants. In order to avoid confusion of meaning, vowels were added to the text and these became standard in Western Judaism by the end of the 10th cent. CE and later throughout Jewish communities.

Massah Near *Horeb; Moses called the place Massah (Hebrew for 'testing') because there in the *wilderness the people put God to the test, complaining about lack of water. Moses struck a rock, and water flowed out (Exod. 17: 7).

master Used in AV and NJB in addresses to Jesus (e.g. Matt. 12: 38) where REB and NRSV prefer 'Teacher'—though REB and NJB use Master in Matt. 21: 3 where AV and NRSV translate 'the Lord' for a different noun in the Greek. A third Greek noun is used for Master (six times in Luke e.g. Luke 5: 5, AV, NRSV, REB, NJB) where there is an overtone of authority.

Mattathias A common personal name in the Maccabean period; the best-known was the priest who initiated the rebellion against the Seleucid rulers in Modein in 167 BCE and whose five sons continued the struggle (1 Macc. 2: 1) against *Hellenization. Two of Jesus' ancestors bore the name (Luke 3: 25).

Matthew One of the *Twelve (Mark 3: 18) whose call to follow Jesus came to him while he sat in his tax office (Matt. 9: 9). In the parallel narrative in Mark (2: 14) the tax official's name is *Levi—perhaps a second name for the same person—though, if the disciples were intended to correspond to the tribes of Israel (Luke 22: 30), then on some reckonings there are thirteen tribes (*Ephraim and *Manasseh, sons of *Joseph, counting as one each) and thirteen apostles (counting Levi and Matthew as two persons). As a tax collector, Matthew was collaborating with the occupying power.

Matthew, gospel of The first book in the NT. It was widely held in the past, as by St Augustine, that Matt. was the earliest of the four gospels to be written and that the gospel of Mark was an abbreviation of it. The more usual view now is that Mark was used by Matt., and in the process Mark's inelegant Greek was polished and Mark's treatment of the person of Jesus developed in a more devotional direction. In addition to Mark, the author of Matt. prefaced his work with the two chapters about the *birth and infancy of Jesus. In the body of the work, Matt. incorporated, in a very orderly fashion, much of Jesus' *teaching that was unrecorded by Mark. Since there is close verbal similarity between Matt.'s discourses and the teaching material also incorporated, in a different way, by Luke, it is generally (but not universally) held that both these evangelists were using a common source, designated *Q.

Date and place of composition. The *Fall of Jerusalem in 70 CE seems to be alluded to as a rather distant event at Matt. 22: 7, and this suggests a date of about 85–90 CE. Social conditions reflected in the gospel support the view that Matt. may have been written in *Antioch in *Syria on the River Orontes, or for a cluster of urban communities in that neighbourhood. The author seems to have been educated in the tradition of Jewish scribes and he brings out of his treasure what is new and what is old (Matt. 13: 52). Possibly Matthew the apostle had founded the Church in the city, which would account for the ascription of the gospel to him from early in the 2nd cent.—but a work so dependent on earlier sources could not have been written by one of the apostles.

Evidently in the area of Matt.'s Church there existed tension between Christians and Jews. From the very beginning of the gospel (2: 1) until its conclusion, the Jewish leaders are branded opponents of Jesus. Christians could expect to be hauled before a local *sanhedrin (Matt. 10: 17), where Jesus might be dismissed as a magician and a deceiver. Controversy with Jews is also the motive for the reply (Matt. 28: 11–15) to current Jewish explanation about the alleged empty tomb. Throughout the gospel the author makes use of his knowledge of the OT to support Christian claims; the *genealogy, for example, taken straight from the OT, is designed to show that Jesus belonged, through *Joseph, to the royal house of *David (Matt. 1: 6, 16). He adopts the philosophy of history that the people of Israel were punished for their *sins by the *Exile and restored in *forgiveness (2 Chron. 36: 17–23).

Characteristics of the gospel. Matt.'s rabbinical background emerges not only in his preference for the term 'Kingdom of heaven', so avoiding, as Jews did, the use of the holy word God, but also in the careful structure of the discourses in five great blocks, each ending with the phrase 'When Jesus had finished these sayings . . .'. The narrative framework echoes the six books of the *Hexateuch: the birth narratives/Genesis; the *baptism in the *Jordan and Jesus' *temptations/Exodus; healing of a leper and an untouchable woman/Leviticus; callings of disciples/Numbers; the *Passion and *Death of Jesus/Deuteronomy; the *Resurrection/Joshua (the entry into the *Promised Land).

The ethical teaching of Jesus, embodied principally by Matt. in the Sermon on the Mount (chs. 5–7) relates the new *righteousness in the light of the coming of the Kingdom to the former obligations of the *Law. The authority which the Law has for Christians (5: 7) has to be interpreted by the overriding injunction to love and therefore stretched out to an ideal not attained in the practices of the *Pharisees (5: 20).

Matthew's theological interests. This gospel became the Church's favourite—the one most often read in the liturgy. Possibly this is because the primary concerns are pastoral and catechetical rather than theological. Nevertheless, there *are* important theological affirmations. There is the repeated refrain, 'that it might be fulfilled which was spoken by the prophets'. This is not meant to imply that the event mentioned took place *in order that* prophecy should be fulfilled, but that things happened *with the result that* OT scripture was fulfilled—even when the OT question was plucked right out of context, as when Hosea 11: 1 is quoted by Matt. 2: 15 to interpret the family's flight into Egypt. Matt.'s theology is that what was said of Israel in Hosea could be appropriately applied to Jesus because he *was* Israel as it was called to be; if Israel was 'recapitulated' in Jesus, then what was said of ancient Israel could now be predicated of Jesus but sometimes in reverse: where Israel was disobedient and failed, Jesus was obedient to the transcendent God. Jesus is said to be endowed with the spirit of God (3: 16) but is not regarded as himself divine.

Matt. is an ecclesiastical gospel. There are traces in the gospel which suggest that the Kingdom is beginning to be interpreted as the Church (as in history it later was). The parables of the Tares (13: 24–30, 37–43) and of the Talents seem to have the situation of the Church in mind: and Matt. alone of the gospels mentions the *word* 'Church' ('*ecclesia*') in connection with Peter as the Rock (Matt. 16: 18; 18: 17). But along with the recognition of the Church is the expectation of Judgement and the *End expressed in the terms of the later biblical literature (Dan., 2 Macc. 7, 1 Enoch) but instead of their accounts of *visions Matt. 24–5 reports Jesus' prophetic discourse.

The presentation of the Person of Jesus. In Matt., Jesus frequently uses the phrase of Daniel '*son of Man' of himself and the *demons recognize him as '*Son of God', which seems indeed to be Matt.'s most prominent title for Jesus (8: 29), and Matt. takes over the *centurion's confession (27: 54) from Mark, except that in Matt. he is joined by his troops in saying the words. There are several other Matthaean touches in the Passion narrative: *Pilate's wife intervenes (27: 19). He affirms his blamelessness for committing Jesus to crucifixion by washing his hands (27: 24), and he admits Jesus' innocence of the charge against him, and passes over responsibility to the Jews. Their enthusiastic acceptance of it for themselves and for their children (27: 25), has often been an excuse for Christian antisemitism. Jesus' resurrection is anticipated in Matt. (cf. 1 Cor. 15: 23 'the first fruits') by introducing into the Passion narrative the legend that 'many bodies of the saints who had fallen asleep were raised' and walked about the city (27: 52), 'after his resurrection'. Other Matthaean features are the stories of the guard posted at the tomb, and of alarming apocalyptic intrusions. None of the Passion, in Matt.'s account, takes Jesus by surprise. He warned the disciples of what was to come (26: 1–2); he is shown to be perfectly aware who was to betray him (26: 25)—both are subtle alterations of Mark.

At the end of Matt. the Risen Lord and the Eleven are on a Galilean mountain. He is the one to whom all authority is given (28: 18) and through his Church he will make disciples in all nations. He is the Son of God, now exalted to the Father.

Matthias Apostle. The defection of *Judas from the *Twelve created a vacancy (unlike a death) and in order to maintain the number twelve intact, corresponding to the twelve tribes of Israel, a successor was chosen. Of two nominees, Matthias was elected, either by lot or, possibly, by votes cast by the existing apostles. The qualifications for office were that those nominated should have accompanied Jesus during his *ministry and been a witness to the *resurrection (Acts 1: 22).

meals Of great importance in Israel: if two strangers met in the desert and shared a meal, an unbreakable bond was established. Often meals were vegetarian, and meat (but only from animals prescribed in Lev. 11 as *clean) could be eaten on special occasions provided that the blood was drained out (Deut. 12: 23). Meals were part of celebrations and festivals, especially at the *Passover (Deut. 16: 1–8), and the laws about food were an essential part of Israelite identity. Banquets are mentioned in the gospels, and it is apparent that a host would issue a preliminary invitation which was followed up by a second when the meal was ready (Luke 14: 16–17). Much care was taken to seat guests in accordance with age and rank. Guests reclined round a table, eating by using the right hand. Jesus clearly enjoyed meals with friends and was accused by enemies of over-indulgence (Matt. 11: 18–19) but meals with disciples symbolized the kind of *fellowship and *joy that would obtain in the age to come (Matt. 8: 11). The image (cf. Isa. 25: 6) of the Messianic banquet (Matt. 22: 2; 26: 29; Luke 22: 30) made every *Eucharist a foretaste of it (Acts 2: 46).

Medes A people who rose to power in the Near East, and captured *Nineveh in 612 BCE, but were conquered by Cyrus and became part of the Persian Empire in 549 BCE. During the Exile they were expected by *Jeremiah (51: 11) to be God's instrument for punishing *Babylon. Dan. 5: 31 and 11: 1 supposes that the Medes fulfilled this prophecy before the rise of the Persians, which is unhistorical. In NT times Media included a Jewish population (Acts 2: 9).

mediator A go-between; in the OT priests fulfil a mediatorial function on behalf of the people by offering *sacrifices to God, for which they were consecrated to their office (Exod. 29). On the Day of *Atonement the high priest entered the Holy of Holies in the Temple, alone, and made atonement for the *sins of the people during the past year. Other OT mediators included Moses in the *wilderness (Exod. 32: 30–4), *Joshua (Josh. 24: 14–28) and *Solomon (1 Kgs. 8: 22–53). In the NT Jesus is named as a mediator. In the epistle to the Heb. Jesus is said to be responsible for a new *covenant between God and humankind (Heb. 8: 6; 9: 15; 12: 24). Christ is the high priest who eternally offers the one perfect sacrifice.

medicine Folk medicine was available (Jer. 8: 22) but knowledge of physiology was rudimentary—the *heart was thought to be the seat of moral *judgement (Job 27: 6; Deut. 2: 30). Diagnosis was haphazard. Disease was regarded as connected with *sin, and good health with good behaviour (Isa. 58: 8). Medicines were simple in character, and the one recommended for the faithful was *prayer (Ps. 38). Recourse to a shrine was common (1 Sam. 1) and bronze serpents were part of the therapeutic apparatus (Num. 21: 9; 2 Kgs. 18: 4). Observation had led to some sound precautions, such as avoiding a second person living in a house with another person with certain kinds of ailment (Lev. 13: 46). Diets were important; *balm of Gilead was applied to soothe pain (Gen. 37: 25). Regulations governing clean and unclean foods probably had their origin in popular observations about personal hygiene and cases of food-poisoning.

By the 2nd cent. BCE Hellenistic theories were displacing ancient Hebrew beliefs about disease and medicine. Instead of a diseased skin being seen as penalty for sin (Num. 12: 10–11), for which the prescribed remedy was penitence, Greek scientific knowledge led to appreciation of the skills of the physician. Whereas *Asa was rebuked for resorting to a physician to treat gangrene in his foot (2 Chron. 16: 12) Ben Sirach urges that 'the Lord has created medicines' and doctors, and it is sensible to use their services (Ecclus. [= Sir.] 38: 1–15). Physicians were available in NT times but were not always very successful (Mark 5: 26). Jesus went out of his way to meet sufferers who were not able to visit the Temple (Mark 14: 3). The NT, however, does not share a former hostility to ordinary medicine (as in 2 Chron. 16: 12); and Luke was appreciated as a physician by Paul (Col. 4: 14).

Mediterranean The only sea the Hebrews knew (Num. 13: 29) and feared (Ps. 104: 25; Rev. 21: 1); they were content to leave maritime trade in the hands of *Phoenicians (1 Kgs. 9: 27). In NT times the Romans controlled the sea and the port of *Caesarea was used by ships which conveyed cargoes to *Rome, and Paul embarked there (Acts 9: 30). Paul's main activities were in places on the *Aegean Sea, a NW inlet of the Mediterranean.

meekness Moses is described as meek (Num. 12: 3), that is humbly submitting himself to the will of God; and it is to the meek in the sense of the poor and deprived that Jesus promises that their present afflictions will ultimately be reversed (Matt. 5: 5).

Megiddo City in the plain of *Esdraelon situated on two important routes for merchants and for armies; one from Jerusalem in the south, and the other from the coast in the west. It is mentioned as a battleground in Egyptian and Assyrian texts, and in the OT Joshua is said to capture it (Josh. 12: 21) and *Solomon to fortify it (1 Kgs. 9: 15 ff.) and *Josiah died there (2 Kgs. 23: 29 ff.). German and American archaeologists have made important discoveries at Megiddo.

In the NT Armageddon (Rev. 16: 16), 'hill of Megiddo', is not surprisingly the symbolic setting for the final conflict of *good and *evil.

megilloth Hebrew for 'rolls'; exegesis by the *rabbis of the Song of Solomon, Ruth, Ecclesiastes, Esther, and Lamentations.

Melchizedek King of Salem (Jerusalem). After *Abraham's victory over Chedorlaomer he was greeted by Melchizedek (Gen. 14: 17 ff.) and in return was given a tenth of the captured booty. Described as 'priest of God most High' who brought out *bread and *wine, Melchizedek became a powerful model in subsequent Hebrew and Christian tradition. David inherited Melchizedek's dynasty (Ps. 110: 4) and Christ inherited his high priesthood, which is superior to that of the Levitical priesthood (Heb. 5: 6 ff.; 6: 19–7: 28). From about 200 CE Melchizedek's offering of bread and wine was taken as prefiguring the Christian *Eucharist.

member(s) Limbs of the human body, used in NT both literally and metaphorically. Individual Christians are members of the *Body of Christ (1 Cor. 12: 27).

memorials Permanent reminders. In the OT stone pillars were erected as memorials of striking events, as of the crossing of the *Jordan (Josh. 4: 4–24). The *Passover celebrated annually was a memorial of the *Exodus from *Egypt. In the NT *Cornelius is told that his prayers and alms have ascended as a memorial before God (Acts 10: 4); that is, what he had done was as acceptable to God as the sacrificial offering made by a Jew in the Temple (Lev. 2: 2; Ps. 141: 2).

The *Last Supper was to be repeated as a memorial of the death of Jesus (1 Cor. 11: 24), a bringing back into the present of that past event, the *sacrifice of Christ, risen from the dead and victorious. Thus, the Christians' celebrations with bread and wine were believed to be the means by which the power of Jesus' accepted sacrifice could be continually made present in their midst.

Memphis Egyptian town on the *Nile (Isa. 19: 13, NRSV; called Noph in Hebrew, hence AV, REB, and NJB, and Hosea 9: 6). It is to be a place of exile for people from Israel (Jer. 2: 14–18).

Mene mene tekel upharsin Daniel interpreted these words (which were perhaps nouns for several weights or coins), which were written on the wall of Belshazzar's palace. They had baffled the wise men of *Babylon (Dan. 5: 25–8). He claimed they meant 'to number', 'to weigh', 'to divide', adding up to a message of doom: the kingdom would shortly be divided between the *Medes and the *Persians.

menorah The seven-branched candelabrum which became the symbol of *Judaism (Exod. 25: 37) and was ceremoniously carried during the triumphal march of *Titus in *Rome after the capture of Jerusalem in 70 CE.

Mephibosheth On hearing the news of the death of *Jonathan (Mephibosheth's father) and *Saul (his grandfather), his nurse dropped the infant Mephibosheth in her haste (2 Sam. 4: 4), after which he was permanently lame. *David treated him generously, even when he was implicated in the plot of *Absalom to seize the throne (2 Sam. 16: 1–4). It is likely that Meribbaal (1 Chron, 8: 34) is another name for the same person; '*bosheth*', meaning 'shame', was substituted for the objectionable word 'baal'.

Mercury In Roman mythology, the gods' messenger, corresponding to the Greek Hermes. At *Lystra Paul was taken to be this god, who had a reputation for eloquence (Acts 14: 12).

mercy In Hebrew the noun *hesed* ('mercy') is also translated *'loving-kindness', and implies the loyalty of God to the covenant. In the NT God's mercy is revealed in the salvation offered through Jesus (Luke 1: 58; Rom. 11: 30–2) and a similar outgoing compassion towards human suffering is shown by Jesus (Luke 17: 13).

mercy-seat A cover on top of the *Ark, described by the *Priestly source as the throne of God, where he is seated in a *cloud (Lev. 16: 2) and where blood was sprinkled by the high priest on the Day of *Atonement (Lev. 16: 14–16). It is the image behind Heb. 9: 3–5.

Merodoch-baladan King of *Babylonia, 721–710 BCE, who sent diplomats to *Hezekiah of *Judah in the hope of securing support against the ruler of *Assyria who had overthrown the Babylonian allies in 710 (2 Kgs. 20: 12–19). The embassy probably came to Jerusalem in 705, which was the year when *Sennacherib succeeded Sargon on the Assyrian throne.

Mesha King of *Moab, renowned as a sheep breeder (2 Kgs. 3: 4), who threw off a long subjection to Israel, about 853 BCE, a success which he attributed to his sacrificing his son as a *burnt offering (2 Kgs. 3: 27). The Moabite Stone, discovered in 1868 and now in Paris, records Mesha's territorial acquisitions.

Mesopotamia The land area between the rivers *Tigris and *Euphrates; the southern part was held by *Babylonia and the northern by *Assyria. The Code of *Hammurabi (1792–1750 BCE) was found on a great stone at Susa, in modern Iran, and is now in Paris.

Messiah Hebrew for 'anointed one', who will be a saviour of the people, and used in OT of both kings and priests, especially of *David and his successors, but also of Cyrus (Isa. 45: 1). In the prophets a future king of eschatological hope is expected to reign with justice and in peace (Isa. 11: 1–5), but the term 'Messiah' is not itself found in this sense in the prophets. The *Dead Sea scrolls refer to the future coming of two anointed figures, one regal and one priestly, in the tradition of *Melchizedek (Ps. 110: 4; Heb. 7: 1), who combined both functions in his person. In the Similitudes of Enoch (Enoch 37–71), from the middle of the 1st cent. CE, there is a Messiah who is also the heavenly *Son of Man. So the fundamental reference is to God intervening in human history by sending his representative, and Christian readers discovered hints in the OT (e.g. Ps. 22: 6–8) that this person would have to suffer.

In the NT the Hebrew 'Messiah' becomes the Greek 'Christos', but references in the synoptics to Jesus as Messiah are scarce (Matt. 27: 17, 22); this rarity seems to reflect the known tradition of Jesus' reluctance to use such a title of himself, though the entry into Jerusalem (Mark 11: 1–10) and the Cleansing of the Temple (Mark 11: 15–19) suggest the notions of delivering his people, as prophesied by Malachi (3: 1). When Jesus was called Christ by Peter, he was rebuked (Mark 8: 29) and told to keep silent about it. Only at the *trial did Jesus reply to the high priest's questions about Messiahship in the affirmative (Mark 14: 61 f.) and in Matt. (26: 64) this is modified to mean 'you are right'—an assent to the question in a form in which Jesus minimizes the importance of the statement about the present: he was, to be sure, the Messiah (he means); but in the near future he would receive a position which would establish his Messianic dignity beyond doubt 'at the right hand of God'—an utterance of blasphemy needing no further corroborative evidence in court. Similarly, the reply of Jesus to *Pilate (Matt. 27: 11) is an admission of Messiahship but without explanation or justification—except in John (18: 36—'my kingdom is not from this world').

When the Church became predominantly *Gentile, 'Christ' lost its original significance of Messiah, anointed one. Gentiles were not much interested in a Messiah who would restore the kingdom to Israel, so Christos became used as an adjective (perhaps confused with Chrestos, pronounced in the same way, meaning 'good') to describe Jesus. Then Christos became a proper name on its own account. Even Paul, Jew though he was, was beginning to use 'Christ' as a substitute for Jesus, or combined with it.

Messianic Secret A theory about the gospels. It was observed by William *Wrede in 1901 that the inconsistencies in the narrative about Jesus in Mark render incredible some accounts of the *life of Jesus. It is not possible to trace how Jesus became during his *ministry increasingly conscious of his Messianic status. Wrede remarked on the oddness that Jesus commanded people he had healed to be silent about it (surely an impossibility! Mark 5: 43). Indeed sometimes (Mark 7: 36) they immediately disobeyed the request.

Wrede explained that Mark was giving not a historical account but a theological interpretation. It was impossible for Jesus' Messiahship to have been accepted until after belief in the *resurrection. Then it was realized that Jesus had been the Messiah all along. But how was it that he was not recognized? The answer was that he had sworn people to secrecy—though the *demons, being supernatural beings, recognized the truth (Mark 1: 24). When Jesus spoke in *parables, Mark's view was that they were deliberately couched in language which concealed the truth from the listening crowd (Mark 4: 12). So Mark's editorial explanation is given as an injunction of Jesus in 9: 9.

It is generally agreed that Mark's gospel *is* a theological interpretation born out of the Church's post-resurrection faith, but Wrede's theory is not acceptable as it stands. The secrecy motif embraces much more than Messiahship (as indeed with the public *miracles). And some apparently Messianic events (e.g. the entry into Jerusalem) attract no command to secrecy: cf. Luke 19: 40. There was no reason in the Jewish upbringing of the disciples for them to have come to regard Jesus as Messiah as a result of the resurrection. Another explanation of the material is that in the 1st cent. CE there were a number of Messianic conceptions on offer (as in the book of Jubilees). If Jesus was to reveal in any way his sense of self-understanding, he was bound to use the categories that were available. It was necessary to disavow, for example, any obviously nationalistic, xenophobic features of Messiahship. The injunctions to secrecy could therefore have been a precaution taken by Jesus, not always with success, to imply that he was a Messiah aiming to save by suffering and service; and he called upon the disciples to tread the same path; which perhaps explains their horror (Mark 8: 32) and foreboding.

metonym A word used in place of the person or thing actually meant: e.g. in Rev. 13: 3 the 'head' refers to the emperor and in 1 Pet. 5: 13 'Babylon' refers to Rome.

Micah, book of One of the OT Minor Prophets. Micah came from the countryside to Jerusalem during the reign of *Hezekiah (Jer. 26: 18) and was a contemporary of *Isaiah. The main part of the book appears to derive from about 720 to 700 BCE and is a denunciation of the wrongdoing in the nation, especially by the ruling classes. Mic. 6: 8 is often quoted as the great summary of prophetic teaching. Injustice will bring God's punishment in the form of military disasters. Micah had the *Assyrian threat in mind. It is possible that there are some interpolations from a later generation: 7: 1–7 has a different style, 7: 8–10 seems to suppose the destruction of the nation by the *Babylonians (early 6th cent.), and 7: 11–20 could have in mind the Return from Exile, and the rebuilding of the walls and the Temple. Mic. 5: 2 is interpreted by Matt. 2: 6 as a prediction of the Messiah being born in Bethlehem.

Micaiah (1) The best-known of several OT persons of this name was a prophet in *Ahab's court (1 Kgs. 22) who advised against a battle to wrest *Ramoth-Gilead from the Aramaeans. (2) A friend of *Jeremiah who heard the prophet's oracles being read aloud by *Baruch; after he told his father Gemariah what he knew, it was arranged for the words to be passed on to the king (*Jehoiakim; Jer. 36: 11–13).

mice Ceremonially *unclean creatures (Lev. 11: 29) for faithful *Hebrews, but allegedly eaten by idolaters (Isa. 66: 17), who will suffer the consequences. Golden mice were placed in the *Ark by the *Philistines (1 Sam. 6) to symbolize the *plague which was afflicting them. The hope was that, if they were carried back to Israel, the plague would cease.

Michael Hebrew for 'Who is like God?' According to Dan. 10: 13, 21; 12: 1 he is an *angel representing Israel. His role grows in the last centuries BCE and he is named as the chief adversary of the devil in the War Scroll of the *Qumran community and in the NT by Jude 9 and Rev. 12: 7.

Michal A daughter of *Saul, given to *David in marriage and his accomplice in his escape (1 Sam. 19: 11–17). She rebuked David for dancing in religious ecstasy before the *Ark of the Lord (2 Sam. 6: 20) when it was brought into Jerusalem. David, offended, refused to give her a child (2 Sam. 6: 23).

Michmash A village about 11 km. (7 miles) north of Jerusalem between *Bethel and *Jericho, where *Jonathan won a great victory over the *Philistines (1 Sam. 14: 13–15). Some of the exiles returning from *Babylon settled in Michmash (Ezra 2: 27).

Midianites An eastern people (Judg. 6: 3) who recur frequently in the OT. A group of them took *Joseph into *Egypt (Gen. 37: 28); the young Moses fled from Egypt to Midian (Exod. 2: 15) and there married the daughter of a priest; later, some Midianites were massacred by Moses (Num. 31: 1–12) and *Gideon defeated them (Judg. 7: 23).

midrash Early Jewish commentary on scripture. The purpose is to bring it up to date for readers of each generation; it is a kind of interpretation, and also a work of *reconciliation—explaining an original narrative by the insights of a later age. Thus, in 2 Sam. (24: 1) it is said that God, in anger, incited David to take a *census of the people, for which David, or rather the people, were punished by a pestilence (2 Sam. 24: 15). But by the time of 1 Chron. (21: 1) the proposal is said to be imparted to David by *Satan, which is more in tune with the theology of the Chronicler.

In the *Qumran community the narratives of the OT were retold and the commentators composed stories based on them, in the manner of preachers. Paul explains in 1 Cor. 10: 11 that scripture has, over and above its original meaning, a derivative meaning, applicable to his own day, which he endeavours to tease out.

In the gospels Matt. is particularly adept with the *midrashic* method. Much of this gospel is a re-writing of Mark, using the methods of *midrash*, but in the case of the *infancy narratives, where he had no predecessor, Matt. uses a variety of OT texts to expound his belief about Jesus as the Christ. Events of the OT that happened to Moses and David, among others, are repeated in Jesus, who is thus the New Moses destined to lead his people into a new *Promised Land. The *genealogy established that Jesus was of the house of David—a 'Son of David' was part of Messianic expectation. These narratives are not therefore historical in the modern sense, though they contain historical material.

mile In Roman measurement, 1,618 yards. Jesus instructed *disciples to go the 'second mile' (Matt. 5: 41) in the *Sermon on the Mount. Jews were obliged to carry Roman soldiers' equipment for distances up to 1 mile; so Jesus' command is an illustration of the behaviour of selfless love in face of the coming of the kingdom.

Miletus A port on the SW coast of Asia Minor of considerable antiquity. Taken by *Alexander the Great (334 BCE), it became a prosperous city. Paul invited the elders of *Ephesus to meet him there when he was on the way to Jerusalem for the feast of *Pentecost (Acts 20: 15–38).

milk *Canaan was seen as a land flowing with milk and *honey (Exod. 3: 8), and milk became a symbol for prosperity (Isa. 60: 16). Paul (1 Cor. 3: 2) and the letter to the Hebrews (5: 12) refer metaphorically to milk as elementary teaching at the beginning of Christian life; and converts are exhorted to crave uncontaminated milk, like new-born babes (1 Pet. 2: 2), which may be an allusion to milk which was given immediately after *baptism, mingled with honey—an echo of the Messianic age foretold by Isaiah (Isa. 7: 15). A cup of milk was also given to initiates in the pagan mystery rites.

mill The Palestinian mill consisted of two large stones put together for grinding corn. One kind was worked by two women (Matt. 24: 41) who sat opposite each other; another kind had a pole thrust through a hole and was operated by an ass, and this may have been in the mind of Jesus (Matt. 18: 6). A millstone was a vital piece of village equipment (Deut. 24: 6) and could even be used as a weapon (Judg. 9: 53). Its weight was proverbial (Rev. 18: 21).

millennium A thousand years. Christian *apologists, explaining the delay in the return of Christ to the world, appealed to Ps. 90: 4—in the sight of God a thousand years are but as yesterday (2 Pet. 3: 8). Rev. (17: 1–18: 24) expected the *End quite soon, after which there would be a millennium. In this time faithful Christians would reign together with Christ (Rev. 20: 1–6), and then *Satan would be allowed a final assault before being banished for ever into the lake of fire and brimstone (20: 7–14).

'Millennium' is often used to refer to the era of total bliss (beyond the thousand years of Rev.) where God reigns for eternity.

millo A fortress: the name derives from the Hebrew 'to fill'. It consisted of an earthen construction held together by walls and on which a building might be erected, such as the house at *Shechem (Judg. 9: 6) and at Jerusalem (2 Sam. 5: 9), which was kept in good repair (1 Kgs. 9: 15; 2 Chr. 32: 5).

mind The English translation of several Hebrew and Greek nouns which denote reflection and intention. Paul speaks of the mind of Christ in willingly accepting human limitations (Phil. 2: 5), and of himself as having the *spirit of Christ (1 Cor. 2: 16). Paul's readers are exhorted to have a similar disposition of humility and a concern for others. But it has to be a mind 'transformed' and renewed (Rom. 12: 2), for left to itself the human mind is far from good (Rom. 1: 28). Mind is not, in Paul, the higher or more estimable part of a human being.

mine There are few mineral resources in Palestine, and consequently few references to mining operations. Copper was extracted (Deut. 8: 9) in quantity (Zech. 6: 1). The fullest account of mining is contained in Job 28: 1–11, where the hidden metals are said to be as difficult to locate as *wisdom (28: 12).

minister One who serves either in a personal or a religious sense: as did Mark to Paul (Acts 13: 5) or priests at the *altar of the Temple (Ezek. 45: 4). In so far as the State was an ordinance of God, its administrators could be his ministers, or servants (Rom. 13: 4). Paul himself was a minister when he established the *collection for the poor Christians in Jerusalem (Rom. 15: 25). The model for NT ministers was Jesus himself (Mark 10: 45).

ministry In the OT there was a ministry of *priests and Levites in the Temple, and in NT times *synagogues had an official ministry (Luke 4: 20; Acts 18: 8). Paul gives a list of ministries operating in the Church and all are equally inspired by the *Spirit: there are *apostles, *prophets, *teachers, *evangelists, and pastors (1 Cor. 12: 28; Eph. 4: 11), though he does not here define duties and qualifications. Later, the duties of *bishops, *elders, and *deacons are more carefully explained in Acts 20: 17, 28; 1 Tim. 3: 1, 8; Tit. 1: 5, 7. The threefold ministry of the Church taking the form of bishops, *presbyters, and deacons later sprang from these roots. The first leaders of the Church, apostles, did not perpetuate their office, but an 'apostolic ministry' of witness to the Risen Christ was continued by local leaders. At first they (e.g. Timothy) acted under the general oversight of the apostles (Phil. 2: 19–24) and there were groups of presbyters at Jerusalem under the presidency of *James. The structure was probably borrowed from the synagogue.

*Timothy and *Titus are regarded in the *Pastoral Epistles as Paul's delegates in ministry and appointed by him by a *laying on of hands (2 Tim. 1: 6). Timothy and Titus in their turn appointed elders (presbyters) in every town (Tit. 1: 5) or bishops (*episcopoi*, Tit. 1: 7; 1 Tim. 3: 1–7), who again must appoint successors (2 Tim. 2: 2). There must be an unbroken succession of apostolic teaching. The duties of deacons are specified in 1 Tim. 3: 8–13.

It is unclear whether the 'bishop' of 1 Tim. 3: 1 is one of the elders of 1 Tim. 4: 14 who 'rule' (1 Tim. 5: 17) or whether the bishop is a single presiding minister. Possibly elders exercised corporate leadership in house *communities while the apostle or his delegate was absent, but eventually the local authority yielded to the leadership of one in the city who had special gifts of teaching. He became the president (*episcopos*) and on him devolved duties (but not of course that of being a witness of the Risen Christ) of the original *Twelve and Paul. In Asia Minor a ministry of presbyters existed (1 Pet. 5: 1, 5) and they formed a corporate defence against the destruction of the Church by persecution. They were under-shepherds under the chief shepherd, Christ; Peter the apostle, a 'fellow elder' had authority to address the Asian elders. Cf. 2 and 3 John.

Thus in the 1st cent. there existed both a local ministry and a general, apostolic ministry.

minuscule A MS written in small, cursive letters, not capitals. The style became normal from the 10th cent. CE; there are a large number of such MSS of the NT.

miracles Writers of the *Bible believed that God who created the world could and did intervene in the lives of people and in the course of nature. Mighty acts of power were performed by God at vital moments in the history of Israel—as at the crossing of the sea during the *Exodus (Exod. 14), which was seen as a confirmation of God's promise to his people. Healing miracles were wrought by God in response to the *prayer of *prophets (1 Kgs. 17: 20–2).

In the gospels the miracles of Jesus continue this tradition of healings by servants of God, who were inspired by the *Spirit. There are resemblances between the miracles of Jesus and those recorded by various charismatic figures in the Hellenistic world, including the Galilean Hanina ben Dosa, in the 1st cent. CE, before the Fall of Jerusalem in 70 CE. Characteristic of the miracles of Jesus (called 'signs' in the gospel of John) is the framework of *eschatology in which they are related; they are indications of the coming of the *kingdom of God; they are fulfilments of expectations of the OT as when the Feeding of the 5,000 (Mark 6: 30–44) echoes the miraculous feeding of Israel in the *wilderness; they are done in a context of *faith and prayer; they represent the conquest of *evil. But it is recognized that others besides Jesus worked similar acts of power (Mark 13: 22; Matt. 12: 27).

The mechanism by which the healings were effected (and there is no reason to doubt the historicity of Jesus' healings, which are told with sobriety and restraint) may well be explicable. For there are no accounts of restorations to wholeness such as the replacing of an amputated limb (unless it is the slave's ear, Luke 22: 51). The readers of the gospels at any rate saw healings as authentic works of God in which believers received the power of the Kingdom which had begun to come already in the person of Jesus. Modern readers might argue that human bodies are more than machines which can be manipulated by medicines and surgery, and positive spiritual and mental influence from Jesus could have been therapeutic.

The ministry of healing continued in the apostolic age. Peter and Paul healed the lame and the sick (Acts 3: 6; 14: 10; 16: 18). Because they were taken out of a lonely and desperate environment and shown friendship and love, it is not surprising if some who were ill felt better.

Miracles of Jesus other than healings may be symbolical rather than historical. The Feedings of the 5,000 and 4,000 (Mark 6: 30–44; 8: 1–10) with the significant numbers of loaves and fish and baskets may point to symbolism—of Jesus as God's agent of compassion for both Jews and Gentiles; the enormous quantity of water available at *Cana for Jewish *purification ritual and changed by Jesus into *wine (John 2: 1–11) may be designed to teach that the Church of Christ has superseded *Judaism and the *synagogue in the purpose of God.

Miriam Moses' and *Aaron's sister, who enjoyed an authority among the people and led the celebrations after the successful *Exodus (Exod. 15: 20). She did not hesitate to criticize Moses (Num. 12: 1, 10), though for this presumption she was afflicted with *leprosy and temporarily excluded from society. She was remembered as a national heroine (Mic. 6: 4).

mirror *Glass; but bronze mirrors were used in the OT period in Palestine.

Mishnah Rabbinic law as collected and organized into categories about 200 CE. It represented the culmination of the oral tradition of centuries. Worked out in Palestine, it became the basis of both the Palestinian and Babylonian *Talmuds, which are commentaries on the Mishnah, and in which some modifications were made. The laws covered both religious and social and economic matters, and in total there are sixty-three tractates. The one most often referred to in NT commentaries is **Pirke Aboth* for its light on *Judaism of the NT period.

Mithraism The cult of the Persian God Mithras (a sun-god), which was established in *Rome by NT times, and it had certain beliefs and rites which resembled those of the Christians: after Mithras had overcome and sacrificed a bull, he ascended into *heaven; *bread, water, and *wine were used by initiates in the rituals; and there was a *ministry of priests. Several German scholars in the early decades of the 20th cent. argued for an influence of Mithraism on the development of sacramental Christianity. The Mithraic festival of the Natalis Solis Invicti on 25 December was taken over as *Christmas Day by the Church in the 4th cent.

Mitylene On the coast of the Greek island of Lesbos; Paul stayed there on his way to Jerusalem (Acts 20: 14).

Mizpah The name in Hebrew denotes a look-out point, and there are five such places mentioned in the OT, of which the best known is that on the border between Israel and *Judah (1 Kgs. 15: 22). After the capture of Jerusalem by the *Babylonians in 586 BCE, Mizpah was established as the capital of the province and *Gedaliah installed as governor until he was assassinated (2 Kgs. 25: 25).

Moab The plain of Moab lay east of the *Dead Sea, south of the River Arnon. According to a scurrilous story (Gen. 19: 30–7) the Moabites owed their origin to an incestuous union between Lot and daughter. There were regular contacts between Moabites and Israelites, and *Ruth the Moabitess became *David's great-grandmother (Ruth 4: 17). There were conflicts with Moab during the reigns of *Saul and David (2 Sam. 8: 2). In the 8th and 7th cents. the Moabites were under the power of the *Assyrians and suffered defeat when the *Babylonians triumphed in the 6th cent. Moabites worshipped the god *Chemosh (Jer. 48: 46), who did not save them, and any future rehabilitation would be seen as the work of Yahweh (Jer. 48: 47).

Moabite Stone Discovered in 1868 and broken into pieces by local tribesmen when they realized its value; but reassembled and now in Paris. The inscription, dated about 830 BCE, in a language related to biblical Hebrew, records military successes against Israel by King Mesha and ascribed to the god Chemosh. A fragment of another stone was found in the country of Jordan in 1958.

Molech A god to whom children were sacrificed at Tophet near Jerusalem; the idolatrous practice is condemned (Lev. 18: 21; 2 Kgs. 23: 10; Jer. 32: 35). Evidently the reforms of *Josiah were not able to eradicate this remnant of heathen Canaanite religion.

molten sea Made by *Solomon (1 Kgs. 7: 23, AV; also called 'brasen sea', 1 Chr. 18: 8, AV; 'bronze sea', NRSV, REB, NJB) and used for the priest of the Temple to wash himself and the vessels before entering the holy place. It stood on twelve oxen of bronze; and it was broken in pieces and carried to *Babylon in 586 BCE.

money When *Joseph collected the 'money' that Egyptians brought as payment for corn (Gen. 47: 14–17), it consisted of metal objects rather than coins, though barter was also a common practice. Animals were used as payment of religious services in the Temple at Jerusalem (Lev. 2: 6–8; Luke 2: 24), but *silver was available during the era of the monarchy, and coins from the time of the Return under the Persians. The exchange of coins became standard practice for commercial transactions when *Alexander the Great had persuaded the civilized world of the advantages of a common currency: silver *shekels no longer had to be weighed if coins were authenticated with a royal stamp. Jewish authorities had been striking their own coins in bronze, by permission of the Persian government, in the 4th cent. BCE, showing the name 'Judah': but in 332 BCE licence to strike coins was restricted to the official mints of Alexander, and a national Jewish currency was not issued until the Maccabean period.

In the NT era, coins were minted by *Herod the Great and bore his name and a date. Roman governors also issued coins, such as the *denarius (Matt. 20: 10), which was a labourer's wage for one day's work, and the amount of the Temple tax. The thirty pieces of silver paid to *Judas (Matt. 26: 15) were shekels equivalent to 120 denarii. The smallest coin in circulation was 'the widow's mite' (Mark 12: 42), the Greek *lepton*, 'a penny' (NRSV, REB, NJB). During the Jewish revolt of 66–70 CE, some Jewish shekels were struck, and when the uprising was crushed coins commemorating the Roman victory showed Jews kneeling in submission bearing palms—the Jewish national emblem.

money-changers Employees of the bureau de change in Herod's Temple. Contributions had to be paid in a special currency since pilgrims came from many areas. Tyrian silver didrachmas were therefore used. This currency had the additional advantage of not being subject to inflation. Officials were employed in the Court of the Gentiles to exchange it.

monotheism The belief in a single God, or a religion affirming that belief, as opposed to polytheism, belief in many deities. There does not seem to be a time when Israelite worship was officially, as portrayed in the sources, other than monotheistic, though historically it may have evolved out of *henotheism such as was typical in the area. The 9th-cent. *Moabite Stone refers to the national god *Chemosh without denying the existence of other divinities. The Hebrews worshipped Yahweh as their national god and there are OT allusions to other nations' gods which imply a belief in their existence, alongside Israel's Yahweh. Such gods, it was held, were worshipped legitimately in their own country (1 Sam. 26: 19). Now and then undisguised polytheism crept in by the monarch's back door, as when Yahweh was allocated a female consort (1 Kgs. 11: 5), and opposite the Temple were built shrines to Kemosh and Milcom. The prophets were zealous in opposition and in the 7th cent. BCE these shrines were demolished by *Josiah. In Deuteronomy, monotheism was linked to strong ethical demands (Deut. 6: 4) and after the Exile there was no looking back: Second Isaiah (*Deutero-Isaiah) accepts that Yahweh is responsible for both *good and *evil (Isa. 45: 7). He can ridicule polytheism without serious comprehension of it but also without any expectation of being contradicted. Nevertheless, even when belief in the one God was fundamental, there remained traces of recognition of subordinate deities (Job 1–2) and the existence of *angels and of *Satan was also assumed.

Monotheism continues to be taken for granted in the NT (Mark 12: 29; 1 Cor. 8: 6), together with the existence of subordinate angels (Heb. 1: 4), and polytheism was blamed for the horrors of immorality (Rom. 1: 24–31). But trouble was imminent when Christians found themselves giving worship to Jesus. Jewish opponents accused them of abandoning monotheism. The Christian answer, after agonizing speculations and deviations, eventually issued in the doctrine of the *Trinity.

months The early Hebrews used a lunar *calendar, and the first of each month was signalled by the new moon. The names of the months were *Canaanite: several are mentioned in the OT (e.g. Ziv, the second month, 1 Kgs. 6: 1; Bul, the eighth month, 1 Kgs. 6: 38). During the Exile numbers were adopted, followed after the Exile by the *Babylonian names which had been learnt; those months mentioned in the OT begin with Nisan (March–April), during which *Passover was celebrated; then Chislev (Neh. 1: 1), Tebeth (Esther 2: 16), Shebat (Zech. 1: 7), and Adar (Ezra 6: 15). When a solar calendar was adopted, the old lunar calendar was retained in some religious circles (e.g. at *Qumran). In the NT there are no names of months.

moon The Hebrews attached special significance to each new moon, which was the occasion for *sacrifices (Num. 28: 11 ff.), and on the first day of the seventh month no work was to be done (Lev. 23: 24 f.). The moon was a celestial object of awe (Job 31: 26) and moon-worship was practised by Israel's neighbours. The modern superstition that the moon affects the mind ('lunatics') can be discerned in the OT (Ps. 121: 6) and in the Greek word translated 'epileptics' (Matt. 4: 24; 17: 15) in modern translation but 'lunatics' by AV.

Mordecai A Jewish hero in the story of *Esther, which illustrates the reversal of fortunes. His execution for lack of respect to

*Haman, favourite in the court of the Persian king *Ahazuerus, was authorized by the king, along with extermination of Jews throughout the empire. Mordecai was to be hanged; gallows were erected. But the Jewish queen Esther intervened; records existed to show the king that Mordecai had once saved him from an intended assassination (Esther 2: 22). Thus informed, the king revoked the execution order served on Mordecai and instead *Haman was hanged on the same gallows. Although fictitious, it is possible that the story may be based on some historical incident. It is celebrated by the minor Jewish feast of *Purim.

mosaics Pictures formed by inserting small pieces of coloured tiles of glass or marble into walls or floors; not a Hebrew skill and not mentioned in the OT, which prohibited all images of God. But some examples of mosaic work were discovered at *Masada from the period of *Herod the Great, and in grand 1st-cent. houses recently excavated in Jerusalem; and Jewish *synagogues in a more liberal age of the 4th cent. CE used the decoration. The best Christian examples of the art are the mosaics in several churches in and near Ravenna, Italy, dating from the 6th cent., which portray a regal, Byzantine, Christ.

Moses Leader of the Israelites. In a way comparable with the legends of other ancient peoples, the story of Moses begins (Exod. 2) with his being hidden at birth to escape a massacre, and then unexpectedly rescued by *Pharaoh's daughter and cared for at her request by a Hebrew nurse, who is in fact the child's mother. Thus Moses was brought up in Egyptian regal surroundings. As the narrative proceeds after the infancy stories, Moses left *Egypt after an altercation with a fellow Israelite, lived in Midian, got married, and was called to be the leader of his people at the Burning Bush (Exod. 3: 1–6). He was equipped by God with powerful weapons against Pharaoh, king of Egypt, who was responsible for reducing the Israelites to slaves; he had the *gift of working *miracles, and the co-operation of his eloquent brother *Aaron to act as his spokesman (Exod. 4: 14). The first gift was exercised by sending *plagues of ever-increasing ferocity upon the recalcitrant Egyptians (Exod. 7–11) culminating in the death of all the Egyptian *first-born sons (Exod. 12: 13), after which Pharaoh was at last coerced into releasing his slaves. They went out into the wilderness under Moses' leadership, where Moses divided the sea (Exod. 14: 21–15: 21), purified water (Exod. 15: 22–5), and guaranteed a victory over Amalek by holding up his hands (Exod. 17: 11–12).

Moses was the people's intercessor to God, and God's *prophet to the people. He gave them the *Law and he himself acted as a judge and arbitrator (Exod. 18). Thus, he gave the people a national identity and was on the point of leading them to the *Promised Land (*Canaan) when he ascended Mount *Nebo in Moab and there died.

The historicity of Moses has been much discussed by OT scholars. Generally, it is accepted that there are three *sources in the narrative known by the symbols *J, *E, and *P, of which the two first are earlier and considered to contain some reliable history; but there is no confirmation of Moses' existence from archaeology or other ancient Near Eastern documents. We are left only with the view held about the internal evidence of the *Pentateuch itself, where there are inconsistencies and doubts. Though few will nowadays be willing to support the traditional view that Moses himself *wrote* the five books of the Pentateuch, there are certainly those who regard the leadership of Moses as too firmly based in Israel's corporate memory to be dismissed as pious fiction.

In the NT Moses is named as the giver of the Law (Mark 7: 10; Rom. 9: 15; 2 Cor. 3: 13), as one whose *faith in God is an example to the Church (Heb. 3: 2; 11: 24), and as the prophet of the *Messiah (Acts 3: 22). In Matt. Jesus is portrayed as the 'second Moses', who proclaims the New Law (Matt. 5–7). At the *Transfiguration, Moses is said to be present and so authenticates Jesus' claim as uttered by the heavenly voice (Mark 9: 7).

Most High Frequently used in the OT for God (as in Pss. 21: 7; 91: 1) and sometimes in the NT, as when Jesus is addressed by the demoniac as 'Son of the Most High God' (Mark 5: 7) and also in *Stephen's defence speech (Acts 7: 48).

mote The translation in AV for a speck of dust in Jesus' *parable about judgmental attitudes (Matt. 7: 3) ('splinter' NJB). The mote is contrasted with a beam.

mountains In *Palestine there are two mountain chains, on each side of the River *Jordan, running from north to south. Trade and communications therefore moved in these directions, and east–west routes were sparse and limited to the few gaps in the mountains. In the OT mountains dominate the history of Israel and are mentioned in many connections—for assemblies (Josh. 8: 33), military sites (1 Sam. 17: 3), and battles (1 Sam. 23: 26); they are also useful for pasture (Ps. 104: 18). The term 'mount' is used for particular peaks within the ranges, and two are especially important for Israel—Mount *Sinai, where the *covenant between God and Israel was enacted (Exod. 19: 23) and the *Law received, and Mount *Zion, regarded as the dwelling place of God (e.g. Ps. 68: 16).

Mountains are held to be awesome and sacred (Deut. 11: 29), which is a continuation of the *Canaanite mythology, whose God *Baal, controller of thunderstorms, was thought to live on Mount Zaphon. For the Hebrews too, God was a God of the mountains (1 Kgs. 20: 23, 28 f.); on a hill was the resting-place for the *Ark (1 Sam. 7: 1) and on a hill ecstatic prophets congregated (1 Sam. 10: 5). Mount Zion, on which the Temple was built, will always be the 'mountain of God' (Pss. 2: 6; 78: 68–9), the 'holy mountain' (Isa. 8: 18). And this perspective about mountains reappears in the NT, notably in Matt. 5–7 where Jesus delivers the *Sermon on the Mount, which is the counterpart of the old *covenant now given to the New Israel as a new manifesto: Jesus is the New Moses who saves people from their servitude. So too, the *glory of Jesus in the *Transfiguration takes place on a mountain (Mark 9: 2) and on a mountain the Eleven are given their marching orders (Matt. 28: 16 ff.).

mourning customs Death in the OT is not a private matter for the bereaved family. *Job's calamities brought in friends and weeping neighbours for a whole week (Job 2: 12–13) and *Jeremiah assumes a corporate mourning ritual which includes cutting off hair and self-laceration (Jer. 16: 6), though prohibited in the *Law (Lev. 19: 27–8). It was obviously contrary to custom that *Ezekiel did not mourn for his dead wife (Ezek. 24: 17). Professional singers sometimes sang dirges (2 Chron. 35: 25). These rites might last for a week (Gen. 50: 10) or three weeks (Dan. 10: 2), but a precise period of time was not laid down. For a full year after the death of a parent the children avoided festive celebrations.

Some of these customs are mentioned in the NT, although Paul discourages excessive grieving (1 Thess. 4: 13); but there was certainly communal mourning round a *synagogue (Mark 5: 38), and Matt. (9: 23) adds the detail that there were flute-players in attendance.

Muratorian canon A list in *Latin of NT books dating from the second part of the 2nd cent. in an 8th-cent. MS discovered by Lodovico Antonio Muratori in 1740. It is usually held to be the earliest known list of writings accepted by the Church as normative. The total of twenty-two books omits Hebrews, 1 and 2 Peter, James, and 3 John, but includes the Wisdom of Solomon and the Apocalypse of Peter (a 2nd-cent. work with a description of heaven and hell, not eventually accepted by the Church).

murder Unlawful killing was prohibited in the *Ten Commandments (Exod. 20: 13), for which capital *punishment was prescribed (Exod. 21: 12). Safe havens, however, were available for those who had killed another accidentally (Deut. 19: 5), subject to enquiry when the offender reached the gate of the designated city (Num. 35: 24). This provision acted as a restraint on tit-for-tat killings and the consequential social unrest. At the Return from Exile the system of asylum was replaced by the judiciary (Ezra 7: 25–6), but under the Romans the right to inflict the death penalty was withdrawn from the Jewish courts (John 18: 31), though there is some evidence that there were exceptions to this regulation. In the *Sermon on the Mount (Matt. 5: 21–2) Jesus extends the law against murder and forbids even anger.

music Hebrews had no talent for the visual arts, but they compensated for that with a variety of singing and instrumental music, and the narrative mentions Jubal the archetypal lyre and pipe player (Gen. 4: 21). There were many opportunities for musical celebrations (Judg. 11: 34; Isa. 5: 12) and people at work would sing (e.g. Isa. 21: 12; Jer. 25: 30), as would soldiers or leaders even after some were slain in battle (2 Sam. 1: 19–27). Music was important in the Temple, and probably some of the psalms were, as tradition claims, composed by King *David against the time when the Temple should be built. David is said to have been an accomplished player of the lyre (1 Sam. 16: 16–18) and a dancer (2 Sam. 6: 14–15). During the centuries of monarchy there was singing and jubilation at a coronation (Ps. 72). After the Return from Exile music was important in the new Temple, and some of the psalms (e.g. Ps. 73) are attributed by their editors to the authors. Other comments are directions about the performance, e.g. 'upon stringed instruments'; and *Selah*, which occurs frequently, probably indicates a pause where perhaps another group of singers might take over. Some psalms (e.g. 13; 20) were intended to be sung antiphonally. The stringed instruments were lyres and harps with ten strings, and wind instruments were flutes and trumpets of ram's horn. There were also tambourines and cymbals of bronze, and bells.

Some of these instruments are named also in the NT, and Jesus and the disciples sang the *hallel* (Pss. 114–18) after the *Lord's Supper. Singing certainly formed part of early Christian *worship (Eph. 5: 19) which was expected to continue in *heaven (Rev. 4: 10).

mustard Said, with some exaggeration, to have 'the smallest of seeds' (Mark 4: 31). It is used in Jesus' *parable to illustrate that the small initial response to the coming of the *kingdom of God is destined to increase dramatically.

myrrh A luxury item produced from a plant grown in Arabia and India. It was a gift brought to the infant Jesus in the narrative of the *Magi (Matt. 2: 11) and was used as a narcotic (Mark 15: 23) and for embalming (John 19: 39). Other uses may have been as an aphrodisiac (Prov. 7: 17; Esther 2: 12), but in *Rome at the hour of judgement no one will buy it (Rev. 18: 13), for there is no market for luxury goods. The later tradition that myrrh symbolized grief or mourning is not biblical.

Mysia Part of N. Asia Minor on the Aegean Sea; Paul passed through two of its cities (*Troas, Acts 16: 8, and *Assos, Acts 20: 13).

mystery A secret in the mind of God which was unknown to human beings until revealed. In Dan. (as also in the *Qumran scrolls) the secret is God's plan for a coming new age when the wicked will be punished and the righteous rewarded. The divine mysteries are sealed in *heaven until made known at the end of this age. In the NT the concept of mystery has taken an additional meaning with the belief that the *Messiah has come and the *kingdom of God has been inaugurated.

The previously hidden divine plan for the world has now been made known to the *Gentiles (Rom. 16: 25–6); it is the mystery of God (Col. 2: 2) or the mystery of Christ (Eph. 3: 4). The Gentiles are now fellow heirs with the Jews in the privileges of the gospel. But whereas this mystery is apparently revealed to all alike in the Church in Col. (1: 26), in the later Eph. (3: 5) the revelation is for apostles and prophets only. Some would take this difference to be an argument for regarding Eph. as unlikely to be from the same hand (Paul's) as Col.

Contemporary with the early Church were the pagan mystery religions in which initiates were admitted by a rite resembling *baptism which promised *immortality. The Eleusinian mysteries were Greek in origin, as were rites of Dionysus, who was believed to have descended into Hades and risen from the grave. Similar mystery religions were founded on the legends of Orpheus, or of Mithras and Adonis, gods who died and returned to life, and were the focus of annual commemorations of their dying and rising, with corresponding hopes offered to the initiates. After the NT era some of the features of these mystery religions may have been incorporated into Christianity and in 150 CE Justin Martyr observed the similarities of Christianity with Mithraism, and regarded them as Satanic counterfeits. But later the word 'mysteries' was used in the Church to designate the *sacraments.

myth In modern parlance, a myth is a *legend or fairy-story unbelievable and untrue but nevertheless disseminated. It has a more technical meaning in biblical studies and covers those stories or narratives which describe the actions of the other-worldly in terms of this world, in both OT and NT. In Gen. the *Creation and the *Fall are myths, and are markedly similar to the creation stories of Israel's Near Eastern neighbours. It was a mistake of expositors and preachers in the past to treat these chapters as 'historical' accounts of the origin of the universe and the cause of original *sin. There are also myths in OT (e.g. Isa. 25: 7) and NT (e.g. Matt. 24: 31) to express beliefs about the end of the world and God's judgement. Some theologians use the term 'myth' to denote a way of expressing intimations of supernatural significance; they are rendered intelligible by means of familiar terms derived from valid natural experience.

Much of the discussion in the second half of the 20th cent. about biblical myths was prompted by *Bultmann's *demythologization enterprise.

N

Naaman An army officer under Benhadad of Syria who had contracted *leprosy and heard from an Israelite slave-girl about the miraculous powers of the prophet *Elisha. Naaman was sent by his king to *Samaria with a letter to the king of Israel (apparently then a vassal of Syria). Elisha ordered Naaman to wash in the River *Jordan, and he was cured. Naaman then vowed to worship Yahweh and, in the belief that a god could only be worshipped on his own soil, carried back two mule-loads of Israelite soil (2 Kgs. 5) to Syria.

Nabataeans Occupants of territory E. and SW of the *Dead Sea, who spoke a form of *Aramaic. They were important in the inter-testamental and NT periods. King *Aretas IV (9 BCE–40 CE) gave his daughter in marriage to Herod *Antipas, who later divorced her in order to marry *Herodias, a union condemned by *John the Baptist (Mark 6: 17–29). The ethnarch of Aretas intended to arrest Paul at *Damascus (2 Cor. 11: 32).

Important excavations have taken place at *Petra, one of the cities built by the Nabataeans, 71 km. (47 miles) from the Dead Sea, and chosen as their capital. Although an earthquake in the 4th cent. CE severely damaged the city, what remains shows evidence of high technical skills.

Naboth Owner of a vineyard which King *Ahab wished to buy because it adjoined the royal property; but when Naboth refused to sell, Queen *Jezebel conspired to bring a false charge of *blasphemy against him, for which he was stoned. Ahab moved in to take possession of the vineyard but was confronted by *Elijah, who predicted disaster for his dynasty and for Jezebel (1 Kgs. 21).

Nag Hammadi A modern town in upper *Egypt. It is the nearest town to *Chenoboskion, where texts relevant to the early Christian Church were discovered in 1945, two years before the accidental discovery of the first scrolls by the *Dead Sea. These twelve papyrus codices are as important for the study of early Christianity as the *Qumran scrolls are for *Judaism. They are Coptic translations of the original Greek; their literary *genres are diverse—sayings, prayers, apocalypses, and epistles—and their theological stances as they wrestle with the problem of evil are also diverse. But notably they (especially the Gospel of Truth) are evidence for a flourishing Christian *Gnosticism in the area, though some of the texts are not Christian at all, and in some there is a strong Jewish input. One work, the *Valentinian Exposition*, is named after the Gnostic teacher Valentinus (a native of Egypt who died in Rome about 165 CE). The codices themselves were probably copied and translated in the 4th cent.

Among the texts is the gospel of *Thomas, which gives alternative versions of some of the sayings of Jesus preserved in the canonical gospels, together with new material. Some scholars have recently argued that Thomas is independent of the NT gospels and contains traditions of considerable historical value.

Nahum, book of Nothing is known about the author whose book is included among the Minor Prophets, except that he was born in Elkosh; but the single theme—the fall of the *Assyrian capital *Nineveh (612 BCE)—permits an approximate date. The prophet is exulting about the impending doom of the empire which more than any other neighbour had heaped sufferings on *Judah and its northern relation Israel; but one section at the end (3: 18–19) is a taunt-song which seems to be celebrating the news that Nineveh has already just fallen. These verses might therefore have been added by an editor to the earlier part of the book which anticipates the fall. The mention of the fall of *Thebes (663 BCE) gives the earliest possible date of composition. The revolt of King *Manasseh and his subsequent pardon (2 Chron. 33) suggests a weakening of Assyria and a climate in which Nahum with his powerful images celebrating God's vengeance over Judah's merciless foe would have escaped retribution.

nails (1) Women captives are to pare their fingernails (Deut. 21: 12). (2) Iron nails used by builders are mentioned by Isaiah (41: 7) and *Jeremiah (10: 4) in their sarcastic description of the construction of idols. Nails were sometimes used by the Romans for fastening the victims of *crucifixion and in 1968 Israeli archaeologists discovered, in a cemetery on the Mount of *Olives, a nail driven through the anklebones of a man crucified in the 1st cent. CE. Evidence that Jesus was nailed rather than tied to his cross is in the request of *Thomas (John 20: 25) to see 'the mark of the nails'.

Nain A village SE of *Nazareth where Jesus encountered a funeral procession and resuscitated a widow's only son (Luke 7: 11–17).

names Most Hebrew personal names in the *Bible are compounds which begin or end with the divine name Yahweh, such as Jehoshaphat, 'Yahweh establishes justice', or Joel, 'Yahweh is God'. Many names of towns and villages were also compounds with the name of God—Beth-el, 'house of God'—or compounded with the place's reputation—Beth-lehem 'house of food'. Some places preserved their ancient *Canaanite name (e.g. *Megiddo; Zech. 12: 11), while others were changed by new rulers, as when *Herod renamed *Samaria as Sebaste. The forms -iah and Jo- are derived from the divine name Yahweh.

Some personal names are changed, indicating the authority of the one who effects the change: *Jacob to Israel (Gen. 32: 28), Simon to Peter (Mark 3: 16). The latter was an unexpected change and it indicated that the very human Simon was to be a 'rock', for Peter = Cephas, Aramaic for 'rock', and under this new name he takes his place in history.

names of God

1. *In the OT. Yahweh*: but when being read aloud the title *Adonai* ('My great Lord') was substituted out of reverence. When vowels were added to the Hebrew text, the vowels of *Adonai* were combined with YHWH to jog the reader's memory to use *Adonai*. (Hence the incorrect English hybrid 'Jehovah'). Some modern Jewish writers observe the traditional reverence by writing 'G—d'.

El: God, in Canaanite vocabulary, but found also in the OT, sometimes in conjunction with another word, e.g. Beth el = House of God.

Elohim: the more common form in the OT; it is plural in form, emphasizing majesty.

El Shaddai: God Almighty (perhaps originally, God of the mountains).

Adonai: My great Lord—used for kings, but after the Exile to replace 'Yahweh' in worship.

Baal: the Canaanite deity to whom the Israelites sometimes turned; for this they were denounced by *Hosea, and other prophets.

2. *In the NT. Theos*: the standard Greek word for God translating in *LXX the Hebrew *Elohim*. Used of Jesus once (John 20: 28).

**Kurios*: the Greek used for *Yahweh in the *LXX, and printed as THE LORD in some English versions. Often used of Jesus (e.g. John 20: 28), but can also be applied to a person in a high position, 'sir' (Matt. 27: 63; John 12: 21). Another Greek noun, *despotes*, meaning Lord in the *LXX is found for God, translated 'Master' (NRSV, Luke 2: 29), and for Jesus at 2 Pet. 2: 1.

**Alpha and Omega*: the first and last letters of the Greek alphabet represent God as the Beginning and the End (Rev. 1: 8) and applied to Jesus (Rev. 22: 13).

The Holy One: Used of God by Isa. in OT and in NT (Rev. 16: 5) but also of Jesus (Mark 1: 24; Luke 4: 34; John 6: 69).

Naphtali A northern tribe taking its name from a son of *Jacob (Gen. 30: 7–8) and occupying a strip of land west of the Sea of *Galilee. *Barak from *Kedesh in Naphtali was inspired by *Deborah to rout the forces of *Sisera (Judg. 4: 6–10). Later Naphtali was invaded by the *Syrians (1 Kgs. 15: 16–20) and by the *Assyrians who deported part of the population (2 Kgs. 15: 29). *Jonathan the Maccabean 160–142 BCE defeated the Syrians at Naphtali, but it was only under *John Hyrcanus (134–104 BCE) that the area was fully restored as Jewish, and as such it was the scene for much of Jesus' *ministry.

Narcissus A Christian in *Rome; possibly a house-church met in his home; members of his household are sent greetings by Paul in Rom. 16: 11.

narrative criticism A development in biblical studies, especially in USA, since about 1970. The main thesis is that readers (e.g. of the gospels) should read the narratives and respond to them as the authors hoped. Historical criticism is not rejected, but the analysis of the gospels into sources and *pericopae, to what is historically probable or theological interpretation, has reached no consensus, so another approach is suggested. Whereas scholars have often identified repetitions in a gospel or inconsistencies or gaps as indicating diverse sources or flawed editorship, narrative criticism invites the reader to assess the work as a whole and to note its stylistic characteristics which resemble those of other literary works with a beginning, a middle, and an end.

Some narrative critics, however, regard the newer discipline as complementary to, not a substitute for, the achievements of previous scholarship. Primarily they are concerned with those elements in the text which are relevant to the plot or theme or storyline: how the text engages the reader in its world and system of values; they note the characteristics and points of view of the narrator and his asides to the reader.

There is no single meaning in a biblical text. The gospel of Luke will be read differently by a black Christian in S. Africa from a white Protestant in the USA. Readers are active, not passive recipients of an unbiased text. In the past, Paul's reference to the female *Junia (Rom. 16: 7) has often been taken to be a man 'Junias' on the ground that Paul could only have accorded apostolic status to *men*—'of note among the apostles'.

Mark contains two accounts of the miraculous feeding of a multitude—of 5,000 (6: 31–44) and 4,000 (8: 1–9) and whereas historical criticism has often taken this as an editorial duplication of a single event, narrative critics read the two stories as we have them. Mark may have intended by the first story to show that Christ was the Saviour of the Jews, and by the second that he was the Saviour of *Gentiles, since the event took place on Gentile soil (7: 31). Narrative critics propose to put the reader in the position of the *intended* readers of the gospel and ponder on the remarkable obtuseness of the disciples who fail to understand, even after the second feeding, what Jesus was teaching and signifying (8: 17).

narrative ethics Some modern scholars urge readers of the *Bible to treat it not as a compendium of doctrine but as a story which by sheer familiarity enables us to shape our own lives. For example, in *liberation theology, the narratives of the *Exodus are held up as a model for oppressed Christians in Latin America with which to confront the authorities and to claim their human rights.

Nathan Of several OT persons with this name, the best known is the prophet in the time of *David. He was consulted about the propriety of building a Temple, and advised David to leave the project to his son *Solomon (2 Sam. 7: 13); it would be a sign of the permanence of the house of David. Nathan rebuked David for arranging the death of *Uriah in battle in order that he could marry Uriah's wife *Bathsheba (2 Sam. 11: 12), which illustrates the religious tradition in Israel of a prophet's privilege of confronting the monarch himself. At the end of David's reign, Nathan was instrumental in organizing the right of succession for Solomon (1 Kgs. 1: 13–14). He is said to have been a palace historian (1 Chron. 29: 29).

Nathanael A *disciple greeted by Jesus as an Israelite without guile (John 1: 47) in an obvious reference to Israel (Jacob) of Genesis, who was notoriously a deceiver. *Jacob dreamt of a ladder from earth to *heaven (Gen. 28: 12) and Jesus tells Nathanael that the *Son of Man is now to be that ladder (John 1: 51). The identification of Nathanael with Bartholomew, though often made, has no basis.

nations Sometimes an alternative English translation for '*Gentiles' (non-Jews), as Matt. 28: 19 and elsewhere.

A regular feature of the OT prophetic books is the 'oracles against the nations'.

natural In Rom. 11: 17–24 Paul explains that *Gentiles have been grafted on to the people of God by using the image of an olive tree. Gentiles are like a new shoot with new life and vigour and, as in arboriculture then and now, the new gave a splendid richness to the failing 'natural' tree, which stood for Israel. Paul is warning the Gentiles that they still depend on the tree of Israel for fullness of life.

nature, law of Both the OT and NT accept that God has laid on the world a certain order and structure and that there are fundamental moral principles given to the *Gentile nations even though they do not have the blessing of the *Torah. There is a law embodied in nature, which the *birds know (Jer. 8: 7) and a *covenant made by God with all the sons of *Noah, the ancestors of the whole human race (Gen. 9: 9–10). The Mosaic covenant was made with a group of those who were already in covenant with God, and these hints of a primeval law were in due course elaborated by the *rabbis. It is possibly what is demanded in the apostolic decree (Acts 15: 29), and what is affirmed by Paul in Rom. 1: 19–21, and 2: 14–15. The good pagan does virtuous acts 'by nature' in accordance with the law of God. Similarly in the prologue of the gospel of John it is said that the Word 'was the light of all people' (John 1: 4): the Word as light was in the whole world, and those who welcomed it included both faithful Israelites (Hos. 1: 10) and also Gentiles who observe the Noachian precepts 'by nature' (Rom. 2: 14–16). All are children of God. Just by being human, all men and women have a capacity for a knowledge of God, and this is apart from, or precedes, the special revelation of God made through Jesus, or the Bible, or the Church.

Nazarene Jesus is so called, probably on account of his coming from *Nazareth in *Galilee, though there are difficulties about the precise meaning of the Greek (which is transliterated 'Nazorean' by NRSV). When Matt. (2: 23) claims an OT prophecy in support, no one has yet discovered any such reference. It may be a punning interpretation of Isa. 11: 1, where a similar Hebrew word means 'branch', and if so Matthew is clumsily applying a possible Messianic prophecy to Jesus.

Nazareth A town in S. Galilee approximately half-way between the Sea of *Galilee and the Mediterranean not mentioned in OT but described as the residence of *Joseph and *Mary (Luke 1: 26 f.) and so of Jesus (Luke 2: 4; 4: 16). Excavations in 1955 revealed that the town was inhabited before the Christian era, and in 1961 an inscription found in Caesarea mentions Nazareth. It was a satellite town of Sepphoris, 6.5 km. (4 miles) away, and had a population when Jesus lived there of about 500. It was assumed in Judaea that 'nothing good' (John 1: 46) could possibly come out of the Graeco-Jewish area of Galilee.

Nazirite In the OT Nazirites were ascetics who had taken a vow, or whose parents had taken a vow on their behalf, as with Samuel and Samson (1 Sam. 1: 1–11; Judg. 13: 7). They were to abstain from intoxicants, to allow their hair to grow long, and to avoid all contact with a corpse. They protested against the moral decline of civilization. *John the

Baptist may have been a Nazirite; and Paul paid the expenses of four men completing vows, possibly Nazirites (Acts 21: 20–30). The resemblance of 'Nazirite' to '*Nazarene' may have carried an overtone of Jesus' consecration from the womb.

Neapolis The port for *Philippi (modern Kavella), where Paul landed (Acts 16: 11) after crossing the Aegean Sea from *Troas.

Nebo (1) A mountain in the land of *Moab which commanded a magnificent view over the *Dead Sea and the *Jordan valley from which Moses was allowed to observe Canaan at the time of his death (Deut. 32: 49). (2) A *Babylonian god; the idol was carried in procession at the New Year festival and is the theme of *Deutero-Isaiah's taunt (Isa. 46: 1). The god's name is incorporated into that of several Babylonian kings.

Nebuchadnezzar Four kings in ancient *Mesopotamia had this name. The second reigned in *Babylon from 605 to 562 BCE. (The alternative spelling Nebuchadrezzar (e.g. Jer. 39: 5, NRSV, REB) is closer to the Babylonian form.) He was a powerful and cruel monarch who defeated *Assyria and *Egypt; in 597 BCE he captured Jerusalem and deported King *Jehoiachin to Babylon and appointed *Zedekiah in his place. When Zedekiah rebelled, Nebuchadnezzar renewed the assault on Jerusalem, destroyed the Temple, and took many of the people to Babylon (586 BCE), though *Jeremiah had urged submission to the Babylonians (Jer. 27: 11).

neck Used by OT prophets in a symbolic way: a neck 'as stiff as iron' described stubbornness (Isa. 48: 4). Breaking the yoke off the neck of Israel and *Judah was *Jeremiah's way of prophesying their future liberation.

Paul praises *Priscilla and *Aquila for risking their necks to save his life (Rom. 16: 4).

Negeb A dry area in the south of *Judah through which were several roads of strategic importance; many villages were built there as defences. At the time of the *Exodus *Amalekites dwelt in the Negeb (Num. 13: 29). From the time of the Exile (586 BCE) it was occupied by *Edomites, and later known to Greek-speakers as *Idumaea (Mark 3: 8).

Nehemiah A post-exilic patriot and hero. Although an OT book bears his name, little is known of him. He was said to be the cupbearer of King *Artaxerxes I of Persia, by whom he was appointed governor of *Judah about 445 BCE. He supervised the rebuilding of the walls, returned briefly to Persia (Neh. 5: 14), and then came back to Jerusalem (Neh. 13: 7).

Nehemiah, book of The last of the historical books of the OT and probably a continuation of the work of the Chronicler. There are several sections: chs. 1–7 (narrated in the first person—cf. the 'we-sections' of Acts) describe Nehemiah's return to Jerusalem and his first governorship. In chs. 8–9 there is an interlude for *Ezra to read the law, which is a continuation of the book of Ezra 7–10 (Ezra may have come to Jerusalem *after* Nehemiah, under Artaxerxes II). Neh. 11–13 continues the account of the rebuilding of the walls under Nehemiah's second governorship, after providing a comprehensive list of officials. Opposition from neighbouring peoples had to be withstood and Jewish laws enforced. Nehemiah lived with a consuming conviction of the providence of God, who guided him.

neighbour Israelites were under an obligation to respect and assist their neighbours (fellow citizens) in the community; it was what was implied by membership of a 'holy nation' (Exod. 19: 6). This was also required by the laws listed in Lev. 19 and Deut. 22. In the age to come at the *End neighbours will live in peace and love each other (Zech. 3: 10), which implies that the laws in the *Pentateuch were often not observed. Jesus quoted Lev. 19: 18 (e.g. Luke 10: 27) but extended loving one's neighbour beyond the commonly accepted limitation that it did not apply to foreigners, still less to enemies. (Although Matt. 5: 43 quotes the OT as ordering hatred of enemies, it is nowhere to be found in the OT. However, it is a fair summary of many passages, such as the merciless injunction of Deut. 20: 16.) The *parable of the *Good Samaritan represents the anti-racist teaching of Jesus; it was shocking to Jesus' contemporaries that a despised Samaritan should have behaved as a good neighbour.

Nero Roman emperor, 54–68 CE; after a reign that began with moderation, Nero became increasingly brutal and irresponsible. In 64 there was a serious fire in a largely residential area; it was where Nero then built himself a palace, and in order to scotch rumours that he had himself caused the fire, he laid the blame on Christians, some of whom were executed by *crucifixion or burning or by wild beasts in the circus. *Tacitus remarks that this cruelty won wide sympathy for the Christians amongst the populace.

It is possible that Peter and Paul may have been among the victims, and that a reference to Nero may be behind the number of the beast, 666, in Rev. 13: 18.

nethinim Employees in the second Temple; Jewish tradition regarded them as the descendants of the *Gibeonites who saved themselves from death at the hands of the invading Israelites by means of a ruse; they were punished by *Joshua to be 'hewers of wood and drawers of water' (Josh. 9: 27). They are supposed to have been in *David's service (Ezra 8: 20). The *Mishnah reports that *nethinim* could marry into the families of *priests and Levites; though this is late evidence, it could be that, from the time of the Return from Exile, the *nethinim* were absorbed into mainstream Jewish life.

nets Fishing nets could be either let down (Luke 5: 4) from a boat, or cast (John 21: 6), or thrown from the shore (Matt. 4: 18). The drag-net (Matt. 13: 47) which collected a variety of species of *fish (there were twenty-four species in the Sea of *Galilee) would have been weighted at the bottom, with the top held up by floats.

new birth A *blessing from God, giving a hope based on Christ's *resurrection (1 Pet. 1: 3). It is the fruit of *baptism. Paul has the same doctrine but a different vocabulary in Rom. 6: 3–4. The terms of 1 Pet. were also common in the pagan mystery religions, and it has been suggested that if 1 Pet. is a late and pseudonymous document, it could have been influenced by these mysteries. Equally, however, the language of new birth is found in the *Dead Sea scrolls, and so there is the possibility of Jewish origin.

New Testament The twenty-seven books comprising the second part of the Christian Bible, not arranged in the chronological order of writing but roughly according to subject matter: first, the narratives about Jesus, followed by the response to him in the life of the primitive Church (the Acts of the Apostles); then the letters of Paul, of which seven are generally thought to be from his own hand; letters by other early Christians; and finally the Revelation to John. This last is sometimes called the *Apocalypse, and is a Christian vision of the future hope, couched in the form of an address to seven Churches in Asia Minor. The name 'New Testament' is an alternative translation of 'New *Covenant', which looks back to the prophecy of *Jeremiah (31: 31–4) of a new relationship between God and people. As the title for a collection of writings the name can be traced to the end of the 2nd cent., and was standard usage by the 4th cent. But the precise contents of the NT took time to be established: there were doubts about 2 Peter (a very late writing), 2 and 3 John, Jude, and Rev. The last two books were also regarded by Martin Luther in the 16th cent. CE and his followers as lacking the authority of the rest, and Luther also gave less value to Hebrews and James. The present list of twenty-seven was set out by Athanasius in Alexandria (367 CE). *See* canon.

All the NT writings are in Hellenistic (koine) Greek. The divisions into chapters date only from the 13th cent., and into verses from the Greek NT of Stephanus in 1551, adopted for the English translation of William Whittingham (1557).

New Testament Apocrypha Early Christian writings partly parallel to canonical books of the NT but not accepted by the Church. Some are complete; of some, only fragments survive; others are known by name only. Those in the gospels category tend to supply information about Jesus that the curiosity of a later generation welcomed, especially about his infancy, or teaching of a Gnostic kind said to be given after the *resurrection. Much of the literature was intended to be edifying, but some seems unpleasing to a modern reader.

In order to gain credibility, some of the apocryphal writings were ascribed to historical personages such as *Thomas—a tendency already to be found within the canonical NT itself, e.g. 2 Peter is certainly not the work of the apostle Peter. There is also a gospel of Nicodemus, which includes the Acts of Pilate; this section marks a further step on the road, apparent in the four gospels, of diminishing *Pilate's responsibility for the *death of Jesus. The perpetual virginity of *Mary makes its first appearance as a doctrine in the 2nd-cent. Protevangelium of James (also known as the book of James), a narrative about the infancy of Jesus.

These apocryphal writings were not so much formally rejected by the Church as never universally accepted. They were suspected of heresy by Church leaders such as *Irenaeus (*c.*190 CE), who was the first to use the term 'apocryphal' or 'secret' of these works. A modern reader finds some of their stories too bizarre to be taken seriously as historical record—as when in the Infancy gospel of Thomas Jesus aged 6 is said to have broken a pitcher and miraculously put it together again. It can be said of the writings that there are a few sayings attributed to Jesus which have the ring of truth. And they certainly put us in touch with popular piety of the early Christian centuries. Many other alleged sayings of Jesus appear in early Christian writings, and even in certain Jewish and Islamic texts.

new year The first day of the seventh *month (Tishri). It was celebrated like a *Sabbath with rest and *sacrifice (Lev. 23: 23–5; Num. 29: 1–6) and blowing of trumpets. There is evidence that the New Year might have been celebrated, alternatively or additionally, on the first day of the month *Nisan (Exod. 12: 2), which was the custom among the Babylonians, where the annual ritual celebrated the renewal of the monarchy for the coming year. Thus there were probably two calendars in the OT, one beginning in the spring, one in the autumn. The New Year of Tishri assumes an autumnal start, but in the spring reckoning that is the seventh month.

Nicanor (1) The Syrian general (1 Macc. 3: 38) charged by *Lysias with suppressing the Maccabean revolt and routed by *Judas near *Emmaus in 166 BCE (1 Macc. 4: 14) and at Beth-horon in 161 BCE (1 Macc. 7: 26, 43). (2) One of the seven men of good repute appointed to administer charity (Acts 6: 5).

Nicolas (or Nicolaus) A *proselyte from *Antioch who was one of the seven men of good standing appointed to supervise the distribution of charity (Acts 6: 5).

Nicodemus A *Pharisee well-disposed towards Jesus, mentioned only in the gospel of John (7: 50–2). He seems to typify a kind of puzzled half-belief which the evangelist would like to convert by a new birth (3: 4, 9); but Nicodemus is perhaps held back by fear (19: 39). He is aware that a sympathetic approach to Christ will attract Jewish hostility. The evangelist portrays Nicodemus as a possible Jewish convert to the Church, with expulsion from the *synagogue as the consequence (John 9: 22).

Nicolaitans A group with Gnostic tendencies living in *Ephesus and *Pergamum who are denounced in Rev. 2: 6 and 15.

Nicopolis A town in Epirus, NW of *Corinth, where *Titus was asked to meet up with Paul (Titus 3: 12); but since this letter is about ministry in *Crete, it is part of the evidence in the case against the Pauline authorship of this epistle.

night Although it is a time for dreaming divine messages (Matt. 2: 12–13, 14), night is, rather, a metaphor for danger (Ps. 91: 5) and grief (Ps. 30: 5), and many psalms speak of deliverance coming 'in the morning'. In the New Jerusalem there will be no night (Rev. 21: 25) and God's people will enjoy his presence all the time.

Nile The great river, over 6,400 km. (4,000 miles) long, which flows from south to north through *Egypt and has been fundamental for the maintenance of life. Its tremendous floods made an unforgettable impression on the *Hebrews, who found nothing like it when they arrived in Palestine. *Pharaoh ordered Hebrew male infants to be cast into the Nile, but Moses was saved (Exod. 1: 22; 2). One of the *plagues turned the Nile into blood (Exod. 4: 9) and in the *Priestly source not only the Nile but all the waters of Egypt (Exod. 7: 19). That the Nile turns reddish and foul is an observed natural phenomenon; within the context of the deep racial memory of the Exodus and its cult and in thankfulness to God, the natural event has become a *miracle.

Nimrod The first great king to be mentioned in the OT (Gen. 10: 8); he ruled an area consisting of the later *Assyria and *Babylonia. Legend had it that he was a mighty huntsman (Gen. 10: 9).

Nineveh The capital of *Assyria, on the eastern bank of the River *Tigris, and once the most powerful city in the world; but it was captured by the *Babylonians and *Medes in 612 BCE. The city's end was expected by *Zephaniah (2: 13–15) and Nahum, whose whole book is aimed against the 'city of bloodshed' (Nah. 3: 1), which will be devoured as the locust devastates the harvest (3: 15). The prophet *Jonah was said to have preached at Nineveh (Jonah 3: 3; Luke 11: 30).

Nisan The first month of the Jewish year (Neh. 2: 1) according to the later reckoning which began the year in the spring, corresponding to March in Europe, during which (on the 15th) the *Passover was celebrated. The name was adopted from *Babylonia after the Exile and formerly was known as the month Abib, a *Canaanite name.

Noah In the *legend of the *flood only Noah and his family of all the human race are said to have survived the deluge (Gen. 6–9), chosen by God because of Noah's goodness of character (though it was not sustained indefinitely, Gen. 9: 21). In the *ark in which the family huddled there were also pairs (the *P source) of other living species (or seven pairs of *clean animals and two pairs of *unclean animals, Gen. 7: 2–3, the *J source). After surviving the deluge, Noah emerged to offer *sacrifice to God and to receive the promise confirmed by the sign of the *rainbow that there would never again be a similar cosmic catastrophe. Noah also received the *covenant with laws (Gen. 9: 17) to apply to the whole future human race. In the NT the waters of the flood (through which Noah was saved) are regarded as a *type of *baptism (1 Pet. 3: 18–22).

Nob A place a little to the north of Jerusalem (Isa. 10: 32) where eighty-five priests lived with *Abimelech. David, in flight from *Saul, was given refuge and food there (the holy bread of the presence). When Saul heard of this through Doeg the Edomite, he ordered the death of the priests, which Doeg carried out (1 Sam. 22: 6–23). Only *Abiathar survived—hence the mistake in Mark 2: 26 (not repeated by Matt. or Luke) which mentions *Abiathar instead of Abimelech.

Nod The land 'east of Eden' where *Cain settled to escape from the presence of the Lord (Gen. 4: 16).

nomads People who for the most part exist on the move, taking sheep, cattle, and asses with them. None of the OT characters are nomads: the *patriarchs led only a semi-nomadic life since their wanderings were interrupted by periods at *Shechem (Abraham, Gen. 12: 6), Gerar (Isaac, Gen. 26: 6), and Paddam-Aram (Jacob, Gen. 28: 5; 31: 17–18). During the *wilderness journeys after the *Exodus, the Israelites grazed their flocks around the oasis at *Kadesh-Barnea.

Nomadic life was harsh; it enjoined a good deal of mutual dependence and of hospitality to strangers; but the austerity was viewed with nostalgia by *prophets who contrasted the simplicity of living in tents and observing the festivals with the luxury of urban culture (Hosea 2: 14 f.) and the *apostasy of *Canaanite religion.

novice Used in AV at 1 Tim. 3: 6 for 'a recent convert' (NRSV, REB) or a 'new convert' (NJB). Nowadays the word 'novice' is applied to a new member of a religious order.

numbers That the *Hebrews were able to count is clear because *David held a *census (1 Chr. 21), though the number of a million (1 Chr. 21: 5, NRSV, REB) or even eleven hundred thousand (NJB) warriors is exaggerated. Calculations were made in tens (based on the number of fingers on the hands), but certain numbers had a special significance beyond the purely arithmetical; one, for the unity of God; seven (being the days of creation) signified completeness; and forty was the period of a generation and signified a large span of time. In the NT and afterwards other numerals became important: one and eight (cf. Luke 9: 28) as being the first day of the week when Jesus was raised from the dead. *Seven remains important, as does seventy: seventy preachers are sent out (Luke 10: 1); Peter is to forgive seventy times seven (Matt. 18: 22); seven assistants are appointed by the apostles (Acts 6). Seven Churches are addressed (Rev. 1: 3). *Twelve is also important as a symbol in the NT, as corresponding to the twelve tribes of Israel (Luke 22: 30).

Numbers, book of The fourth book in the OT; the title derives from the *census (1: 1–46). The literary sources *J, *E, and *P are all found in Num., and behind them were numerous local traditions which they assimilated. Probably the book as finally edited has conflated a variety of tribal memories into a unity, accepted as its own by the whole nation. The book therefore represents Israel's interpretation of her past.

The historical narrative of Numbers, round which other literary types have been collected, takes the progress of the Israelites from the entrance into the *wilderness, through the wilderness from *Sinai to Transjordan; it records conquests in that region, and finally the approach to *Canaan. These events cannot be regarded as strictly historical reminiscences since Num. in something like its present form was probably compiled much later, about 550 BCE.

Nun (1) *Joshua's father (Exod. 33: 11). (2) The fourteenth letter of the Hebrew alphabet.

Nunc Dimittis The first words in *Latin of the canticle based on *Simeon's words (Luke 2: 29–32) in praise of having seen the beginning of the deliverance of his people and the *salvation of the *Gentiles. These verses are incorporated into the evening office of some Churches.

Nympha(s) A Laodicean man or woman greeted by Paul (Col. 4: 15) who hosted meetings of the local Christian community.

oaks Powerful trees which grew extensively in OT times, and those on the plain of Sharon were comparable to the great forest of *Lebanon (Isa. 35: 2). Some oaks were invested with a kind of sanctity (Gen. 12: 6, NRSV, NJB, which REB gives as '*terebinth*').

oath(s) A sanction employed by religious and secular authorities to ensure that a person, especially a witness in court, shall tell the truth. Oaths were part of Israelite culture: the name of God is invoked and divine punishment is expected if the truth is distorted—it is profanation (Exod. 20: 7). The practice continued in NT times (Matt. 23: 16–22), but in the *Sermon on the Mount Jesus explained that in the *kingdom of God such expedients belied relationships of *truth and *love (Matt. 5: 34). The epistle of James urges believers to renounce the use of oaths (5: 12).

Obadiah, book of One of the Minor Prophecies in the OT, though nothing is known of the prophet himself. It is the shortest book in the OT; perhaps placed fourth in its group in order to follow immediately the reference to *Edom in Amos (9: 12). It is a sustained execration of the Edomites for joining the enemies of *Judah in sacking Jerusalem. The situation seems to be the siege of Jerusalem by the *Babylonians in 586 and the bitterness of the prophecy is aggravated by the kinship of the Edomites to the people of Israel (Obad. 12); the Edomites were descended from *Esau (Gen. 36: 9), the brother of *Jacob (Israel), and they occupied the land later called *Idumaea. The freshness of the denunciation argues for a date of composition not far off 586 BCE, and certainly before Edom was occupied by the *Nabataeans in the 4th cent.

Obad. does not have any hint of the duty of a *Gentile mission, as Jonah does; but the threat of vengeance which is its theme is ascribed to the hand of God (Obad. 21), who is absolute Justice (Obad. 8), rather than being wrought by the injured people.

Obed Father of *Jesse and grandfather of *David (Ruth 4: 22); this otherwise undistinguished character is significant as an ancestor of Jesus (Matt. 1: 5).

obedience The verb 'to obey' translates the Hebrew 'to hear' (as Isa. 42: 24), or 'to observe' the commandments (Exod. 16: 28). Disobedience is punished; Moses' failure to maintain God's holiness among the Israelites deprived him of his greatest wish—to enter the *Promised Land (Deut. 32: 51). In the NT the obedience of Christ to God is celebrated by Paul (Phil. 2: 8) and Christians are called to obedience to Christ (Heb. 5: 9). There is also a proper obedience to the State (Rom. 13: 1–5), and there is a hierarchy of obedience within a household (Eph. 5: 21–6: 9; 1 Pet. 2: 13–3: 7).

oblation A cereal offering presented to God (Lev. 2: 4) and denounced by *prophets (Isa. 1: 13) when unaccompanied by social righteousness. In Christian usage, the word was applied to Christ who offered himself in *sacrifice. Later the word was used of the *Eucharist as a commemoration of Christ's self-offering, and to the elements of *bread and *wine used in the service.

offence Paul writes that Christ crucified is an offence or stumbling-block (Greek, *skandalon*) which arouses opposition among Jews (1 Cor. 1: 23; Gal. 5: 11).

offering, burnt The most usual type of OT *sacrifice (Lev. 1), when blood of an animal was poured on the *altar by a high priest (Lev. 3: 5). The purpose was to win the favour of God.

offering, guilt A private *sacrifice of a ram to atone for an act of desecration (Lev. 5: 15). It is not clear how this differed from the 'sin offering'.

offering, peace The slaughtered animal was shared as meat between the priest (who had the right thigh), his colleagues (who had the breast), and the offerer, who ate the rest within two days (Lev. 7: 15; 19: 6–8).

offering, sin Used for eliminating impurity from the *sanctuary at festivals (Num. 28: 15) and offered at a private ceremony when a prohibition had been inadvertently infringed.

officer The word is used of a variety of civil servants and military and judicial personnel (Num. 11: 16) or officials in the Temple (Jer. 20: 1). In the NT an officer is a kind of policeman (Luke 12: 58) attached to the Temple (John 7: 32; 18: 12).

Og King of Bashan, the last survivor of the Rephaim (Deut. 3: 11); defeated by Moses (Deut. 3: 3) in a vital victory long celebrated in Israelite worship (Ps. 135: 11).

oil Olive oil was important in Israelite economy (Num. 18: 12), used for light (Exod. 27: 20), cooking (Ezek. 16: 13), and healing (Mark 6: 13), and as hair conditioner (Ps. 23: 5). *See* anointing.

ointment (1) Olive oil mixed with fragrant spices (Esther 2: 12). (2) The 'costly ointment' poured on Jesus' head at *Bethany (Matt. 26: 7) was of 'nard' (Mark 14: 3), an aromatic oil from India, alleged to be worth 300 *denarii (John 12: 5). An alabaster jar would have contained about an ounce, not the 'pound' of John 12: 3, which is an exaggeration.

old age The elderly were revered in the Near East for their accumulated *wisdom (Job 12: 12) rather than resented as a burden on society (cf. Isa. 3: 5). Length of years was a sign of God's reward for keeping his laws (Deut. 30: 19 f.), though the OT is realistic about the ills of the aged (1 Kgs. 1: 1 ff.; Gen. 27: 1). One of the Ten Commandments required respect for parents (Exod. 20: 12), and this is more likely to be addressed to those with elderly parents than to children.

Old Testament Term used by Christians for the first part of the *Bible, though many scholars prefer to use the term 'Hebrew Bible' as a more neutral description. It is always regarded as authoritative (except by certain heretics, e.g. *Marcion), though the precise contents of the OT have been disputed. In NT times the *Samaritans accepted only the first five books of the OT, and at the Reformation Catholics accepted all the books contained in the *LXX, while Protestants accepted the Hebrew *canon and relegated the inter-testamental books contained only in the Greek LXX to the *Apocrypha.

The Hebrew Bible is divided into the *Law, the *Prophets, and the *Writings:

The Law (*Torah) reached its final form about 400 BCE and is the *Pentateuch: Genesis to Deuteronomy.

The (Former) Prophets consist of the historical books which record activities of prophets—Joshua, Judges, 1 and 2 Samuel, 1 and 2 Kings.

The (Latter) Prophets are Isaiah, Jeremiah, Ezekiel, and the twelve Minor Prophets (Hosea to Malachi).

The Writings comprise the Psalms, Job, Proverbs, and five Rolls (Ruth, Song of Songs, Ecclesiastes, Lamentations, and Esther), Daniel, Ezra–Nehemiah, and Chronicles.

The Greek OT, the *Septuagint (*LXX) included additional books and compiled them in a different order. The historical books are grouped together, so that after the *Pentateuch, the order is: Joshua, Judges and Ruth; the books of Samuel, Kings, and Chronicles; 1 Esdras (not in Hebrew OT), 2 Esdras (Ezra–Nehemiah in Hebrew); Esther; followed by Judith, Tobit, and 1 and 2 Maccabees (not in Hebrew); then 3 and 4 Maccabees. The next section consists of the poetical books: Psalms, Proverbs, Ecclesiastes, Song of Songs, Job, Wisdom of Solomon, Ecclesiasticus (Ben Sirach), Psalms of Solomon. The Prophetic Books consist of: the twelve Minor Prophets, followed by Isaiah, Jeremiah, Baruch, Lamentations, Letter of Jeremiah, Ezekiel, Susanna, Daniel (including the Prayer of Azariah and the Song of the Three Children), and Bel and the Dragon.

The OT is more of an anthology than a unified document with a single theological outlook. The great insight of modern critical scholarship has been the recognition of different strands in the historical books and the overall theological stamp which has been imposed upon them at a date far later than that of the events described. The OT was compiled over a period of more than a thousand years. A few pieces of *poetry, such as Judg. 5, may have originated before 1000 BCE, while Daniel is as late as 165 BCE. It is one task of critical study to reconstruct the course of the history of Israel, and the records of neighbouring peoples and other discoveries of archaeologists have been important; and a few clues are provided by the books themselves about the time and place of their composition.

Old Testament in the New Testament It was fundamental to the first Christians, being Jews, that their belief in the Messiahship of Jesus should be demonstrable from the OT, the 'oracles of God' (Rom. 3: 2). Hence the frequent citations of OT in the NT. Much of the NT argumentation follows the procedures of contemporary *Judaism, such as **midrash*: the teaching about the *Bread of Life in John 6 : 25–59 is based on an OT text resembling Exod. 16: 4. The gospel of Matt. is especially emphatic in encouraging its readers to understand how Jesus was the fulfilment of the destiny of Israel. Thus Matt. 2: 13–15 describes how *Joseph took the family to *Egypt so that Jesus would escape the wrath of *Herod. This is a most unlikely piece of history, since the obvious way to flee would have been into the hills of *Galilee or the desert of *Judaea or to the east, not to make an enormous trek to Egypt. But Egypt was the place where the nation of Israel had sojourned and later returned to Palestine; Israel was God's 'son' (Exod. 4: 22). So Matt. cites Hos. 11: 1 ('out of Egypt I have called my son') to prove a point. Just as Israel was God's son (and was disobedient), so Jesus is God's obedient son. In Matt. 5–7 Jesus bases his moral teaching on the OT and stretches it to the limit, and beyond, of human practicability.

All the gospels use the OT to show how the *life, *death, and *resurrection of Jesus are fulfilments of the OT; especially important are Ps. 110: 1 and Ps. 118. The use of '*Son of Man' derives from Dan. 7: 13, and Isa. 53 is quoted to explain Jesus' suffering, while Zech. 12: 10 is used as a prophecy of the *crucifixion (John 19: 37). It was less easy to find allusions in the OT to resurrection, but Hosea 6: 2 may have been used, though it is not quoted in the NT, and possibly Jonah 1: 17 (Matt. 12: 40). *Typological exegesis is used by Paul (e.g. 1 Cor. 10: 1–13) and *allegorical interpretation (Gal. 4: 21–31). The epistle to the Hebrews is one long exposition of the superiority of the new *covenant to the old.

olive An evergreen tree found all over Israel. The fruit was crushed and squeezed and the oil stored in vats.

Olives, Mount of A hill about 1.6 km. (1 mile) long to the east of Jerusalem, separated from it by the Kidron valley, on which *Gethsemane (Hebrew for an 'oil-press') was situated. Jesus resorted to it for *prayer (Mark 14: 32) and stayed at *Bethany on the south side (Matt. 21: 17). The mount was a place of *eschatological associations: from it Jesus foresaw Jerusalem's destruction (Mark 13: 3 ff.) and Luke pictured the *ascension and promised return as taking place there (Luke 24: 50; Acts 1: 12). It had been prophesied that on the day of the return of the Lord to Jerusalem he would stand on the mount (Zech. 14: 4), and this was applied to Jesus.

omega The last letter of the Greek alphabet; used symbolically with *Alpha as a title of Jesus (Rev. 22: 13), implying that he is Everlasting, and the contemporary of all the generations of mankind.

Omri Proclaimed king of Israel after a coup in 885 BCE (1 Kgs. 16: 15 ff.); he established his capital at *Samaria (1 Kgs. 16: 24), appropriately for a new dynasty, and *Assyrian inscriptions long regarded Israel as 'Omri's country'.

Omri was astute in his dealings with Israel's neighbours and gave his country prosperity; but his alliances (he gave his son *Ahab in marriage to the Phoenician *Jezebel) allowed pagan practices into the land, for which he is roundly condemned by the Deuteronomic editor of 1 Kgs. 16: 25–6.

On Heliopolis, 'city of the sun', in N. Egypt, with a huge temple and several obelisks. *Joseph married the daughter of the high priest of On (Gen. 41: 45). In the *LXX it is named as one of the cities built by the Israelites (Exod. 1: 11). The book of Jer. (43: 13) predicts that *Nebuchadnezzar will break the obelisks of Heliopolis, and *Ezekiel foretells that its young men will be slain (30: 17).

Onan A son of *Judah who was condemned for failing in the duty of a brother. When his brother died, Onan wilfully disobeyed the *Levirate law (Deut. 25: 5–10) which required him to make good his deceased brother's lack of offspring (Gen. 38: 6–10), by practising *coitus interruptus* (?), and, according to the narrative, was punished by death (38: 10).

Onesimus A slave who absconded from his owner Philemon and sought sanctuary with Paul, who wrote to Philemon urging him to treat Onesimus gently. From the epistle to Philemon, verse 10, it would seem that Onesimus became a Christian believer, and even a leader in the Church (Col. 4: 9). Scholarly speculation has suggested that he became bishop of *Ephesus, and that there he gathered together Paul's surviving letters and wrote the epistle to the Ephesians as an introduction to the collection.

Onesiphorus Mentioned by 2 Tim. 1: 16–18 as a friend of Paul's from *Ephesus who helped him at *Rome.

onyx One of the precious jewels on the high priest's breastpiece (Exod. 28: 20) and considered as valuable as gold (Job 28: 16). It is also one of the twelve jewels set in the wall of the heavenly Jerusalem (Rev. 21: 20).

Ophir A place of uncertain location, but accessible by ship (1 Kgs. 9: 28); it was famous for gold which was mined there (1 Kgs. 10: 11).

oracles Messages from deities, as from Apollo at Delphi in classical Greece, where the priestess uttered after inhaling vapours. Kings and generals sought advice at such *sanctuaries and paid well when the advice was favourable. In the OT oracles by means of lot (**Urim* and **Thummim*) were known (1 Sam. 14: 41). Oracles uttered by the OT prophets in the 8th to 6th cents. BCE were preserved in writing and attained recognizable forms such as *parables and songs, e.g. Isa. 13. In the NT messages from God guided *Joseph (Matt. 1: 20; 2: 13) and the early Church (Acts 10: 10). An oracle was conveyed to Paul after he had appealed three times to the Lord about his thorn in the flesh (2 Cor. 12: 9).

oral tradition Material passed on from generation to generation by word of mouth. There was a long process of gestation before books of the OT reached their final written form. In ancient Israel pieces of *poetry, stories of heroes, *legends about founding figures, and songs of celebration circulated amongst groups and tribes. Repeatedly accounts of the momentous events of the *Exodus and the wilderness journeys were transmitted to descendants and these were followed later by accounts of the early kings and the prophets; they all gathered embellishments and reinterpretations in the process. Then written collections were assembled—of laws, of prophetic utterances, stories about particular individuals. These writings in turn became the raw material for the recognizable written sources which can now be identified in the *Pentateuch. Some modern scholars, especially in Scandinavia, where comparisons have been made with the Norse epics, have argued for a very long period of oral tradition; but it is more usually held that written sources go back to about the 10th cent. BCE.

The NT also had a period, much shorter, during which stories about Jesus and the first Christians were passed from person to person, and group to group, as the needs of teaching and controversy required. It meant that much material that was not needed simply perished and that what survived has passed through the modifications and interpretations that the occasions demanded, though no doubt a limit was set to development by reason of the retentive memories of those days. Sayings of Jesus were assembled and may have been strung together and used when the gospels were written. Matt. and Luke contain *parables and shorter sayings of Jesus in their gospels, though their original contexts may have been lost. Mark compiled a *Passion narrative with great care, to which he attached the *eschatological predictions and affirmations of ch. 13, and prefaced all this with a rapid account of Jesus' journeys, conflicts, choice of disciples, *miracles, and parables expounding and proclaiming the advent of the *kingdom of God.

Oral tradition was not a Christian innovation. It was an established medium of Jewish scribes and *rabbis. It was necessary to update the written injunctions of the *Torah, so there were many legal stipulations (*halakah*) and homiletic expositions of the OT narratives (*haggadah), and this oral transmission continued for several centuries until finally established in the *Mishnah.

orator The word translated 'orator' by AV (Isa. 3: 3) is better rendered 'enchanter' (NRSV, REB) or 'soothsayer' (NJB). The 'orator' Tertullus (Acts 24: 1) is better described as an 'advocate' (REB, NJB).

ordain, ordination (1) In the OT God is said to 'ordain' (AV) or 'establish' (NRSV) the government of the Universe (Ps. 8: 3) and its geographical divisions (1 Chron. 17: 9). This means that God 'orders' or 'appoints' or 'institutes' or 'invests'. (2) Similarly, he 'ordains' certain persons with authority, and they may delegate that authority to others, as Moses to *Joshua (Num. 27: 18–23), by means of the *laying on of hands. In the NT God institutes governments (Rom. 13: 1) and is said to ordain in advance events concerning *salvation (Rom. 8: 29–30). (3) God also ordained certain persons to perform functions: *Jeremiah was to be a prophet (Jer. 1: 5). But the idea of ordination to a function appears more strongly and frequently in the NT: the Twelve were ordained or charged to preach and to heal (Mark 3: 14–15) and elders (*presbyters) were ordained or appointed by Paul and *Barnabas (Acts 14: 23), since Paul himself had been ordained an apostle (Gal. 1: 15). As an apostle, he ordained others to lead local communities (2 Tim. 1: 6).

The *Pastoral Epistles represent an early stage in the institutionalizing of ordination and *ministry. The intention is to continue the ministry of the apostles and to remain faithful to their *teaching. In ordaining, the Church, under the inspiration of the *Holy Spirit (1 Tim. 4: 14), provides for the continuing proclamation of the gospel and for service to the community in the name of Christ. The laying on of hands is the outward sign of the *gift of the Spirit.

Paul writes that *Titus was 'appointed' (2 Cor. 8: 19, NRSV, REB; 'elected', NJB; 'chosen', AV) by the Churches, and the Greek verb literally has the meaning of stretching out the hand: possibly therefore a description of the similar appointments—of the Seven (Acts 6: 6), of Barnabas and Paul at *Antioch (Acts 13: 3).

ordinance Translation (especially in AV) of a variety of Greek words: simple commandments of the Mosaic *Law (Eph. 2: 15); commandments of God (Rom. 13: 2); requirements of law in general (1 Pet. 2: 13).

Origen (185–254 CE) Theologian, born in Alexandria and imprisoned during the persecution of Decius (250 CE). Origen was a prolific writer but attracted opposition for his belief that all creatures will ultimately be saved. He was first and foremost a biblical scholar and wrote commentaries on Matt., Luke, and Romans, and on other books, but these works have not survived. He held that scripture was to be understood under three senses: literal, moral, and allegorical. He wrote a reply to one of the most perceptive pagan opponents of Christianity *Against Celsus* (247 CE).

Orion In Greek mythology a giant and a huntsman who was killed by the goddess *Artemis and then placed amongst the stars with a girdle, a sword, a lion's skin, and a club, according to Homer. *Amos (5: 8) and *Job (9: 9) refer to the creation of this constellation of seven stars shaped, it was held, like a man with a narrow waist, and in Job 38: 31 God asks whether a mere man can loose Orion's girdle and so usher in the winter season (Job 38: 29–31).

ornaments Sometimes these were worn as charms to protect against evil forces (Gen. 24: 53); sometimes as adornments, forbidden in times of mourning (Exod. 33: 4–6). Women's ornaments consisted of necklaces, bracelets, rings, brooches, and anklets (Isa. 3: 16).

orphans Children without parents. The *Hebrews were exhorted to look after the fatherless (Exod. 22: 22:) for whom God particularly cared (Deut. 10: 18; Ps. 146: 9). *Jeremiah rebuked those who violated this law (Jer. 5: 28; cf. Ezek. 22: 7); and the same care is expected of Christians (Jas. 1: 27). Of all people, widows and orphans were most at risk in both OT and NT times. They turned for support to kings (Ps. 72: 1–4, 12–14) but were often rebuffed (Jer. 5: 28).

orthodoxy 'Right thinking'; used of mainstream Christian thought as opposed to *heresy; thus, the gospel of John is orthodox if it maintains that Jesus was truly human while also being *Son of God, whereas *docetism is heretical. However, in a narrower sense, Orthodoxy has come to be used particularly of those autonomous Churches of the East who are in communion with the See of Constantinople.

orthopraxis 'Right-doing', the complement of *orthodoxy (cf. 1 John 2: 3–6). In Christianity right belief about Jesus must be validated by imitating Jesus, including his suffering (cf. Luke 9: 23). The condition of coming to the light is to do the works of God—a Johannine theme emphasized in modern *liberation theology. It is often held that Judaism, which has no formal creeds, is a religion of orthopraxis—though it is interesting that traditional Jews are regularly described as 'Orthodox'.

ostraca Potsherds: broken pieces of clay pottery used in the Near East and *Egypt for writing brief messages and receipts. Being almost imperishable, many of these have been discovered by archaeologists. A notable discovery, in 1935, included some twenty-five of the 'Lachish Letters' to the army commander (Yaosh) who was defending *Lachish; they came from a worried Judaean captain (Hoshaiah) when the *Babylonians were approaching the city shortly before 587 BCE (Jer. 34: 7).

oven A device made for cooking food. Fuel to heat ovens was placed either round the clay chamber, which was usually open at the top, or inside. Ovens were primarily for baking, but could also serve as stoves on which pots could be placed. The metaphorical use of the word denotes the fate of the wicked, who will burn as an oven, on the Day of Yahweh (Mal. 4: 1). In Matthew's *Sermon on the Mount (Matt. 6: 30) Jesus argues that if God has given some natural beauty to grass which is cast into the oven as soon as it is grown, surely he must care much more for human beings?

Oxyrhynchus A site in *Egypt, south of Cairo, excavated at the turn of the 19th–20th cents. by British archaeologists who found a vast quantity of Greek papyri. They consisted of personal and business notes, official documents, and literary texts, including portions of the NT and of the apocryphal gospel of *Thomas. Many date from the 3rd and 4th cents. CE and are housed in libraries round the world, including American universities.

Those that have been published shed considerable light on the language of the NT writers.

ox(en) A castrated bull used throughout the OT period for ploughing (1 Kgs. 19: 19) and hauling (Num. 7: 3), and also as *sacrifices (1 Kgs. 8: 63). There were some humane regulations: an ox and an ass were not to be unequally hitched together (Deut. 22: 10) and they were to be rested on the *Sabbath (Exod. 23: 12). When treading out the grain, oxen were not to be muzzled (Deut. 25: 4). This last prohibition is quoted by Paul (1 Cor. 9: 9) with the surprised comment, 'Is it for oxen that God is concerned?'

P

P Symbol used by OT scholars to designate the Priestly source or Priestly Writer who is regarded by the majority of OT scholars as being one of the four main sources of the *Pentateuch. P material is recognized by a concern for ritual and the prominence accorded to *Aaron. It has a doctrine of *creation according to which God has control of all the nations of the world but in which *Abraham and his offspring have a special role. P offers *genealogies and accounts of cultic institutions which are put back into the period of the *Exodus and the settlement in the *Promised Land, though in fact P was probably compiled (but the date is in dispute) in the exilic century (6th cent. BCE).

palace There is plenty of archaeological evidence that royal residences were richly decorated and well fortified. The OT gives details of *Solomon's palace (1 Kgs. 7: 1 ff.), which contained a splendid banqueting hall 45 m. (150 feet) long made of wood from cedars of Lebanon. In the Portico of Judgement (or Hall of the Throne) was the king's throne (1 Kgs. 7: 7). Other royal palaces are mentioned in Esther (1: 5) and Daniel (4: 29). In a *parable of the Messianic age which Jesus proclaims, the stronger one will prevent his palace being invaded by the powers of darkness (Luke 11: 21–2).

palaeography From Greek words 'old' and 'writing'; it is the rigorous discipline of deciphering ancient handwriting on *codices and papyri. The styles of writing determine the dates of MSS. So palaeography is an essential stage in textual *criticism of the NT.

Palestine The traditional extremity of biblical Palestine was *Dan in the north and *Beersheba in the south (1 Sam. 3: 20), which are about 240 km. (150 miles) apart. There are two inland seas (Sea of *Galilee and the *Dead Sea) and the River *Jordan joins them, flowing through a deep valley, often flooded. The Sea of Galilee (also known as Gennesaret, or Tiberias) is about 21 km. (13 miles) long by 12.8 km. (8 miles) across, and has an abundant stock of edible fish. But the Dead Sea, much bigger, is in very inhospitable country, below sea level. The River Jordan rises in the Lebanon mountains and is fed by numerous small streams, such as the Cherith (1 Kgs. 17: 1–7).

Two valleys run across the centre of the country from the Mediterranean Sea on the west to the River Jordan—the plain of *Megiddo (or Esdraelon) and the valley of *Jezreel.

The main range of mountains runs from north to south just to the west of the Jordan, from upper Galilee to the desert. Beyond the mountains and by the Mediterranean, south of the plain of Megiddo, are the fertile plains of Philistia and Sharon, but at the northern end the plain is interrupted by the Carmel range of hills, in the middle of which is a pass at Megiddo. Here was the vital point to control military and commercial traffic between *Egypt and *Syria and the east. It was the scene of great battles, as when *Josiah tried to resist the Egyptians (2 Kgs. 23: 29 f.).

Other mountains mentioned in the Bible are *Hermon, a few miles north of Dan, *Tabor, west of the Sea of Galilee, and *Sinai, south of Palestine and north of the *Red Sea. The wilderness round Sinai is linked to Beersheba by a desert area known as the *Negeb.

palimpsest A parchment MS on which a text has been written over a previous writing which can sometimes be recovered by ultraviolet photography.

palm A tall, hardy tree with a remarkable capacity for survival in a difficult climate (Ps. 92: 12–14). The fruit (dates) was valuable and the leaves were gathered and strewn as a symbol of *joy (Mark 11: 8). They became a national emblem of the Jews. Jesus' arrival in Jerusalem was marked by the waving of palms (John 12: 13).

palsy Used in AV for paralysis, loss of muscular power, and cured by Jesus (Mark 2: 11). The 'withered hand' or atrophied hand (Mark 3: 1) was probably also paralysed.

Pamphylia An area in SW Asia Minor in which Jews were living in the 2nd cent. BCE (1 Macc. 15: 23) and which was included in Paul's itinerary. At Perga John Mark left the party (Acts 13: 13) and at *Attalia Paul and *Barnabas boarded a ship for *Antioch (Acts 14: 24–6). Pamphylia was raised to provincial status by the Romans in 74 CE.

Paphos City established in the 4th cent. BCE on the SW coast of Cyprus. Paul and *Barnabas met the Roman proconsul *Sergius Paulus, who was astonished at the teaching about the Lord (Acts 13: 12) when Paul caused the magician *Elymas to be struck blind.

Papias Bishop of *Hierapolis (140 CE); quoted by the historian Eusebius as having been once told by 'the Elder' that Mark, not himself a follower of Jesus, acted as a translator or interpreter for Peter. Because he often listened to Peter's preaching, Mark was able to write down what he remembered, though admittedly this was not 'in order'. These words were the basis which established the tradition of the connection of Mark's gospel with Peter. Though it was not first-hand *eyewitness testimony, it was enough to establish the authority of this gospel even after Matt. and Luke became more popular.

Papias also wrote that Matthew wrote down 'sayings' of Jesus in Hebrew, and it has been speculated that he was referring to what scholars nowadays call '*Q'; but evidence that such a source has existed in Hebrew (if it existed at all) is entirely lacking. Perhaps Papias meant that Matthew compiled a convenient anthology of OT proof texts for Christian preachers.

papyrus A plant which flourished near the delta of the *Nile in *Egypt (Job 8: 11), used for making writing material. The stem was cut into sections, split open, and sliced into strips which were then laid flat and a second layer superimposed crossways before the sheets were pressed together. Sheets could be stuck together to form a roll, or (by the 2nd cent. CE) a *codex (a book with pages). Documents made of papyri survived well in the dry climate of *Egypt, and fragments of the Greek Bible written in the 2nd cent. CE have been found.

parables *Teaching by means of a comparison; stories of varying length containing a meaning or message over and above the straightforward and literal, with an element of metaphor. In the teaching of Jesus brief aphorisms (e.g. Matt. 24: 28) jostle with quite long parabolic teaching (e.g. Matt. 25: 1–13) and they were not necessarily delivered 'to the crowd'. They may often have been spoken in more intimate groups in conversation. There were well-disposed and interested or curious persons who might invite Jesus to supper (Luke 14: 15 ff.).

For centuries there was a tradition of interpreting Jesus' parables as *allegories. *Augustine gave every character and action in the parable of the *Good Samaritan a symbol standing for an article of Christian *faith—the inn represented the Church, the innkeeper is Paul, and the two *denarii stand for the two gospel *sacraments. Modern critical scholarship has tried to think its way back to Jesus' intentions and audiences. The first shots in the modern debate were fired by Adolf *Jülicher in 1888: his book on the parables maintained that the essence of a genuine parable was that it has but one single, discoverable point—'Now is the day of *salvation', for example. After a century of further discussion Jülicher's principle has had to be somewhat modified. There are allegorical elements in some parables (e.g.

the Wicked Husbandmen, Matt. 21: 33 ff.) and in the interpretations of them offered (e.g. of the Sower, Mark 4: 13–20). It is often suggested that some parables as recorded look back on the *ministry and *death of Jesus as past and interpret them. As such, they are creations of the early Church in a continuing conflict with *Judaism. Yet they were valid interpretations for the readers of the gospels, and reinforce the sense of crisis with which both the contemporaries of Jesus and the readers of the gospels were wrestling: after the seed-time the harvest, which is always the present.

Some of the parables in the Bible (of the vineyard, Isaiah 5: 1–6; Matt. 13: 24–30) suggest a kind of 'nature theology'—an argument for the existence and activity of God from regular and welcome aspects of nature. Yet on the other hand agricultural processes or human affairs may be described in absurd ways or with a good deal of contrivance—the unjust judge of the parable (Luke 18: 2–5) is not a normal character, nor is the employer who breaks the accepted rules of wage negotiations (Matt. 20: 1–15). Such parables express the contrast of *grace and nature; the sovereignty (kingship) of God is *not* like the judiciously regulated contracts of employer and employee. The loan of a thousand talents which is waived is a fantastic sum; and it is surprising that the whole body of ten virgins went to sleep (Matt. 18: 24 ff. and 25: 5). There is a paradox. The parables spoken by Jesus were to teach and to persuade. They rarely did. And that became a puzzle to the early Church: how was it that the Messiah visited his people and yet his own failed to recognize him? It was the contention of William *Wrede and other NT scholars that Mark had an explanation for the people's failure—he gives it in Mark 4: 11–12 (quoting Isaiah): Jesus was aware that he was Messiah (Wrede's theory went) but he kept it a secret until it dawned on the *disciples after the *resurrection: the parables had been deliberately couched in terms which would for the time being conceal their meaning.

Wrede's theory of the *Messianic Secret is not widely accepted today as he propounded it; and yet there was a kind of secret from the start surrounding the ministry of Jesus. The Greek word *parabole* can be translated 'riddle', and so Mark 4: 11 could be rendered: 'but to those outside everything has the nature of a riddle'; the meaning of the Kingdom is veiled to outsiders but is being disclosed to the disciples by God's revelation. To outsiders it is veiled by the very purpose of God. So the secret belongs to the gospel itself: it is an integral characteristic of God's act in Christ, comparable to the miracles, which could only be recognized by a God-given faith. It is a principle of the NT understanding of Jesus that he does not compel belief by an irrefutable argument; faith and responsibility are left to people themselves because Jesus is committed to freedom. People were not bludgeoned into belief then, nor are the readers of the gospel today. However, the translation 'riddle' is not suitable for every parable of Jesus, and many of them seem straightforward and likely stories designed to urge the audience to do something, or change their attitudes, in a fundamental way.

Paraclete A Greek noun meaning 'one called to the side of' and therefore an 'advocate'; a term especially liked by the gospel of John for the *Spirit (John 14: 16–17; 16: 7–15, cf. 1 John 2: 1) whom Jesus promises that he will send to the *disciples after he has gone. He is the Spirit of Truth; he is the 'Comforter' (AV) who through the disciples will continue Jesus' own work and reveal new teaching (John 14: 26; 16: 12–14) that was not able to be assimilated until after the *resurrection. Other translations indicate other functions attributed to the Spirit: 'advocate' (NRSV, John 14: 16) expresses the role of a counsel for the defence in court; 'counsellor' (RSV) is used for the Spirit who helps disciples in the absence of Jesus.

paradigm Term formerly used in *Form Criticism for the story of an event or saying of Jesus used in the oral period, before the tradition had been assembled in the gospels (e.g. Mark 12: 13–17), but the word is sometimes used now in biblical studies for the methods and assumptions of particular approaches. Thus the shift from a literal to a critical approach, or from historical to literary criticism, will be described as a 'paradigm shift'.

paradise The Greek term used (Gen. 2: 8) in *LXX for the garden of *Eden; originally a Persian word. It became associated with the time of primeval bliss and so also of the destination of the righteous in the age to come. Paul uses the image when he haltingly describes his mystical experience of elevation to the 'third heaven' (2 Cor. 12: 1–4), and Jesus promised this reward to one of the two felons convicted with him on crosses (Luke 23: 43). The Church in *Ephesus is also promised a reward for faithfulness in the paradise of God, where its members will eat from the tree of life (Rev. 2: 7).

paralysis The usual modern translation of the AV *palsy (Acts 8: 7).

Paran A place in the *wilderness, just south of *Canaan. After the Israelites left Mount *Sinai (Num. 10: 11–12) they encamped at Paran. Spies were dispatched from here to investigate the land of Canaan.

parchment A kind of refined leather used in Palestine during the Roman era as a more durable material for writing than *papyrus, as used by the *Qumran community.

pardon One of a group of words with similar meanings—such as *forgiveness and *reconciliation. The OT contains numerous promises that God offers pardon where he finds repentance (2 Chron. 7: 14; Ps. 86: 5) and elaborate provision for pardon by God was made in *sacrifices and by preaching of the prophets (Ezek. 33: 10–16). The message is continued by Jesus (Luke 15), but Paul introduces a new dimension by proclaiming Christ crucified as the basis for obtaining pardon. The NT Church understood the new covenant in Jesus' blood in their celebrations of the *Lord's Supper to be for the forgiveness of *sins (1 Cor. 11: 25): it led to reconciliation and peace with God (2 Cor. 5: 18–21; Rom. 5: 1).

parenesis The Greek term for sections of the epistles dealing with moral exhortation, such as Phil. 1: 27–2: 18.

Parmenas One of the Seven appointed to administer *charity (Acts 6: 5); the name is Greek; he was perhaps a Hellenistic Jew.

paronomasia A rhetorical device, especially in the OT, using two words similar in sound or appearance. It is difficult to reproduce this Hebrew play upon words, but one suggestion for the pun in Isa. 7: 9 is 'If you are not firm in *faith, you will not be firm at all'. Proper names provide fertile ground for paronomasia: 'Isaac' is a play on the Hebrew verb 'to laugh', as is explained by Gen. 21: 6.

parousia Greek for 'presence', and used in the NT for the coming of Christ within an *eschatological complex including *judgement and *resurrection—though the word can also denote simply the 'presence' of an apostle among his flock (2 Cor. 10: 10).

The *parousia* of Christ must be understood against a Jewish background of the national hope of a future vindication. It is expressed in Dan. 7: 13 in terms of a *Son of Man (meaning Israel) triumphing over beasts (meaning the Gentile nations). The Christians took over Jewish *apocalyptic imagery and transformed the eschatological dimension in that the main elements of the coming of the *kingdom of God had already been realized in the *life of Jesus and in his *death, which had been vindicated by the *resurrection (1 Cor. 15: 23–4). But the Temple still stood in Jerusalem although Jesus had foretold its destruction (Mark 13: 2). As long as it survived, the heart of the Jewish system was being perpetuated. When it was destroyed by the Romans in 70 CE, Christians could claim that its place was taken by the crucified and risen Son of Man, the new focus of divine presence.

The hope which Paul expected of a coming *day of the Lord was not for the end of the world and of time (2 Thess. 2: 2) but for the judgement on mankind (1 Cor. 4: 5) and the final conquest of evil (1 Cor. 15: 24–5). Then the *resurrection of Jesus would be consummated by the resurrection of all the faithful and a renewed creation—a kingdom of peace, justice, and love (Rom. 14: 17). Thus Paul and his contemporaries awaited the destruction of the Temple; after 70 CE this historical event of apocalyptic significance became a further vindication of Jesus' message. The fulfilment of what he had predicted for the near future guaranteed the fulfilment at the ultimate. And what still remained to come in the future was the *End, when Christ would hand over the kingdom to God the Father (1 Cor. 15: 24). It would constitute the *parousia*. It could happen suddenly (1 Thess. 5: 2) but, apart from 2 Pet. 3: 3–4, there was no reported dismay or despondency about any delay. Certainly the return of Christ at some time was hoped for, and by some groups has been in the past, and still is, awaited at a precise date in the near future, but that was not the mainstream Christian conviction. The return of Jesus (Acts 1: 10 ff.) is regarded as part of the ultimate future when the whole created order will be renewed; then evil will be judged and defeated; and in such a scenario there would have to be a place for Jesus. But for the early Church the most important part of the process had already occurred with Jesus' resurrection. The timing of the remainder was unimportant.

Parthians A people of Iran, of Scythian origin, and very warlike. By about 130 BCE their empire extended from the *Euphrates to the Indus. By 53 BCE they were threatening Syria and Asia Minor but were checked by the Romans under Ventidius in 38 BCE and by the victory in 20 BCE under Augustus. The Parthians had a brief intervention in the affairs of *Judaea when they supported Antigonus II in Jerusalem until *Herod was installed as king by the Romans. Some Parthians were in Jerusalem for *Pentecost (Acts 2: 9).

particularity The view of the *Bible that God chose a historical people (Israel) and (in the NT) finally one historical person by whom to effect his purpose. The phrase 'the scandal (stumbling-block) of particularity' has been used by modern theologians for the view that the Christian *faith cannot be reduced to religion in general or the uniqueness of Christ compromised.

Pasch(al) The *Passover. A false etymology in the early Church connected this Hebrew word with the Greek verb *paschein*, meaning to suffer. Because of its theme of Christian suffering, it has then been suggested that 1 Peter originated as a homily at an *Easter (Paschal) *baptism (1 Pet. 3: 21).

passion In NT studies the reference is to the sufferings and *death of Jesus, and the Passion Narratives cover the whole period from the *Last Supper (Mark 14), the arrest, the *trials, the *crucifixion, and the burial, as recorded in the four gospels; but the only actual mention of Jesus' 'passion' or 'suffering' (NRSV) is in Acts (1: 3).

The four narratives reveal a tendency to diminish *Pilate's responsibility for the execution of Jesus, at the expense of the Jewish authorities. It is held that this development reflects the situation of the Church after 70 CE in its relationships both to the Roman Empire and to *Judaism. It was desirable for the Church not to appear subversive in the Empire; it was also imperative to establish the claims of the predominantly *Gentile Christian community as the covenant people who had superseded the Jews. The view, however, that Jesus died by stoning, the Jewish method of execution (Lev. 24: 14), is implausible; the crucifixion (the Roman penalty) is far too strongly attested.

Matt. (27: 24–5) introduces the action of Pilate asserting his blamelessness by washing his hands, while the Jews exclaim that they are willing for the blood of Jesus to be on them and their descendants. In Luke, who has an additional trial scene, before Herod, Pilate three times protests Jesus' innocence, and the *centurion at the cross repeats this (Luke 23: 47). In John (19: 12)

Pilate says he intends to release Jesus, and is only frustrated by the threats of the Jews to delate him to Rome. But the acme of Pilate's career in Christian exculpation was achieved when he was canonized by the Coptic Church, which commemorates his *martyrdom on 25 June.

Passover The annual Jewish festival, held on the 14th of the month Nisan, which commemorated the saving events of the *Exodus from *Egypt; as an institution it may have already existed in pre-Israelite *Canaan as an agricultural festival in the first month of the year, and this was absorbed into the life and ritual of the incoming Hebrews. There seems to be a record of this combination when the Israelites, after entering Canaan, are said to have kept the Passover (Josh. 5: 10) and on the following day to have eaten the produce of the land (5: 11).

The late Pentateuchal source *P puts the Passover rites back into the *wilderness of *Sinai (Num. 9: 3, 5) but this assertion is too late to be regarded as historically reliable. From the time of *Solomon (1 Kgs. 9: 25; 2 Chron. 8: 13) the Passover ritual was established, and in the reforms of *Josiah (2 Kgs. 23: 21–3) there were to be regular Passover festivals in Jerusalem. Passover lambs were either roasted (Exod. 12: 8–9) or boiled (Deut. 16: 7) and were eaten at a solemn meal in the evening at which the departure from Egypt after the Lord 'passed over' the Israelite males but destroyed their Egyptian contemporaries was remembered. The feast of *Unleavened Bread (as in Josh. 5: 11) was maintained for a full week from the Passover. The feast was of immense significance for the Jewish sense of identity as a people; it united successive generations in a family. In the NT it has significance because of the *Lord's Supper as a Passover Meal, according to the synoptists, held by Jesus with his disciples in Jerusalem. They followed the traditional ritual with the first and second cups of wine and unleavened bread, but a new significance was given to the meal by Jesus by associating it with his coming death. NT writers refer to Christ as a 'passover lamb' (1 Cor. 5: 7) and the acclamation by *John the Baptist to Jesus as 'Lamb of God' (John 1: 29) is held by some commentators to indicate that this gospel, which connects the death of Jesus with the timing of the slaughter of the passover lambs in the Temple (John 19: 14), had a theology of Jesus as the True Paschal Lamb who brings about the final release of his people from the slavery of sin.

Pastoral Epistles The three letters of the NT placed after 1 and 2 Thess. Since 1726, when Paul Anton called the two NT letters to Timothy and that to Titus 'Pastoral Epistles', the name has stuck. It is appropriate on account of the concern they exhibit for a certain number of pastoral situations in the Church. There is a threat of persecution (2 Tim. 4: 6–8) and of apostasy by Christians (2 Tim. 4: 10), yet Christians are urged to be law-abiding citizens (Tit. 3: 1). There is no call for revolution. Within the Church the author is worried by heretical tendencies of a *Gnostic character, which obliged him to insist on the goodness of God's *creation (1 Tim. 4: 4) and the genuine humanity of Christ (1 Tim. 3: 16).

The three epistles' claim to be written by Paul has been much disputed in NT scholarship for about 200 years, but the traditional view, mentioned by *Irenaeus at the end of the 2nd cent., still has a few supporters. Maintaining Pauline authorship has to contend not only with the epistles' refutation of Gnosticism—which marks them as later than Paul's generation—but also with the difficulty that the organization of the *ministry reflects an era later than Paul's. Style and vocabulary are also said to be uncharacteristic of Paul. Some differences could be explained either by Paul's use of a secretary or by the sheer versatility of the man.

The balance of evidence is that the Pastorals are pseudonymous, though fragments of Paul's own letters may have been incorporated, e.g. 2 Tim. 4: 9–18. A suggested date for the composition of the epistles would place them about 130 CE, since they are probably known to Polycarp (in about 135 CE), but a date near the end of the 1st cent. is also possible in that the kind of church order in the epistles reflects that of 1 Clement (95 CE). A tentative suggestion for their place of composition would be *Ephesus, where Timothy is said to reside (1 Tim. 1: 3).

The epistles could be fairly described as a manual for the management of the household of God. They reflect the structures and authority of an ordered society and the ethics of the surrounding Graeco-Roman world; women are regarded as subordinate (Tit. 2: 5), but on the basis of 'the word of God'.

patience God's reasonable anger with his sinful people is patiently restrained (Isa. 48: 9), and this is a characteristic that should be imitated by Christians confronted with human failings (Matt. 18: 26; Eph. 4: 2) and an uncertain future (Jas. 5: 7).

Patmos A volcanic island in the Aegean Sea, SW of *Miletus, where John the Divine was exiled (Rev. 1: 9) and where he had his visionary experiences.

patriarch(s) The principal ancestors of the Israelites—*Abraham, *Isaac, and *Jacob, whose wives *Sarah, *Rebekah, and *Rachel, bore sons through whom the nation continued. Some scholars do not regard the patriarchal narratives as history at all but cast them in a *genre more like *poetry, no longer relying on disputed archaeological evidence which placed the ancestors within established data of Mesopotamian culture over the wide period 2000–1200 BCE. However, the narratives may have reached their final form as late as the generation of the Exile (*c.*500 BCE) and so have been subject to much editorial construction. It is not therefore easy to disentangle tribal memories from a theological interpretation which imparted a sense of national identity from the basis of a core of individual histories.

'Ancestors' is used rather than 'patriarchs' in much scholarly literature because of the important role of women in Genesis.

patristics The study of the works of the Christian Fathers (theologians from *c.*100 to 700 CE). Their theological data were in the *Bible; but the task of giving coherence and credibility fell to the following generations as they confronted contemporary philosophies and religious speculations. They tried to do justice to the complex data of the NT, to the felt necessity of offering worship to Jesus, while maintaining his authentic humanity. A further task was to preserve continuity of Christian belief with the OT and Jewish monotheism while also being open to the contemporary intellectual scene in the Hellenistic world. The Apologists (120–220 CE) made bold use of the **Logos* concept and argued that Christ, the *Logos* of the Johannine Prologue (John 1: 14), was himself the Reason at the heart of the cosmos and a kind of Omega point of all intellectual aspirations.

Paul, the Apostle A formidable theologian—sometimes even regarded as the founder of Christianity—whose correspondence occupies a substantial part of the NT. There is no serious doubt about Paul's authorship of the epistles to the Romans, the Corinthians, the Galatians, the Philippians, and Philemon; and most scholars accept the first to the Thessalonians and Colossians. The epistle to the Ephesians is sometimes regarded as a compendium of Pauline theology rather than a letter from Paul's own hand, and the *Pastoral Epistles would not generally be used as evidence either for Paul's thinking or for his imprisonment after the end of Acts 28. The epistles were written for particular occasions and were dictated to a secretary, such as Tertius (Rom. 16: 22), and are therefore first-hand witnesses to the life and tensions of the apostolic Church, as well as giving autobiographical information about Paul. The Acts of the Apostles, by contrast, is a secondary piece of historical writing. But neither epistles nor Acts tell us what Paul looked like, though an unreliable work of the 2nd cent. (the Acts of Paul and Thecla) describe him as short, bald, bow-legged; of vigorous physique, with meeting eyebrows and a slightly hooked nose. From references in the epistles, especially 2 Cor., we deduce that Paul was a masterful and proud teacher, and yet could be humble, exacting, and irritable, though also forgiving and magnanimous. He attracted the loyalty, even love, of disciples (Gal. 4: 15) and had an intense pastoral concern for his communities; he was anxious to promote reconciliation amongst the disaffected. But by Jews he was hated for his betrayal.

Although the letters were non-literary in the sense that they were not intended for publication or for posterity, Paul was certainly not uneducated. He was born about 10 CE into an orthodox Jewish family (Phil. 3: 5–6) at *Tarsus. He left home at a young age for Jerusalem (Acts 22: 3) and was trained by *Pharisees. He became a persecutor of Christians and as such was on the way to *Damascus (34 or 35 CE) when he received his call (or experienced a 'conversion', Gal. 1: 23) to follow Jesus as *Messiah and take the gospel to the *Gentiles. Paul was equipped for this by his knowledge of Greek, even though his thinking has a strongly Hebraic tone.

Paul was a man of some social standing as a citizen both of the Greek city of Tarsus and of *Rome itself (according to the Acts; Paul's Roman citizenship is never mentioned in his epistles). The latter conferred privileges, such as exemption from degrading punishments and the right of appeal to the emperor in the case of capital charges. Paul's Roman citizenship was from birth and was probably a reward given to his father by the emperor, which passed on to the son. According to *Josephus, many Jews possessed Roman citizenship. Not surprisingly, Paul maintains that the power of (Roman) government was God-given (Rom. 13: 1–7). It is possible that a directory of citizens was kept at Rome. Paul's marital status is unclear: it has been held, variously, that he was a celibate, a widower, or legally separated from a wife.

Paul had two names: the Hebrew, *Saul, after the first Hebrew king, and Paul, which is Latin (= 'the little one') and was probably adopted because of its sounding rather like Saul. He is depicted as a man of wealth, able to pay the expense of a Nazirite vow (Acts 21: 24) and to excite *Felix's expectation of a bribe (Acts 24: 26). He receives preferential treatment from time to time and the author of Acts seems keen to create an impression that Christian belief and good social standing were entirely compatible. The fact that Paul worked with his hands, making tents, did not relegate him to the artisan class, since it was the practice of rabbis to teach without charging fees and to have a trade by which to live. It is clear that, unlike Jesus, Paul was essentially a townsman, for he refers to the theatre (1 Cor. 4: 9), commerce (2 Cor. 1: 22), shops (1 Cor. 10: 25), and games (1 Cor. 9: 24). After Paul, the Church continued to thrive in cities while paganism prevailed in the countryside.

The view that Paul rather than Jesus was the founder of Christianity as a new religion, separate from Judaism, is based on the contrast between Jesus' preaching of the Kingdom with Paul's religion of redemption in which Christ's death and resurrection are the centrepieces of a new mystery cult. *Baptism is the means of entry, and the defilement of human *sin is removed as the initiate enters into communion with Christ's act of *salvation. The message of Jesus about the Father has been transformed into a religion appropriate for the Graeco-Roman world in which the heavenly Father plays only a minor role, and the *Judaism of Jesus is stripped in Paul's religion of its Jewish basis of the *Torah.

Part of the argument for Paul as the genius of imagination and poetry, who contrived to present the Mediterranean world with a new mystery religion round the *death and *resurrection of Jesus rests on his comparatively infrequent reference to the *life and *teaching of Jesus. However, there *are* references to Jesus' *birth (Gal. 4: 4), to his obscurity and poverty (2 Cor. 8: 9; Phil. 2: 7), to his teaching on marriage (1 Cor. 7: 10), on the practice in antiquity approved by Jesus (which Paul did not avail himself of) that ministers of a religion should be financially supported by their fellow believers (1 Cor. 9: 14); to the Last Supper (1 Cor. 11: 23 ff.); to the recommendation to be imitators of Jesus (1 Thess. 1: 6); to the meekness and gentleness of Christ (Rom. 15: 2–3). But the main interest of Paul is in the *crucifixion and in Jesus' being raised by God on the third day (there is no explicit allusion to the empty tomb). Paul taught as one who believed that God had acted decisively in the *Cross and Resurrection; he interpreted the present situation in the light of those *eschatological events. Both Jesus and Paul had the same hope for final *salvation but each viewed it from his own standpoint: Jesus was the bearer of salvation; Paul was the messenger of what had taken place.

When Paul retreated to Arabia (Gal. 1: 17), he perhaps needed to think through the meaning of the transforming experience on the Damascus road. He knew himself to be now a slave of Jesus Christ (Rom. 1: 1).

The Cross with its saving power was central: Paul had been a persecutor of Christians because he saw them as adherents of a false *Messiah (Gal. 1: 13) who had been cursed by God (Gal. 3: 13). After the Damascus road experience he believed that Jesus was the exalted Lord in glory. Such a dramatic change of direction made him more aware than most of the mercy of God (1 Cor. 15: 9–10); he had been chosen; he was inevitably led to a new doctrine of 'election' and in Rom. 9–11 he argues that election cannot be in virtue of physical descent from *Abraham (9: 7) and practice of good works; in Rom. 8: 38 he declares that God can even turn evil and sin to his own purposes.

Paul's sense of having failed God just when he had supposed he had been working on God's behalf made him, as a Christian, deeply conscious of guilt, and he repudiated the legalism which had laid down in precise terms exactly what was right and wrong. So (Phil. 3) all the values of his past life and the merit it had accumulated, and the conceit of salvation by birth into the *covenant people, lost their value because they had seduced him to self-confidence. From these convictions stemmed Paul's hostility to the campaign of certain Jewish Christians to insist that obedience to the Law could be combined with faith in Christ. For Paul, those two kinds of *justification were incompatible and he could not accept his opponents' demand that his Gentile converts should first accept the requirements of the Jewish Law (by being circumcised and observing the food regulations) before being baptized (Gal. 2: 21). Paul's belief that God had sent Christ to save the world and his conviction that he had been called to proclaim this gospel far and wide made him assert that these events would not have been necessary if the means of salvation were already generally available in Judaism. If his Gentile converts were to accept the Jewish-Christian proposition, it would be equivalent to renouncing Christ. A precondition of *circumcision and *Sabbath observance before being admitted into a Christian congregation would spell the end of Paul's mission. He urged ex-Jews in the Roman Church to throw in their lot with the ex-pagans. Emotional and legal ties with Judaism should be renounced in a united community (Rom. 15: 7). But Christians' release from the obligations of the Law, along with separation from the institutions of Judaism, does not imply that they can live without moral restraint: becoming 'one person' in Christ produced the 'fruit of the spirit'. Dying with Christ led to a share in his risen life, as a new person. Paul's anxiety to preserve the unity of the Church was one motive which inspired his *collection among the Gentile Churches for Christians in Jerusalem. He wanted the Jerusalem leaders to accept the validity of his Gentile mission, in which pagans were baptized without submitting to the Jewish Law. No doubt he also thought that the financial aid would assist his cause.

Paul's encounter with *Gallio at *Corinth (Acts 18: 12–17) can be dated in 50–1 CE, and his execution in *Rome probably in 62 CE, possibly in 64 CE. *See* Nero.

Repeatedly in Church history it has been the thought of Paul, especially in the letter to the Romans, that has inspired creative new movements: Augustine, Luther, Wesley, *Barth.

Paulus, Sergius Proconsul of *Cyprus, 47–8 CE, who was much impressed by Paul (Acts 13: 7). The name suggests that he belonged to an aristocratic family.

pavement *Pilate's judgement seat on a stone pavement, called *Gabbatha in Hebrew (John 19: 13) or Aramaic; it was either adjacent to *Herod's palace or in the courtyard of the fortress *Antonia on high ground, north of the Temple, where a Roman gaming board has been discovered carved into the stone.

peace In Hebrew, *Shalom*; used as a greeting between friends (Ezra 5: 7) and of the absence of hostilities between nations (1 Kgs. 5: 12). It is a *gift from God when it comes (Isa. 54: 10) and if deluded prophets cry 'Peace, peace', when there is no prospect of it, they will be punished (Jer. 6: 14–15). Ultimate peace belongs to the Messianic age (Isa. 9: 6).

In the NT 'peace' means not only harmonious relations among nations (Luke 14: 32) but also the state which should exist both in the Christian communities (Rom. 14: 19) and also towards outsiders (Heb. 12: 14). The death of Christ established peace between God and humanity (Col. 1: 20) and between Jews and *Gentiles (Eph. 2: 14). Yet before this ultimate good is fully realized, the immediate effect of Jesus' message may have to be conflict; a *disciple's own family may be estranged, and could be the price of not being alienated from Jesus (Matt. 10: 34–9).

pearl Used as an ornament in NT times, although discouraged as ostentatious amongst Christian women (1 Tim. 2: 9) but appropriate for the gates of the New Jerusalem (Rev. 21: 21). In one of Jesus' *parables, the kingdom of heaven is compared to a pearl of great price, or value, found by a merchant (Matt. 13: 45–6).

Pekah King of Israel 737–732 BCE after murdering Pekahiah; he was himself murdered by *Hoshea. During his reign he allied himself with *Syria in war against *Assyria and endeavoured to coerce *Judah into the conspiracy. Instead *Ahaz of *Judah, contrary to the advice of Isaiah (Isa. 7: 4), sought the help of Assyria, which obliged by promptly invading Israel.

Pelethites Members of *David's bodyguard (2 Sam. 8: 18), who probably lived along the coastal plain of Palestine.

Pella A city of the *Decapolis (Mark 5: 20) captured for the Romans by Pompey in 63 BCE, about 29 km. (18 miles) south of the Sea of *Galilee and east of the *Jordan. It has been suggested that on the basis of Mark 13: 14 Christians made a mass exodus in 66 CE, leaving Jerusalem to escape the devastation of the imminent siege, and that it was to Pella that they went. But against this is the sense of Jesus' words: they envisage a flight from a rural area in sudden unpremeditated urgency. (Although there is no archaeological evidence at Pella of Christian occupation, the Christians' escape to this Greek city is recorded by the historian Eusebius in the 4th cent.)

pen Two kinds are mentioned in the Bible; a sharp instrument which cut letters in stone (Job 19: 24) and one cut from a reed, with a knife (Jer. 36: 23), used for personal letter-writing on *papyrus (3 John 13).

penny Used in NRSV, REB, NJB, where AV has 'farthing' (Matt. 5: 26) as the translation of the Greek *kodrante*, the smallest coin. The Greek *assarion*, or Latin *as* (Matt. 10: 29) was a Roman copper coin worth 1/16 of a *denarius, and is also translated 'penny' by NRSV, REB, NJB, where AV again renders 'farthing'. In Matt. 20: 2, 9, AV translates the Greek *denarion* by 'penny', where NJB simply uses the Greek word, while NRSV and REB deal with the problem of inflation by describing the wage in the text as that usually paid to a labourer for a day's work.

Pentateuch The first five books of the OT, traditionally ascribed to Moses, though by modern scholarship regarded as an editorial compilation of *sources from different dates. The books are also known as the *Law of Moses (*'Torah' in Hebrew). The *Samaritans held to these five books only as scripture, in their own edition.

The Pentateuch is the holiest part of the Bible for Jews; the Torah scroll is kept in a place of honour in *synagogues.

Pentecost In the OT, the feast of Weeks (Exod. 34: 22), an agricultural festival on which work was forbidden (Lev. 23: 21). Later, the festival was incorporated into Israel's national consciousness of its history and became associated first with the *covenant made with *Noah (Gen. 8: 20–2) and then with the giving of the *Law on Mount *Sinai. Because the interval between the first *Passover and the arrival at Sinai was reckoned as fifty days (Exod. 19: 1), the feast became established after that interval, as is seen from the NT (Acts 2: 1–42), where the Greek word 'Pentecost' is current. The narrative in Acts 2 resounds with echoes of the giving of the Law at Sinai. First there is the contrast, for Christians, as Paul taught, now live under the *Spirit rather than the Law. Secondly, there is the resemblance: the visible *gift of the Spirit was in the form of tongues of fire, and the author of Acts is recalling the descent of the Lord in fire on Sinai (Exod. 19: 18). Thirdly, the 'harvest' of 3,000 converts recalled the harvest of the old OT agricultural context of the feast (Exod. 23: 16). The gift of *languages, as the **glossolalia* is interpreted, marks the reversal of the scattering of nations at *Babel (Gen. 11: 1–9). Therefore Pentecost continued to have special significance in the Church (1 Cor. 16: 8; Acts 20: 16) as the festival of unity (1 Cor. 12: 13), a foretaste and promise of a unity among all the nations of the world through the gospel (Acts 1: 8; 10: 45).

Penuel The site east of *Jordan of *Jacob's wrestling with a man (Gen. 32: 22 ff.), from whom he secured an angelic blessing.

people In the OT, the term designates the whole body of Israelites who are distinct from the nations of the world (Num. 23: 9). This is the case in the NT (Luke 1: 68; Rom. 11: 1–2) but it is soon used for the *new* people of God, that is, the Christians (Tit. 2: 14; 1 Pet. 2: 9). The 'people of the land' came to be a derogatory term for the ignorant and non-observing Jews after the Return from Exile (Ezra 10: 2; John 7: 49), though earlier it had apparently referred to a more influential group in society (2 Kgs. 23: 30).

Peor A mountain in *Moab where the people of Israel began to revere the pagan god (*Baal) of Peor and suffered for it (Num. 25: 5). *Balak, king of Moab, took Balaam there (Num. 23: 28) that he might curse his Israelite enemies, but instead he prophesied that Moab would be crushed.

perdition In the NT used in AV of the destination of those condemned at the judgement to destruction, e.g. of the *antichrist (2 Thess. 2: 3).

Perea Area east of the *Jordan where *John the Baptist preached repentance. Its people were mostly Jewish in the 1st cent. and together with *Galilee it formed the tetrarchy of Herod *Antipas. In order to avoid passing through the unfriendly territory of *Samaria Jesus would have passed through Perea (Luke 9: 51–3).

perfection In the OT perfection is ascribed to God (Job 37: 16) and in the NT to persons, signifying maturity (1 Cor. 2: 6). The command in the *Sermon on the Mount to be perfect (Matt. 5: 48) is changed in Luke 6: 36 to 'be merciful'. In Matt. it is part of the new ethic of the kingdom, such as non-retaliation, in which the conditions of the kingdom are set out as ideals.

Perga A city in *Pamphylia in Asia Minor, where John Mark left Paul and *Barnabas (Acts 13: 13).

Pergamum One of the cities whose Church is addressed in Revelation (2: 12); capital of the Roman province of *Asia, and the site of '*Satan's throne' (Rev. 2: 13), which perhaps means that it was a centre of *worship of the Roman emperor.

pericope A term used in *Latin by *Jerome for sections of scripture and taken over by *Form Critics to designate a unit, or paragraph, of material, especially in the gospels, such as a single *parable, or a single story of a *miracle.

Perizzites A group of inhabitants of *Canaan (Gen. 15: 20) who were defeated by

*Joshua (Josh. 11: 3, 23), as forecast by Moses (Deut. 20: 17).

persecution Active hostility to a group, other than military action, or mob violence against an individual (as Acts 7: 57–8). The Jews suffered persecution when *Antiochus Epiphanes (175–164 BCE) attempted to absorb them into his Hellenistic culture and condemned to death (1 Macc. 1: 41–61) those who refused to cease practising the rite of *circumcision and the food regulations. In the NT persecutions are both prophesied (Matt. 5: 10–12) and described, but they come more from Jewish antagonism than official Roman policy, or at any rate according to the reports in the Acts. It is clear from 1 Pet. 4: 16 that Christian faith could lead to suffering, but it is not certain that this verse implies that being a Christian was itself an indictable offence. There was, however, a persecution against Christians by *Nero after the fire in Rome of 64 CE; and correspondence between the governor in Bithynia, Pliny the Younger, in 110 CE and the emperor *Trajan suggests a formal state policy to get rid of uncompromising Christians; the similarity of that situation with 1 Pet. 4: 15–19 is an argument (though not a conclusive one) for putting the date of this epistle into the early years of the 2nd cent. Actual experience of concerted persecution does seem to be implied by Rev. 13 through the threats of the emperor *Domitian (81–96 CE), who wished to be called 'Lord and God'.

perseverance Rarely mentioned in the Bible as a virtue but perseverance in *prayer is commended (Eph. 6: 18); it is the theme of the *parable of the unjust judge (Luke 18: 2–8).

Persia A vast empire once including part of Greece and reaching India in the east. It impinged on the Jews with the conquest of *Cyrus over the *Medes (550 BCE) and his decree authorizing the exiles to return from *Babylon to Jerusalem, a decree reaffirmed by *Darius I (Ezra 5: 1–6: 15). The legendary story of *Esther is set in the reign of *Ahasuerus, or *Xerxes (Esther 1: 1–2). The empire was destroyed by *Alexander the Great (*c*.330 BCE). The Persians, regarded as barbarous orientals in many Greek sources, are always treated favourably in the OT.

The religion of *Zoroaster (7th cent. BCE) had its origin and base in Persia; its belief was that the god of Light, Ahura Mazda, had overcome Ahriman (darkness) and this dualism has been thought to have had an influence on *Judaism, as did also its developed system of *angels. It was a religion which was tolerant of other cults, and the Greek Apollo and the Jews' Yahweh were among the deities recognized.

Pesher Hebrew for 'commentary' and particularly used for commentaries on the OT in the *Dead Sea scrolls, which looked for hidden meanings in the text which were seen to apply to and to justify the community's way of life. The NT use of OT texts has some similarity with this method: over and above the original sense, a passage is said to have a special meaning for the present time (e.g. 1 Cor. 10: 11). Another, similar, method of interpretation was *midrash* ('study'), which was essentially oral exposition in the synagogue to elucidate difficulties.

Peshitta A Syriac text of the Bible, of which the NT portion is by some said to be the work of Rabbula, bishop of Edessa 412–35 CE, by others to be dated in the 4th cent. The text lacks Rev. and four epistles.

Peter One of the first of Jesus' *disciples to be called; originally *Simon, but given the name Peter (Greek, 'a rock') or *Cephas (Aramaic for 'rock', John 1: 42), he was a Galilean fisherman, owning a boat which was a handy pulpit for Jesus (Luke 5: 1–3) and a house at *Capernaum (Mark 1: 29–31). He is called 'Son of John' in John 1: 42 but 'bar' = 'son of' 'Jonah' in Matt. 16: 17—perhaps because in Matt. 16: 16 he had spoken out in the manner of the sign of the prophet *Jonah (Matt. 16: 4). He became a leader among the group and came nearer than most to recognizing Jesus' *Messiahship (Matt. 16: 16–17). But when he deplored the notion of Jesus' suffering, he was sternly rebuked (Mark 8: 33). He was present at the *Transfiguration (Mark 9: 2). During Jesus' *trial, Peter three times denied any association with him, but after the *resurrection he was accorded a special appearance of Jesus (1 Cor. 15: 3–5). He became willing to admit *Gentiles into the Church (Acts 10–11). But under pressure he reneged, and Paul's anger against Peter for going back on his 'liberalism' is a main theme of Gal. 2. However, Peter is mentioned in more friendly terms as a visitor to *Corinth (1 Cor. 1: 12) and there is a reliable tradition that he lived in *Rome, wrote 1 Peter to Churches in Asia Minor, and died as a martyr in 64 CE. Excavations at the Vatican seem to show that there was a monument from the early part of the 3rd cent., if not earlier, to mark the place where he died, if not his actual tomb.

Peter, first epistle of The twenty-first book of the NT. An epistle of great beauty dealing with *baptism and suffering, with appropriate ethical exhortations to different classes of converts. Some think it may have originated as a homily at *Easter, the time for baptisms, in *Rome, which is concealed under '*Babylon' in 1 Pet. 5: 13 (cf. Rev. 17: 5–6). Then, with an introduction (1: 1–2) and conclusion (5: 12–14), it served as a letter to encourage Christians possibly facing oppression (4: 12). They are marginalized citizens, mere 'resident aliens' (Greek, *paroikoi*) in Asia Minor. Their conversion to Christianity has evidently increased the local hostility towards them. They have become a sect with a culture which separates them from the surrounding society. Members of the Church belong to an alternative family (1 Pet. 4: 17), though they are warned not to withhold their proper duties to the State (1 Pet. 2: 17). By their example they will win over unbelievers (2: 11–12). Authorship by Peter depends on the view taken of the reference in 4: 15–16 to the criminal offence of being a Christian. Some scholars regard this reference as so like circumstances detailed in correspondence between the younger Pliny and the emperor *Trajan that it must belong to the early part of the 2nd cent. The language about being reborn (1: 3) is similar to the propaganda of the pagan mysteries, which could also imply a 2nd-cent. date. Other scholars defend the Petrine authorship of the letter. They point out that it may owe its felicitous style to the literary ability of Silvanus the *amanuensis (5: 12). The references to persecution seem to denote the hostility of neighbours and families (3: 6) rather than official imperial policy.

Peter, second epistle of A late and pseudonymous epistle that hovered on the edge of the *canon, but eventually the claim to be a second letter from Peter (2 Pet. 3: 1) prevailed. Its late composition is indicated by the references to the epistles of Paul which have become collected and widely read (2 Pet. 3: 15–16) and by the incorporation of the epistle of Jude in 2 Pet. 2: 1–18. Christians of the first generation have died (2 Pet. 3: 4).

The epistle attempts to refute heretics who are evidently embarrassed by Christian apocalyptic *eschatology. They argue complacently that the delay of the *parousia* shows that there will be no judgement at all. The writer replies that just as God once destroyed the world by the *flood, so he can equally terminate it, by fire (3: 5–7), notwithstanding the promise of the *rainbow (Gen. 9: 16). There will be a divine judgement; and the apostle's presence (1: 18) at the *Transfiguration gave him authority to preach the *parousia*. But in this defence of traditional Jewish-based ideas, the epistle employs Hellenistic language, and so represents an interesting stage in the transition of Christianity to another culture.

Peter, gospel of An apocryphal work of the 2nd cent., known to *Origen, antisemitic in tone, and docetic in *Christology. A section of it survives in a MS of the 8th cent. CE, now in Cairo.

Petra Capital of *Edom and later of the Nabataean Kingdom, SE of the *Dead Sea, in modern *Jordan, probably the Sela of OT (e.g. 2 Kgs. 14: 7). Petra has impressive ornamental monuments, including a temple, carved out of the sandstone.

Pharaoh Hebrew for the title of the kings of *Egypt, who were regarded as divine and ruled absolutely, with the help of an enormous corps of civil servants; these were supervised by the head of the service, the vizier, which was the office pictured as being held by *Joseph (Gen. 41–7). The Pharaoh of the *Exodus who allowed the Israelites to leave *Egypt (Exod. 5–12) is often identified as Rameses II (1279–1212 BCE) in the light of excavations carried out in his store cities of Pithom and Rameses (Exod. 1: 11).

Other Pharaohs mentioned in the OT include (1) Shishak (945–924 BCE), who invaded Palestine (1 Kgs. 14: 25–6); (2) So (perhaps 727–720 BCE), to whom King *Hoshea begged for help when he rebelled against *Assyria (2 Kgs. 17: 4); and (3) Neco (610–595 BCE), who slew King *Josiah in battle at *Megiddo in 609 BCE (2 Kgs. 23: 29).

Pharisees From a Hebrew root meaning 'separated'; so, according to Origen and other Fathers, Pharisees were 'those who separate themselves'. It is held, but disputed, that they were the Hasideans of 1 Macc. 2: 42; 7: 12 f. (*see Hasidim*). They are mentioned by *Josephus as one of the Jewish 'sects' together with *Essenes and *Sadducees.

Pharisees appear frequently in the NT. Paul describes himself as a former Pharisee (Phil. 3: 5) and as such observed the Law dutifully and attacked the followers of the Nazarene Jesus who were posing a challenge to his Pharisaic tenets and way of life. But Paul does not refer to Pharisees in his letters either as special opponents or as leaders of the Jewish community; it is possible that they did not live in Galatia—the area to which Paul's most polemical anti-Jewish letter was addressed.

Much that is written in the gospels about the Pharisees reflects controversies between Church and synagogue and does not accurately represent the conditions prevailing in Palestinian society before 70 CE. Of the four gospels, it is Mark which gives the most reliable picture. In that gospel, the Pharisees are brought into the narrative only when Jesus is in Galilee or 'beyond the Jordan'—except at Mark 12: 13. There are disputes about fasting (2: 18), Sabbath observance (2: 24), and divorce (10: 2). In Jerusalem (12: 13) Pharisees are joined with supporters of Herod Antipas and are thus shown to be a well-connected political group who are determined to defend that kind of Jewish community which Jesus threatened and which could be outside Pharisaic control. The scribes (e.g. Mark 12: 28) were not identical with Pharisees, but because they had social control the Pharisees associated with them. Scribes were students and expositors of the Law, whereas Pharisees were concerned with people's performance of the Law.

The Pharisees of Matt. are the victims of Church/Synagogue controversy after 70 CE. They are the zealous advocates of a Judaism which contrasts with the interpretation of the Law ascribed to Jesus. Pharisees are contrasted too with the standards expected of Christian leaders in the second Christian generation. They are viciously attacked for failing to be sincere practitioners of their own religion (Matt. 23: 3) and they are unhistorically identified with the scribes (23: 13, 23, 25, 27, 29).

Luke is more careful: he distinguishes Pharisees from the scribes (Luke 11: 37 and 45). He is also rather ambivalent about Pharisees, for on the one hand there are friendly Pharisees who invite Jesus to meals (11: 37; 14: 1) and in Acts (15: 5) there are Christian Pharisees, in line with Luke's view that there is a continuity between Judaism and the Church; it is Luke who repeatedly mentions Jerusalem as the place of revelation. And as members of the Sanhedrin Pharisees are sympathetic when Paul is on trial (Acts 23: 9). On the other hand, Pharisees are rich, whereas Luke emphasizes Jesus' preference for the poor (6: 20); Pharisees preside over dinner parties, whereas Luke records Jesus' compassion for beggars who eat the scraps which fall off the table (16: 21).

The gospel of John portrays the Pharisees as constantly suspicious of Jesus; in union with the high priests they are Jesus' adversaries. They appear to be government officials and teachers of the Law. They are located in Jerusalem (John 1: 19–28) where residents of the city turn to them as bureaucrats charged with matters of public order (John 9: 13).

The evidence of Josephus, the gospels, and post-70 CE rabbinic writings, in the light of critical scrutiny, yields an impression of Pharisees as a literate group, socially above the peasant class but below and dependent on the governing class; subordinate officials, educators, and judges. They appear in every era from the *Hasmonean period until the destruction of Jerusalem in 70 CE. They could be regarded as a sect with gradualist reforming intentions, Fabian rather than revolutionary, and seeking alternatives in accordance with divine revelations. They welcomed novel beliefs such as resurrection of the dead and had a strong interest in updating laws regarding tithing, ritual purity, and Sabbath observance. They were less attracted to regulations surrounding the Temple, which was more of a Sadducean preserve.

Philadelphia One of the Seven cities whose Churches are addressed in Rev. (1: 11; 3: 7), in western Asia Minor (modern Turkey). It was important for trade and communication but was subject to earthquakes. The Church there had remained faithful during persecution from a faction of Jews.

Philemon, Paul's epistle to The shortest of Paul's extant letters; it is an appeal to Philemon, written while Paul was in prison, begging Philemon to be merciful to the runaway slave *Onesimus, who had been converted to Christianity (verse 16) and sought refuge with Paul. Onesimus had come with stolen property (verse 18) and the prospect of returning was appalling. The penalties for slaves who absconded were severe and Paul not only pleads for gentle treatment by Philemon but also seems to hint that Onesimus might be spared for service to himself (verse 13)! Philemon is urged to go the extra mile (verse 21, cf. Matt. 5: 41), and not to exact the penalties (which might include death) which under Roman law he was fully within his rights to impose. A bishop of *Ephesus of the name Onesimus (which means 'useful', hence Paul's *paranomasia in verse 11) is mentioned by *Ignatius (107 CE), and the theory has been proposed, but not generally accepted, that Onesimus was responsible for making a collection of Paul's epistles.

Philemon lived at *Colossae, and some have held that this is so much nearer to Ephesus than *Rome that it was more likely that the letter was sent together with the epistle to the *Colossians (with which it is closely associated) from Ephesus; but the only evidence for Paul's imprisonment in Ephesus are the vague generalizations in 1 Cor. 15: 32 and 2 Cor. 11: 23. It has also been suggested that the slave owner was actually Archippus (verse 2). What is, however, perfectly clear from the letter is that Paul is trying to involve the whole local Church (verses 1 and 23), for the risks being taken by Onesimus and urged on Philemon, and indeed accepted by Paul, were very considerable. It is also clear that the institution of slavery as such is not condemned, though the common Christian faith of slave and owner requires a totally transformed relationship.

Philip A name borne by many biblical characters, from Philip, father of *Alexander the Great, onwards (1 Macc. 1: 1; 6: 2). (1) *Herod the Great had a son, by Mariamne, called Herod Philip, who married *Herodias and had a daughter, *Salome; this last married Herod Philip II, who was the son of Herod the Great by Cleopatra of Jerusalem. This Philip was tetrarch of *Ituraea and *Trachonitis (Luke 3: 1) from 4 BCE to 33 CE. (2) One of the Twelve who followed Jesus was named Philip (Mark 3: 18; John 1: 43) and was in Jerusalem after the *Resurrection (Acts 1: 13). (3) Another Philip ('the evangelist') was one of the seven appointed to administer the Church's charity (Acts 6: 5) but afterwards he extended his ministry (Acts 8: 29) to baptizing a God-fearing Ethiopian. Later Philip lived at *Caesarea (Acts 21: 8–9) with four daughters who prophesied.

Philippi A city in NE Greece named after Philip II of Macedon. It was refounded by Mark Antony as a Roman colony for veterans of the victorious army (41 BCE), augmented by further settlements in 31 BCE and governed as a Roman provincial municipality. The mixed population included Jews (Acts 16: 13). Converts were made by Paul, who paid several visits to the city and wrote his affectionate letter (54 CE). In about 125 CE a letter of Polycarp of Smyrna was written to the Church and is extant.

Philippians, Paul's epistle to the In the NT, the sixth of Paul's epistles. The Christian community at *Philippi in *Macedonia was highly regarded by Paul and this letter includes a warm appreciation of their financial contribution for his needs (Phil. 4: 15–20). There seems to have been plenty of exchange of news between Paul and the Church, both by correspondence (Phil. 3: 1) and also through personal visits of *Timothy and Epaphroditus (Phil. 2: 19–30). This has suggested to some commentators that Paul's imprisonment where he was writing was more likely to have been in Ephesus than in distant Rome, in which case (at Ephesus) it would have been in about 55 CE. The references to the '*praetorium*' (1: 13) and 'Caesar's household' could apply as well to Ephesus as to Rome. But in fact there is no clear evidence for an Ephesian imprisonment. *See* *Philemon.

The epistle to the Philippians includes the important Christological section, 2: 1–11, which is also very difficult to interpret. Possibly it was a hymn already in Christian use which Paul has incorporated here as a poem of praise for the Christ who had voluntarily renounced his equality with God, accepted the agony of the Cross, and was at length crowned by God to rule over the whole universe. It is a passage seized on by a few modern theologians to explain a doctrine of the *incarnation: Christ 'emptied himself' of such divine attributes as omniscience and so in his human person had all the limitations in knowledge of a 1st-cent. Jew. This is a theory now generally discarded, for it seems to posit a Christ who retained some divine attributes but relinquished others. The difficult expression in 2: 6 is interpreted by NRSV as meaning that Christ did not regard his equality with God as a status to be 'exploited'.

Behind this account of the pre-existence and humanity of Christ may lie the portrait of the Servant of the Lord of Isa. 40–55, especially Isa. 52: 13–53: 12, who suffered undeservedly but was vindicated by God and in the event (his death) he justified many and himself carried their iniquities. What Christ did should be an example to the Philippian Christians to renounce self-centredness and give themselves to others (2: 4).

Philistines A sea-going people who came originally from *Crete (= Caphtor, Amos 9: 7) but after defeat by the Egyptians, they settled along the sea-border of the southern part of Palestine (which takes its name from them). They became neighbours and enemies of the Israelites. The five main cities were *Gaza, *Ashdod, Ashkelon, *Gath, and Ekron. The Philistines and the Israelites fought for the available land; after the monarchy was established with *Saul as king, Israel tended to prevail, in spite of setbacks, but it was *David who claimed the final victory (2 Sam. 8: 1). The Philistines became vassals of *Assyria in the 8th cent. and of Egypt in the 7th, but finally the Babylonians conquered them (604 BCE) and deported the people (Jer. 47).

The popular reputation of the Philistines for cultural insensitivity is not borne out by archaeological discoveries of their artefacts from many centuries; these suggest an aesthetic appreciation richer than that of their Hebrew rivals.

Philo A contemporary of Jesus and Paul (20 BCE–*c*.50 CE). Though a prolific writer in Alexandria, Philo shows no knowledge of Christianity. He was an apologist for *Judaism in a rich Hellenistic culture and his aim was to combine a non-rabbinic Judaism, often using *allegory, with Platonic philosophy. His use of the **Logos* concept is reminiscent of the gospel of John, and there may also be affinities between Philo's thought and ideas in the epistle to the Hebrews. Later Christian theologians in *Alexandria (*Clement and *Origen) were more clearly influenced by Philo.

philosophy There has always been a tendency in Christian thought which is suspicious of philosophy and human reasoning. Paul (1 Cor. 1: 16 ff.) denigrated *wisdom: it had no power to save, compared with the 'foolishness of the cross'. In modern times the classical expression of the view that God is known only by his self-revelation in Christ and not as a concept demonstrable by philosophical argument was in Karl *Barth's commentary on Paul's epistle to the Romans (1919; ET 1933).

However, a more conciliatory *exegesis of the Bible took contemporary philosophy seriously at Alexandria. Clement and Origen admitted crudity and lack of credibility in biblical narratives, though they were opposed by those like Tertullian who could see only a literal meaning in the texts and asked 'What has Jerusalem in common with Athens?'

Phinehas (1) A grandson of *Aaron, who ministered before the *Ark of the Covenant (Judg. 20: 28). (2) The younger son of *Eli, a priest (1 Sam. 1: 3) who failed in his duties (1 Sam. 2: 13) and was killed in battle by the *Philistines (1 Sam. 4: 11).

Phoebe A woman deacon at *Cenchreae near Corinth commended by Paul (Rom. 16: 1 and possibly the bearer of the letter).

Phoenicia Coastal territory north of Israel with *Tyre and *Sidon as the principal cities. There were commercial links with Israel (2 Sam. 5: 11; 1 Kgs. 9: 26 f.) and these led to the infiltration of Phoenician *Baal worship (1 Kgs. 18: 19; Isa. 65: 11; Jer. 32: 35). Phoenician art and writing were highly developed.

phoenix In classical mythology, a bird resembling an eagle; according to one legend, at the age of 500 years it built a funeral pile composed of spices, lay upon it, and died, but out of the ashes a new body arose. Hence the phoenix became a symbol of the *resurrection in Christian art.

The name was given to a harbour in *Crete (Acts 27: 12).

Phrygia A section of the Roman province of *Asia, except for part of its eastern area which in 25 BCE was incorporated into *Galatia. The native people practised orgiastic worship of a female deity but the cities of *Laodicea, *Colossae, and Pisidian *Antioch had Greek-speaking communities, among whom Churches were established.

phylacteries Small leather cases, mentioned in the *Law (Exod. 13: 9; Deut. 6: 8). They were worn in NT times by *Pharisees and some others and were attached to the forehead and the left arm of males when reciting daily prayers. Inside the cases were texts from the *Pentateuch. *Scribes and *Pharisees were alleged to make a display of exaggerated piety by making their phylacteries more conspicuous than was necessary (Matt. 23: 5).

physicians In OT times there were professional healers in Palestine but they were not well regarded (Job 13: 4), even in the 1st cent. CE (Mark 5: 26). But Ecclus. [= Sir.] 38: 1 ff. urged respect for their services, so that it was not sarcastic to call *Luke a 'beloved physician' in Col. 4: 14, and Luke has traditionally been held to be the author of the third gospel and the Acts.

piety A word used of personal *faith and behaviour (1 Tim. 5: 4) which includes *repentance and *prayer and the practice of Christian *love.

Pilate, Pontius *Prefect, also called, less accurately, procurator, of *Judaea from 26 CE, commander of the army and chief justice. He appointed the high priests and had little sympathy with his Jewish subjects, having been associated with Sejanus, an antisemitic conspirator in *Rome. He seized funds from the Temple to build an aqueduct; for cruelty in *Samaria he was recalled to Rome in 37 CE. In the gospels he is portrayed as doing his best for Jesus but as unable to resist the demands of the Jews, who could have denounced him to the emperor *Tiberius. Later Christian *antisemitism tended to exonerate Pilate altogether.

An inscription bearing his name was found at Caesarea in 1961.

pilgrimage A journey to a shrine or place of religious importance. The practice has its roots in the OT, when *Abraham visited Mount Moriah (Gen. 22). In the course of time journeys to the Temple became a part of Jewish life (hence Luke 2: 41 ff.). In the 4th cent. CE a Spanish abbess, Etheria, made a pilgrimage to Jerusalem and other holy places and her account in Latin described the kind of *worship she found there, as well as enabling various sites to be identified.

The NT sometimes regards human life as a pilgrimage; Christian believers are temporary dwellers on earth who are travelling towards an eternal destination (Heb. 11: 13 ff.).

pillar In the OT both memorials and also the focus for religious devotion. Made of stone or wood, their capitals would often be elaborately decorated. *Jacob erected a pillar at *Rachel's grave (Gen. 35: 20). Moses set up *twelve pillars (one for each tribe) round an *altar in the *wilderness (Exod. 24: 4). *Joshua caused twelve 'stones' (obelisks?) to be erected in the *Jordan, and later removed to *Gilgal (Josh. 4: 9, 20). Because of the association of pillars with the agricultural fertility religion of the *Baals, the prophets condemned cultic pillars (Hos. 10: 2; Mic. 5: 13), but the two free-standing pillars outside the Temple, called *Jachin and *Boaz, certainly had a religious significance, unlike the pillars of a Greek temple, which were integrated with the structure and had a strong aesthetic appeal.

Symbolically, Hebrews in *Egypt are a pillar (Isa. 19: 19), as are *James, Peter, and *John in Paul's snide description (Gal. 2: 9). The Church is a pillar of truth (1 Tim. 3: 15).

pinnacle The highest points of the Temple roof ('parapet', REB, NJB), possibly overlooking a precipitous valley, mentioned as the imaginary scene of Jesus' second *temptation (Matt. 4: 5).

Pirqe Aboth Hebrew for 'Sayings of the Fathers'; and the earliest section of the *Mishnah, the oral *Law, put into written form in the middle of the 2nd cent. CE. There are resemblances between some of the sayings found in *Pirqe Aboth* with sayings of Ben Sirach in Ecclus. A verse in *Pirqe Aboth* (1: 2) is a fair summary of Ecclus.: 'By three things the world is sustained: by the Law, the service in the Temple, and by deeds of loving-kindness'.

Pisgah A mountain in *Moab east of the *Dead Sea from whose height Moses viewed the land beyond the *Jordan (Deut. 3: 27) and promised its slopes to the tribe of *Reuben (Josh. 13: 20).

Pisidia Southern area of Asia Minor and in NT times part of the Roman province of *Galatia. Its principal city was Antioch, where there was a Jewish population. Paul and Barnabas travelled in the area and met a hostile reception (Acts 13: 14–51; 14: 24).

pit Wide-ranging term covering holes for burials (2 Sam. 18: 17) and traps (Ps. 7: 15), as well as the abode both of the dead (Job 33: 18) and of the devil (Rev. 9: 1, 20: 3). The owner of a pit who failed to cover it would be liable to pay compensation (Exod. 21: 33–4) in the event of an accident.

Pithom A city which the *Hebrews in *Egypt were employed to build (Exod. 1: 11).

pity The kindness of a person towards another in trouble (Mark 8: 2); this may be compared to the pity which God has for his creatures (Mic. 6: 8) which he will exercise at the *Last Judgement (Matt. 25: 31–46).

plagues Various diseases and pestilences are called plagues in the OT; they are often seen as divine punishment (Deut. 28: 21). But pre-eminently it is the ten plagues visited on *Egypt that were ingrained in Hebrew memory or folklore (Exod. 7–12). The narratives are a combination of the *J and *P sources, the latter emphasizing the drama as a contest between God and *Pharaoh. All except the last were regular and natural phenomena in Egypt but are grossly exaggerated in the biblical version. The Hebrews saw in them the providential care of their God Yahweh, as in Ps. 78: 43–51. Even the last plague, the death of the *first-born, could have been an infectious illness which was confined to the Egyptians.

plain, the cities of the Five cities to the SE of the *Dead Sea (*Sodom, *Gomorrah, Admah, Zeboiim, and Zoar) where there was a battle between kings (Gen. 14). If there is a historical basis to the story, the cities (Gen. 13) were probably *around* rather than *in* the Dead Sea plain, according to archaeologists.

plain, the Sermon on the Luke's much shorter version (6: 17–36) of Matt.'s *Sermon on the Mount. The reference to Jesus' descent to a level plain seems to be a contrived emendation of Matt.'s 'on a mountain' (5: 1), since preaching on the level would not in itself call for such an explicit comment. To the *beatitudes, Luke has added corresponding Woes, and it is characteristic of him that he draws attention to Jesus' attitude to wealth and poverty.

plants Regarded in the OT as gifts of God to be valued and not consumed selfishly or greedily (Gen. 1: 29–30; 3: 18), especially the vine (2 Kgs. 19: 29). In the NT plants and planting are used metaphorically—of God's work (Matt. 15: 13) and people's (Matt. 21: 33–41), and of a stage in the growth of a Church (I Cor. 3: 6–8). *Palm trees were especially important, and flourished in adverse circumstances. In 71 CE the Romans depicted a palm tree on coins commemorating the defeat of the Jews in the recent war.

Platonism After the death of Plato (347 BCE), his philosophical teachings were continued into the Hellenistic era by his followers and were an influence on *Judaism in the Book of Wisdom. Mankind is given immortality because human beings are made in the 'image' of God's own eternal self (Wisd. 2: 23).

In the NT there is a possible influence of Platonism (through *Philo) on the epistle to the Hebrews. In 8: 5 the author uses the idea of a heavenly temple having its earthly, imperfect, counterpart. The Levitical priests minister in what is only a copy or shadow of the heavenly places, where Jesus exercises his priesthood. It is thought that these concepts of this epistle show awareness of Plato's theory that the 'forms' or 'ideas' laid up in heaven are the perfect realities of which there are imperfect reflections on earth.

pledge A form of protection for a lender. There was a legal procedure in Israel governing loans. The lender might accept a piece of property pledged as guarantee for repayment, but if this was the warm cloak needed for the night, it was to be returned before nightfall (Exod. 22: 26).

Pleiades Stars in the constellation Taurus, mentioned with Orion as part of God's creativity (Amos 5: 8) but which to Job seems to make a mockery of his alleged justice (Job 9: 9, 15). In Job 38: 31 God replies to his complaint and accuses Job of arrogance. Job has no power to alter the patterns of the constellations.

pleonasm The repetition of an idea both in Hebrew *poetry, as Ps. 8: 4, and in the gospels: e.g. Mark 7: 21; but when Matt. follows Mark, the superfluous words 'from within' are omitted by Matt. 15: 19, thus eliminating Mark's pleonasm.

pleroma Greek for 'fullness', and used in Eph. 1: 23 and Col. 1: 24; it can have either an active or a passive sense 'that which fills up' or 'that which is filled up'; therefore the meaning is either that the Church (Eph. 1: 22) completes Christ, or that the Church is filled by Christ. The word (which the epistles may have hijacked from *Gnostic terminology) may refer to Christ as both over the Church, as its head, but also within the Church, which is his Body.

Pliny 'The Elder' (23–79 CE), author of many works, including a Natural History; he died when Vesuvius erupted over Pompeii. Pliny 'the Younger' (b. 61 CE) was the nephew of the former; as governor of Bithynia he wrote a celebrated letter in 110 CE to the emperor *Trajan asking advice on how to deal with Christians in his province. He mentioned their morning worship of Christ 'as a god' and wondered whether their refusal to pray to the emperor's statue or to honour the pagan gods was treasonable. *See* Peter, first epistle of.

plough When the Israelites settled in *Canaan, they were obliged to learn agricultural arts and sought help from their neighbours (1 Sam. 13: 20). Ploughing took place in the winter in preparation for planting seed. Jesus used ploughing as a metaphor for total dedication to the cause (Luke 9: 62).

pneumatology That area of theology which deals with the doctrine of the *Holy Spirit (from the Greek *pneuma*, meaning 'breath' or 'spirit').

poetry 'All good poetry is the spontaneous overflow of powerful feelings' (William Wordsworth, 1801).

The distinction in modern literature between prose and poetry is difficult to apply to the *Bible, but there is a tradition that regards certain OT books—the Psalms, Proverbs, Job, Lamentations, the Song of Solomon (or Song of Songs)—as poetry; and modern translations commonly print them in verse form. But many other examples of poetry are claimed by modern scholarship. Not having the same recognizable characteristics intrinsic to much Western poetry, of metre and rhyme, Hebrew poetry requires other criteria for recognizing it. There is a terseness of style; there is sometimes an ambiguity of meaning when a strictly literal description is replaced by figurative language; and the words used demand a response from the listener. The order of words may be unusual or they may be repeated as a kind of refrain (Isa. 5: 25; 9: 12, 17, 21; 10: 4). There are to be sure some instances of rhyming, by the use of suffixes; Jer. 12: 7 is cited as an example. And the relationship between Hebrew poetry and *music indicates that it had a kind of metre which was fluid and flexible, not unlike the free rhythm of Gregorian plainsong of the Latin Church. (Some modern responsorial chants for the psalms developed in France using the rhythm of ordinary speech, stylized to some extent, may convey the flavour of Hebrew recitation.) When Jerome (342–420 CE) was working on his *Vulgate, he thought that Hebrew poetry was usually composed in hexameters, like Greek and Latin verses, but modern scholars are inclined to hold that Hebrew poetry had no system of regular metre, as did Greek or Latin poetry; yet the music must have required some conventions of rhythmic stress.

The frequently (and perhaps misleadingly) mentioned use of 'parallelism' is a convention about the ordering of words in Hebrew poetry. In some cases there is a second line which extends or echoes the theme of the first (e.g. Ps. 104: 28). In Job 10: 12 two masculine nouns in the first line are paired with two feminine nouns in the second. Elsewhere a second idea is brought in by way of contrast with the first, but not always in strict line-by-line parallelism, e.g. in the Song of Deborah (Judg. 5: 4–5; 26–7). It was water that Sisera requested: she gave him milk. She brought him curds on a plate; but in her hands she held a tent-peg and a mallet; the parallelism and the terseness seem to be there, but they are not clearly marked out. It is present also in the book of Job, though with complex variations (e.g. 28: 12, 20).

Poetry is found in connection with worship in all eras of Israelite history. There are sayings uttered by priests in the Temple, e.g. the threefold blessing of Num. 6: 24–6, and there are hymns sung at Passover, e.g. the celebration of victory over the Egyptians and the rule of God as king in Zion in Exod. 15, which must derive from Temple worship rather than from any historical exodus when Zion (Jerusalem) was unknown. Other fragments of war poems were woven into liturgical forms (e.g. Hab. 3). Funeral dirges and laments have an additional force from the parallelism, as in Jer. 9: 1 and 12: 8, 10; and liturgical poems of thanksgiving, though the rhythm and parallelisms are less clear, have been adapted by editors to fit into pieces of narrative, e.g. Jonah 2: 2–9, where appropriate references to the sea are followed by a promise to make vows in the Temple, so affirming a theological point relevant to the editor's own time. As the God of Jonah could rescue him from the deep, so also he could rescue the people of Nineveh from unbelief into repentance.

There is very little poetry in the NT (though the Prologue of John 1 is similar to the poetic parallelism of passages in the Wisdom literature, e.g. Prov. 8: 22–36 and Wisd. 7: 22–8: 1). But a number of Jesus' aphorisms contain parallelisms, e.g. Matt. 11: 30, Luke 16: 10. In some cases, the second line makes a contrast with the first, as in Matt. 8: 20 ('Foxes have holes, and the birds of the air have nests; but the Son of Man has nowhere to lay his head'). The canticles of the Lucan *infancy narratives (the *Benedictus, *Magnificat, and *Nunc Dimittis) resemble Hebrew poetry. The Greek poet Aratus is quoted by Paul in his speech at Athens (Acts 17: 28) and Menander in 1 Cor. 15: 33, though the latter quotation rather looks as if Paul had not actually read the lost play *Thais* from which the words are taken; the Cretan poet Epimenides (or, another suggestion, Callimachus' *Hymn to Zeus*) is quoted by Titus 1: 12.

Pontus Roman province along the coast of the Black Sea in which there were several Jewish communities. Some Jews from the region were in *Rome for *Pentecost (Acts 2: 9), and *Aquila was born there (Acts 18: 2). Christians in Pontus are addressed by 1 Pet. 1: 1, and advice on their treatment was sought by the provincial governor *Pliny in letters (110–12 CE) to the emperor *Trajan. The 2nd-cent. heretic *Marcion, who wished to detach the Church from the OT, was a shipping magnate in Pontus.

poor Those existing at the bottom of a society's economic scale. In the OT there is recognition of injustice and inequalities. These impose strong duties on the better-off to relieve those who have lost out in ruthless competition, perhaps even to the point of selling themselves into slavery (2 Kgs. 4: 1; Amos 2: 6, 7). The prophets denounced conditions which accelerated the divisions between rich and poor (Isa. 3: 14–15) and reminded the nation that God hears the cry of the poor (Ps. 12: 5; Isa. 10: 2). The *Law made provision for the poor (Exod. 23: 11) and imposed restrictions on a farmer's unbridled greed (Lev. 19: 9–10).

Charitable giving to the poor was, however, circumscribed by the conviction that poverty was a punishment for wickedness (Prov. 13: 18). There was no systematic help for the army of beggars in NT times, but Jesus' concern for the poor is particularly clear in the gospel of Luke. In the epistles Paul more than once asks his Churches to support his collection for the poor members of the Church in Jerusalem (Rom. 15: 26; 2 Cor. 8–9), who were impoverished after spending all their capital (Acts 4: 32–5) and were severely hurt by the universal famine (Acts 11: 28) of, perhaps, 46 CE.

Such sympathy for the poor was alien to contemporary Graeco-Roman ways of thinking: gifts were certainly sometimes made by the well-to-do but from motives of self-interest such as honour or prestige. The élite in society regarded the majority with contempt and were apt to parade their superiority. The Church's ideal of holding together in one congregation all strata of society (Paul's aim in 1 Cor.) must have been regarded as extraordinary, and Luke 6: 35 quite outlandish.

porch A forecourt with pillars in *Solomon's house (1 Kgs. 7: 6); and in *Herod's Temple at Jerusalem a colonnaded portico or passageway was known as Solomon's porch (Acts 3: 11).

Potiphar A high official of *Pharaoh (Gen. 39: 1). The patriarch *Joseph prospered in *Egypt and was promoted in the household of Potiphar. Potiphar's wife accused Joseph of sexual harassment and he was temporarily imprisoned (Gen. 39: 6–20).

Potiphera An Egyptian priest whose daughter was given in marriage to *Joseph (Gen. 41: 45).

potter The craft of the potter was both ancient and essential, and archaeological discoveries have yielded important clues for dating the occupation of cities, as far back as the Neolithic Age at *Jericho. In the *Bible the earliest mention of pottery is when *David was brought vessels full of grain (2 Sam. 17: 28) and from then on techniques of workmanship improved. There is a brief description of a potter at work in Jer. 18: 3 f. By NT times the potter worked with two wheels joined by a shaft, rotating the lower wheel with his foot. The clay used contained sand and vegetable matter and was purified before being thrown on to the wheel and moved as the wheel spun. The jar or bowl (Matt. 26: 23) or basin (John 13: 5) was baked or dried in the sun. God is likened to a potter (Isa. 29: 16; Rom. 9: 20–4) and human beings should not argue with what he chooses to make.

Potter's Field A cemetery for *foreigners near Jerusalem, bought with the thirty pieces of silver which had been given by the priests to *Judas as a reward for betraying Jesus and which he threw back at them when he realized that what he had done was irreversible; hence the field was also called the Field of Blood (Matt. 27: 8). Matthew highlights the event by quoting an incident in Jeremiah (incorrectly—it is mostly from Zech. 11: 12–13) in accordance with his policy to link events in the life of Jesus to apparent OT prophecies. The story takes a different form in Acts 1: 18–19, where it is Judas himself who purchases the property.

pound The translation of *mina* (Luke 19: 13) as 'pound' (NRSV, NJB) is said (NRSV marg.) to be what a labourer would earn in three months.

power The power of God was shown in nature (Jer. 10: 12) and in the workings of history (Exod. 15: 6). In the NT it was manifest in Jesus' *healings (Mark 6: 2), in the descent of the *Spirit at *Pentecost (Acts 1: 8), and in the apostolic preaching of the gospel (Rom. 1: 16).

praetor A Roman magistrate with office in the judiciary, the army, and the civil service. The position was a step in the career ladder, leading to appointment as consul.

praetorian guard A seasoned unit of the Roman army (Phil. 1: 13; perhaps also Phil. 4: 22).

praetorium The governor's residence in Jerusalem (Mark 15: 16. REB, NJB, NRSV marg.) and *Herod's palace at *Caesarea (Acts 23: 35, NRSV marg, NJB).

praise In the OT and NT *worship, the act of ascribing *glory to God (Ps. 113; Acts 2: 47).

prayer The act of communicating in words or in silence with the transcendent God. Conversations between God and men are reported in the OT (e.g. *Abraham, Gen. 15: 1–6; Moses, Exod. 3: 1–4; 33: 11; *prophets, 1 Sam. 3: 4–9). OT prayer includes petition, *intercession, *confession, and *thanksgiving, and set hours and days are prescribed for prayer. In the NT Jesus is reported to have prayed to his Father frequently and he gave the '*Lord's Prayer' to the disciples (Matt. 6: 9–13; Luke 11: 2–4). The epistles teach that prayer to God is offered through Christ (Rom. 1: 8). NT prayers include *praise (Acts 2: 47), thanksgiving (1 Cor. 14: 16–17), and petition (Phil. 4: 6). Prayer is not regarded as a method for compelling God to act but for asking that his will be done and his kingdom come.

preaching A fundamental method of communication in the NT (though it continues the 'announcement of good tidings' in the later OT prophets, Isa. 40: 9). It was said both of *John the Baptist (Mark 1: 7) and Jesus (Luke 4: 18) that they preached their messages, and Paul regarded preaching as his main work (1 Cor. 1: 17). The Greek word commonly used for the preached sermon is *kerygma*, which is sometimes distinguished from the other kind of NT communication, teaching or instructing new members, for which the Greek word was *didache*.

Jesus' good news is succinctly given (Luke 4: 18–19) as good news to the poor, release to captives, sight to the blind, liberty to the oppressed, and was the proclaiming of 'the acceptable year of the Lord'. The good news of Paul was centred on the *death and *resurrection of Christ 'in accordance with the scriptures' (1 Cor. 15: 3–8).

predestination God's pre-arranged intentions for the destiny of human beings. Because of the history of this concept in Christian theology (e.g. by Augustine and Calvin), it is associated with the idea of an inscrutable God who from eternity had decided that some are to be 'saved' and others 'condemned'. However, the scriptures do not so much affirm predestination as a rational deduction within a philosophy of theism as give expression to a glorious sense of God's graciousness. God has a plan to deliver human beings from the power of sin—this is predestined: people may freely accept or reject this offer (Phil. 2: 12). Paul exclaims in wonder that he has been chosen by God and that the Church has been chosen, but he nowhere says that those who have rejected faith are thereby damned.

Special interest attaches to Paul's understanding of God's purposes for Israel (Rom. 9–11). If by his *covenant God seemed to predestine Israel to a perpetual role in his purposes, does this mean that he is fickle and has, now that Christians are the true children of *Abraham (Gal. 3: 7), broken his promise? Paul declares that God is faithful; he has a purpose for Israel, even for Israel's unbelief. Israel has been the vehicle for the Gospel and in the end the conversion of Gentiles will persuade Jews also to believe, out of sheer envy (Rom. 11: 11).

pre-existence A NT view that Christ already existed, in eternity, before he became man. He was altogether outside time. This view is expressed in the gospel of John (17: 5) and by Paul (Col. 1: 15 ff.). It was their experience of a divine dimension in Jesus that made them declare that he must always and eternally have existed. It could be argued on the basis of the prologue to John's gospel that the **Logos* was pre-existent and that when Jesus was born (John 1: 14) the *Logos* was then identified with the Jesus of history.

prefect The title of the Roman official of the equestrian order who ruled *Judaea from 6 to 41 CE. From 44 to 66 CE he was a 'procurator'. 'Prefect' is the title accorded to Pontius *Pilate in the inscription from Caesarea, but he is called a procurator by *Josephus. The two terms are almost synonymous.

preparation In NRSV, NJB, and AV (John 19: 31) this was the day before a religious festival, or any *Sabbath, when it was necessary to get things ready on account of the law ordering cessation from all work on the Sabbath itself.

presbyter Greek for *'elder', a Christian leader, probably modelled on the *synagogue precedent. Each Church had a group of presbyters, as also at *Qumran, and *Paul and *Barnabas appointed them in every Church they founded (Acts 14: 23). In due course, one of the group enjoyed a pre-eminence. In some cases the 'elder' is also known as a 'bishop' (Tit. 1: 5–9). *See* ministry.

pride Although regarded as the root of *sin, because it excludes other people, and God, nevertheless the NT more often praises humility than castigates pride. Paul persuades himself that his only boasting is in the *cross of his Lord (Gal. 6: 14).

priesthood of Christ Because priesthood existed for the purpose of offering *sacrifices, Jesus is regarded as a priest because he has offered a sacrifice (himself) once for all. This has rendered the OT sacrifice system obsolete, together with the Temple priesthood. The OT model for Christ's high priesthood is that of the priest-king *Melchizedek (Heb. 7: 26).

priesthood of the Church Because Christians are united by *baptism and *Eucharist to Christ, it is appropriate to regard the Church as itself priestly, (1 Pet. 1: 9–10), and all participate in the mystery of Christ (1 Cor. 10: 16). The priesthood is corporate; in the NT individuals are not called priests.

Priestly narrative One of the sources in the compilation of the *Pentateuch, commonly designated by the symbol *P; the author(s) came from priestly circles (6th cent. BCE) and had a special interest in *genealogies and a concern with cultic practices which it was hoped would be restored with the Return from Exile.

priests and Levites In the OT the priests were intermediaries between the people and their all-powerful God. Their main function was to offer *sacrifices, with a subsidiary duty until the *Exile, of teaching the *Law. They were also in charge of **Urim* and **Thummim*, the sacred lots, carried by the priests in a front pocket (Exod. 28: 30). During the period of the Judges, priests of the tribe of *Levi were particularly valued (Judg. 17: 10). When *worship was centralized at Jerusalem, local shrines were closed and many Levites who were not descendants of *Aaron had no employment (Deut. 12: 12 f.), and were not accepted at Jerusalem as true priests. They were suspected of having become tainted with *Canaanite practices in their rural shrines, and so in the Temple were relegated to an inferior status (Ezek. 44: 10–14). Priests were superior as being descendants of Zadok (Ezek. 48: 11). Post-exilic literature (Mal. 2: 4–7) suggests that both priests and Levites were accorded a common descent from Levi and as such the Levites, though distinct from priests, had important duties at the services (1 Chron. 16: 4–27) and in teaching (2 Chron. 17: 7–9). In the NT the distinction between priests and Levites is maintained (Luke 10: 31, 32).

Chief priests, together with *elders and scribes, comprised the Jerusalem *Sanhedrin (Mark 14: 55) before whom Jesus appeared before being sent for trial by Pilate. The 'chief priests' were members of priestly families from whom the 'high priest' (Caiaphas, Matt. 26: 3) was chosen. In the NT Church the whole body of Christians were regarded as a 'royal priesthood' (1 Pet. 2: 9) not in the sense of a ruling or tyrannical priesthood but as a body charged, like a king, to mediate blessings to the world and to offer spiritual sacrifices.

prince In the OT, not only the heir to the throne, but also a man of eminence (an official, NRSV in Esther 1: 14); in the Messianic era there will be a 'prince of peace' (Isa. 9: 6).

Prisc(ill)a One of Paul's colleagues. The longer version of the name is preferred by Acts, but Paul (Rom. 16: 3; 1 Cor. 16: 19) uses the shorter (Latin) name for this apparently well-connected leader of the Church, who seems to take precedence over her husband *Aquila. Both were expelled from *Rome under the edict extraditing Jews in 49–50 CE and moved to *Corinth and to *Ephesus.

prison Among the prisons mentioned in OT were *Jeremiah's (Jer. 32: 2, 8) and King *Jehoiachin's (Ezek. 19: 9). In the NT *John the Baptist was kept in the mighty fortress of Machaerus (according to *Josephus), and Peter (Acts 12: 3 ff.) and Paul (Acts 21: 34) were both held in Jerusalem, probably in the Antonian fort. Paul was also imprisoned, with chains attached, in *Rome (Acts 28: 16).

proclamation The preaching of the gospel of Christ in words (Gal. 1: 8; 1 Cor. 9: 14; Phil. 1: 17), or the *Eucharistic celebration (1 Cor. 11: 26). Proclamation was usually followed by *teaching and instruction in the elements of the *faith, or creed, though what Jesus did and taught in his ministry may well have been included within the basic proclamation.

proconsul A Roman provincial governor, such as Sergius *Paulus (Acts 13: 7) and *Gallio (Acts 18: 12), appointed for one year by the senate.

procurator A Roman administrator who acted on behalf of the emperor, subordinate to the legate. *Pontius Pilate is sometimes referred to as the procurator of *Judaea (26–36 CE), but the more correct translation of the Greek is governor, as given by AV, NRSV, REB, and NJB (Matt. 27: 11), as also for *Felix (Acts 23: 26) and *Festus (Acts 24: 27). Pilate does seem to have had the status of *prefect, which was the title used, according to an inscription, from 6 to 41 BCE. The governor was called 'procurator' from 44 to 66 CE. The frequently applied title procurator to Pontius Pilate is due to the historian *Josephus, who did not always give in Greek the exact translation of Latin titles. (From 41 to 44 CE Judaea was governed by Herod *Agrippa I.)

prolepsis Reference to a future event as though already accomplished, as Matt. 23: 38, or to a past event as being still anticipated, as Luke 19: 44. The latter is also regarded as a **vaticinium post eventum*.

promise God is said to be able to accomplish what is promised (Isa. 55: 10 f.); all that is promised in the OT is according to Paul fulfilled in Christ (2 Cor. 1: 20).

Promised Land *Canaan—the destiny of the Israelites after the *Exodus. Although God had created the whole world (Ps. 95: 4 ff.; Isa. 40: 28), he had set apart a special land for the descendants of *Abraham (Gen. 12: 2; Deut. 26: 5 ff.). It was not a land to be ruthlessly exploited (Jer. 2: 7), nor was this land essential for worship of *Yahweh (Jer. 29: 4 ff.). Similarly, no specific 'promised land' was needed for the new Israel, the Christian Church, which is not confined to any particular place or country for its worship (Jas. 1: 1). Modern Zionists regard the State of Israel as falling within the ancient promise of a land.

pronouncement stories Used by English *Form Critics to designate short illustrative stories of an event in the gospels and reflecting the catechetical interests of the early Church e.g. Mark 3: 1–6; 12: 13–17.

prophetess Female prophets in the OT performed functions similar to those of male prophets. Deborah was a judge who led people into battle (Judg. 4: 1–10) and *Huldah received enquiries from King *Josiah about the will of God (2 Kgs. 22: 14–20) with regard to the book of the law found in the Temple. The *infancy narrative of Luke (2: 26–38) recounts that a prophetess, *Anna, greeted Jesus as the *Messiah. A false prophetess, reviled as '*Jezebel' (after the heathen wife of King Ahab of Israel) is said to be tolerated by the Church of Thyatira (Rev. 2: 20).

prophets 'Prophets' in English Bibles usually renders Hebrew *'nebi'im'*, though other terms are used. Thus there were 'seers' (Hebrew, *ro'eh*, a seer, 1 Sam. 9: 9), who passed on messages from God received in dreams or divination. There were also sorcerers and soothsayers who engaged in necromancy and whose activities are condemned as pagan practices (Deut. 18: 10–11). They also claimed to predict future events, and in this they were indistinguishable from traditional characteristics of the prophets of the Lord as recorded in the stories of the books of Samuel and Kings. But in the Deuteronomic tradition in the mid-6th cent. BCE it is Moses who is regarded as the prophet (*nabi*) *par excellence*. He is the man of authority who conversed with God 'face to face' and who wrought signs and wonders (Deut. 34: 9–12), and looked to the future (Deut. 18: 18). Miracles continued to be one of the authenticating marks of prophets such as *Elijah and *Elisha in the 9th cent. BCE (though before then *Nathan was recognized as a prophet in the court of King *David (2 Sam. 7) about 1000 BCE). But also the commonly held view that prophets were above all inspired to foretell future events is based on the Deut. text. A true prophet was recognizable if his predictions were fulfilled (Deut. 18: 22), which could be a long time ahead. Prophecy of divine punishment, for example, is seen by the Deuteronomist as fulfilled by the destruction of Jerusalem (586 BCE), which was not an incomprehensible disaster, because it happened in accordance with prophecy. Prophets also played a regular and respected role in Israel's religious and social life. Although they were sometimes threatened (Amos 7: 10–13; Jer. 26: 8) or ignored (Isa. 6: 9 ff.), they were on the whole accorded remarkable toleration even when their utterances were feared and resented. Several enjoyed royal protection (Isaiah, Huldah, Jeremiah); some, under the early monarchy, had support by membership of a prophetic guild (2 Kgs. 4: 38). Some may have been held in awe on account of the ecstatic and paranormal phenomena and frenzy (1 Sam. 10: 6) which were characteristic and, in some cases, even suspected of being an indication of possession by an evil spirit (1 Kgs. 18: 26–9; Jer. 29: 24–8). At any rate there were times when prophets were struck with aphasia (Ezek. 3: 26) and hallucinations (Jer. 4: 19), and some were conspicuous for wearing distinctive dress (2 Kgs. 1: 8). There is no evidence that prophets had disciples who were responsible for collecting their masters' utterances (the only possible reference to such a custom, Isa. 8: 16, is difficult to interpet) but Jeremiah had a secretary (Jer. 36: 4), and someone at any rate collected the teachings of Isaiah for preservation and for reinterpretation and additions during the Exile.

Although prophets engaged in normal community life (Amos 1: 1) and constantly expressed praise for Jerusalem (Isa. 2: 3) or were even employed in the Temple services (Ezek. 1: 3), they uttered their prophetic oracles, sometimes reluctantly, when the word of the Lord compelled them (Jer. 1: 6). They were often fiercely critical of the sacrificial system (e.g. Isa. 1: 10 ff.; Amos 5: 24–5) when it was divorced from moral responsibility. They were not, however, opposed to such worship in itself, as is evident from the wealth of liturgical material in the prophetic literature (e.g. Isa. 38 and Habakkuk 3). Isaiah had his great vision in the Temple and there discovered that holiness rebuked the sinfulness of himself and society (Isa. 6: 5).

Prophets were often critical of royal courts and institutions and policies, and were then felt to be very threatening by the Establishment. At the same time it was part of the prophetic convention to utter strong denunciations when appropriate against the nation's enemies, even if the crimes had been committed some time ago (Amos 1: 3–2: 6) and the enemies would never hear them. But the people of Israel would hear them, which then rendered the denunciation of Israel itself all the more telling. Prophets were upholders of social justice; they condemned owners of property (Mic. 2: 2), men who were persistently drunk (Joel 1: 5), and women who adorned themselves excessively with fine clothes and jewellery (Isa. 3: 6–23). They warned that punishments were in store for the nation's iniquities, though in the end, since God would never be unfaithful to his *covenant promises—there remained the consolation of a strong hope. When the prophets condemned religious abuses (Isa. 1: 9–17) and social vice (e.g. Isa. 5: 8), and gave political advice (Isa. 7: 4; Jer. 32), there were differences of emphasis between prophets in what they proclaimed. But all accepted the basic doctrine that Israel was chosen by God as his special people, and that this had moral consequences. The prophets were one of the influences on the ideology of the covenant eventually formalized by the *Deuteronomist school. This conviction enabled the people's faith to survive the hardships of the Exile and the hazards of the Return in the 6th cent. BCE. After the prophets of that era (*Haggai, *Zechariah, and *Malachi) prophecy disappeared, under a cloud of suspicion and antipathy (Zech. 13: 2–6). Thereafter characteristic communication between God and the people was in the form of *apocalyptic, though there is no clear line of demarcation between that and prophecy. The prophets' hope of ultimate *salvation combined with their threats of judgement, which is apparent in a late oracle such as Isa. 58: 1–12, became part of the content of apocalyptic expectations for the *day of the Lord.

When *John the Baptist began to preach and baptize by the River *Jordan, this was hailed as a revival of long defunct prophecy (Matt. 3: 4).

Prophecy was not a uniquely Israelite

phenomenon. Discoveries at *Mari have uncovered letters (18th cent. BCE) which describe activities of religious men similar to that of the OT prophets. Within Israel there could be 'false prophets', as when Hananiah was challenged by *Jeremiah (Jer. 28); the truth could only be discovered by the people who heard their dispute by waiting to see what happened (Deut. 18: 21–2). Nevertheless unfulfilled prophecies are faithfully recorded: Micah predicted in the 8th cent. BCE that Jerusalem would fall (Mic. 3: 12), whereas his contemporary Isaiah predicted that the city would be saved. It would seem that over the years these prophets acquired a kind of unique authority. The non-fulfilment of a single utterance was less important than the whole context of the nation under hostile oppression. Seen like that, prophecies were validated.

There are prophets in the NT who had a kind of specialized office (Eph. 2: 20) and they can be traced up to the second half of the 2nd cent. Then they disappeared, possibly because a rash of false and spontaneous prophecies made them suspect to leaders of the Church. Paul is certainly not very enthusiastic about this *ministry: utterances were to be tested (Rom. 12: 6; 1 Cor. 12: 3). But it was more acceptable than unintelligible speech, and is mentioned in every list of spiritual gifts. It was regarded as helpful in building up a community of faith (1 Cor. 14: 1–5), but, unlike OT prophecy, it seems that only exceptionally (Acts 11: 28) was NT prophecy associated with foretelling the future.

prophets, books of In the Hebrew Bible the books of Joshua, Judges, 1 and 2 Sam., and 1 and 2 Kgs. are known as the Former Prophets and are all works influenced by the Deuteronomic editors. They have one consistent aim: to show that obedience to the terms of the covenant is rewarded and disobedience punished. The Latter Prophets are Isaiah, Jeremiah, Ezekiel, and the Twelve Minor Prophets, which were once contained in a single scroll; but not Daniel, which was assigned to the Writings. These great prophecies all belong to the period 750–350 BCE.

propitiation The means of warding off the justifiable anger of God. Often used in AV when 'expiation' would be a sounder translation, being the means by which an offender's guilt can be removed or forgiven. Expiation does not enter the realm of divine wrath as needing to be appeased. AV translates 'propitiation' in Rom. 3: 25, Heb. 2: 17, 1 John 2: 2 and 4: 10 of the sacrifice of Christ; NRSV has a 'sacrifice of *atonement' and NJB 'a sacrifice for *reconciliation', and REB 'the means of expiating *sin'. The modern translations are intended to make it clear that the NT does not assert that an angry God is placated (propitiated) by the death of Jesus. Rather, expiation is directed towards sinners by removing their guilt.

proselytes Converts to *Judaism, who had gone farther than the mere sympathizers called 'God-fearers' or 'worshippers of God' (Acts 13: 16; 16: 14). Proselytes had taken the final steps for which *circumcision, *baptism, and a sacrificial offering were required. They were generally as opposed to Christianity as born-Jews, but some pagans who were attracted to the ethical monotheism of Judaism but repelled by its ceremonial and legal restrictions welcomed the Gospel. Christianity spread rapidly among these, causing resentment amongst Jews who regretted the loss of potential proselytes (Acts 18: 12–17; cf. Matt. 23: 15).

prostitution Part of *Canaanite fertility religion which the Israelite followers of *Yahweh sought to suppress, though individual Hebrew women were paid for sexual favours and were not stigmatized. The most famous was *Rahab, who received the Hebrew spies (Josh. 2) and who appears in the *genealogy of Jesus (Matt. 1: 5).

Proto-Luke An earlier form of the gospel of Luke. Several British scholars in the period between the two world wars argued that the gospel of Luke was compiled in three stages: the earliest form (Proto-Luke) consisted of *Q + *L; to this compilation most of Mark was added, and the whole was finally prefaced by the *infancy narratives (1 and 2). It is a theory now generally discarded.

proverbs Short, sometimes witty, comments on human life and manners, deriding fools (Prov. 1: 7), or noting consequences (Jer. 31: 29) or the absurd (Amos 6: 12). They are maxims to express *faith and right conduct. The Hebrew word can also mean a 'taunt-song' (Isa. 14: 4).

Proverbs, book of A collection of pithy sayings and moral advice for the instruction of the young, traditionally ascribed to *Solomon, the archetypal wise man, but in fact dating in its final form from the post-exilic period. The book is written in poetic form —that is, with regular patterns and vivid images—and emanates from a fairly established group in society, anxious to maintain stability and tradition. These teachers of the Wisdom School were sometimes in conflict with the more radical proclamations of the prophets (Jer. 9: 23); they were from the same school as Ben *Sirach and the author of Job and were intellectuals but not in the Greek tradition of speculative philosophers; Hebrew wisdom was exemplified in practical skills, knowledge about how best to manage one's life and about the purpose of life. Hence 'wisdom' came to be personified as an attribute of God (Prov. 8: 22) and eventually the concept was available to the Church as it struggled with its *Christology (Col. 2: 3).

providence The belief that over all the vicissitudes of mankind and the evolution of the universe there is a conscious purpose of God for good. There is a reason, however opaque, for everything. God is responsible not only for making but also for preserving the created order, and history is a process moving ever onwards towards an end, which God foresees (Ps. 104: 31–3). In the NT the providence of God is demonstrated above all in Jesus, who is the guarantor that promises of God from of old are being fulfilled and will lead to *salvation for mankind (1 Pet. 1: 3–9).

provinces Geographical areas of administration. In the OT, provinces of the Persian Empire are mentioned in Ezra (7: 21) and the book of Esther reports that each of the Persian provinces had its own governor (Esther 1: 22). Under the Romans provinces were created from NW Africa to *Mesopotamia. Senatorial Roman provinces mentioned in the NT (created between 146 BCE and 27 BCE) are: *Macedonia, *Achaia, *Asia, Bithynia, *Crete and Cyrenaica, *Cyprus. Imperial provinces are: *Syria (which included the sub-province of Judaea), *Galatia, *Pamphylia and *Lycia, *Egypt and *Cappodocia. The former group were governed by consuls, chosen by lot; the latter by lieutenants of the emperor himself and were regions where disorders were anticipated, and these were created between 64 BCE and 30 BCE.

Land taxes were collected by officials appointed by the emperor, customs taxes by *publicani*, who had bought the franchises and who collected the maximum they could squeeze from travellers.

Psalms In the OT, a collection of 150 self-contained poems and prayers without narrative contexts, traditionally ascribed to King *David, but regarded by modern scholarship as mostly anonymous compositions of various dates and for undeclared occasions, which have to be inferred from the content.

The psalms have devices and techniques that are apparent in other Hebrew *poetry, including parallelism, though there is no uniform system and the presence of a traceable metre is disputed. Some of the few psalms having a regular rhythm are Pss. 5, 29, 46, and 117. Ps. 29 (an early psalm, resembling literature found at *Ugarit) contains parallelism in the first two verses. Other poetic devices include the use of repetitions and plays on words (e.g. Ps. 93: 4), refrains (Ps. 136), and acrostics (Ps. 111). Ps. 119 consists of twenty-two sections, each of eight lines, which begin with the same letter; the initial letters for each stanza are in turn the twenty-two letters of the Hebrew alphabet.

Psalms were composed for a variety of purposes over a long period of the nation's history, certainly up to the Exile in Babylon (Ps. 137), and perhaps even to the Maccabean struggle (Ps. 44). But most modern scholars have been concerned less with possible historical settings than with the literary forms of psalms. Many are hymns celebrating deeds for which God is praised—the Creation (Ps. 8), the Exodus (Ps. 114), and the long course of events from Abraham to the invasion of Canaan—which presupposes a knowledge of Gen. and Exod. (Ps. 105) and so is probably a post-exilic composition. Another category is that of laments, sometimes in a personal form (Ps. 22) and sometimes in a communal cry of distress (Ps. 44). There are also personal thanksgivings, such as for recovery from illness (Ps. 30).

One group of psalms relates to the king of the line of David but it is uncertain whether those were composed during the time of the monarchy when a king on the throne was being honoured or, less probably, whether the psalms had in mind some future idealized king through whom divine blessings would be channelled to the nation. The king is regarded as a son of God, and there are references to his anointing and coronation (Pss. 2, 72 and 101). Some scholars believe that there was an annual enthronement festival. A royal wedding is the theme of Ps. 45, and Ps. 110 (which was to be an important text for the early Christian Church—e.g. Mark 12: 35–7) reflects the priestly role of the king in cultic ceremonies, acting on behalf of the whole people. The cultic context of these psalms is shown by Ps. 132, where the king and the *Ark are associated in a thanksgiving for Jerusalem. Choral recitation in the Temple cult is suggested by psalms which contain antiphonal elements: Ps. 118 with liturgical instructions (verse 19) may have been for king and people to recite in turn; Ps. 15 was probably sung by pilgrims entering the building, perhaps at the feast of ingathering at the beginning of the year; they would be welcomed and invited to seek admission; they would respond by framing ten conditions (verses 2–5b); and then the priests pronounced a blessing (verse 5c). However, such is the diversity of forms that it is necessary to be cautious about the actual occasions for composition and use. Because there were changes in the cult over so many centuries scholars are reluctant to pronounce confidently about a context beyond that of liturgy in general for the Temple. Terms which appear, such as **Selah* (e.g. 3: 2), of uncertain meaning, and directions for the choir master at the head (e.g. Pss. 8–14) indicate the use of psalms with music in items of a service, such as a procession, or a blessing, or a petition.

The continuing liturgical use of the psalter argues for its having been specially put together for practical use in the Temple worship. Another view, however, proposes a literary emphasis: that it was compiled for teaching or theological purposes, expounding the everlasting kingship of God. On the latter view, the psalter is considered as a whole rather than as a collection of miscellaneous units, and the critical focus is more on the readers than on the original speakers of individual psalms in worship. But whether the psalter assumed its final shape through priests of the Temple as a book of hymns and prayers, or whether it was an anthology compiled by scribes, it is certain that the psalms have been an essential ingredient of both Jewish and Christian worship. Many psalms are among the Dead Sea discoveries at *Qumran, and in the NT there are many quotations from the psalms. For example, sermons reported in Acts (e.g. 2: 25 ff.) quote Pss. 16 and 132 in support of belief in the resurrection of Christ, and Ps. 118 is regarded by the gospel of Matt. (21: 9) as foretelling Jesus to be the expected deliverer, leading a victory procession. By the 2nd cent. CE, the psalms were an established element in the Christian liturgy, and in due course became part of the daily office in religious houses whose members, both monks and nuns, were required to know the Latin psalms by heart before the invention of printing. It was a fixed text into which successive generations imported new meanings.

An additional psalm, Ps. 151, is found in early Greek versions, and was also in use at Qumran. It is held to be canonical by some Orthodox Christians in the East. It celebrates

David's killing of Goliath and thus taking away 'reproach from the children of Israel'. In one English version (NRSV) Ps. 151 is included with the Apocrypha.

Psalms of Solomon Eighteen psalms not included in the *canon and only known in Greek and Syriac translations, composed in the middle of the 1st cent. BCE. They speak of a Messianic king, Son of *David, who will be God's instrument in the establishment of his kingdom on earth through the destruction of God's enemies.

Pseudepigrapha Jewish Writings not included in the OT or *Apocrypha, published under assumed names. They give information which is important for understanding the background of the NT. Prophecy had ceased, and interpretations of the will of God were offered in apocalypses which tried to reconcile the prophetic hopes and promises with the realities of national sufferings. Some of these writings emerged from within Palestine: the Psalms of Solomon, the Testaments of the Twelve Patriarchs, and the book of Jubilees. These works contain legends and alleged revelations; the Martyrdom of Isaiah explains that the prophet was sawn in two (cf. Heb. 11: 37); the book of Enoch is a vision of judgement and a prophecy of events down to the Messianic age (cf. Jude 6), and the *Assumption of Moses (which is referred to in Jude 9) has a view of world history. The Apocalypse of Ezra and the Apocalypse of Baruch offer hope of future Messianic blessings to those suffering in the present. Writings from the Hellenistic environment commend the *Law to *Gentiles. The Letter of Aristeas (*c.*100 BCE) and the Sibylline Oracles, written in Alexandria (*c.*140 BCE), invite Gentiles to abandon their pagan worship; 4 Macc. is not concerned with the Maccabean war but with the sufferings of Jews and the use of reasoning. These works share a spirit of universalism different from the post-exilic nationalism which marks some of the literature of that epoch.

pseudonymity It became a convention in the period between OT and NT that works should be published under the authorship of a fictitious name, such as Moses or *Enoch, or the sons of *Jacob. Such a name imparted a certain authority to the writing. In the OT the book of Daniel is pseudonymous, and in the NT it is widely held that 2 Peter and Jude are pseudonymous, and also the *Pastoral Epistles. Literary standards of the time did not necessarily condemn pseudonymity as deceitful.

Ptolemais The name given to the seaport Acco (Judg. 1: 31) by the Egyptians in the 3rd and 2nd cents. BCE and one of the towns which mobilized an army against *Judas Maccabaeus (1 Macc. 5: 15). Paul broke his journey to Jerusalem at Ptolemais and spent a day with the Christian community there (Acts 21: 7).

Ptolemy The kings who ruled Egypt from 323 BCE after the death of *Alexander the Great took this name; their aim was to introduce Greek culture and learning at *Alexandria (where there lived a large community of Jews). In the 2nd cent. BCE there was considerable strife within the dynasty, with the Seleucids of Syria, and finally with the Romans. The last of the line was Cleopatra VII, married to Mark Antony, who gave her his territory in Palestine; but when Augustus defeated their combined navies at Actium in Sept. 31 BCE Cleopatra committed suicide.

publicans Tax collectors in the Roman Empire (Latin, *publicani*). The system was to offer contracts to capitalists, who then hired natives of the region to obtain the money: they were able to do this in excess of the amounts transmitted to Rome. There was thus considerable scope for theft and for greed, with the result that in Palestine the publicans were a despised class. *Zacchaeus (Luke 19: 1) and his colleagues were hated both for their probable corruption and also for their collaboration with 'unclean' Gentiles (Matt. 9: 11). Jesus rejected the customary contempt and said that some publicans and *harlots were believers who would enter the Kingdom (Matt. 21: 31). Zacchaeus is cited as a model of penitence (Luke 19: 8).

Matthew, one of the twelve disciples of Jesus, was a publican (Matt. 9: 9).

Publius The chief official in Malta, the local representative of *Rome, whose father was suffering from gastric fever and was healed by Paul (Acts 28: 7–8).

punishment, everlasting The ultimate destiny of the irredeemably wicked. The ancient Hebrews regarded **sheol*, the abode of the dead, as a place of weariness and forgetfulness and not of punishment or retribution for earthly sin. The concept began to emerge later of a punishment where 'their worm would not die and the fire would not be quenched' for the wicked, after death (Isa. 66: 24), and at a *resurrection of all the dead (Dan. 12: 2) the wicked would be condemned to everlasting contempt. The valley of Hinnom, notorious for child sacrifices (2 Kgs. 23: 10) became a symbol for everlasting punishment (2 Esdras 7: 36) and, now called *Gehenna, a fiery punishment is envisaged there for the unrepentant wicked (Matt. 5: 22). Much elaboration followed in apocalyptic literature. In modern theology, the idea of everlasting punishment is generally regarded as inconsistent with the concept of a God of love. It is held either that in the last resort the love of God will prove universally irresistible and all will be saved—there is no hell—or that it is possible to be so irredeemably evil that a person forfeits the gift of eternal life and ceases to exist at all. 'Hell' in this case is the total absence of the presence of God. The scriptural basis for universalism is Rom. 11: 26, 32. The gospel of John suggests that *hell is a state in this present life of total opposition to God, whereas eternal life is a quality of abundant life available here and now (John 5: 24).

purgatory The interim state, according to Catholic belief, for those of the dead who must be punished or purified before being fit for the *Vision of God. The biblical basis for the doctrine is in 2 Macc. 12: 39–45 and 1 Cor. 3: 11–15, but there was elaboration in Church tradition. Prayers for the dead are a prominent feature in the worship of both Eastern Orthodox and Roman Catholic Christendom; and many non-Catholic Christians teach the probability of an intermediate state, after death, of preparation and growth.

purification Rites necessary for Hebrew men and women in the frequent events of their contracting a kind of ritual impurity, as in menstruation or by touching a corpse. It was the duty of a priest to remove the traces of impurity by means of a *sacrifice at an *altar (Lev. 14: 25). The subject is mentioned by John 2: 6, where it is explained for *Gentile readers. At John 3: 25 there is a difficulty about the precise Greek reading, and the discussion between *John the Baptist's disciples and a Jew over a purification has been thought by some scholars to indicate that the Baptist's disciples sided with other Jews against Jesus in requiring strict adherence to the purificatory laws.

Purim A Jewish festival. According to Esther 3: 6 all Jews were to be exterminated in a holocaust. The date was to be chosen by lot (Hebrew, *purim*), but before this took place Queen *Esther, a Jewess, was able to foil the schemes of the antisemitic chancellor Haman. Purim is therefore observed as an annual holiday.

purity The avoidance of ritual contamination. Important to the Hebrews, for impurity effectively prevented much religious and social activity. Impurity could be 'caught' by contact with the dead, from prohibited foods, from contact with bodily fluids (e.g. Lam. 1: 9) and various diseases. Lev. and Num. describe the laws regarding purity, and the *Dead Sea scrolls contain regulations which were designed to maintain within the community a rigorous standard of ritual purity. The purpose of purity regulations is to guard the group against what seems alien and to reinforce the sense of being the chosen people of God, as well as to establish moral purity.

In Christian theology the Jewish laws on purity have been abrogated and their requirements of moral purity strengthened. Jesus embraced what was taught should be avoided on the ground that moral insights and strengths overrode traditional codes (Mark 7: 12), but it is unlikely that he swept aside the whole Jewish system. The words ascribed to him in Mark 7: 19 surely come from Mark rather than from Jesus; if this principle had been known when Paul was battling on behalf of 'his gospel' (Gal. 1: 8), he had only to appeal to it to settle the controversy.

purple A royal colour, found in a fluid derived from sea creatures on the Mediterranean coast of Palestine. Purple-dyed cloth was used for priestly vestments (Exod. 25: 4) and a purple cloak was mockingly thrown round the 'King of the Jews' (John 19: 2, 5) before Jesus' *crucifixion. Paul encountered a seller of purple at *Thyatira (Acts 16: 14).

purse The purses of the OT (e.g. Isa. 46: 6) were small bags, often poorly made (Hag. 1: 6). Because they often allowed coins to drop out, they were an apt symbol for an inflationary economy. They were carried in the girdle or belt (Matt. 10: 9). Judas had charge of 'the common purse' (John 12: 6, NRSV, REB; 'common fund', NJB).

Puteoli The port (near modern Naples), captured by the Romans in 338 BCE, where Paul landed after his long sea voyage (Acts 28: 13) which ended with a fast spurt in the final two days. He was welcomed by a Christian group, presumably connected with the Church of Rome, 140 miles distant by road.

Pyrrhus Father of Sopater of Beroea, who was one of the companions of Paul on his final journey to Jerusalem (Acts 20: 4). It does not seem that Pyrrhus was himself one of the party.

python The name in Greek for a mythical serpent at Delphi, supposedly slain by the god Apollo. 'Python' was then applied to persons at the oracle regarded as inspired by the god on account of words uttered without rational control—possibly they were ventriloquists. At *Philippi, Paul exorcized a woman with such a spirit of divination—in Greek 'a pythoness' (Acts 16: 16).

Q The symbol (said to have been adopted as being the initial letter of the German *Quelle*, source) for that material common to both Matt. and Luke of about 230 verses, mostly sayings of Jesus, which is not from Mark. It is assumed that there was a stock of Jesus' *teaching which was preserved in an early community. Some modern advocates of the Q hypothesis suggest that it shows Jesus to have been a teacher of aphoristic, quasi-Gnostic, quasi-Cynic wisdom, but that, when edited into the gospels of Matt. and Luke, it was given a change of direction and emphasis. Even so, many of the sayings in Q do resemble some ancient Near Eastern Wisdom writings. In due course Q had become available to these two evangelists and each incorporated it (in different ways) into their gospels. The classical exposition of this solution to the 'synoptic problem' by the English scholar B. H. *Streeter (1924) held that Q was a now lost written document. There is, however, a small group of NT scholars who reject the Q hypothesis altogether on the ground that, where the three gospels run in parallel, Luke sometimes follows Matt. rather than Mark. This suggests the possibility that Luke knew and used Matt. to supplement his use of Mark, rather than the hypothetical, non-existing Q.

Defenders of the Q hypothesis argue, for example, that Matt. 3: 7–10 and Luke 3: 7–9 is a Q passage. Matt. has deliberately introduced the *Pharisees and *Sadducees in verse 7 (omitted by Luke) to provide a polarized audience; Luke is less polemic: but opponents of the Q hypothesis suggest that Luke has omitted the opposition simply because he is less hostile to those groups than Matt., and preferred to introduce 'the crowds' (verse 7) because, in his theology, the Baptist was preparing the whole nation for the coming of Jesus. And so, on this view, Luke knew and altered Matt.

queen In Israel there was no right of succession to the throne by daughters of a deceased king, or by his widow; but a queen mother, Athaliah, ruled for seven years when her son died (2 Kgs. 8: 26; 11: 1–4). A foreign queen, of *Sheba, visited *Solomon and was impressed by his wealth (1 Kgs. 10: 6). *Candace, queen of the Ethiopians (Acts 8: 27) was a queen mother and effective ruler: 'Candace' was a generic title borne by other queen mothers. In the ancient world the queen mother was often a much more prominent figure than the king's wife.

queen of heaven Various goddesses in the ancient world were accorded this title, even at Jerusalem before the Exile. *Jeremiah regarded the fall of Jerusalem as just punishment for this *idolatry (Jer. 44: 17, 23). It was also the title of *Artemis (Diana) at *Ephesus (Acts 19: 27).

Quest of the Historical Jesus, The The title of Albert *Schweitzer's work (ET 1910) which traced the course of investigation by (mostly) German scholars into the presentation of Jesus in the gospels. The starting-point was in the late 18th cent. and ended with an account of William *Wrede's argument (1901) that Mark's gospel was not an objective report of historical facts but a theological narrative with a particular theory—that the people failed to recognize Jesus as *Messiah because he kept it a secret; the Church came to accept the belief only after the *resurrection. Schweitzer challenged this view, and all those which preceded it in his chronicle, and instead posited an *eschatological Jesus, one who believed himself to be the future *Son of Man—Messiah—who would descend on the *clouds as judge.

The quest continued after Schweitzer with the *Form Critics, who maintained that the Christ who was preached in the early

(*Gentile) Church was not the historical Jesus but the Christ of *faith. For the traditions encapsulated in the gospels had been shaped by the Church, and sayings attributed to Jesus were the utterances of Christian prophets. After the Form Critics, however, a 'New Quest' was prepared to find authentic words and deeds of Jesus, and this line was reinforced by much new work on the Jewishness of Jesus and the appropriateness of such titles as *Son of Man, *Son of God, and *Christ to the Jesus of history.

The continuing quest has been developed by scholars with a special interest in sociology, and much greater knowledge of the social environment of Jesus and its various renewal movements has to some extent undermined previous historical scepticism. It is possible to compare Jesus with other Jewish teachers of his time. He differed from contemporaries (such as Hanina ben Dosa) by virtue of his combination of 'new teaching' (Mark 1: 27; 2: 12) and actions that were a threat to the authorities (such as the cleansing of the Temple) and by his *miracles of *healing and *exorcisms. Controversy surrounded the issue whether Jesus' works were wrought by someone possessed by *Beelzebul (Mark 3: 22) or by the *Spirit (Mark 3: 29–30), and the survival of this unattractive story would appear to be authentic historical reminiscence. Thus, there is some confidence in the achievements and methods of the historical-critical approach, though no consensus of results can be affirmed.

Qumran The modern Arabic name for the site of the monastery on the *Dead Sea, 14.4 km. (9 miles) south of *Jericho, accommodating a group usually regarded as *Essene in character. The scrolls were discovered in caves nearby in 1947 and following years, and the site of the monastery was excavated. It was occupied by a community from the middle of the 2nd cent. BCE until it was overrun by the Romans in the war of 66–7 CE.

quotations The *faith of the NT is undergirded by constant references to the OT—250 direct quotations and many more allusions. It was important to maintain the continuity of the new *covenant with the old and to justify the Christians' claim to their having superseded the Jews as a chosen people. The OT is usually quoted in the *LXX version and without any necessary relationship to the historical context of the original. Such quotations are especially frequent in Matt.; thus the *infancy narratives are set in a context of fulfilment: *Hosea had once mentioned how God had called his 'son' (Israel) out of *Egypt at the *Exodus (Hos. 11: 1, quoted by Matt., unusually, from the Hebrew), and when Jesus was taken as a refugee to Egypt and brought back (Matt. 2: 21) this was a fulfilment of prophecy. The OT story of Exodus, crossing the water, wanderings for forty years in the *wilderness, and the giving of the *Law at *Sinai is understood typologically in the gospels. What happened to Israel, which was a disobedient people, happens by contrast to Jesus, who is obedient, and so God's plan is fulfilled. There are therefore many OT allusions and quotations in Matt., including one difficult to recognize (Matt. 2: 23), and in the other gospels.

Paul's many quotations from the OT are usually from the *LXX text. Sometimes they are quoted because the words are required in a straightforward literal sense (Rom. 3: 10–17) and sometimes he adapts an OT text to a subject to which it bears no obvious relation (1 Cor. 9: 9). Other OT quotations are used *typologically. *See* Old Testament in the New Testament.

R

R The conventional symbol for 'Redactor', an editor or collector of material of books of the *Bible to whom we owe the completed form, and who was inspired by his own beliefs and theological convictions.

rabbi In 1st-cent. CE *Palestine in the time of Jesus the word was a form of address, equivalent to 'sir' but by the end of the century and in rabbinic literature it is used for a teacher, and this development is reflected in the usage in the four gospels.

The word occurs in Mark (9: 5; 11: 21; 14: 45) as a polite, even reverential, form of address when spoken by Peter. In Mark 10: 51 the form Rabbouni (NRSV marg.) is used by the blind man who then follows Jesus 'on the way'. There is a suggestion of the greatness of Jesus.

In Matt. only Judas addresses Jesus with this greeting and the sense is plainly hostile (26: 25, 49). For in this gospel rabbis are among the adversaries of Jesus and they are painted in unfriendly colours: they like to receive deferential treatment and to be called 'sir' and they are assuming the role of teachers (23: 8) of the *Law. Matt. keeps Jesus at a distance from them; he is not to be associated with them and the hateful *Pharisees, and his disciples are forbidden to arrogate to themselves the title (23: 8).

The word is not found in Luke, but it is quite frequent in John where Jesus is addressed by the word at 1: 36–8, 49 by two disciples and by Nathanael. The evangelist explains that the Hebrew word means 'teacher', and in 3: 2 Nicodemus calls Jesus a rabbi because he can perform 'signs'. Members of the crowd call him rabbi after the feeding of the 5,000 (6: 25). When Mary Magdalen encounters the risen Jesus, she first mistakes him for a gardener and calls him 'sir' (Greek, *kurie*) but when she recognizes Jesus he is Rabbouni, 'my Teacher' (20: 16).

The use of the word in Mark and John conveys a suggestion of Jesus' public ministry and puts him on the same level as the scribes, with whom he debated (Mark 12: 28).

Rabshakeh An emissary from *Sennacherib to *Hezekiah of *Judah demanding tribute (2 Kgs. 18: 17). This was duly paid, but *Isaiah urged the king not to offer total surrender. In fact, the *Assyrian soldiers were afflicted with a *plague which obliged the army to withdraw.

raca Aramaic for 'fool', and a term of abuse probably forbidden by a Jewish rabbinic source known to Matt. The word occurs nowhere else in the NT and the text (Matt. 5: 22) appears to be in a confused state. It is difficult to interpret. It hardly seems appropriate that the use of '*Raca*' should incur the judgement of the *Sanhedrin, whereas uttering much the same kind of abuse is to be punished with a 'hell of fire'—a punishment hardly mentioned at all in Mark and Luke, and never in John. The general tenor of the passage is probably that while a court of law can take cognizance of external acts, internal dispositions, such as anger, will come under the ultimate judgement of God.

race Athletic games were an important part of Greek cultural life; they included races on foot and in chariots, and were staged in honour of the gods. As pagan customs they were abhorrent to orthodox Jews when they were promoted by Hellenists in Jerusalem (2 Macc. 4: 7–12).

However, the metaphor of racing is used by Paul without any pejorative sense (1 Cor. 9: 25; Phil. 1: 30) and no doubt he had often been a spectator. It was a form of intense striving and a good model for Christian self-discipline and effort. A picture of cheering witnesses in the stadium watching the races is offered by the epistle to the Hebrews (12: 1–2) by way of encouraging fellow Christians on the course of life.

Rachel Daughter of *Laban; *Jacob's second wife (Gen. 29: 17) and mother of *Joseph and *Benjamin; she died while giving birth to the latter (Gen. 35: 18). Previously infertile, she had pleaded with Jacob to take her maid Bilhah as another wife, in order to bear him sons (Gen. 30: 4).

Rad, Gerhard von German OT scholar, 1901–71. He wrote important works on Genesis and OT theology in which he maintained that the historical narratives, much of whose information was reliable, provided the basis for a theological account of divine intervention in Israel's history. *See also Heilsgeschichte*.

Rahab (1) A prostitute with a house conveniently built into the wall of *Jericho (Josh. 2) who entertained the Israelite spies and helped them to escape their pursuers. In return, she and her household were spared when the city was later captured and sacked. In the NT Rahab is regarded as a heroine (Heb. 11: 31 and Jas. 2: 25) and is included by Matthew (1: 5) in the *genealogy of Jesus. (2) A mythical monster killed by God in order that out of chaos God could create the world (Isa. 51: 9; Ps. 89: 10). It becomes also a symbolic name for *Egypt (Ps. 87: 4; Isa. 30: 7).

rain Because it was not totally predictable (Amos 4: 7) and could fail altogether, rainfall was much prayed for, looked for, and even induced (1 Kgs. 18: 45); it has been suggested that *Elijah's action in bowing to the ground with his face between his knees was a piece of sympathetic *magic; he was trying to make himself look like a *cloud. Such rain-making was practised amongst ancient peoples. Early rain in Palestine fell in the autumn, with the main rainfall from October to March, but there could be prolonged droughts when the rains failed (1 Kgs. 17: 7; Joel 1: 20). The amount of rainfall in the south was considerably more than in the north of the land.

rainbow Said to be a sign of God's promise after the *flood not to destroy the earth again (Gen. 9: 12–13) and its beauty is to be praised (Ecclus. [= Sir.] 43: 11); it is part of the *glory surrounding God (Rev. 10: 1).

Ramah Several elevated sites have this name in the OT: (1) A place 8 km. (5 miles) north of Jerusalem where *Deborah was a judge (Judg. 4: 5) and which later was destroyed by *Asa of *Judah (1 Kgs. 15: 22). It was on the *Assyrians' route towards Jerusalem (Isa. 10: 29). (2) The early home of *Samuel (1 Sam. 1: 19; 7: 17), in *Ephraim. *Saul visited Samuel there (1 Sam. 9: 6) and *David was given safe lodging there (1 Sam. 19: 18).

Rameses The area in *Egypt where *Joseph installed his family (Gen. 47: 11); it takes its name from the town which was later built there by Israelite slave labour under another Pharaoh, Rameses II. It became the delta residence of the dynasty.

Ramoth-Gilead A town east of the *Jordan designated as a place of refuge (Josh. 20: 8) and given to the *Levites (Josh. 21: 38). It was frequently contested between Israel and *Syria and changed hands several times; during a battle at Ramoth-Gilead King *Ahab of Israel was killed (1 Kgs. 22: 1–40). *Elisha sent a prophet to anoint *Jehu king of Israel at Ramoth-Gilead (2 Kgs. 9: 3).

ransom The word used in the *LXX for the sum of money to secure the liberation of a slave (Lev. 25: 47–55). This is the background for the phrase about Jesus in Mark 10: 45b: 'the Son of Man came to give his life a ransom for many'. Here the thought extends the preceding sentence (45a) that the *Son of Man came not to be served, but to serve. The sufferings of this one man are used by God to bring benefit to others: 'many' means 'all'.

In 4 Macc. 17: 2 ff. the Jewish *martyrs are said to have become a ransom for the nation's sins: they suffered on behalf of others—but as representatives, not substitutes. The martyrs represented Israel. There is also a recollection in Mark 10: 45 of Daniel 7, where a Son of Man is vindicated and pronounced the victor. But Mark 10: 45b is to be understood in the light of 45a; victory is achieved only after suffering and service.

Raphael An *archangel sent to cure Tobit of blindness (Tob. 11: 7–8) and to be the companion of Tobias, son of Tobit. The development of belief in angels (Tob. 12: 15) is a feature of literature of this period and Raphael is also mentioned in 1 Enoch (10: 7).

Ras Shamra The modern name for the ancient *Ugarit on the coast of *Syria and the scene of excavations since 1928. Five phases have been identified, of which material from the Late Bronze Age (1500–1200 BCE) with *cuneiform texts is especially interesting. Those with an alphabetic script are in Ugarit, a language akin to Hebrew, and include literary *myths and *legends not unlike some of the stories in the OT.

raven A raven was sent out of *the ark by *Noah (Gen. 8: 6), but it was regarded as ceremonially unclean in the *Law (Lev. 11: 15, in the *Priestly source). The prophet *Elijah was brought bread by ravens when he lived east of the Jordan (1 Kgs. 17: 6).

Reader-Response Criticism The recognition that modern readers of a biblical text come to it with a variety of experience and assumptions which affect their appreciation of the narratives. Biblical scholars, especially in America, have argued that in addition to the rigorous examination of the text to discuss the theological ideas in the writings and the accuracy of the history and the identity of the authors, the part played by the readers of the books is also important; for, as with a sermon, the words are lifeless until the person reading or hearing begins to ask questions. What are the hidden assumptions behind the description of some incident? Why in the NT is there so little personal information e.g. about Jesus, or Peter? Much has to be supplied by the reader to evoke a response (John 20: 31), though the response of a modern reader may be one of scepticism or contempt, quite different from that expected from the 'implied reader' whom the author had in mind. For example, throughout the OT the writers celebrate the victory of the *Exodus, and the destruction of the pursuing Egyptians. But a modern reader (a Palestinian, perhaps) would be more likely to identify himself with the helpless, drowning soldiers of *Pharaoh, while a Western feminist reader would deplore the assumed masculine hegemony in OT and NT. She would bring new insights to the text which she approaches with her own integrity and assumptions which she will not abandon.

That a Levite could allow his concubine to be appallingly abused in order to fulfil obligations of hospitality to a man (Judg. 19: 24) has been regarded as barbaric and primitive, but a feminist reader also repudiates, and persuades others to repudiate, the shocking complacency of the masculine narrator.

The situations of readers of the gospels and of their communities has influenced the narratives. Matt.'s additions to Mark reflect the life and experiences of his community. The behaviour of the Twelve as they oscillate between faith and unbelief and obedience and failure is just what Christians knew in Matt.'s Church, and they received encouragement from the story. Like the women at the tomb (Matt. 28: 8–10) they knew both fear and joy; like the Eleven on the mountain of Galilee (Matt, 28: 17) they have known both faith and doubt.

reaping Corn is harvested and stored in barns (Luke 12: 18). Hence the metaphorical use of reaping for judgement at the *End (Matt. 13: 24–30; Rev. 14: 15–16) and the reward for good deeds (2 Cor. 9: 6) and punishment for evil (Gal. 6: 7–8).

Rebekah Wife of the patriarch *Isaac and mother of *Jacob and *Esau, her twins born after twenty years of marriage. She contrived to get Jacob, her favourite, to receive the dying Isaac's blessing at the expense of the first-born Esau (Gen. 25: 20–26; 27); this seems to be a legend from oral tradition told to explain the historical relationship of the *Edomites (descendants of Esau) to the Israelites: they were an older nation, related to the *Hebrews and settled on their southern border, and often in conflict.

recapitulation A term used by some of the Fathers, especially *Irenaeus (130–200 CE), taken from Eph. 1: 10, where it is said that God summed up everything in Christ. The entire work of God in *redemption, and the long process of hopes and promises of the old *covenant, is completed in Christ.

recension A revision of a MS or group of MSS by an editor.

Rechab One of the supporters of *Jehu's coup (2 Kgs. 10: 15). His followers came to Jerusalem (*c.*816 BCE) to escape from the *Babylonians, but in the city they continued to live in rural and teetotal simplicity. *Jeremiah contrasted their fidelity to their chosen ways with the religious disloyalty to their God of those who had always dwelt in Jerusalem (Jer. 35: 14–16).

reconciliation An improved relationship between individuals or groups who were previously estranged (Matt. 5: 24) and in particular the relationship created by Christ's work of *redemption between God and mankind. It is God who effects this reconciliation (2 Cor. 5: 18) even when mankind was at enmity with God (Rom. 5: 10). The process included, above all, the death of Christ on the *cross (Col. 1: 22), and the result is a peace with God (Eph. 2: 14) and the inauguration of a 'new creation' (2 Cor. 5: 17) of cosmic dimensions (Col. 1: 20).

A *ministry of reconciliation is entrusted, so Paul believes, to himself (2 Cor. 5: 19); therefore he travels as an ambassador of Christ to 'sell' the gospel without charging fees (1 Cor. 9: 18).

Redaction Criticism The elucidation of the theological outlook of the editors (redactors) of OT and NT books. After the *Pentateuch had been analysed into its several sources (*J, *E, *D, *P), it was some time before scholars took an interest in the scribes or archivists who must have combined the material. But if the history of Israelite religion could to some extent be discovered by Source and *Form Criticism, it was eventually asked whether the later stages of Israelite institutions could be explored through the apparent intentions of the editors who stitched the raw materials together. Increasingly the editors were seen as creative writers with convictions and discernible concerns. The work of Redaction Critics was aided by the existence of 1 and 2 Chron., which are patently a revision by redactors of the books of Samuel and Kings; so having the first draft as it were of the history and comparing it with the Chronicler's work, critics could recognize his distinctive theological standpoints, which emphasized God's promises to the line of David, and the central place in the nation of the Jerusalem Temple.

Another example: Redaction Critics have found a pattern in the book of Judges which has been imposed on the material by a theologically informed redactor. It is a pattern of people's rebellion, followed in turn by oppression, penitence, and then deliverance by a judge, a military hero who brought some sort of cohesion into the scattered people, and gave them 'rest' for twenty, forty, or eighty years. The redactor has woven together a miscellany of stories into a unity to illustrate the providence and patience of God.

Redaction Criticism of the NT was developed in Germany after the Second World War and in Britain and North America during and after the 1950s. As with OT work, it accepts the principles and necessity of *Source Criticism and recognizes the importance stressed by *Form Critics of the oral period in which traditions about Jesus were passed on in the Church. But it rejects the description of the synoptic gospels as mere collections of fragments. The gospels, in Redaction Critics' view, are sophisticated works with plans, presuppositions, and motifs; although the gospels may sing in harmony, they do not sing in unison. Mark is a creative author in his careful selection of the available material—e.g. there are *twelve healing *miracles of Jews, and one (the Syro-Phoenician woman's daughter, 7: 24–30) of a *Gentile. Matt., in the *infancy narratives, using the Jewish method of commentary called **midrash*, regards it as important to show for his community how Jesus fulfils Messianic expectations in the OT; he enlarges the gospel of Mark and arranges his material in blocks against an OT background. Luke shows that in his day Christianity was now at home in the Roman Empire, where tiled houses (Luke 5: 19) have replaced the miserable dwellings in Palestine with roofs of straw (Mark 2: 4), and the Church is conscious of living the Christian life 'daily' (Luke 9: 23) in the Spirit-filled community (Acts). The evangelists' distinctive theological outlooks can be discerned by the changes they have made in vocabulary, style, and structure to the sources being used. For example, Matt. 4: 12–11: 1 has made variations to the order of Mark, which he was using, to suit his careful design. More recently, many Redaction Critics have insisted that the evangelists' stances are also revealed by what they have retained from their sources.

redemption Deliverance by a 'redeemer' from some evil by paying a price. Slaves could be redeemed (released) by that means. If an ox gored someone to death, the irresponsible owner was liable to execution, but a ransom could be paid to rescue him (Exod. 21: 30). The kinsman who has the right to redeem by payment is known as the *go'el* in Hebrew, and the same word is used of Yahweh as the deliverer of his people (Isa. 41: 14). The epistles claim that Christ redeemed people 'from the present evil age' (Gal. 1: 4) or bondage to the devil (Heb. 2: 14–15). This redemption is both already effected (Rom. 8: 29) and yet also awaiting final completion with the destruction of death (Rom. 8: 23). As in the OT, NT writers also mention that God has redeemed a people for himself (Rev. 5: 9), and the death of Christ is proclaimed as redeeming mankind from the consequences or punishment for their sins (Gal. 3: 13).

There was a doctrine of a Redeemer in *Gnosticism. He was thought to descend in disguise to the earth from a spiritual realm to enlighten mankind with some knowledge (*gnosis*) of their identity. Gnosticism in a Christian setting did not regard the work of Christ as delivering from sin but rather as leading people to self-realization.

Red Sea The sea which divides Africa from Arabia, and always an important trade route. *Solomon kept a naval force at Eziongeber (1 Kgs. 9: 26–8; 10: 11, 22) more successfully than *Jehoshaphat (1 Kgs. 22: 48), whose ships were wrecked there.

The 'Red Sea' which the Israelites crossed after leaving *Egypt is properly translated 'Sea, or Lake, of Reeds' and may refer to the marshes of Lake Timsah which are now part of the Suez Canal (Exod. 14: 18). *See* Exodus.

reed Shallow waters in *Palestine as in the *Jordan valley produced several kinds of reed which were used in writing (3 John 13). Cf. 2 Kgs. 18: 21. Reeds were the vegetation motif on the first coins issued by Herod *Antipas, who founded his capital city on the border of the Sea of *Galilee, where reeds were prolific. Jesus' reference (Matt. 11: 7) to a 'shaken reed' may be a title of ridicule and recognition for Herod Antipas, who had a reputation as a waverer. It is a reference that could only have been appreciated in Palestine, and therefore would appear to be an authentic utterance of Jesus.

Reformation Religious movement in Europe at the end of the Middle Ages. Although there were differences in the interpretation of scripture among the Protestant Reformers, Luther, Calvin, and Zwingli, they were agreed that the inter-testamental (deuterocanonical) works included in the *Apocrypha were not to be regarded as part of scripture (and Luther also rejected the epistle of James from his inner *canon as lacking explicit reference to Christ). They also insisted that scripture should be regarded as canonical in the original languages of Hebrew and Greek. This principle demoted the Latin Vulgate and its occasional mistranslations. But it was fundamental to the Reformers that scripture should be available in the vernacular, and that Christian beliefs and traditions were subject to the overruling authority of scripture.

refuge Moses designated three cities east of *Jordan and three west of Jordan (Deut. 4: 41–3; Josh. 20: 7) as *sanctuaries for those who had accidentally killed a person and whose relatives were seeking revenge.

regeneration From the Latin, meaning 're-birth'. The idea, but not the word, is found in the OT, in the *prophet's promise of future *blessings at the End (e.g. Ezek. 36: 25–6). In the NT that time is said to have arrived in the *life, *death, and *resurrection of Jesus (Matt. 19: 28), and Christians are 'born anew' (1 Pet. 1: 3), 'saved by the washing of regeneration' (Tit. 3: 5, AV; 'through the water of rebirth', NRSV). The ground of regeneration is the work of Christ, but it is to be apprehended through *faith and *sacraments.

Rehoboam Son of *Solomon. He succeeded to the throne but of Judah only because the northern tribes rejected Rehoboam's proposed tax increases and accepted *Jeroboam as king. He reigned probably from 926 to 910 BCE, and in 920 failed to repel an Egyptian invasion (1 Kgs. 14: 25–8); this the Chronicler regarded as divine judgement on his disloyalty to Yahweh (2 Chr. 12: 5).

Reimarus, Hermann Samuel (1694–1768) German biblical scholar in Hamburg. As a deist he rejected the historical worth of events in the *Bible ascribed to supernatural agents, such as *miracles. He was reckoned by Albert *Schweitzer to be the pioneer in the *'quest for the Jesus of history'.

religion In general, a complex of beliefs, cultic practice, and ethical demands in a system related to God or gods. More narrowly, the reference is to *Judaism in Acts 26: 5, but the same Greek noun is also used in Col. 2: 18 of the 'worship' (NRSV) of *angels. Christian faith in 1 Tim. 3: 16 ('religion', NRSV) is denoted by a different Greek word. Two sorts of 'religion', bad and good, are contrasted by James (1: 26, 27).

remembrance In the OT remembrance is especially ordained to recall the saving acts of the works of God (e.g. Deut. 5: 15); but people could be remembered after death by a memorial *pillar (2 Sam. 18: 18), as could events of history (Josh. 22: 10). In the NT the words of Jesus are to be remembered (Acts 20: 35), but above all it is his sacrificial death which is to be held in remembrance (1 Cor. 11: 23 f.) and his living presence renewed. The repeated words in connection with the cup 'as often as' (11: 25, 26) indicate that for Paul this rite was enjoined by Jesus to be continued.

remnant That part of a group which has escaped (e.g. exile, Jer. 24: 8) or will not be destroyed (Amos 5: 14–15) or which has remained faithful under trials. It was characteristic of Isaiah's teaching that a remnant of the nation would repent, and in Micah (4: 7) the remnant consists of those to be redeemed. On the other hand, *Jeremiah predicts untold misery for any remnant that survives, yet if blame lies with the nation's rulers a remnant may be saved (Jer. 23: 3–4), though not if they flee to *Egypt (Jer. 42: 15–16). *Ezekiel too foresees total destruction (5: 1–4), yet prays for the safety of a remnant (11: 13). In the NT Paul rejoices that there is a remnant of the Jewish people who have turned to Christ 'by *grace' (Rom. 11: 5).

Renan, Joseph Ernst (1823–92) French biblical scholar. He wrote a *Life of Jesus* (1863) which stripped him of any divine or supernatural attributes.

repentance A change of mind or behaviour; this may be attributed even to God in OT (1 Sam. 15: 11). It was demanded of the people whose repentance was often merely formalized in cultic actions and as such condemned by prophets as inadequate and empty (Amos 4: 6; Hosea 6: 4; Isa. 1: 10–17) and without the radical change demanded by the *Law. The hope lay in the possibility that one day God would give his people a new *heart (Ezek. 36: 26–31) and there would be *forgiveness to all who repented (Isa. 1: 18–19). In the NT repentance is called for by *John the Baptist (Matt. 3: 9–10) and is to be validated by *baptism. The call is repeated by Jesus (Luke 5: 32) and sometimes Paul (e.g. Rom. 2: 4) and in Rev. (2: 5).

Rephaim Inhabitants of *Transjordan before the arrival of the Israelites (Deut. 2: 11). The name is also applied to the 'giants' (NRSV) from Philistia (2 Sam. 21: 16; 18, 20, REB), and to giants who exist as 'shades' (Isa. 26: 14) in the underworld but with their earthly status intact (Isa. 14: 9).

reprobate 'Reprobate silver' (AV, Jer. 6: 30) means silver rejected when tested, and is applied metaphorically to the wicked. Paul uses the same adjective to describe the unrighteous (AV, Rom. 1: 28) which NRSV translates 'debased', REB 'depraved', NJB 'unacceptable'.

rest *Sabbath, as the day of rest, was formally grounded in the account of God's resting after the work of *creation (Gen. 2: 2); and the necessity of rest both for animals (Ezek. 34: 14) and people (2 Sam. 4: 5) was an established part of Hebrew life. In the NT the promise of God at the end of earthly life is named as rest (Heb. 4: 1; Rev. 14: 13). The idea is that the Sabbath rest not only commemorated the past work of God but also looked forward to the as yet unfinished work of Christ. The Hebrews failed to achieve the fulfilment of their covenant life (Exod. 33: 12–14) in fellowship with God. NT writers offer this promise to believers in Christ.

restitution The practice ordered of making good an act of evil, such as theft (Exod. 22: 1, 4). If an ox is killed, five are to be paid for in restitution. In addition, a criminal must offer a *guilt offering (Lev. 6: 1–7). In the NT the tax collector *Zacchaeus promised restitution after he had met Jesus (Luke 19: 1–10).

restoration In the NT the concept is of putting things in order. *John the Baptist (in the role of the new *Elijah) has done that (Mark 9: 12), in the sense of having accomplished his mission successfully. The writer to the Hebrews (Heb. 13: 19) hopes to be 'restored' to the readers, either by being released from prison, or perhaps by a simple return visit.

The idea of a 'universal restoration' (Acts 3: 21) when God will re-establish the whole cosmos to its initial state of perfection (Rom. 8: 21) is expressed in terms of a new *heaven and a new earth (Rev. 21: 1) and the *salvation of all (Rom. 5: 18).

resurrection There was a variety of beliefs about life after death held both by people generally and by intellectuals in the biblical eras. The ancient *Hebrews rejected both the *Canaanite *Baal worship, which included in the cult the annual dying and rising again of the god, and also the Greek notion of the inherent immortality of the 'soul'. But the NT concept of resurrection has only the barest hints in the OT. The idea of a hopeless shadowy existence in *sheol* (e.g. Ps. 88: 3–5) gave way only gradually to a richer conception of life after death. Job 19: 25–7 is searching for a more satisfactory view which would conform to the Hebrew sense that the human body, part of God's creation, was 'very good' (Gen. 1: 31), and therefore life without 'body' was incomplete and unsatisfying. Moreover, while existence in *sheol* might be a fair reward for the wicked (Ps. 49: 14), surely the faithful deserved something better? So there is a promise of resurrection for Israel as a nation (Isa. 26: 19); Yahweh's loyalists who have suffered will rise to an appropriate reward (Dan. 12: 2) and apostates will endure shame and everlasting contempt. In 2 Macc. there is hope for the resurrection of those who suffer (7: 9), and in the time of Jesus this was also the view of the *Pharisees (but not the *Sadducees) and of Jesus himself (Mark 12: 18–27).

The resurrection of believers is part of Paul's hope for all believers at the end of history. He anticipates a complete transformation of the whole human person (1 Cor. 15: 53–5).

resurrection of Jesus There are predictions of Jesus' own resurrection in Mark 8: 31; 9: 31, and 10: 34, but the precision of 'on the third day' argues against the authenticity of these sayings, which are more likely to have been the Church's elaboration of an expectation expressed by Jesus of some sort of ultimate vindication. In the gospel of John (11: 1–44) the raising to life of *Lazarus serves as a kind of prediction of Jesus' own resurrection. It signifies the power of resurrection which the Father has given to Jesus. Cf. John 12: 24. But the raising of Lazarus, like that of the daughter of *Jairus (Mark 5: 42–3) was a resuscitation. They returned to their ordinary, mortal existence, until their eventual natural deaths. Jesus' risen life is depicted in the NT as transformation into a new bodily life: his resurrection was an *eschatological rather than historical event.

That God raised Jesus from the dead is a triumphant belief that rings through almost every one of the NT books but is beset with problems. For no act of a transcendent God can be open to scrutiny as though it was just another event in the course of history. It was not an observable occurrence. Hence the gospel of Mark ends at 16: 8 without any account of the act of resurrection or of an appearance of the risen Jesus. There are, however, several ways of affirming the Christian belief in the resurrection of Jesus on the basis of the NT evidence. Much depends on the evidence of the earliest witness available to us: Paul. In 1 Cor. 15: 3–8 he provides a summary of what he had personally and authoritatively instructed that Church about 50 CE. Within two decades of the first *Easter Paul was proclaiming Jesus' resurrection. Moreover, Paul paid a visit to Jerusalem not later than 35 CE, soon after he received his call on the Damascus road (Gal. 1: 18–19) and there conversed with Peter. It is reasonable to assume that the earliest traditions about the resurrection were not passed over in silence when they met. Paul therefore received confirmation of his own experience of his encounter with the risen Christ; and he met the infant Church of Jerusalem which was the living testimony to the truth of Jesus' resurrection. His own experience on the Damascus road was regarded by him as similar to other resurrection appearances (1 Cor. 15: 8). For unlike Luke (24: 43) Paul appears to deny the physicality of Jesus' risen body (1 Cor. 15: 20 and 50).

The gospels in fact develop a second kind of resurrection *apologetic: the tomb was empty on *Easter morning. A widely held view is that the gospels' accounts are late and designed to support the existing belief based on the witnesses to the appearances. The gospels have too many inconsistencies (e.g. the number of women at the tomb, and of '*angels'; the purposes of the women's visits, and what they did about the message they were to deliver) as well as improbabilities (the rising to life of Jewish heroes to walk the streets, Matt. 27: 52–3; and 27: 62–6) to be accepted as reliable, historical records. Luke's account of appearances in which Jesus was given food to eat and was presumably supposed by Luke to be wearing clothes but nevertheless was able to pass through closed doors (24: 36)—and yet was unrecognized by his two companions who conversed with him on the road to *Emmaus (24: 16)—seems to be more theological interpretation than historical narrative. The meal at Emmaus was reminiscent of the *Lord's Supper; later, in Jerusalem, he ate a piece of fish (Luke 24: 42); Luke is emphasizing the continuing identity of the risen Jesus with Jesus of the ministry. His renewed 'flesh and bones' (24: 39) express a theological conviction of the enduring value of the human body and all its experiences. Thus it is the considered view of much modern scholarship that Paul's early account is to be accepted as the more reliable expression of early Christian belief rather than those of the gospels.

A radical view held by several NT scholars

(notably Rudolf *Bultmann) is that there were no objective visions of the risen Jesus. The historical truth is that there were a number of 'individual risings to new life'. The disciples were brought to new and radical decisions about themselves. By themselves they knew they could do nothing. Now they could truly be alive. For disciples, then and now, the stories of the resurrection of Jesus are to be understood as Christian proclamation to give assurance of the victory of the Cross. In itself the crucifixion, an event of history under Pontius Pilate, is not a *saving* event. It only becomes that when it is preached and believed, and the resurrection is an invitation to contemplate the triumphant significance of the Cross, in faith, and experience its liberating power. Then indeed we know ourselves as we truly are. However, Bultmann's understanding of the resurrection (which is in terms of *existentialist philosophy) fails to account satisfactorily for the new power which seized hold of the disciples. 'Risings to new life' were not what the resurrection of Christ actually meant, but are implications or consequences of resurrection belief which came to be held not by a number of individuals but a number of groups (1 Cor. 15: 3–8).

A defence of the historical value of the story of the empty tomb has been made. It admits the inconsistencies between the four accounts but suggests that if there were no discrepancies in the years when the traditions were taking shape it would lead to the suspicion of a carefully constructed Christian deceit. Granted that there are legendary accretions in the narratives, nevertheless it is interesting that Paul's account, which many scholars accept, mentions that Christ was buried and was raised (1 Cor. 15: 12, 20) which *implies* that the tomb was empty; and 'the third day', instead of the *rabbis' expectation of a general resurrection of the nation on a *seventh* day, adds to the evidence: if the tradition was false, Christians would have chosen another day. Paul, staying in the city for two weeks some six years after the crucifixion, had been in a good situation to hear about the tomb. The announcement conveyed by women, which could have been humiliating for the disciples, is a further piece of good evidence. It is also claimed that the resurrection faith could hardly have survived in face of a tomb containing the body of Jesus—a faith which the Jews could have instantly refuted by producing the body.

There are, however, difficulties about this defence of the reliability of the empty tomb narratives. Mark ends at 16: 8 with the women running away in fear, and not telling anyone what they had seen. This would seem to be an explanation by Mark to account for the story of the empty tomb *not* being generally known. It was not part of the earliest *kerygma*. Matt. expands Mark, but his additions read like the story of the Church and the Jews in the second half of the 1st cent. rather than an account of the first *Easter. Luke's narrative is historically questionable: he provides an impossible sequence of events for a single day (24: 29)—a return of 12 km. (7 miles) to the city in the dark, an appearance to the disciples, a walk to *Bethany, and prayer in the Temple (24: 53). If the empty tomb narratives were a late piece of Christian apologetic, then it was too late for the Jews to produce the body in order to refute the story.

The resurrection of Jesus could be regarded by Christians as God's ratification of the values of Jesus the teacher and man of holiness. After the resurrection Jesus, according to the gospels, was recognized by disciples as the one they had already known, who had gone about doing good (Acts 10: 38).

The appearances are located by Mark and Matt. in Galilee ('of the Gentiles', Isa. 9: 1) wherein Jesus' main ministry lay, but in Jerusalem according to Luke–Acts to comply with Luke's presentation of Jerusalem as the place of revelation: Luke writing for Gentiles is concerned to remind them that *salvation is 'from the Jews' in the first place.

retrojection The possibility that Christian experience in the early Church has been put back by the writers of the gospels into the lifetime and *ministry of Jesus. The gradual exoneration of *Pilate as the principal agent of Jesus' *crucifixion and its transfer to the

Jewish leaders reflects the experiences of the Church in the Roman Empire seeking to establish favourable relations with the authorities. The reference (John 3: 23) to *Aenon, which is difficult to identify, may be an account of the activity of followers of the Baptist still surviving in a later generation and resident in that place.

Reuben Son of *Jacob and *Leah (Gen. 29: 32), and ancestor of the tribe of Reuben. The story of Reuben, the first-born son whose behaviour (Gen. 35: 22) led to his losing the right attaching to that position (Gen. 49: 3–4) is in fact the story of the tribe told in the person of the founder: once pre-eminent, the Reubenites (Gen. 49: 4) lost their power to the tribes of *Judah and *Joseph. Reuben is mentioned first in the blessing of Moses (Deut. 33: 6) but in the Song of Deborah (Judg. 5: 15–16) Reubenites refuse to join the campaign against *Sisera and are reproved. Settled in the *Transjordan with the tribe of Gad (Num. 32: 33), the Reubenites may have become absorbed into that tribe (2 Kgs. 10: 33) until overcome by the *Assyrian invasion of 732 BCE.

revelation New knowledge imparted without the deductive reasoning of the recipient. As such, it is a divine *gift (Matt. 16: 17) but it could be passed on to others (Gal. 1: 15–17). The mode by which it is conveyed might be in visions (Jer. 1: 11–13) or through an understanding of events in history (Ps. 111: 6).

Revelation, book of (the Revelation to John) Also known as the *Apocalypse (Greek = 'revelation'), a similar kind of writing to the OT book of Daniel. From early in the 2nd cent. it was ascribed to John the Apostle, thought also to be the author of the fourth gospel, but most modern scholars cannot reconcile the barbaric Greek of Rev. with the fluent prose of the gospel and regard Revelation as the work of a recognized prophet in the Church, writing for his contemporaries about the year 95 CE.

The book has often been an embarrassment to the Church. Its bizarre imagery has been incomprehensible or misunderstood. Apocalyptic sects have used it as a handbook to predict the future. Many Christians have found the apparent gloating over the defeat of the Church's enemies to be morally repulsive (Rev. 18: 6–7).

An important principle of interpretation is to see what messages Rev. was giving out to the Churches for whom it was written. References to *Rome are clear: the 'seven hills' of 17: 9; and '*Babylon' of 16: 19 stand for Rome (as in 1 Pet. 5: 13), the power which destroyed the Jerusalem Temple in 70 CE as did the Babylonians in 586 BCE. The seer writes to encourage Christians under the threat, or experience, of persecution, which is usually thought to be an escalation by the emperor *Domitian (81–96 CE) of the sporadic outbursts of persecution, such as that by *Nero, whose persecution in 64 CE seems to be remembered in Rev. 13 and 17. The book is thus more hostile to the empire and the surrounding culture than was Paul (Rom. 13: 1, 7). It condemns *idolatry ('fornication', Rev. 17: 2) and the abuse of power in this world, and looks forward in hope to the coming of the Kingdom with a new heaven and a new earth. At the time of writing the power of evil represented by Rome is symbolized in Rev. by the traditional images of beasts, and those forces of chaos are in rebellion against the Creator. But he will preserve his faithful people in their adversity. The visions of Rev. consist of five series and they have a common pattern of persecution of the faithful (as 6: 9–11), followed by the judgement on the nations (6: 12–17) and then the victory of the Lamb (Christ) and the *salvation of his followers (7: 9–17). This present life for Christians is therefore one of struggle, suffering, and death but with the hope of ultimate *joy and *peace where evil is subdued and 'there is no more sea' (Rev. 21: 1). The theme of the book is sustained by formidable symbolisms: *twelve months, twelve signs of the zodiac (a mistake for the correct number of thirteen), twelve tribes and apostles; there are seven planets, days of the week, colours of the *rainbow—all numbers signifying completeness. But the number of the antichrist is six (one short) and Satan is a treble six (666), and the name *Nero Caesar if written in Hebrew letters would make 666. There was a terrifying rumour that Nero, who committed suicide in 68 CE, would return from the dead (Rev. 13: 3).

The clearest sections of Rev. are the first three chapters, with messages to seven Churches in Asia Minor. Their situations are well known to the writer, and the seer's exhortations are designed to fit each Church's current position. Where they have stood firm, praise is given (to *Smyrna and *Philadelphia); where they are failing, they are rebuked—especially *Laodicea. Modern readers of Rev. may not be able to take literally what is prophesied, though there exist today fundamentalist and conservative Christians, as there have often been in the past, who believe the book to be a prediction of the end of the world in this generation. Some have even anticipated the battle of *Armageddon (Rev. 16: 12–16) to be fought with the mass destruction by nuclear weapons. There is certainly in the seer's words the faith that the salvation of Jesus, the Word of God (19: 13), can be available to *every* generation. Every Christian can follow Jesus and through suffering like his win through to victory (2: 7, 17, 26). We shall all be judged by our deeds (20: 12–13). And there is a role for the Church—that of mediating God's forgiveness and urging repentance (3: 7–9). In this way the Lord comes on his day (1: 10) year in, year out, in preaching and in *Eucharist, and the New Jerusalem descends from God not at the end of all time but whenever a *martyr wins a crown (3: 12; 21: 2, 10).

reverence The attitude of awe before God or in the presence of the holy, as in Isa. 11: 2–3; Rom. 8: 15. The opposite is blasphemy or sacrilege (Mark 13: 14).

revised versions of the English Bible The 'Revised Version' (RV) by a group of scholars in England was a conservative revision of the AV in the light of abundant new textual evidence. The NT was published in 1881; the OT in 1885. The fidelity of the translation to the original made RV the standard Bible for students, but it was not well received by the public at large, who found it wooden and pedantic. Not surprisingly, when it was adapted for the American public as the Standard Version, there were many alterations.

The AV of 1611 was itself a revision of the Bishops' Bible (1568), which was a revision of the Great Bible (1549), Coverdale's revision of the Matthew Bible (1537).

After 1881, the American Standard Version, which incorporated the preferred alterations to RV, appeared in 1901. The RSV NT, a revision of the Standard Version, was published in the USA in 1946, followed by the OT in 1952, and the Apocrypha in 1957. These immediately enjoyed a high reputation for students both in America and Great Britain. A committee continued to watch over the RSV and eventually the NRSV was published in 1989, embodying the latest knowledge of MSS and using inclusive language. NRSV represents the end term of a long process of revision since the 16th cent.

The Revised English Bible (1989) is a 'substantial revision' of the New English Bible (NT, 1961; OT, 1970) which was a completely new *translation. *See also* Bible, English translations of.

reward There are many promises in the NT of a reward not only in this life but also in the world to come—'a hundredfold now . . . and in the world to come *eternal life' (Mark 10: 29, 30); give to the poor, and there is to be treasure in *heaven (Luke 18: 22). Complementary to the promise of rewards is the warning of punishments (Matt. 22: 13). However, the rewards are promised by Jesus (Mark 8: 35) to those who follow him from motives other than for the sake of the rewards, such as 'for my sake and the gospel's', and it is possible that the references to reward are a recognition that the ideal of self-forgetfulness is not obtainable by obsessive self-scrutiny and discipline. Reward was not the goal of ethical goodness; neither was its possibility to be deliberately suppressed.

rewritten Bible Term used for writings which amplify, modify, or in some other way revise existing books of the OT, thus making them more relevant or acceptable to a later generation of readers. Thus 1 and 2 Chron. go over the books of Samuel and Kings from a post-exilic standpoint, which is primarily to enhance the standing of *David. Ecclus. [= Sir.] 44–9 rewrites much biblical material to emphasize the dignity of the priesthood; the book of Jubilees in the 2nd cent. BCE supplements the narratives in Gen. about *Noah and *Abraham and revises the OT chronology; the Temple Scroll from *Qumran contains instructions about *worship not contained in the OT. The Jewish *Book of Biblical Antiquities* of the 1st cent. CE contains additional psalms and affirms a doctrine of *resurrection. *Josephus' *Antiquities* (94 CE) rewrites much of the history, as do the **targums* for use in the *synagogues.

Rezin King of *Damascus who paid tribute to *Assyria in 738 BCE but later joined an alliance which included Israel to resist Assyria; but *Ahaz of *Judah, supported by Isaiah (Isa. 7: 1–10) rejected his overtures even when he attempted to impose his will by force. Rezin was executed by the Assyrians (2 Kgs. 16: 5–9) when the alliance was crushed.

Rhegium Port on S. Italian coast. Paul's journey to *Rome was held up for one day at this port because of the need to wait for a favourable wind (Acts 28: 13). The city was noted for its Greek culture, but the emperor *Augustus augmented its population with veterans from his fleet.

rhetoric The art of persuasion by public speaking. This was studied intensively in the Hellenistic and Roman world, and the gospels and the epistles of Paul reflect some of the techniques that had to be mastered. Amongst the teaching methods of Jesus, his use of *parables was pre-eminent. The gospels employ the standard rhetorical device of comparison: Jesus is shown to be greater than Moses (the greatest of the Hebrews), and *John the Baptist, the Temple, *Jonah, and *Solomon. The purpose is to enhance the significance of Jesus. Paul's epistles, though written for particular needs and with characteristics not unlike many of the personal notes and business letters discovered in *papyri from the Egyptian desert, are not without rhetorical devices, e.g. Philemon 11, 2 Cor. 6: 4 ff.; 2 Cor. 11: 22 ff.

rhetorical criticism Examination of the devices employed to create a literary unit out of a piece of prose or poetry. Study of biblical books to assess them from this point of view asks how they accomplish their purpose in communication to the reader. Is the purpose achieved? What categories of material are used? What are the occasions which elicit the rhetorical units? In classical rhetoric a speech was divided clearly into three parts, and it has been suggested that Paul wrote Gal. in such a scheme. His letters were to be read aloud and to communicate a message; they were more than mere business letters such as were discovered in the papyri in Egypt by G. A. *Deissmann, even though some of the same vocabulary is used.

Rhoda A member of the Jerusalem household of *Mary, the mother of John Mark; Rhoda was so overwhelmed with joy when she went to the door after Peter's release from prison (Acts 12: 12–16) that she forgot to open it.

Rhodes Island off SW Asia Minor. Paul passed it on the way by sea from *Ephesus to *Tyre (Acts 21: 1). At one time the island was an independent republic with extensive commercial enterprises, and its own navy, but its prestige and wealth were diminished under Roman rule.

riddle A saying which is deliberately made obscure, as when *Samson proposed the riddle of the lion and the honey (Judg. 14: 12–19). The word could also encompass messages difficult to interpret; such riddles were not used by God when speaking to Moses (Num. 12: 8). In the NT some of Jesus' *parables could be described as riddles (Mark 4: 11–12).

righteousness In Hebrew, the quality of being found right or innocent; it is what describes a good king (Isa. 32: 1) or a fair partner or neighbour (Amos 5: 6–7). God is righteous because he observes his *covenant, delivering Israel from enemies and offering hope for the future (Jer. 23: 5). In the NT, the word righteousness occurs frequently in Matt., where it means ethical righteousness, doing the will of God (Matt. 5: 6, 10), which is less radical than what was written by Paul, who asserted that righteousness is not only right conduct before God but a right relationship with God. The initiative is from God; it is received in *faith; it issues in right conduct (Rom. 3: 21–6). Righteousness is ascribed to Jesus in 1 John 2: 1 in the sense of a total conformity to the will of God.

right hand In OT the hand or arm of the Lord is a way of expressing his saving power. The king of Israel was said to be at the right hand of God when he was enthroned (Ps. 110). The ascension of Jesus to the right hand of God (Heb. 1: 3) implies that the pioneer of our *salvation (2: 10) now possesses sovereignty: God reigns and judges through Jesus.

Rimmon The name of a Syrian god worshipped by *Naaman in *Damascus (2 Kgs. 5: 18) but not in the end exclusively, since he had to share temple space with Yahweh after Naaman had been cured of *leprosy by *Elisha.

Rimmon was also the name of a place south of Jerusalem (Zech. 14: 10) in which some of the returning exiles settled.

rivers Some streams (wadis) which are dry in summer are incorrectly called rivers (Deut. 2: 37). They are violent in winter and their torrents are used as an illustration of the power of enemies (Ps. 124: 4). Genuine rivers are mentioned in the Bible: the *Jordan in Palestine, the *Nile in *Egypt, and its delta, and also the *Euphrates, which is known as 'the great river' (Deut. 1: 7). The 'river of Egypt' (1 Kgs. 8: 65, AV) is not the Nile but the wadi Arish, and is rendered 'brook' (RV), 'wadi' (NRSV, REB), and 'torrent' (NJB); it is the traditional boundary of the *Promised Land.

roads Essential for trade and travel in *Palestine; roads went from north to south along the coast, and further inland between the hills, with branches into *Galilee. Roman highways were kept in good repair and greatly facilitated Paul's journeys, especially the Via Egnatia from Dyrrhachium on the Adriatic coast across *Illyricum, *Macedonia, Thrace, *Thessalonica, *Berea, *Amphipolis, *Neapolis, and *Philippi on the Aegean. Paul entered *Rome along the *Appian Way.

rock Rocks are mentioned in the Bible as places of refuge (Prov. 30: 26) and as hiding-places from God (Isa. 2: 19). Moses hid himself among rocks lest he should have to face God (Exod. 33: 20–3). And he struck a rock to provide water (Exod. 17: 6), an incident used typologically by Paul, who interpreted the rock as Christ (1 Cor. 10: 4). Rocks provided solid foundations for building (Matt. 7: 24–5), and so the term was a suitable description for Peter (Matt. 16: 18), upon whom the Church was to be built.

Romans, Paul's epistle to the Written in *Corinth, probably in 57 CE, to a Church which Paul had never yet seen but which he was hoping to visit. The letter was to prepare the ground for such a visit. Paul wrote to give careful exposition of his understanding of the Christian Church in relation to the Jews, with whom God had made a covenant: the Jews were 'the chosen people', and many former Jews were by now members of the Church in Rome, which had possibly been presided over by no less a personage than Peter. The leading apostle had perhaps come to Rome shortly before Paul was writing (he was to die under the emperor *Nero in 64 CE), so Paul's mission would be delicate, even though it would seem that in his own mind (Rom. 15: 24) his main plan was to go on beyond Rome and, with Roman help, to proceed to the west. If Paul's epistle to the *Philippians was written from Rome, it provides evidence (Phil. 4: 22) that Christians had already infiltrated into *Caesar's palace itself. The Church had travelled from Jerusalem to the capital of the empire already, and the apostle to the *Gentiles had better catch up!

The theme of the epistle is the *Salvation offered by God to all humanity through the *faith of the gospel. Paul expounds this in three directions: first, as a kind of statesman, in which he examines the present position and future prospects of *Judaism; then, as a theologian, he reflects on the bankruptcy of Judaism, and the remedy for it; and finally, as a man of action, he indicates his intentions and plans.

He is aware (Rom. 14–16) that there existed two groups in the Roman Church: Jews who became (Jewish) Christians and ex-pagans. He wanted the former to realize that Gentile Christians, who did not observe the Jewish *Law, represented the heart of the Church. The ex-Jews should worship alongside the Gentiles and sever all their emotional and legal links with the *synagogue before Paul arrives. Inevitably then he has a problem: that if Jesus is the Messiah, what is to be affirmed about the revelation given in the OT? If God is now calling the Gentiles, is he not being unfaithful to his promise to the Jews? This is the question addressed by Paul in Rom 9–11. In fact, he claims, the OT itself shows that God chooses people quite freely and irrespective of natural descent or race.

Paul must also of necessity give a full treatment to his views, previously adumbrated to the Galatians, of the Law (*Torah). It is at once 'holy, righteous and good' (7: 12), yet Christians are discharged from it (7: 6), and they are thus liberated because salvation is through Christ, and not by means of food-laws, Sabbath observance, and *circumcision. All are justified only by faith, and no one can boast about achievements. Obedience to the Law is not the decisive criterion of one's standing with God, and therefore Christians 'have died to the law' (7: 4). The criterion of righteousness is faith in Christ, which has taken the place of the Law at the centre of a believing life (10: 4). The Law was not so much bad in itself, therefore (it is 'holy'), but the human heart had twisted it into badness, notably in the command not to covet.

Paul is aware that his strictures on the Law might lend substance to the charge that he is indifferent to morality. So he spends four chapters (12–15) in relating 'his gospel' (Gal. 1: 9) to the duties of everyday life in Church and society, in which the predominant motive must be love; and love's obligations are even more exacting than those of the Law.

The epistle appears to come to an end with ch. 15, making ch. 16 into an appendix, which has led some commentators to suggest that the epistle was a circular letter, with appropriate greetings added at the end to each recipient Church, but the evidence for this is insufficient. A similar doxology also occurs at the end of ch. 11.

It was for a long time fashionable among commentators to interpret Romans in the same terms as Martin Luther had done—namely that Paul was opposing the view supposedly held by the Jews that good deeds done as required by the Law gained merit in the sight of God, and so a man was justified by his works. Luther urged that Paul, on the contrary, said that acceptance by God, and being acquitted at the judgement, was in virtue of grace from God and solely by faith on the side of man. Luther had in mind Roman Catholicism and the strenuous discipline of the anxious asceticism of his life as a monk. He felt liberated by reading Romans and Galatians. But in truth Luther's was not the situation of the 1st cent. Paul's attack on the Law is not because it induces self-righteousness through accumulating merit, but because it has been displaced in the scheme of salvation by Christ, and for that reason Jews and Gentiles are in the same boat. It is in that sense that Paul can claim that 'he upholds the law' (3: 31): the law places the Jews in the same position as Gentiles—both are guilty before God; and Jews and Gentiles in the Church are one in Christ (3: 21–30). There can be absolutely no question of converted Gentiles first submitting to the Law by circumcision before being baptized. Jewish Christians should sever all relations with the synagogue and join in worship and social life with Gentile Christians on Pauline principles (15: 7).

Paul ends the letter with a premonition of failure: the Jews may frustrate his plans; his offering may not be acceptable (15: 31). And according to the Acts the apostle did reach Rome, but in chains (Acts 28: 16).

Rome The capital of the Roman Empire, traditionally founded in 753 BCE. In NT times it had a population of about one million, mostly crowded into tenements, though the wealthy lived in pleasantly heated villas. The emperor *Augustus undertook a magnificent building programme for which materials were imported from Africa, Greece, and Asia Minor. Such commercial enterprise attracted to Rome a cosmopolitan mixture of customs, languages, and religions, though Greek was the language of communication. *Latin was spoken only by the social élite.

There were enough Jews in the city by 49 CE for the emperor Claudius to issue an edict expelling them on account of disturbances (see Acts 18: 2), which probably revolved round the arrival of Christians. The latter were numerous and unpopular enough to be blamed by *Nero for the calamitous fire of 64 CE, which conveniently laid waste a part of the city which he needed for his own building. The Christian community nevertheless grew and both *1 Peter and *1 Clement were written from Rome.

roofs The roof of a flat building was to be provided with a parapet to prevent an accidental fall (Deut. 22: 8). This is an example of OT humane legislation. Roofs were useful amenities to a house (for observation—by David, 2 Sam. 11: 2; for guests—Elisha, 2 Kgs. 4: 10). In Palestine of OT times roofs were simple constructions; branches laid horizontally supported by columns were covered with earth or straw (Mark 2: 4), though in cities of the Roman Empire houses had roofs of tiles (Luke 5: 19).

Rosetta Stone Found by French soldiers at Rosetta on the west bank of the *Nile in 1799 and surrendered to the British in 1801; it contains inscriptions in Egyptian and Greek written at the beginning of the 2nd cent. BCE, and the parallel Greek version made it possible to decipher ancient Egyptian hieroglyphics. It is now in the British Museum in London.

Rufus ('Red') Possibly the same man mentioned both in Mark 15: 21 as a son of *Simon of Cyrene and greeted by Paul in Rom. 16: 13 along with his mother. The connection is sometimes advanced as one link in the chain of evidence associating the gospel of Mark with *Rome.

ruler of the feast 'Master of the feast' (REB) was the master of ceremonies or 'chief steward' (John 2: 8, NRSV) who regulated the arrangements for drinking and music.

ruler of the synagogue The 'leader of the synagogue' (NRSV, Luke 8: 41) was an official appointed by the elders to look after the building, its contents, and its arrangements for *worship. He was probably not the same person as the 'attendant' of Luke 4: 20, but in Luke 13: 14 it is the ruler who was outraged about the undignified disturbance caused when Jesus healed a crippled woman on a *sabbath. Sometimes there were several persons responsible for the *worship (Acts 13: 15).

Ruth An outsider, a *Moabitess, who as a widow married into the tribe of *Judah at *Bethlehem. Her son was the grandfather of *David, and therefore Ruth appears in the *genealogy of Jesus (Matt. 1: 5).

Ruth, book of A charming OT story, placed between the books of Judges and 1 Sam. as being a bridge between the two epochs in Israelite history—the judges and the monarchy. Ruth, a Moabitess, cared for her widowed mother-in-law Naomi, who in turn was instrumental in finding a husband for Ruth, whose husband had also died. The man chosen was from the tribe of *Judah, by name *Boaz (Ruth 2: 1), who was prominent and rich. The son born was called *Obed (Ruth 4: 17) and was the grandfather of David. Because of the Moabite involvement, it has often been supposed that the book was written as a universalist tract (in the same spirit as Isa. 56: 6–8) in protest against narrow nationalism in the time of Ezra and Nehemiah after the Exile, when intermarriage with foreigners was prohibited (Neh. 13: 23–8). On the other hand, is it likely that at such a time after the Exile someone should deliberately ascribe a foreign ancestry to King David? Thus there is also an opinion that the book should be dated earlier—between the 10th and 8th cents. BCE, and not as late as 400 BCE—and could be a timeless story about vulnerable people in society.

S

sabachthani An Aramaic word, preserved at Mark 15: 34, uttered by Jesus from the *cross, quoting Ps. 22: 1, meaning 'you have forsaken me'.

sabaoth *Hebrew for 'military hosts'. Used in AV, RV at Rom. 9: 29, which is a quotation from Isa. 1: 9. The meaning is Lord of the hosts (of *heaven), as in Ps. 89: 6–8. The mercy-seat of the *Ark was regarded as the throne of the invisible God, a kind of general of the armies of Israel, seated on this throne. It was therefore important for the Israelites to take the Ark into battle (1 Sam. 4: 4). The term also occurs in Jas. 5: 4.

Sabbath A word meaning cessation (from work) and so the weekly day of rest, fundamental in Israelite life, sanctioned by God's rest from the work of *creation (Gen. 2: 1–3) and accepted as such in all parts of the OT (though not mentioned in the Wisdom literature). Legislation for the Sabbath is outlined in Exod. 20: 8–11; 31: 12–17; 34: 21; Deut. 5: 12–15. It was a day for rejoicing (Hos. 2: 11) and for visiting the Temple (Isa. 1: 13). Before the Exile the discipline was not absolute, and it is recorded (2 Kgs. 11: 5–9) that Athaliah was arrested and executed on a sabbath. Indeed the regulations could hardly go back to the time of Moses, since they assume agricultural society, not a heterogeneous collection of nomads. The prophets *Jeremiah and *Ezekiel insisted on strict observance of the Sabbath (Jer. 17: 19–27; Ezek. 20: 11–24) and it was an important aspect of Nehemiah's discipline after the Return (Neh. 13: 15–23). From then on it was an increasingly visible part of the Jewish sense of national identity. Early in the Maccabean struggle, Jews preferred to die rather than fight (1 Macc. 2: 38), but they recognized that if they continued that policy they would become extinct (1 Macc. 2: 41), and they therefore resolved to defend themselves on the Sabbath.

A Sabbatical Year was ordered every seventh year when the land was to lie fallow (Lev. 16: 31; 26: 34, 43).

Jesus worshipped in the *synagogue on the Sabbath (Luke 4: 16) but the tradition of his ambivalent attitude to the Sabbath (Mark 2: 28), combined with the belief in his *resurrection on the third day (our Sunday), soon turned the first day of the week instead of the last into the Christians' day of liturgical observance (1 Cor. 16: 2). Nevertheless in modern popular non-Jewish usage 'Sabbath' is often made to refer to Sunday rather than Saturday.

Sabbath day's journey A Jew was permitted to travel 2,000 cubits on the Sabbath (Exod. 16: 29 and Num. 35: 5), about 1.2 km. (3/4 mile), and the Mount of *Olives was within this distance from Jerusalem (Acts 1: 12).

sackcloth A coarse material made of goat's hair (Rev. 6: 12), uncomfortable as a garment and inexpensive to buy. It was worn by those mourning for an individual (Gen. 37: 34) or for the nation (Esther 4: 3), or as a sign of *repentance (Matt. 11: 21).

sacraments From the Latin '*sacramentum*', meaning an oath, as taken by men joining the Roman army. Already by the time of the governor *Pliny (112 CE) the term is used of Christian practices—he misunderstood their sacraments to be oaths by which they promised not to commit crimes. The Latin *Vulgate translated the *Greek *mysterion* by *sacramentum*, which led to *Baptism and *Eucharist being designated sacraments. Other rites were added to these sacraments by the mediaeval Church, but the Reformers restricted the number to the two most clearly mentioned in the NT (Matt. 28: 19 and 1 Cor. 11: 23–5). The biblical references to other rites such as *confirmation are less clear.

It is generally held that in the Church baptism corresponds to the OT rite of initiation (*circumcision), and Eucharist is related to the OT festival of redemption, *Passover.

sacrifices An essential part of Israelite and Jewish life and culture, as in other religions with which anthropological study has made interesting comparison. They were devised primarily as a means for dealing with *sin. Offences which needed expiation were not necessarily those committed deliberately; they all still required the appointed 'guilt offering' (Lev. 5: 17–19). The animal sacrificed was killed as a substitute for the human transgressor who otherwise could have died. The offerer laid his hand on the head of the victim in order to identify himself with it (Lev. 1: 4) and offer himself to God. Amos (5: 21–7) and Isaiah (1: 10–20) repudiate the practice of sacrifices if taken as easy means of access to God without obedience to his will.

The NT adopts the sacrificial principle in order to explain the *death of Christ; he died to atone for our *sins (1 Cor. 15: 3), by his blood (Rom. 3: 25), and 'for our trespasses' (Rom. 4: 25). Jesus' death is in Paul's mind when he writes that one man's act of *righteousness leads to acquittal (Rom. 5: 18). Jesus, who was without sin (2 Cor. 5: 21; cf. Heb. 4: 15), took the place of sinful human beings (2 Cor. 5: 21; Rom. 8: 3). Paul does not, however, labour the sacrificial metaphors, and prefers to base a Christian's relationship with God on his theory of participation in Christ by sharing sacramentally in his death and *resurrection. *See* offering.

Sadducees A minority group in 1st-cent. CE *Judaism. They were traditionalists in *faith and practice. It was important in their eyes to be faithful to the terms of membership of the Jewish nation by observances of the written *Law and participation in the Temple cult, but they held that this was compatible with submission to the Roman occupying power. They opposed armed conflict and preparations for it.

Sadducees rejected the *Pharisees' concept of the oral law as being no less valid than the written. They were therefore loyal adherents to what had been established by *Ezra at the time of the Return from Exile in Babylon (5th cent. BCE). Compromises with the Romans meant that the high priesthood was a government appointment and the Temple was overseen by the Roman military. Thus, though at a price, Israel was enabled to maintain its relationship with God and atone by *sacrifices for its *sins. But novel beliefs in the immortality of the soul or *resurrection of the body were rejected by Sadducees. After the destruction of the Temple in 70 CE the Sadducees no longer existed. According to one theory, they were *Zadokites who had controlled the Temple for several centuries.

saints Those who are faithful to God and love him (Ps. 31: 23) and are dedicated to his service (Dan. 7: 27) as the 'holy ones'. In the NT the 'saints' are Christians as distinct from non-believers (1 Cor. 6: 2); they belong to Jesus Christ and are God's beloved (Rom. 1: 6–7), and are members of particular *communities, e.g. Jerusalem (Acts 9: 13) or *Lydda (Acts 9: 32). In Rev. (17: 6) they are *martyrs for the *faith.

Salamis The main port of *Cyprus, on the eastern coast, named after the Greek island. Paul and *Barnabas stayed there and preached in the *synagogues (Acts 13: 5). The city had a fine natural harbour and there was much commercial activity.

Salem The Hebrew word for '*peace' (Heb. 7: 2) as a place is normally connected with Jerusalem (Ps. 76: 2).

Salim *John the Baptist moved from *Bethany (John 1: 28) to *Aenon near Salim (John 3: 23), a place difficult to locate but a reasonable suggestion puts it in *Samaria. The assertion, 'much water there', implies that it was away from the River *Jordan. Jesus is said to remain in *Judaea—a period of his ministry unrecorded in the synoptic gospels.

Salmone A cape on the NE of *Crete round which Paul sailed on his way to *Rome (Acts 27: 7), now known as Cape Sidero.

Salome (1) Possibly the wife of *Zebedee and mother of *James and *John (Matt. 27: 56; cf. Matt. 20: 20), one of the women from *Galilee who was present at the *crucifixion (Mark 15: 40) and came to the tomb (Mark 16: 1) and received the message that Jesus had been raised. (2) Daughter of *Herod (not Philip) and (Mark 6: 22, NRSV, marg.) of Herodias whose second marriage was to *Herod Antipas, tetrarch of Galilee (Mark 6: 17), who is therefore both uncle and stepfather to Salome. Her dancing pleased Herod, who promised he would give her anything she requested; prompted by her mother she asked for the head of John the Baptist (Mark 6: 24). Salome's *name* is not mentioned in the NT; she married Herod Philip, tetrarch of *Trachinitis.

salt Either extracted from the *Dead Sea and allowed to dry out, or mined in the adjacent rock. It was an essential preservative and salt became synonymous for a power for good—hence the *disciples could be called 'salt of the earth' (Matt. 5: 13).

Salt, city of (or **Salt Town**) A town given to the tribe of *Judah (Josh. 15: 62; REB, Irmelach), considered to be Khirbet *Qumran, where the community which wrote the *Dead Sea scrolls was living.

salutations Greetings. In the OT they are lengthy and formal (as Ps. 118: 26) and prolonged (cf. 2 Kgs. 4: 29) or obsequious (Dan. 4: 9). An extended greeting is implied in the command to the *Twelve on their missionary journey (Matt. 10: 12–13). 'Go in peace' (1 Sam. 1: 17) were the words of farewell.

salvation Initially, God's intention to rescue from danger; later his promise to establish his Kingdom. God's *gift to his *covenant people of help when they confront an enemy (1 Sam. 7: 8 etc.) or, later, God's resolve to establish his reign (Isa. 52: 10). In the NT Jesus (the name is from the *Hebrew word for 'save') brings salvation, by means of the reign or *kingdom of God (Matt. 1: 21; 21: 31–2; Luke 19: 10; John 4: 42; Phil 3: 20). But *faith 'saved' the woman with the haemorrhage (Mark 5: 34) in the sense simply of restoring her to health.

The *death and *resurrection of Jesus are the decisive moments according to the NT in the scheme of salvation. Those who have faith receive salvation; they are reconciled to God in the present (Rom. 13: 11–14) and saved from the perils of future *judgement (1 Thess. 1: 9–10). The experience of possession by the *spirit is a foretaste of future *joy of salvation in the kingdom (Rom. 8: 23; Eph. 1: 14).

Samaria The capital of the separated northern kingdom from the time of *Omri (*c.*870 BCE, 1 Kgs. 16: 24) until 722 BCE when the *Assyrians captured it. The city was well fortified and some of its inhabitants were affluent (Amos 6: 4–6). The hill country round Samaria was occupied by the Joseph tribes at the time of the settlement, but after the Assyrian conquest the local population was largely deported (2 Kgs. 17: 6) in exchange for foreign colonists probably in 720 BCE. This at any rate was the Jewish explanation of the beginning of antipathy between *Judah and Samaria, exacerbated by the rebuilding of Jerusalem after the Exile (Ezra 4: 8–24). On the other hand, the post-exilic antagonism of the Jews could have influenced the hostile account of the origin of the Samaritans. During the *Hellenistic period (325–63 BCE) there was further disruption and the Samaritans rebuilt the city of *Shechem and a temple to Yahweh on Mount Gerizim. In the *Maccabean period the territory was taken by *Judaea. The Romans gave the territory to *Herod the Great in 30 BCE, from whom it passed to *Archelaus (4 BCE–6 CE). Samaria as a district is mentioned in the NT (e.g. Matt. 10: 5; Acts 8: 4–25).

Samaritan Pentateuch The sacred scripture of the *Samaritans (who still exist near the modern city of Nablus) consisted of a version of the Hebrew *Pentateuch. Other OT books were not accepted as scripture by them.

Samaritans The people settled by the *Assyrians in the district of *Samaria (according to 2 Kgs. 17: 29) and who were alleged by Jews to practise a form of *Hebrew worship contaminated by combination with their previous cult. In NT times Samaritans were despised by Jews as *foreigners (Luke 17: 18) though in fact they still had much in common with Jews. While the Samaritan Bible consisted only of the *Pentateuch, the group claimed to observe it more strictly than the Jews, especially in the regulations for the *Sabbath. The Samaritan temple was on Mount *Gerizim and was staffed by priests (John 4: 20). The common heritage of Jews and Samaritans combined with the history of friction and dissent adds to the piquancy of Jesus' friendliness towards them (Luke 17: 18; John 4: 7) and the astonishing anti-racism of the parable of the *Good Samaritan (Luke 10: 33). In the expansion of the Church from Jerusalem to *Rome, the Samaritans occupy a midway position between the evangelization of Palestinian Jews and Hellenistic Jews on the one hand, and *Gentiles on the other (Acts 8: 12). In spite of persecution and the political and military upheavals of Palestine, a small Samaritan community has survived to modern times.

Samos An island in the Aegean Sea, SW of *Ephesus, visited by Paul briefly before he came to *Miletus (Acts 20: 15). Jews had been living on the island for some time (1 Macc. 15: 23); it had been part of the Roman province of *Asia since 129 BCE.

Samson One of the Israelite judges, of the tribe of *Dan (Judg. 13: 24–16: 31). He was a Nazirite, but sat lightly to his vows (Judg. 14: 8–10). He carried on a mostly individual battle against the *Philistines. The stories of Samson's erotic attachment to women are legendary. They include the discovery by Delilah of the secret of his astonishing strength, which lay in his uncut *hair. When it was shorn, he was captured and blinded, but in a last display of strength, as his hair grew, he seized the pillars which supported the house where the Philistines were celebrating, and died together with his mocking captors (Judg. 16: 4–31). The story of Samson was used by Handel for his oratorio, and there are many great pictures (by Rubens, Rembrandt, Reni) in the art galleries of London and other cities.

Samuel Son of Elkanah and Hannah (1 Sam. 1). He was promised by his mother to the service of the Lord at *Shiloh, where he grew up. He became a national leader (1 Sam. 7: 13–14) and judge (1 Sam. 7: 15–17), and anointed Saul as the first king of Israel. After his death, his ghost was invoked by Saul, who hoped to hear from him a prediction of victory over the Philistines (1 Sam. 28).

Samuel, books of Originally, with 1 and 2 Kgs., one book in *Hebrew but divided first into two, then in the *LXX into four; this arrangement has been generally continued. The books were named 'of Samuel' because Samuel is the first major character to appear. In the Hebrew *canon they are numbered among the books of Former *Prophets. The work has been compiled and edited by the Deuteronomic school, probably about 560 BCE, but many units of material have been utilized. There are local traditions preserved at shrines (e.g. Mizpah) and also from a group deriving from *Shiloh (1 Sam. 1–3), some of which concern the fortunes of the *Ark. Some of the stories, such as the account of *Absalom's rebellion (2 Sam. 15–18), may reflect popular *legend rather than recorded history.

There seem to be two sources describing the accession of *Saul: in one he is presented unsympathetically, the prophet Samuel is the hero, and God regrets having established the monarchy; a second set of narratives (1 Sam. 9: 1–10, 16; 11: 1–15) offers a more attractive portrait of *Saul—he is a charismatic figure of great promise, and excels in battle.

Some of the narratives about *David—e.g. the account of the *Ammonite war in

2 Sam. 10: 1–19—read like a contemporary description. Other David stories must have been assembled by a collector not later than the time of *Solomon. The Succession Narrative (most of 2 Sam. 9–20, and 1 Kgs. 1–2) describes the intrigues of David's sons and the final triumph of the Solomon party. There are also three poems in 2 Sam.: the elegy over the deaths of Saul and *Jonathan, David's last words, and a variant of Ps. 18 which is inserted into 2 Sam. 22.

The *Deuteronomic editors have stamped their theological interpretation on 1 and 2 Sam. The affairs of the nation have been guided by God, especially through David. But even David committed *sins; and for these he was punished, as was *Eli, at the beginning of the book. But the nation prospered and expanded under David because he was chosen by God and responded. He captured Jerusalem and under him it began to be the centre of the nation's worship. It was there God had chosen 'to put his name'.

Sanballat Governor of *Samaria in the second half of the 5th cent. BCE. When the Jews had returned from the Exile in *Babylon, the walls of Jerusalem were rebuilt by *Nehemiah, and this incurred the wrath of Sanballat (Neh. 2: 10). He probably had ambitions to extend his rule into *Judaea and therefore alleged that Nehemiah was planning to rebel against the Persians.

sanctification A process of making holy by separation. In the OT people and places are sanctified, i.e. reserved to the Lord, by means of sprinkling of sacrificial *blood. The persistent sinfulness of the people required their ritual sanctification if they were to be the people of God's name. The *Spirit of the Lord would give them a 'new *heart' (Jer. 31: 33; Ezek. 11: 19). In the NT Jesus has been sanctified or consecrated by the Father and sent into the world (John 10: 36), and the Son has himself consecrated *disciples (John 17: 17–18). But Paul teaches that for the mission to which disciples are sent, they are to live a life of sanctification (Rom. 6: 19) producing the fruit of the spirit (Gal. 5: 22–5).

In later theology, sanctification is the name given to a process which begins with *baptism, continues in the life of *faith supported by the sacraments, and is eventually to be completed at the final *judgement (Eph. 4: 30).

sanctuary A place of holiness or security. (1) The Israelites trekking through the *wilderness believed that the Lord was present with them in a portable 'tent of meeting' (Exod. 25: 8 ff.; 29: 4) and in the Temple at Jerusalem, when *Solomon had built it. The Presence was located in the *Holy of Holies, the innermost section, which only the *high priest was allowed to enter. (2) Six *cities were set aside as places of sanctuary or refuge (Josh. 20: 7–9) for persons fleeing from avengers after an accidental homicide.

sand The vast quantities of particles of sand afford many apt allusions in both OT and NT. The descendants of *Jacob are to be as numerous as the sand of the sea (Gen. 32: 12), and Paul quotes *Isaiah's prediction that of this great number only a remnant will be saved (Rom. 9: 27). Sand, though heavy (Job 6: 3), is dangerously deceptive when wet and a building on it will collapse (Matt. 7: 26).

Sanhedrin From a Greek word meaning a council of leaders. In NT Palestine there were various local councils of seventy-one (or twenty-three) members (Mark 13: 9; perhaps 15: 43). But in the gospels the reference is usually to the supreme court in Jerusalem. Its members were chief priests and *elders who assembled with the goodwill of the Roman authority, which found it convenient to use the local élite for various matters of administration. They conducted the preliminary *trial of Jesus (Matt. 26: 59). It is uncertain whether these leaders included *Pharisees as well as the aristocratic *Sadducees; probably they were present, and part of the strategy of the *high priest at the trial was to unite the parties in a unanimous condemnation of Jesus. At the trial of Paul (Acts 23) both Pharisees and Sadducees were involved.

Sapphira Wife of Ananias (Acts 5: 1–11); both died from shock when confronted by Peter about a case of fraud.

sapphire The second jewel in the *walls of the new Jerusalem (Rev. 21: 19); and in *Ezekiel's *vision the heavenly *throne resembled sapphire (Ezek. 1: 26).

Sarah Wife of *Abraham; her striking beauty made Abraham fear the jealousy of powerful men (Gen. 12 and 20). Because of prolonged infertility Sarah persuaded Abraham to beget a child by her maid *Hagar, but later Sarah (also called Sarai—both names mean 'princess') was promised a son in her old age and *Isaac was born. The story of Hagar's use by Abraham and subsequent expulsion with her son *Ishmael is given an allegorical interpretation by Paul in Gal. 4: 21–31.

sard, sardin, sardius A red stone, possibly a kind of carnelian (Ezek. 28: 13, NRSV) which adorned the high priest's breastpiece (Exod. 28: 17) and the wall of the new Jerusalem (Rev. 21: 20).

Sardis Capital city of *Lydia in the Roman province of Asia; its Church was one of the seven addressed in Rev. (1: 11; 3: 1, 4) and rebuked as moribund apart from a small nucleus of the faithful, not enough to diminish the influence of the imperial cult which had a centre in Sardis.

Sargon King of *Assyria, 722–705 BCE; he defeated an alliance of Israel and *Syria, with Egyptian support, and took *Samaria (720 BCE). Some of the population (27,000 was the Assyrian claim) was deported. Sargon continued campaigns against enemies on all sides, but throughout *Judah remained loyal and was spared.

Satan In Hebrew, 'the adversary'; and as such he appears in Job as one of the sons of God, delegated to inform God about human frailties. Later he becomes a malevolent being (1 Chron. 21: 1), and the enemy of God over Jesus (Luke 22: 3). He thus becomes the *Devil (*diabolos* in Greek), identified with *Beelzebub (Matt. 12: 26). Satan is said to have a battalion of demons who may enter human beings (Luke 9: 42) and may tempt them to evil (Jas. 2: 19). In the gospel of John there are *no* lesser demons and Jesus works no *exorcisms; the principals alone, Jesus and the Devil, engage in conflict (John 13: 2) but eventually according to Rev. Satan will perish in a lake of fire (Rev. 20: 10–15).

satrap The name of Persian governors or viceroys of provinces. In Esther 3: 12 AV translates the word 'lieutenants', and in Dan. 3: 2 'princes'. Although the OT suggests that *Darius appointed 120 satraps for his kingdom (Dan. 6: 1), the Greek historian and traveller Herodotus, writing in the 5th cent. BCE about the Persian War, mentions that there were *twenty*.

satyr The translation in NJB at Lev. 17: 7 for 'goat-demons' (NRSV), '*demons' (REB), '*devils' (AV); the same Hebrew noun is used in Isa. 13: 21 in connection with the imminent downfall of *Babylon. 'Goat-demons' (NRSV) are also part of the Lord's day of vengeance (Isa. 34: 14).

In Greek mythology satyrs were nature-deities, and thought to have bristly hair, pointed ears, horns projecting from the forehead, and the tail of a goat. So the translation of the Hebrew *sa'ir* as satyr has a certain appropriateness.

Saul (1) The first king of Israel. He came from the tribe of *Benjamin, and was probably chosen as king on account of his military leadership (1 Sam. 14: 47–8). There are two accounts in *1 Sam. of the institution of the monarchy; in one Saul is well regarded, but in the second he is overshadowed by the prophet *Samuel and loses God's favour, which is transferred to *David. The main enemies during Saul's reign were the Philistines, who were defeated at *Michmash (1 Sam. 14: 31). In later life Saul was tormented by an evil spirit and obsessed by jealousy and suspicion, and he visited his hatred on David. In the feud that developed between them, David twice spared Saul's life (1 Sam. 24: 6). On Mount *Gilboa Saul died in battle (he fell on his own sword, 1 Sam. 31: 4), along with his son *Jonathan (1 Sam. 31: 2), who had been the close friend of David (2 Sam. 1: 26). (2) The Hebrew name of *Paul of Tarsus.

saviour Used in the OT of God (e.g. Isa. 43: 3), who delivers Israel out of the Exile, and of Jesus in the NT in a similar sense (of deliverance) (Luke 2: 11). It is used of Jesus as the saviour of the world (John 4: 42) and occurs frequently in the *Pastoral Epistles (e.g. 1 Tim. 1: 1) but otherwise rarely in the NT. The title was commonly applied to pagan deities and also to ruling 'benefactors' (Luke 22: 25), such as Ptolemy I and Roman emperors. The *salvation which Jesus is believed in the *Pastoral Epistles to offer is deliverance from sin. As a Christological title, it may have been influenced in the later NT writings by current Hellenistic usage.

savour, sweet Used in AV of offerings which are acceptable to God, especially the fat parts of a *sacrifice (Lev. 3: 5). Modern versions prefer 'pleasing odour'.

scandal A literal translation of the Greek noun, which means something which causes offence, such as the *Cross, which is a 'scandal' to the Jews (1 Cor. 1: 23). *Satan was a 'scandal' to Jesus in tempting him to sin (Matt. 16: 23). Modern theologians have spoken of the 'scandal' (offensiveness) of the uniqueness of Jesus. More often the word is used today of the disasters caused to a community by unacceptable behaviour of its members.

scapegoat In English usage, a person who bears the blame which properly falls on others; in OT (Lev. 16: 8, 10) a goat was chosen 'for *Azazel' (the name of a *demon) and instead of being sacrificed it was released into the desert, symbolically carrying with it the sins of the people to a place of no return. It is a theme not used in the NT; it might have been expected at Heb. 9: 7. But, later, it becomes part of Christian piety (e.g. epistle of Barnabas, 7: 7–11).

scarlet The colour used to indicate honour (2 Sam. 1: 24); in mockery, for Jesus at his *trial (Matt. 27: 28); or reverence, for the tabernacle furnishings (Num. 4: 8). A scarlet cord outside the house of *Rahab in *Jericho was a warning to the invaders that it was to be protected (Josh. 2: 18).

sceptre Part of the king's regalia; a short rod taken to symbolize authority (Isa. 14: 5; Esther 4: 11) and applied to Israel (Num. 24: 17, NRSV) as exercising the kingship of God, and to *Judah (Ps. 60: 7).

Sceva An otherwise unknown and possibly self-styled Jewish high priest whose seven sons fell foul of Paul at *Ephesus (Acts 19: 14) when they attempted unsuccessfully to imitate Paul's power of *exorcism.

schism From the Greek word meaning a split or division. In the NT (1 Cor. 12: 25) Paul uses the word when he deplores factions in the Church. From the 2nd cent. (Irenaeus) it is used of a formal separation from the Great Church.

school(s) The so-called 'schools of the *prophets' were not institutions but rather groups of *disciples who preserved the words of their masters. Education in the wider sense was important in Israel both for the national religion and for politics and commerce, and the OT itself often refers to writing (e.g. the chronicles that were used in the compilation of the books of Kgs.) and the duty of reading the *Law (Deut. 31: 12–13), all of which presumes a process of instruction. In the 2nd cent. BCE the evidence of Ecclus. [= Sir.] 51: 23 is that there was a school in Jerusalem. The *Qumran community must have had a system of teaching and the rabbis who preserved the traditions after the fall of Jerusalem in 70 CE produced a massive body of literature. No doubt the *synagogue network was a fundamental agent of Jewish education, which centred on the Torah. *Rabbis were not paid for teaching.

Paul had been educated but he nowhere describes his schooling. There is a mention (Gal. 3: 24) of a schoolmaster or custodian; he is the slave who accompanied a boy to and from school and may even have done some home tutoring, and it could be that some of Paul's readers had experience of such guardians, for the early urban Christian communities came from all classes of society.

Schweitzer, Albert (1887–1965) Born in Alsace; NT scholar who held that Jesus believed that he was the Messiah-designate who was to herald the coming of the kingdom of God in the near future. Schweitzer held that both traditional and liberal interpretations of Jesus overlooked the fundamentally apocalyptic nature of Jesus' *teaching. His most famous book was *The *Quest of the Historical Jesus* (ET 1910, with a commendation by the Cambridge professor F. C. Burkitt). Schweitzer was also a doctor of medicine and founded and ran a hospital in Lambarene, for which he raised funds in Europe by organ recitals of the music of Bach, of which he was a notable interpreter.

scorpions Members of the spider family equipped with a terrible stinging power in their tails and therefore feared in themselves (Deut. 8: 15), but they were also symbols of intense pain (Rev. 9: 3), of evil (Luke 10: 19), and of *punishment more severe than whipping (1 Kgs. 12: 11).

scourging *Punishment by a lash of leather thongs inflicted on Paul several times (2 Cor. 11: 24–5; Acts 16: 22–3; 22: 24). Jesus was scourged either as a warning (Luke 23: 22) or as a severe punishment before *crucifixion (Matt. 27: 26).

scribes From the Latin word meaning 'writers'. The learned class of many ancient societies, responsible for administration and diplomacy, as was *Baruch, *Jeremiah's secretary who wrote down his words. *Ezra, who had been a high official in the Persian civil service, was well versed in Israelite law (Ezra 7) and this tradition was continued through the Maccabean age (167–63 BCE) into NT times.

As a group, scribes belonged to different parties, as is implied by Mark (2: 16), who mentions those who belonged to the *Pharisees. In the gospels they are presented as opponents of Jesus, associated with the *high priests and *elders (Mark 11: 27) and hostile to the Church (Acts 4: 5; 6: 12). Their work was to pass on their knowledge of the *Law and of the interpretation of scripture; they were active in the Temple. Jesus' *teaching is said (Mark 1: 22) to have possessed an authority unlike that of the scribes, meaning that he could not only interpret existing law but actually make law.

scripture From the Latin for 'writing'. In *Judaism the term 'scripture' was eventually restricted to those books in the OT which were recognized as canonical, though the content of the *canon was disputed. Christians used the Greek OT and this was 'scripture' for them (as 1 Cor. 15: 3), but in due course Christian writings also became 'scripture' as the notion of the canon was formulated.

scrolls Rolls of *papyrus (which was grown in *Egypt) used for writing. They were sometimes made of leather, from which mistakes were more difficult to erase, and this was the preferred material for the scrolls of the Dead Sea *Qumran community. Pens were made from rushes.

The Church adopted the *codex in preference to scrolls.

Scythian A people of notorious cruelty (2 Macc. 4: 47) who threatened in turn both the *Assyrian and *Persian empires, identified with the Ashkenaz (Jer. 51: 27).

Sea The Mediterranean Sea is called the Sea of the *Philistines (Exod. 23: 31) and the Great Sea (Num. 34: 6), but the other seas known to the Jews were the Sea of *Galilee and the *Dead Sea; in the forecourt of *Solomon's Temple the great *laver or basin was called the *molten or bronze sea (1 Kgs. 7: 39).

In the *Ras Shamra texts, the sea and the storm-god, Baal Haddu, did battle, and so caused the alternating seasons, and it is suggested that there are traces of this *mythology in the OT where God seems to feature as a storm-god raging against the sea, for which the Hebrews had a kind of horror (Isa. 27: 1; 51: 10; Ps. 89: 9; 93: 4; Job 26: 12; cf. Rev. 21: 1).

seal In use from the earliest OT times as a means of authenticating documents and ensuring ownership. They were often made from small semi-precious stones, oval in shape, with a name carved on the seal so that, when it was pressed in wax or clay, an impression remained. Letters were sealed (1 Kgs. 21: 8); a *covenant was sealed (Neh. 9: 38); and a metaphorical use of sealing is used by Paul, who defined the *circumcision of *Abraham as a seal of the *righteousness that he had by *faith (Rom. 4: 11). Christians are sealed by the *Holy Spirit (Eph. 1: 13). Christ is held to reflect the nature of God as closely as the impression made by a seal (Heb. 1: 3).

seat, chief The place indicating special honour. The chief seat (AV, REB), or 'best seats' (NRSV), 'front seats' (NJB; Mark 12: 39) were much sought after at the right of the host. In the *synagogues the front seats enabled the occupants to face the rest of the congregation in a position of eminence—though it is not certain whether this had become the custom by the 1st cent. CE.

Second Coming The *parousia*. In the gospels Jesus is said to predict a day when the *Son of Man will come. But it will be only after sufferings and conflicts (Mark 13: 30). Then it will be the time of *judgement (Matt. 13: 41–3) and of the final coming of the Kingdom (Mark 8: 38–9: 1). No date can be given (Acts 1: 7). The term *parousia* covers the whole range of expectations about the *End, which the first Christians may have expected in the near future (Mark 13: 30); hence the death of some of them came as a shock. Paul explained that when Christ came those already dead would be raised (1 Cor. 15: 20–3). But the NT does not employ the precise term 'second coming' of Christ; and his 'coming again' (John 14: 3) or 'appearance a second time' (Heb. 9: 28) are to be understood in their contexts. Christ does come again, and again, not as Judge but as Saviour, for all mankind, but the final coming stands for the fulfilment of all God's work in Christ and a kingdom of love and peace and joy 'where *righteousness is at home' (2 Pet. 3: 13, NRSV).

second death In a *vision of the day of *judgement all the dead will rise (Rev. 20: 12) and the wicked will be thrown into the lake of fire (Rev. 20: 14) to suffer their second death. Jesus gave a warning that God has the power to destroy both soul and body in *hell (Matt. 10: 28).

secret Or 'mystery', something hidden which should be made known (as Dan. 2: 29–30). In the *teaching of Jesus, *parables are that which discloses the secret of the Kingdom to believers; but the parables are only *riddles to those outside (Mark 4: 11). In the epistles of Paul, the 'mystery now revealed' is the good news of Christ (Col. 2: 2) to the *Gentiles (Eph. 3: 8–9).

see There is a kind of seeing with discernment (Mark 4: 12, quoting Isa. 6: 9), but also a merely external observation of a *miracle or marvel without discernment of its significance (Mark 15: 32; John 4: 48). That is to say, there can be physical sight, merely; or there can be (John uses another Greek verb) spiritual perception too (John 14: 19 and 16: 16 ff.) which is the work of *faith (John 20: 29). The request of the Greeks to 'see Jesus' (John 12: 21) shows that they are within reach of faith and that this faith is a possibility open to all races. However, whatever degree of sight people may be given now, it will be exceeded in the Kingdom (1 Cor. 2: 9; Heb. 11: 1). It is suggested that 'no one has ever seen God' and he can be 'known' only through his Son (John 1: 18), but in general this Christian exclusivism is expressed more modestly: it is 'the *glory of God' which can be 'seen' (John 11: 40), or 'knowledge', rather than 'sight', 'face to face' (1 Cor. 13: 12).

seed Used literally of plants or corn (Deut. 11: 10), as in Jesus' *parables (Mark 4: 26–9; 30–2), but also used of human descendants (Mark 12: 19, AV, RV; modern versions interpret as 'children' or 'heir'), especially of *Abraham (Rom. 4: 13, AV, RV; modern versions prefer 'descendants'), meaning Israel.

seer A visionary through whom divine revelations were believed to be channelled, such as *Samuel (1 Sam. 9: 5–21).

Seir The *Edomites' territory, SE of *Canaan, and mostly mountainous. The Israelites were refused passage through the area as they journeyed towards the plains of Moab (Num. 20: 18).

selah Hebrew word found in certain of the psalms and in Habakkuk. In AV it is incorporated into the text e.g. Ps. 88: 7; Hab. 3: 9), but in NRSV and REB the word is italicized and put in the margin. Its meaning is uncertain but NJB adopts the view that the word is some kind of instruction to readers or singers, and translates it as 'pause', in the margin.

Seleucia The port for the *Syrian city of *Antioch which was 26 km. (16 miles) to the east; the River *Orontes was 8 km. (5 miles) to the south. Egyptians and Seleucids wrested control of it from each other in the 3rd and 2nd cents. BCE (1 Macc. 11: 8) but in 63 BCE it was granted the status of a free city in the Roman Empire by Pompey, which was its status when Paul, *Barnabas, and *Mark embarked there on a ship bound for *Cyprus (Acts 13: 4).

self-control The opposite of drunkenness, in the corresponding list of vices (translated 'temperance' by AV, Gal. 5: 23); and for Paul not so much a human achievement as a *gift of the *Spirit. Leaders of the Church such as *apostles (1 Cor. 9: 25) and *bishops (2 Tim. 1: 7) are nevertheless enjoined to exercise this gift.

semantics The examination of what words mean; a form of scholarship much applied to the *Bible in the second half of the 20th cent. Account is taken of the context and situation of words and phrases; the same word may have different connotations in various biblical passages (*elohim*, which can mean 'God' or 'gods', is such a word in Hebrew) or a word may change its meaning as between its being uttered by a *prophet and subsequently incorporated in the book of his oracles.

semiotics The study of the system of signs; one of the *synchronic methods of analysis. It claims that words do not have any meaning in themselves, but only in the context in which they appear; and hence the study of the function of contrasts in language. It is not interested in what might have been the process of composition of a text. The discipline arises from the observation of social behaviour and a network of relationships and conventions; a certain gesture or word receives its meaning from its context and not by reason of its inherent nature. Applied to biblical studies, semiotics is the attempt to discover the conventions by which the reader understands what is signified; to discern what significance a particular narrative conveyed to the first readers. Evangelists and readers shared common conventions. There existed rules and structures in discourse which enabled sayings and events transmitted during the 'oral period' to be captured in writing and made intelligible. Thus it was possible both for a writer to use shared conventions to manipulate the readers and for readers to infer quite different meanings from the same text e.g. of *parables. This could be considered to be as valid a method of studying as traditional historical criticism. A group can discuss together how a parable was created and proclaimed, and what its purposes and effects were and are.

Semites *Assyrians and *Babylonians and others who lived round the ancient Fertile Crescent, and their modern descendants, both Jews and Muslims. The name is derived from *Shem (son of Noah, Gen. 6: 10).

(In modern parlance, since the late 19th cent., including NT scholarship, the term '*antisemitic' is used more restrictively to describe opposition to *Judaism.) Groups of languages spoken by peoples of that region are known as Semitic, classical Hebrew being just one. The particular Semitic contribution to civilization was the development of an alphabet.

semitisms Hebrew or Aramaic words or phrases which have influenced the Greek in the NT and which seem to reflect the *Aramaic speech of Jesus and the first Christians, such as **Abba* ('Father', Mark 14: 36) and **maranatha*, 'Come, our Lord' (1 Cor. 16: 22).

It has been argued that the first two chapters of the gospel of Luke have a strong Semitic flavour. Poetic and symbolic language of the OT is employed: Luke 1: 46–55 recalls the style and thought of 1 Sam. 2: 1–10.

Sennacherib King of *Assyria, 705–681 BCE. He led several campaigns against *Babylon, to the south, but the main campaign in the west was in 701 BCE, after *Hezekiah king of *Judah had organized an alliance of coastal states with some Egyptian support. The Assyrian response was swift and decisive; the coalition collapsed and Hezekiah only saved Jerusalem by paying tribute (2 Kgs. 18: 14–16). Isaiah had counselled defiance. The account in 2 Kgs. 19: 32 ff. that the siege of Jerusalem was suddenly lifted is not corroborated from Assyrian sources. Sennacherib was assassinated by his sons (2 Kgs. 19: 37).

Septuagint Denoted by the symbol *LXX, and taking the name from the *legend that this translation into Greek of the Hebrew OT was undertaken at *Alexandria by seventy (or seventy-two) Jewish scholars in as many days. The work was done in the time of Ptolemy II Philadelphus (285–246 BCE), but there were repeated revisions and also further translations (of Aquila, Theodotion, and Lucian). The LXX is a valuable check on the accuracy of the official Masoretic Hebrew text, which has also undergone revisions. There are 2nd-cent. BCE MSS fragments of the LXX among the *Dead Sea scrolls. The LXX became the Bible of the early Christians; it included some books not contained in the Hebrew (the *Apocrypha) while other books (e.g. Jeremiah) were shorter than in the Hebrew.

sepulchre Used in AV, RV for the tomb of Jesus (Mark 15: 46). Hence the church of the Holy Sepulchre on the supposed site in Jerusalem. The emperor Constantine built a church there in 335 CE; it was destroyed in 1009 CE. The present church, smaller than its predecessor, dates from 1149 CE.

seraphim Angelic beings such as those seen in *Isaiah's *vision of the Lord of *hosts (Isa. 6: 1–8). They touch the prophet's *lips with a hot coal from the *altar, thus cleansing him of *sin. These fiery creatures are described as having six wings: two to cover their faces, two to cover their feet (i.e. genitals), and two for flying.

Sergius Paulus *See* Paulus, Sergius.

Sermon on the Mount The discourse of Jesus in Matt. 5–7 placed, according to Matt.'s scheme, to correspond to Moses' receiving the *Ten Commandments on Mount *Sinai. It was first given this title by St Augustine (about 392 CE). Some of the material is replicated in Luke 6: 20–49, which is shorter and said to be delivered on the level ground. Luke does, however, add to his four Beatitudes four corresponding Woes, relevant to Christian life in the Church, which is a typical theme of Luke: notice the addition of the word 'daily' at 9: 23 (it is not in Mark 8: 34).

Matt.'s Sermon is a compilation of Jesus' sayings edited by the evangelist. They were delivered on various occasions and as now arranged they present Jesus as the New Moses, setting out the New Righteousness for his disciples. The Sermon is an integral part of the whole gospel but can be examined as a distinct unit. The first part (Matt. 5: 1–12) consists of the nine Beatitudes, or Blessings, showing that in the *kingdom of Heaven (or God) there will be a great reversal of values. The beatitudes are pronounced in the indicative form, not the imperative. They are statements about a kind of life entirely independent of a particular community, Jewish or Christian. They apply to everyone. The disciples are addressed (5: 1) but so are the crowds (7: 28). The imperative is implied—disciples at any rate ought to imitate that ideal form of living wherever it is found. There follows a series of practical examples setting out the ideals of the Kingdom in terms of the conditions of this life: murder, adultery, oaths, and animosity are all forbidden by the absolute standards of the Kingdom. There follows (6: 1–7: 12) a collection of Jesus' *teaching on *righteousness towards God; generosity and self-discipline; undetectable fasting and discreet almsgiving; perfect trust and non-judgemental attitudes. The centrepiece of the section is the Lord's Prayer (6: 9–13), and it ends with the *Golden Rule.

Interpretation of these injunctions has been split. On the one hand are those (predominantly Catholic) who have regarded them as commands to be taken as they stand; for with the help of divine *grace, they are able to be put into practice, at any rate by those with a special vocation to do so, such as members of religious orders. Others (mostly Protestant) have regarded the sermon as giving ideals to be striven towards; for a requirement to love as God loves (5: 46) is an impossible demand for perfection (Matt. 5: 48) which accuses the readers of their imperfections and which already Luke in his version (6: 36) modifies to the more practicable command to be merciful as God is merciful.

One section of the sermon (Matt. 5: 17–48) contrasts the righteousness demanded by the *Law with that put before the disciples of Jesus in a series of 'antitheses'. They are in an extreme form, to goad Matt.'s readers into actions inspired by love (Matt. 22: 39), but they represent the heightening of the Law and not its abolition (Matt. 5: 17).

serpent First appears in the *Bible in the *J story of the *creation (Gen. 2: 4–3: 24), as an explanation of human sinfulness. The serpent (not yet identified with *Satan or the devil) is the tempter. *Aaron was endowed with a power to change his staff into a serpent in order to bring pressure on *Pharaoh (Exod. 7: 10). Moses had a 'bronze serpent' which possessed properties of healing (Num 21: 4–9), which is referred to again in John 3: 14–15. Paul also had an encounter with a snake, supposed to be poisonous (Acts 28: 1–6), though poisonous snakes are not now thought to have existed in *Malta.

Servant of the Lord In *Deutero-Isaiah, four passages are known as 'Servant Songs': Isa. 42: 1–4; 49: 1–6; 50: 4–9; 52: 13–53: 12 and have sometimes been regarded as by a different author from the rest of the book, into which they were inserted later. The theme is of the undeserved suffering of the servant up to death itself as the means of taking away the nation's *sins. Controversy surrounds the identity of the servant. Was he an individual or a group? One view is that he might have been *Jeremiah. Or does he represent the whole people of Israel, or the faithful remnant of the nation? The picture of a sinless, persecuted man (Isa. 53), who died on behalf of others, has always been a particular focus of commentary, but with the other passages the Servant presents a composite figure who embodies the prophet's hope of a future national redemption.

The first Christians saw in the Servant Songs a prophecy of their *Messiah. His *suffering and *death they believed were in accordance with those scriptures and were not arbitrary or unanticipated (Luke 22: 37; Mark 14: 24). In particular, Isa. 53 is used by Paul to explain Christ's atoning work (Phil. 2: 6 ff.) and early reported sermons also rely on the Servant Songs for the OT evidence that the Messiah was, in the providence of God, to be a suffering servant figure (Acts 3: 13, 26; cf. 1 Pet. 2: 21 ff.; 3: 18).

servant, service The translation 'servant' spares the harshness of the Hebrew and Greek words for a slave, and modern translations often prefer the reality. As such, the OT 'servant' possessed no legal rights though he was admitted to membership of the *covenant people (Gen. 17: 12) and observed its basic laws (Deut. 12: 18).

In the NT Jesus speaks of people as servants of God, i.e. wholly given to his service. Jesus can himself accept this role of extreme humility (Mark 10: 45). Matt. (12: 18–21) quotes the first Servant Song (Isa. 42) as a portrait of Jesus, and Paul's words in Phil. (2: 7) seem to echo Isaiah; so Christians are also to be servants of each other (John 13: 14) because they are servants of Christ (1 Cor. 7: 22).

Seth The third son of *Adam and *Eve (Gen. 4: 25) and regarded in the narrative as heralding a more respectable line than that of the criminal Cain. With the birth of Seth people began the custom of calling on the name of Yahweh (Gen. 4: 26), and it is through the line of Seth that the human race was propagated (Gen. 5: 7).

seventy Seventy *elders accompanied Moses (Exod. 24: 9); and seventy *disciples were sent out by Jesus (Luke 10: 1–7); seventy years was predicted to be the period of the Exile (Jer. 25: 12), though it turned out to be less. Seventy years was regarded as a whole human lifespan (Ps. 90: 10). Peter was told that we are to forgive seventy times seven (Matt. 18: 21–2). Thus, seventy indicates completeness, as being a multiple of seven (the number of days in the week), which is therefore itself significant; seven men were to help the apostles (Acts 6: 1–6), and seven Churches are addressed in Rev. 2–3.

seven words The sayings of the crucified Jesus as recorded in the four gospels. They are of forgiveness, consolation, love, and desolation (Luke 23: 34—though omitted by some MSS; John 19: 26–7; Matt. 27: 46); but also of thirst, of triumph (John 19: 29, 30), and a quotation from Ps. 31: 5 commending his soul to God (Luke 23: 46). However, these seven 'words' cannot be regarded as a historical sequence; they represent the theological interpretations of the *evangelists. They are alternative explanations of the event: the cry of desolation expresses a sense of the total abandonment by God (Mark 15: 34) but is changed by Luke, who replaces the terror of Mark's narrative by a note of serenity supported by its being a quotation from Ps. 31: 5. The notion of the 'seven words' from the cross became widely established after Roman Catholic Jesuits instituted a popular *Good Friday service based upon them, following the Lima earthquake of 1687, which continued in regular use well into the 20th cent.

sexuality Sexual differentiation is explained in the *creation narratives as part of God's design for humanity. It is good; and it is necessary for the continuation of the race. It is expressed in OT narratives in polygamy (1 Kgs. 11: 1–3), concubinage (Gen. 16: 1–4), and Levirate marriage (Deut. 25: 5–10). Divorce was permissible (Deut. 24: 1–4); homosexual acts were not (Lev. 18: 22). The emphasis throughout was on the production of future generations and the birth of children showed the divine favour (Ps. 127: 3–5). But the Song of Songs is a witness to the worth of erotic love in itself without reference to marriage.

The NT *teaching is conditioned by an expectation of the possible end of the world and the consummation of the Kingdom, which lent a negative view on human, temporal relationships which could distract from the imperatives of expectation (1 Cor. 7: 31). However, the *Gnostic view that the material world is evil and the body and its passions to be denied is firmly rejected (Col. 2: 20–3; 1 Tim. 5: 23; Heb. 13: 4), but Paul is aware that passions do not have unrestricted moral licence (Rom. 1: 24–7).

shadow The Jewish *Law offers only a shadow of the reality to come in Christ (Heb. 10: 1); it can give only an insubstantial expression of future heavenly promises. The author's thinking here resembles that of the Alexandrian Jewish philosopher *Philo, who combined *Judaism with Platonic ideas. In the OT the disappearing shadow is a reminder of the transience of human life (1 Chron. 29: 15—David's prayer).

Shallum Numerous men have this name in the OT, of whom the best known are (1) the king of *Judah (609 BCE), son of *Josiah (Jer. 22: 11); his father was killed by Necho king of *Egypt, who within three months deported Shallum (called Jehoahaz) to Egypt (2 Kgs. 23: 30–4); (2) the husband of the prophetess *Huldah (2 Kgs. 22: 14).

Shalmaneser (1) Shalmaneser III was king of *Assyria, 858–824 BCE; he defeated *Hazael of *Damascus in 841 and accepted tribute from *Jehu of Israel, as recorded on the Black *Obelisk, now in the British Museum in London. (2) Shalmaneser V, 727–722 BCE, put down a rebellion of *Phoenicia, Israel, and Egyptian supporters; he imprisoned *Hoshea king of Israel (724 BCE) and besieged *Samaria, which fell in 722.

shame A sense of disgrace. Various customs in NT times were held to be shameful, such as short *hair worn by women or long hair on men (1 Cor. 11: 6, 14); the real shame in NT is the guilt attaching to *sin (Phil. 3: 19). However, there was a shame which Christ himself bore on the *cross (Heb. 12: 2) on our behalf (2 Cor. 5: 21), which is cause enough for the apostle to 'boast' (Gal. 6: 14).

Shammai A rabbinical *teacher in the time of Jesus. He gave a rigorous interpretation of the *Law, contrasting with the more relaxed and liberal views of *Hillel. Those of the Shammai school were often in conflict with the Roman authorities, and after the fall of Jerusalem in 70 CE were much diminished in influence and numbers. The dispute about *divorce (Matt. 19: 1–9) must be understood against the background of rabbinic controversy about Deut. 24: 1. It puts the Christian position in the Shammai camp but with an exceptive clause (19: 9) which greatly liberalizes its rigorism.

Shaphan Served in the court of King *Josiah (2 Kgs. 22: 3–20); and when the law book, thought by many scholars to be part of Deuteronomy, was discovered in the Temple, he gave the news to the king and was dispatched to consult the prophetess *Huldah. Shaphan saved the prophet *Jeremiah from execution (Jer. 26: 24) and provided accommodation in which the words of Jeremiah could be studied by *Baruch (Jer. 36: 10–12).

Sharon An extensive but not altogether level plain in the north of Palestine used as pasture land (1 Chr. 27: 29) and known for its beautiful wild flowers (S. of S. 2: 1).

Shear-Jashub Hebrew for 'A remnant shall return'; the name given to one of *Isaiah's sons as a symbol of his message (Isa. 7: 3) that in the kingdom a mere remnant (those faithful to their God) would survive.

Sheba, queen of Ruler of the Sabeans (SW Arabia), who visited King *Solomon and was impressed by his *wisdom (1 Kgs. 10: 1 ff.). She may also have had commercial interests in mind. Jesus contrasted her journey 'from the ends of the earth' with the indifference to his teaching of his own generation (Matt. 12: 42).

Shebna King *Hezekiah's under-secretary of state (2 Kgs. 18: 18), who was sent to obtain the advice of *Isaiah at the time of the Assyrians' campaign. It seems that he had previously held the higher office of steward and that Isaiah had urged his replacement by Eliakim (Isa. 22: 20).

Shechem A city north of Jerusalem controlling a vital trade route, and so very prosperous. It was visited by *Abraham (Gen. 12: 6) and it was the burial site of *Jacob (Josh. 24: 32) and the place where *Joshua renewed the *covenant (Josh. 24). It was destroyed by *Abimelech, self-styled king (Judg. 9: 45), but rebuilt, and became the place chosen by *Rehoboam for his coronation; but it was in the northern area which revolted against him soon afterwards, and Jeroboam I of Israel used Shechem temporarily as his capital. The city survived successfully until the *Assyrian invaders again destroyed it; once more it was rebuilt and became a religious centre in 350 BCE for the *Samaritans. It was destroyed again when *Samaria was subdued by John Hyrcanus in

107 BCE, and again rebuilt. It survives today as part of the city of Nablus.

sheep Domesticated by the time of *Jacob (Gen. 30: 37–42); sheep provided meat, wool, hide, and milk, and so were fundamental to human existence. Their shepherds were in constant search of suitable pasture (1 Chron. 4: 39–40). In the NT sheep are often used as a metaphor: thus, the people of Israel are compared to sheep without a shepherd (Matt. 9: 36) and Jesus is 'a good shepherd' who gives his life for his sheep (John 10: 7–9) but, also, he himself is compared to a sheep that is led to the slaughter (Acts 8: 32).

Sheep Gate On the north side of the Temple in Jerusalem and built after the return from Exile (Neh. 3: 1). It was probably the best way into the city for sheep to be driven for the *sacrifices. The gate was apparently near the pool of Bethesda (John 5: 2), though the Greek text here is uncertain (the word 'gate' is absent in the Greek).

shekel In OT times the shekel was a unit of weight—about 2/5 of an ounce or 14.5 grams of silver. In NT times it was a silver coin of the same weight.

shekhinah The Hebrew term (not itself in OT) for the important idea of God's dwelling in the midst of his people, usually located in the *Tabernacle or the Temple, and described as his '*glory' (Exod. 40: 35). There were theological difficulties in reconciling the idea of the localized presence with that of the one God who could be encountered throughout the world; but the concept provides a way of holding that God is both transcendent and immanent, and it became available to Christians who struggled to relate their Jewish monotheism with a doctrine of the *incarnation of Christ. The concept underlies the doctrine of the *Word (John 1: 14), the narratives of the *Transfiguration (Luke 9: 31) and the *Ascension (Acts 1: 9), and the promise of the presence of Jesus where 'two or three are gathered in my name' (Matt. 18: 20) 'to the end of the age' (Matt. 28: 20).

shema The first Hebrew word (meaning 'Hear!') of the fundamental expression of Jewish faith in the oneness of God, as given in Deut. 6: 4. It was called by Jesus the first commandment in the law (Mark 12: 29).

sheol Hebrew term for the lower parts of the earth or underworld to which the departed were thought to be consigned (Prov. 9: 18). There they continued in gloomy insubstantiality (Ecclus. [= Sir.] 17: 27–8) which was foreshadowed by the deep depressions in present existence. Therefore restoration to health and vitality was a cause of thankfulness to the Lord (Ps. 30: 3).

shepherd The importance of shepherds in Israel lay in the dependence on sheep of both the religious cult and the national economy. Sheep in Palestine relied totally on the protection of their shepherd, who rescued them from lions and bears (1 Sam. 17: 34–5; Amos 3: 12), checked their numbers (Jer. 33: 13), or failed to do so (Ezek. 34: 8); and searched for any that were missing (Matt. 18: 11–14). The image of the shepherd occurs frequently in the gospels (e.g. Luke 15: 3–7), and Jesus is the 'good shepherd' (John 10: 1–19) or the 'great shepherd' (Heb. 13: 20).

Shesh Bazzar A trusted servant of *Cyrus who brought back to Jerusalem the sacred vessels of the Temple in 538 BCE. They had been removed to *Babylon (Ezra 1: 7) at the time of the Exile.

shewbread (or **showbread**) Bread made without yeast. Instructions are given in the OT (Lev. 24: 5–9) for *twelve loaves of unleavened bread to be placed in the Temple as an offering to the Lord. On the *Sabbath the priests renewed them, consuming the old loaves. David, in an emergency, once ate this bread at *Nob (1 Sam. 21: 4–6), an incident cited by Jesus to justify an infringement of the *sabbath law by his *disciples (Matt. 12: 3–4).

shibboleth In English usage, a test word or opinion which gives away one's party or nationality; hence also a doctrine once regarded as essential but generally abandoned as old-fashioned. The word in Hebrew was the password demanded by *Gileadite soldiers (Judg. 12: 6) to pick out the *Ephraimite enemies who were physically unable to sound the initial consonant.

Shiloh North of Jerusalem and *Bethel and west of the River *Jordan, Shiloh was a religious centre of the tribes in the 12th cent. BCE at the time of the distribution of land (Josh. 18: 9). It was a *sanctuary with a priesthood and *Eli and his sons ministered there until the *Ark was captured by the *Philistines (1 Sam. 4).

Shinar A plain in *Babylonia (modern S. Iraq), where the tower of Babel was supposed to have been built (Gen. 11: 2). Some of the Israelite exiles were to be rescued from there (Isa. 11: 11). In *Zechariah's seventh *vision the wickedness and guilt of *Judah is sealed in a *basket and carried to Shinar to be deposited in a temple of its own (Zech. 5: 11).

ships Seafaring was not a skill of the Israelites; it was the sailors of *Hiram, king of *Tyre, who transported the materials for *Solomon's Temple (1 Kgs. 5: 9), and in the NT Paul's journeys across the Mediterranean were made in trading vessels from Egypt (Acts 28: 11), not without misadventures (2 Cor. 11: 25). Jesus' disciples had fishing vessels on the Sea of *Galilee (Mark 1: 19–20). One such vessel of that time was discovered by archaeologists in 1986 and has been carefully preserved.

Shittim A place in *Moab, east of the *Jordan, where *Joshua was proclaimed as Moses' successor (Num. 25: 1; 27: 12–23) and spies were dispatched to investigate the Promised Land (Josh. 2: 1).

shophar A musical instrument made from a ram's horn (Josh. 6: 4–13) used as a call to arms (Judg. 3: 27) and in music (Ps. 98: 6), and still in use in Jewish worship.

Shunem A town in the territory assigned to *Issachar at *Shiloh (Josh. 19: 18), near Mount *Gilboa (1 Sam. 28: 4). It was a favourite stopping-off place for *Elisha, and there the prophet once revived the son of his hostess from apparent death (2 Kgs. 4).

Shushan Susa, NRSV, REB, NJB: the capital city of the kings of Persia (Esther 1: 2) where *Daniel saw a *vision in the third year of the reign of King *Belshazzar (Dan. 8: 2).

Siccuth This is the reading in some English versions (e.g. RV) at Amos 5: 26, following a mistake in the Hebrew text for Sakkuth, a Mesopotamian star-deity; the idols of Sakkuth and Kaiwan were being carried in processions in syncretistic worship, which the prophet condemns. These gods were connected with the planet Saturn.

Sidon A coastal city of *Phoenicia often associated with *Tyre (Jer. 25: 22) and 35 km. (25 miles) to the south of it (in modern Lebanon). In NT times it was known as a place of sophistication and wealth, and Jesus apparently travelled in that area (Mark 7: 24), as also did Paul (Acts 27: 3).

siege Military operation to surround a community. It was a common experience in the ancient world for towns to be surrounded by an enemy and all supplies cut off. The defenders built walls and gates to resist the enemy, and sometimes were successful. *Samaria held out against the *Assyrians for three years, until 722. The siege against Jerusalem in 701 was lifted and the city remained free until the Babylonian siege of 597 when *Nebuchadnezzar attacked it (2 Kgs. 25: 1). Some of the Roman siege works against *Masada are still visible.

signs Remarkable events which witness to the will of God, such as the *rainbow (Gen. 9: 12) or the *plagues wrought by Moses and *Aaron in *Egypt (Exod. 10: 2). In the NT the Jews demanded miraculous signs from Jesus (Matt. 12: 38), which he refused. In John's gospel the *miracles are called signs (*semeia* in Greek) which are intended to produce *faith (John 20: 30–1). Jesus would therefore be accepted as the one who performs the works of God, which are the eschatological acts of *salvation (John 5: 20–3).

Sihon King of the *Amorites who tried unsuccessfully to prevent the Israelites traversing *Transjordan on their way from the south (Num. 21: 23–4), a victory long celebrated in Israel (Ps. 135: 11).

Silas Latinized as Silvanus; one of the delegates from Jerusalem to the Church in *Antioch (Acts 15: 22–35). After the rupture with *Barnabas, Silas became Paul's principal companion (Acts 15: 41–18: 5). He assisted Paul in writing or composing his letters (e.g. 1 Thess. 1: 1) and also shared in the writing of 1 Peter (5: 12), which some scholars regard as the means by which the authorship of Peter, at one remove, can be defended.

Siloam A pool of water in Jerusalem which served as a reservoir. It was connected by a tunnel to the spring of *Gihon and an inscription was discovered by two boys in 1880 which explains that the tunnel was dug from each end; the Hebrew script suggests a date not long before 701 BCE, when *Sennacherib laid siege to the city; so the tunnel would have been part of the defence system. *Isaiah mentions that its waters 'flow gently' (8: 6), and there is a record in John 9: 7 of Jesus healing a blind man by means of this water. The tower of Siloam (Luke 13: 4) which so disastrously collapsed was part of the fortifications.

silver Mentioned frequently in the *Bible; *Joseph in *Egypt had a silver cup (Gen. 44: 2); silver was used for articles needed in the Temple (Neh. 7: 71), for the purchase of goods (Josh. 24: 32), and for decorating idols (Isa. 2: 20). In the NT silver is either mentioned bitterly, as the thirty pieces awarded to *Judas for betraying Jesus, or in deprecation as a corrupting influence (Jas. 5: 3) or as an unnecessary luxury (Matt. 10: 9).

Simeon (1) Son of *Jacob and *Leah (Gen. 29: 33), and regarded as founder and ancestor of the tribe Simeon which united with *Judah at the time of the conquest of *Canaan (Judg. 1: 3–17) but was regarded by the Chronciler as part of the northern kingdom (2 Chron. 15: 9). (2) An elderly man who took the infant Jesus into his arms in the Temple (Luke 2: 25–35) and praised God that he had seen *salvation brought to Israel. (3) The Hebrew name of *Simon (Peter), found in 2 Pet. 1: 1.

Similitudes of Enoch Chapters 37–71 of 1 *Enoch. The term '*Son of Man' occurs in this book, but its date is difficult to determine and therefore it is impossible to argue whether Enoch influenced the use of the term in the gospels.

Simon Several men in the NT bear this name, taken perhaps from Simon Maccabaeus (142–134 BCE). (1) Simon Peter, leader of the Twelve (Mark 3: 16). (2) Another member of the Twelve, 'the Zealot' (Luke 6: 15). (3) A brother of Jesus (Mark 6: 3). (4) A leper in Bethany (Mark 14: 3–9). (5) A Pharisee (Luke 7: 36–50). (6) A man from *Cyrene who, on entering Jerusalem, was conscripted to carry Jesus' cross (Mark 15: 21). (7) *Simon Magus (Acts 8: 9 ff.). (8) A tanner in *Joppa with whom Peter lodged (Acts 9: 43).

Simon Magus A magician who held people spellbound in *Samaria (Acts 8: 9 ff.). He was baptized by *Philip (Acts 8: 13), after which he in turn was no less astonished by Philip's signs and wonders, and was filled with such envy that when Peter arrived Simon sought to obtain a similar power by a financial deal, which Peter repudiated (Acts 8: 20). Despite his repentance he acquired a bad reputation. He was regarded as a Gnostic leader and the offence of 'simony' is named from him.

sin In the OT 'sin' denotes what is unacceptable (to God or humanity), not necessarily disobedience to God, or rebellion against him (as it is in 1 Kgs. 8: 50); not to be equated with crime, which is an offence against society. Anything wrongly related to God is sin (Rom. 14: 23). *Idolatry is therefore the supreme sin (Rom. 1: 23). In the NT one Greek word has the meaning 'missing the mark' (John 8: 46; Rom. 5: 12). Another means 'lawlessness' (2 Cor. 6: 14) or 'moral depravity' (1 John 3: 12). When sin came into the world, it provoked God's wrath (Gen. 3: 24) and it is explained that the fact of death is to be ultimately attributed to sin (Gen. 2: 17). Everybody is infected by sin and its guilt, of pride rather than *faith (Rom. 5: 12 ff.). The condition is irreversible by human effort (Rom. 7) and it is only by God's initiative that any change is possible (John 3: 3 ff.). Sin was in principle conquered by the *life, *death, and *resurrection of Jesus and we are released from its suffocating power by our being united to him by faith and *baptism.

Sinai Identified since the 4th cent. CE (but other locations are proposed) as the peninsula which joins *Africa and *Asia; on the north is the Mediterranean, on the south the *Red Sea; much of it is desert, though the area has always managed to support a smallish nomadic population. The Israelites trekked through the *wilderness of Sinai when they left the slavery of *Egypt. At Mount Sinai, in the south of the peninsula, Moses had witnessed the *Burning Bush (Exod. 3: 1–2). It was to this mountain that the Hebrews escaping from Egypt at the *Exodus made their way. This was the sacred mountain where Yahweh was believed to dwell, and it was natural that because they associated him with their rescue from slavery they would proceed to Sinai before going on to the promised land of *Canaan. At Sinai the people received the *Law (Exod. 20–3) and the *covenant was ratified. This 'mountain of God' is also known in the *E and *D sources, as Mount Horeb. The mount became so much identified with the revelation to Israel that when Paul discusses the relationship of the new covenant to the old, he makes Sinai in his *allegory stand for the old system of *Judaism (Gal. 4: 24–5).

Sinaiticus, Codex Denoted by the Hebrew letter *aleph. A 4th-cent. CE MS of almost the whole *Bible in Greek, together with two other works; written in four columns to a page, it was found by the German scholar Constantin von Tischendorf (1815–74) by chance in the course of two visits to the monastery of St Catherine on Mount *Sinai in 1844. It was later presented to the Tsar of Russia, and after the Bolshevik revolution passed into the hands of the new regime, by whom it was sold to the British Museum in London in 1933.

Sirach, Son of (in Hebrew, Ben Sirach) The author of the inter-testamental book *Ecclesiasticus, or the Wisdom of Sirach.

Sisera Commander of an army opposing Israel (Judg. 4–5). The decisive battle was fought at Taanach; a sudden storm of rain and rising flood waters were Israel's salvation and Sisera's forces were routed. Sisera sought refuge in the tent of *Jael, a Kenite woman, who pretended to offer the proper hospitality but killed him while he slept (Judg. 4: 17–22; 5: 27).

Sitz im Leben A term employed by German *Form Critics and one difficult to translate exactly into English. It denotes the social context or 'life setting' in which a narrative emerged. The point being made is that particular items in the OT can only be understood when they are related to the culture and social life of ancient Israel. Before the sources which comprise the *Pentateuch were written, they were transmitted orally, probably within the context of worship (their 'Sitz im Leben'). The character of the social situation determines the style of the communication. In NT scholarship critics try to relate sayings of Jesus both to the *Sitz im Leben* of the Church in which they were transmitted and to the *Sitz im Leben* of Jesus' ministry in which they may have originated. It has been suggested that sayings of Jesus can only be regarded as authentically his when they can be slotted into neither the experience of the early Church, nor to that of 1st-cent. CE *Judaism. Others regard this criterion as excessively sceptical, on the ground that the teaching of Jesus, as a Jew of the 1st cent., must have reflected that of contemporary Judaism.

slander Defamation of other people by lying speech; forbidden in OT (Lev. 19: 16) and in the Church (2 Cor. 12: 20).

slavery Absolute property of a person by a legal owner. Men captured in battles were often enslaved (1 Kgs. 9: 21) and debtors who defaulted, together with their families, could be enslaved by the lender (2 Kgs. 4: 1) or they could sell themselves (Lev. 25: 39). Humane laws mitigated the severity of slavery—e.g. the *Sabbath rest (Exod. 20: 10), the *jubilee (Lev. 25: 13), and the right to asylum of foreign fugitive slaves (Deut. 23: 15–16). In the NT slaves are mentioned as belonging to the Christian community, and some may have expected that liberation would be granted by Christian owners if the slaves were baptized. But this did not seem necessarily to follow (1 Cor. 7: 21). Paul, however, does put on subtle pressure and requests Philemon to treat his slave *Onesimus as a brother (Philem. 15–16).

sleep Used both literally (e.g. 1 Sam. 3: 2 ff.) and also metaphorically for idleness (Prov. 24: 33 f.), spiritual laziness (Eph. 5: 14), or death (1 Cor. 15: 20, AV, RV, NRSV marg.).

Smyrna A city on the west coast of Asia Minor (modern Turkey) where one of the Churches addressed by Rev. (1: 11) was established. It was in conflict with the local Jewish community (Rev. 2: 9–10). *Ignatius, on his way to martyrdom in *Rome in 107 CE, called at Smyrna, where Polycarp was bishop, and subsequently *Ignatius wrote letters both to the Church of Smyrna and to its bishop.

snakes Common in Palestine, and often mentioned in the Bible from Gen. 3: 14 onwards. There existed cobras and vipers. In *Malta Paul is said to have been bitten by a snake (Acts 28: 3). Not surprisingly, snakes feature also as symbols of *evil (Ps. 58: 4) and the *Devil is called 'that old *serpent' (Rev. 20: 2.). Nevertheless snakes are credited with an enviable *wisdom (Matt. 10: 16).

snares Traps which held the legs of birds or animals. Used in OT times (Amos 3: 5), they became a symbol for death (Ps. 116: 3) and in NT for wickedness (1 Tim. 3: 7).

snow Common in mountainous parts of Palestine but rare elsewhere, though mentioned in the OT (e.g. 2 Sam. 23: 20 and Jer. 18: 14). It was regarded as the ultimate in whiteness (Isa. 1: 18 and Matt. 28: 3).

So King of *Egypt who received messengers from King *Hoshea of Israel (2 Kgs. 17: 4), which provoked the *Assyrian assault on *Samaria (725 BCE). No Egyptian source mentions this ruler, and the accuracy of the reference has been widely suspected.

sociology The study of human communities. It is increasingly important in OT and NT studies to determine the structure and changes in the society of ancient Israel and in the development of Christian groups in the Roman Empire. A particularly interesting feature of Israelite society has been the transition from the tribal structure of the period of the *Judges (who were heads of powerful families and who acted as arbitrators in quarrels), up to the time of the early *monarchy, and the examination of this system has assisted in the understanding of prophecy. Like the judges, the early prophets were also heads of guilds, as was *Elijah.

A lineage-based social system, in which there was much inter-tribal friction (Judg. 19–21) gave way to the monarchy under Saul. Pressures from outside and economic necessities within forced the independent units into accepting it; but the divergent attitudes to the monarchy in the narratives reflect opposition on the part of those losing out in the power struggle. A strong merchant class emerged under *Solomon (965–924 BCE) and the power of the central government grew. Prophets could oppose the political and religious powers of the kings (1 Kgs. 12–14) because they were able to feel detached from the Establishment. Nevertheless *Isaiah, who was at home in the royal court (Isa. 7: 3–17), could both criticize the king and also uphold popular morale in the state when it was under assault from *Assyria in 701 BCE. The Exile saw the end of the system of central government in Jerusalem. After the Return from Exile there was a brief resurgence of prophecy. Malachi combined concern for the cult with a call for social justice (Mal. 3: 5), but thereafter groups that regarded themselves as deprived sought solace in apocalyptic fantasies (e.g. Dan. and 1 Enoch)

NT scholars have noted that primitive Christianity involved some breaks with the established norms of society in the Roman Empire: Jesus' demand that the disciples abandon family relationships if they follow him (Mark 10: 29) is one such social surprise. In the Pauline Churches the abolition of social barriers promoted social mobility (cf. Rom. 16: 23) amongst those on the margin either economically or religiously. The Pauline denigration of Jewish food laws and the law of *circumcision was especially an affront to Jews because it threatened the basis of the people of the *covenant. It was therefore essential for Christians to establish an alternative basis for the claim to be the people of God and to affirm the solidarity of the new groups. The controversy of Paul in Gal. and his theological argument in Rom. can be understood in sociological terms. It was necessary for the success of the Gentile mission that the new Christian Churches should separate from *Judaism and the *synagogue and the demands of the *Law. Paul was trying to persuade Jewish Christians in *Rome to worship alongside the *Gentile converts (Rom. 15: 7).

Much space is devoted in Mark to narratives about purity, *hospitality, and meals, and in John to the demonstration that Jesus' *miracles are 'signs' of heavenly food and divine light that are available to those who enter the new *community. Mark's community needs to understand itself both against the Jews, in the matter of purity, and in relation to the empire, to have self-confidence as a mere sect in the midst of mighty pagan powers. So Christians were fortified by nursing the hope that beyond the present suffering lay divine deliverance, as it did for Jesus (Mark 13).

The study of social relationships has thrown light on the progress of a small group on the margin of rural Jewish society in Palestine into a world-wide Church consisting of related communities in many urban centres. Paul's letters to *Corinth show that the Church there comprised members of different classes, both sexes, and varied occupations. In that Church was taught an ideal of equality which was unknown in the Graeco-Roman world, and this affirmation of each individual's worth acted as a strong means of recruitment. Cf. Gal. 3: 28.

Sodom One of the cities of the plain (Gen. 19: 28) where *Abraham's nephew *Lot decided to live (Gen. 13: 8–13), near the *Dead Sea; it was said to be a place of great wickedness (Amos 4: 11; Matt. 10: 15), though there is no evidence that this wickedness was 'sodomy'—an offensive description of homosexual acts.

sojourning *Foreigners who stayed for a considerable time in another country were said to be sojourning, as did the tribes of Israel in *Egypt (Gen. 47: 4, AV; 'reside as aliens', NRSV), or Elimelech in *Moab (Ruth 1: 1, AV).

soldiers A frequent word in the Bible for men equipped with weapons some of which are described in 1 Sam. 17: 5–7 and Eph. 6: 11–17—the latter being a metaphorical use for a Christian's spiritual armour.

Professional soldiers were employed by *David (1 Sam. 22: 2) and subsequent rulers. In NT times Roman soldiers were a familiar sight throughout the empire. The guard of soldiers offered by *Pilate to the *Sanhedrin (Matt. 27: 65) were presumably Roman soldiers, but it is unlikely that they would have reported to the chief priests (Matt. 28: 11). Indeed the whole account of the guard at the tomb is implausible; if the *disciples were unaware of any prediction about Jesus rising on the third day, it is unlikely that the *Pharisees knew about it and remembered it.

Solomon Successor of King *David, and his son by *Bathsheba; installed on the throne through the intervention of *Nathan the prophet (1 Kgs. 1: 23). He reigned from about 962 to 922 BCE. He was an administrative reformer (1 Kgs. 4: 7–19) and international activist. He fortified Jerusalem and entertained 700 wives and 300 concubines. Solomon acquired a reputation for sagacity (1 Kgs. 4: 30), and the book of Proverbs is traditionally ascribed to him. Although Solomon built the first Temple, he also permitted local shrines at 'high places' and in his international policy he provoked hostility from three neighbouring peoples, who enlisted help from *Egypt (1 Kgs. 11: 21). This religious syncretism and foreign opposition led to enemies both within and outside the kingdom and contributed to its disruption soon after Solomon's death.

Solomon's porch (AV); 'Solomon's portico' (NRSV, REB, NJB) formed part of the reconstruction of the Jerusalem Temple by *Herod the Great. Jesus went to it (John 10: 23) and so did the apostles after the resurrection (Acts 3: 11; 5: 12).

Son of David A Messianic title of which the first documented occurrence is in the Pss. of Sol. (17: 21); used of Jesus (e.g. Mark 10: 47; Matt. 1: 1; 21: 9). It was part of Jewish Messianic expectation that the *Messiah should be a descendant of King *David, Israel's ideal king. Matt. emphasizes this claim: Jesus accepts the people's acclamation of him as such when he visits Jerusalem; and he confounds his critics when they question him about this at the climax of controversy in Jerusalem (Matt. 22: 41–6). The title is used nine times by Matt. and seems to be part of his answer to Jewish objections, when the gospel was compiled, about the Church's claims for Jesus. The child Jesus was a Son of David (2: 2); as Son of David he rode into Jerusalem (21: 9); but he is never addressed as such by *Pharisees. He does not fit into current expectations of the Son of David. The liberation he offers is not like that of King David but of healings and love. The title does emphasize the Messiahship of Jesus but in Matt. it is a modification of Israel's hopes. Indeed it is used of him by a Gentile woman (Matt. 15: 22).

Son of God A title of the king in the OT (e.g. Ps. 2: 7), and in the plural form it is applied to faithful Jews (Wisd. 9: 7). It is a Messianic title in the *Dead Sea scrolls and in 4 Ezra (7: 28). Although Christians attributed divine Sonship to Jesus as *Messiah after the *resurrection, the tendency was to push its usage ever earlier—to the *baptism by *John (Mark 1: 11) and the *Transfiguration (Mark 9: 7), or even before his birth. According to Paul (Rom. 1: 4 and also John 1: 18) Jesus was designated 'Son of God'. This could mean that a pre-existent divine being entered human life and functioned in the person of Jesus. At any rate according to Matt. 1: 18–25 the beginning of Jesus' life was an act in which God took the initiative. A saying is recorded in which Jesus refers to God as his Father (Mark 13: 32). At the end of the gospel (Matt. 27: 54, following Mark 15: 39) the *centurion makes the Christian confession of *faith. If he in fact spoke such words, he may have said, 'He is a son of God' (there is no definite article in the Greek) meaning a divine being in the pagan sense; the emperor *Augustus had the title 'Son of God' after his adoptive father Julius Caesar had been deified by decree of the Roman senate in 42 BCE. For the Christian readers of the gospels the importance was in the words being an unsolicited witness to what they themselves believed. It is a Christological title in the epistles of Paul (e.g. Gal. 4: 4), and Heb. (1: 5) quotes Ps. 2: 7. Unambiguous divine status was accorded to Jesus by the Council of Nicaea (325 BCE).

Son of Man A common phrase used on the lips of Jesus frequently in all the gospels, but not elsewhere in the NT (except Acts 7: 56 and Rev. 1: 13); in Greek, a curious translation of the Aramaic *bar nasha* (= in Hebrew *ben adam*). In the OT it is the form of address by God to the prophet *Ezekiel, and in Ps. 8 it is used of mankind in general as pre-eminent in the created order; in Dan. 7: 13 f. the phrase symbolically represents the people of Israel, contrasted with the beasts who stand for the *Gentile nations. Later, in the *Similitudes of Enoch, the Son of Man is a supernatural figure, the agent of *salvation and of *judgement.

In the course of the many discussions of

this term, it has been suggested that Jesus was not referring to himself at all by this title and that it was attributed to him by the post-resurrection Church. However, the lack of its use outside the gospels suggests that it was avoided because it was regarded as sacred to the utterances of Jesus himself. Other scholars have distinguished those sayings which refer to Jesus' present life and imminent death from the sayings which anticipate the future Kingdom and the place of the Son of Man in it. Most German scholars take the view that 'Son of Man' could only have the latter, apocalyptic, significance; it was part of Jesus' expectations of the future. Sayings which do not fall into this category are creations of the Church. Yet another view, increasingly accepted, is that Jesus *did* use the term and did mean himself by it, but that it was an elusive, roundabout, deliberately enigmatic way of saying: 'I, being the man I am'. Matt. certainly intends 'Son of Man' to mean 'Jesus' for he sometimes substitutes 'I' for 'Son of Man' (Matt. 10: 32 for Mark 8: 38). For him, it is a title of special dignity. Jesus is no ordinary man (Matt. 16: 13), and when the passion is complete ('from now', Matt. 26: 64) the Son of Man will be enthroned.

Such a view maintains the authenticity of the 'present' sayings, while allowing that the apocalyptic sayings affirming the future glorification of the Son of Man may be creations of the Church as part of mainstream Christian interpretation.

Song of Songs (REB, NJB) or **Song of Solomon** (AV, NRSV) A poetic book in the OT attributed to *Solomon but possibly written in the later part of the 6th cent. BCE. Other dates ranging over several centuries have been proposed. It is a series of love poems in dialogue form between a man and a woman who both use powerful and moving imagery. The conventional interpretation in *Judaism and Christianity has been to regard the Song's erotic vocabulary as allegorical: it is said to express the love of God for his people (Israel, or the Church), but the book is not mentioned in the NT.

Most probably the exuberant lyrical poems were written to celebrate human love. There are beautiful and sensuous metaphors and similes to extol the bodies both of women (e.g. 7: 1–6) and of men (5: 10–16). Mountain goats and pomegranates and nectar; liles and spices and sapphire, are among the images with which the lovers entice each other to the vineyards (7: 12).

Song of the Three Children An addition to Dan. after 3: 23 found in the *LXX and included there in Roman Catholic Bibles (e.g. NJB, 'the story of Azariah') but in the *Apocrypha by others. It was probably an independent composition, and included in Dan. as a suitable insertion in the 2nd cent. BCE. As a prayer of penitence, it seems to reflect the desecration of the Temple by *Antiochus Epiphanes in 167 BCE and accepts it as a *punishment for Israel's *sins. Verses 35–66a have been incorporated as a canticle known as Benedicite in some Christian liturgies, e.g. in Books of Common Prayer of 1549 (as an alternative in Lent), of 1552 (as an alternative on any day), and subsequently in later editions. At the Savoy Conference in 1661 the Presbyterian delegation asked that it should be omitted, as coming from the non-canonical *Apocrypha, but they were overruled.

Sons of God Or 'children of God'. In Gen. 6: 1–4 'Sons of God' united with 'daughters of men' to produce a race of supermen who were presumably destroyed by the *Flood. The expression, meaning supernatural beings, is also used at Deut. 32: 8, in *LXX, and Job 38: 7 (NRSV marg.). In Dan. 3: 25 'a son of the gods' (NRSV marg.) appears in the furnace as a fourth figure.

In the NT 'sons' or 'children of God' (men and women alike) are believers in Christ (Gal. 3: 26; 1 John 3: 1).

Sons of the prophets A term for a prophetic group, such as that from which Amos (7: 14) kept his distance.

soothsayer Those who claim to foretell the future; the OT is hostile to them (Jer. 27:

9). They are denounced by *Micah (5: 12) as akin to sorcerers.

Sopater Son of *Pyrrhus, one of Paul's companions on his journey to Jerusalem (Acts 20: 4), possibly to be identified with the Sosipater who sends greetings to *Rome (Rom. 16: 21, there said to be a relative of Paul).

Sosthenes (1) Ruler of a *synagogue at *Corinth and assaulted by the angry crowd when he failed to win the case against Paul. *Gallio the proconsul refused to intervene (Acts 18: 17). (2) Mentioned as associated with Paul in writing to the Corinthians (1 Cor. 1: 1) and possibly the former ruler of the Corinthian synagogue.

soteriology The doctrine of *salvation, of which the basis is found both in the OT and the NT. The Israelites were saved from *slavery in *Egypt; in Christian experience there is need for liberation from the slavery to *sin and death. OT images and expectations were adopted and reinterpreted by Christians, including the sacrificial language of the OT. Paul (Rom. 3: 25) asserts that God put forward Christ as a *sacrifice of atonement by his blood; this has sometimes been understood as meaning that God needed to be propitiated by the sacrificial blood of his Son. Such an interpretation, however, seems incompatible with justice—that someone innocent should be punished in place of a criminal. Others understand NT soteriology in terms of God's love and grace: God in Christ went to such lengths to manifest his love, and so win the response of repentance from human beings. The salvation promised by Christ is, as it were, a liberty to captives (Luke 4: 18), and some modern theologians would include in the captivity the enslaving forces of economics and conflict.

soul Used In the OT of the principle of life, as when God breathed into Adam (Gen. 2: 7), and thus 'soul' referred to the whole living human person. Later the book of Wisdom (9: 15) introduces the Greek view, in which the 'body' is distinct from the imperishable 'soul'. In the NT 'soul' means 'life'. So NRSV and REB of Luke 12: 20, but AV, RV, and NJB have 'soul'. The Greek idea of an immortal soul different from the body has come into the NT at 1 Pet. 1: 9 and retained a place in Christian theology. Paul's reference to 'spirit and soul and body' (1 Thess. 5: 23) does not imply a threefold division of the human person: 'soul and body' means a living, not a dead, body.

sources

1. *The OT*. Although the *Pentateuch reads well as a coherent and continuous document, closer examination reveals that it contains a number of sources. Some of these are named: e.g. 'the book of the wars of the Lord' (Num. 21: 14–15). Other items in the OT seem also to have existed independently, such as the lament of David (2 Sam. 1: 18). But critical scrutiny of the text of the *Pentateuch has disclosed clear evidence that several formerly separate sources now form the book we have today. It was noticed, for example, that the *creation narrative in Gen. closes at 2: 4a and another parallel narrative, different in style and content, takes over at 2: 4b. Similarly, in Gen. 6: 5–9: 19, the story of the *Flood, there are not only differences in style, but also contradictions and inconsistencies—e.g. seven pairs of animals are to be taken into the *ark (Gen. 7: 2) but only one pair in Gen. 6: 19.

The widely accepted hypothesis to account for these features in the Pentateuch fastens particularly on the different names used for God—*Yahweh* and *Elohim*. The sources have been given the symbols *J, *E, *D, and *P. J from Jahveh (the German form for Yahweh) is thought by many scholars to derive from the south of the country in the 10th cent. BCE; E (from *Elohim*) from the north in the 8th cent. BCE; D = Deuteronomy, associated with King *Josiah's discovery of the law book in the Temple (622 BCE); and P ('Priestly') is dated to the *Babylonian Exile or just after with a strong interest in the ritual of the *sanctuary and the emphasis on offerings being made by *priests of the tribe of *Levi. This four-document hypothesis is attributed to the German scholars *Graf and *Wellhausen, and it remains a springboard for all OT research. Nevertheless much of the hypothesis has been disputed; perhaps the most vulnerable part is the identification of a source E. It is held by some that the different names for God are not sufficient ground for positing different literary sources. Elsewhere in the OT it is widely agreed that Isa. 1–39 is an editorial compilation of sources, and certainly distinct from Isa. 40–66.

2. *The NT*. It is generally accepted that the close verbal relationships between the *synoptic gospels demand a hypothesis about the use of sources. Luke (1: 1) actually explains that the author knew of several predecessors. Although there is no universally agreed solution to the *synoptic problem, it is usual to hold that Mark was the first gospel to be written, and that Matt. and Luke embodied Mark in their longer works. The remaining material common to Matt. and Luke is explained either by their both having access to the same source ('*Q') of tradition, or (a minority view) that Luke knew Matt. as well as Mark.

There is a minority view that the gospel of John is a compilation of pre-existing sources, e.g. of the Signs or the Discourses, arranged under the heading of the Feasts. A pattern on the whole work has been imposed by an editor. But an objection to this view is that the work as a whole has a consistent and recognizable style. It is, however, probable that the author of John had available written sources of material, such as the accounts of the Feeding of the Multitude, used in John 6: 1–21.

sowing The scattering of seed. The practice assumed in the *Bible is that after seed of wheat and barley was scattered broadcast over the land it would be ploughed in (Matt. 13: 3–8).

Spain The Spanish peninsula was inhabited by several fierce and courageous tribes and it was not wholly subdued by the Romans until the time of the emperor *Augustus. He established some fifty colonies in the country, so that it became very Romanized and in time produced Latin poets and three Roman emperors. It was a natural missionary destination for Paul (Rom. 15: 24, 28), but whether or not he ever reached it is uncertain. The result of his appeal to *Caesar is unrecorded and the last years of his life are the subject of speculation and legend.

spices Perfumes and aromatic oils were sold and bought in *Tyre (Ezek. 27: 22), no doubt partly as luxuries, but also used for embalming and anointing a corpse (Mark 16: 1).

spikenard An expensive Indian plant; the value of the amount poured over the head of Jesus by the woman of *Bethany (Mark 14: 3) was equivalent to wages for almost a whole year.

spinning The preparation of fibres. Before weaving could take place, fibres of wool, flax, or hair (of camels or goats) had to be washed, combed, and spun. A mass of fibres was first hooked on to a distaff (a wooden rod) and then drawn out as thread and attached to the spindle (Prov. 31: 19), which was wound on until the spindle could hold no more. Next, the spun yarn was woven on a loom by means of intersecting flexible material. The shuttle, formed in the shape of a boat, held the bobbin of weft thread and was passed through the space made by keeping the warp threads apart (Job 7: 6). Finally, the weaver had to push each new thread down against the existing fabric with a batten, and on an upright loom the web was woven from the top down, which would have been the case with the seamless tunic of Jesus (John 19: 23).

Weaving was done by both men and women (Exod. 35: 35; 36: 6; Acts 18: 2–3).

spirit The Hebrew *ruach* means wind (Exod. 10: 13) or breath (Gen. 6: 17) or divine power (Ezek. 37: 9 f.) All imply something awesome. In the NT the Greek *pneuma* has a width of meaning—the spirit would come as fire and *judgement (Matt. 3: 11 f.), its coming on the *disciples at *Pentecost resembled fire (Acts 2: 3) but the sound was like wind (Greek, *pneuma*). Prefaced by the adjective 'holy', the reference is to the divine Spirit, or Spirit of God. As such it is not mentioned very much in the synoptic gospels, since in Christian belief this Spirit was not given before the Lord's *resurrection/glorification (John 7: 39). But at a few significant and exceptional moments the Spirit is said to come—at Jesus' conception (Luke 1: 35) and *baptism (Luke 3: 16 f.). He was driven by the Spirit into the *wilderness (Mark 1: 12). The Spirit is promised to the disciples (Mark 13: 11; Luke 11: 13) and the promise is fulfilled for the assembled Church at *Pentecost (Acts 2) and on individuals at the beginning of their new life (Gal. 3: 2 f.; Rom. 8: 9 ff. Cf. John 3: 3 ff.). The coming of the Spirit brought *joy (a word which sometimes seems equivalent to Spirit, as in Acts 8: 39) and a variety of gifts (1 Cor. 12: 4 ff.) for building up the Church. The Spirit reveals new truth to every generation (John 14: 26; 16: 12 f.) and it is by the Spirit that the believer is enabled to recognize who Jesus is (1 Cor. 12: 3) and to follow his example (2 Cor. 3: 18).

spirits in prison The dead in *Hades before the time of Christ. Belief in the *resurrection of Christ led Christians to speculate about his whereabouts between the *crucifixion and his resurrection on the third day. One suggestion was that he went to preach to those who had died and who were in the place appointed by God for the departed (1 Pet. 3: 19). This could be a *mythological way of affirming that Christ's *salvation was universal—it stretched both backwards in time and forwards. The 'descent into *hell' became a clause in the earliest Christian creeds.

spiritual gifts Endowments given by the Spirit. All Christians should show the fruit of the Spirit (Gal. 5: 22–3), but the gifts bestowed by the Spirit are adapted to each individual, and are listed by Paul in Rom. 12: 6–8 and in 1 Cor. 12: 8–10, 28–30. They comprise gifts of speech, both coherent utterance and also the unintelligible speaking in *tongues (though some present might have the complementary gift of interpretation); gifts of service and administration; and gifts of *healing. A further gift mentioned by Paul is apostleship, which builds up the Church by leadership and pastoral care. In later epistles (Eph. 4: 11 and 1 Pet. 4: 10–11) the gift of speaking in tongues (**glossolalia*) is omitted, possibly because experience taught the Church that words understood by all were preferable to sounds intelligible only to God (1 Cor. 14: 1–5).

spitting The worst kind of insult was to spit into someone's face (Num. 12: 14; Matt. 26: 67), but sometimes Jesus spat in the performance of healings (Mark 7: 33).

star of Bethlehem The star which led the *Magi or Wise Men to Bethlehem to pay homage to the infant Jesus (Matt. 2: 1–12). In so far as the story has any basis in historical fact, the 'star' is more likely to have been a conjunction of the planets Jupiter and Saturn in 7 CE, which could have alerted astrologers ('magi' has this meaning in Greek) to an important event. It is, however, strange that the Magi should only have travelled under the stars by night over such country; if they were indeed being led, it is odd that they needed to ask the way at the court of King *Herod (Matt. 2: 2). It is better to regard Matt.'s narrative as a **midrash* based on the reference to a star in Num. 24: 17 and to the coming of *Gentiles to Jerusalem in Isa. 60: 3.

In that case the story is an expression of the universal *salvation offered by Christ, proclaimed also at the end of the gospel (Matt. 28: 19). It could also be Matt.'s way of showing that the superstitious arts of astrologers yield to the truth and rationality of Christianity, which Luke has demonstrated by his account of the burning of magical books by converts at *Ephesus (Acts 19: 19).

stars In the ancient world of the Near East and in Graeco-Roman societies stars were held to be supernatural beings, and astrologers flourished, as in *Babylonia (Isa. 47: 13).

Astrology was rejected by *prophets (Jer. 10: 2) as an unhealthy foreign influence. When it appeared it was condemned, e.g. by *Amos (5: 26) and suppressed by King *Josiah (2 Kgs. 23: 4–14); it had been introduced during the reign of *Manasseh (687–642 BCE, 2 Kgs. 21: 3–5).

state A political *community organized under one government. As the biblical historians looked back, they understood Israel to have been a theocratic state; that is, the power of the monarchy was limited by the tradition of divine laws and the voice of the *prophets. Church and State were, so to speak, one community. In the era of the NT the Christian Church was a tiny group of communities, closely related to each other, and united by the authority of an apostle or his delegates, suspected by Jews, and denied privileges extended to Jews in the pagan Roman Empire. Christians, whether free citizens or slaves, were members of this great state and owed it responsibilities, as is recognized by Rom. 13: 1–7 and 1 Pet. 2: 13–17. The NT thus sets itself against political anarchy, but the courageous words of Peter that the disciples must obey God rather than men (Acts 5: 29) show that there are limits to Christian subservience to the secular authorities—reached by the time Rev. 13 was written and the Church was enduring persecution.

In subsequent centuries these two NT attitudes to the State have been in tension. It was especially so in the Protestant Churches, where the State was seen as a divinely ordained barrier against the all-pervasiveness of *sin, that obedience to the State was most stressed, though since the disastrously obsequious behaviour of the Protestant Churches during the Nazi regime (1933–45) in Germany, a much more critical stance has become predominant.

stater A Greek silver coin; its value could be reckoned as the wages for four days' labour. It was worth four *drachmas and would have been the appropriate tax payable by two persons to the Temple as required by the *Law of Moses (Exod. 30: 11–16). Refusal to pay it would have been regarded as implying withdrawal from the life of *Judaism, and would have been a controversial issue for Jewish Christians after their conversion. The *miracle story of Matt. 17: 24–7, in which a fish is caught with a stater in its mouth, though hardly historical, was a recognition for Jewish Christians who read Matthew's gospel that they had been permitted to pay the tax to the Temple while it was still standing as instructed in the Law (Exod. 30: 13; 38: 26). But after its destruction in 70 CE the tax was transferred to the temple of Jupiter in Rome. This was the situation when the gospel of Matthew was written.

The words of Jesus 'then the sons are free' from the obligation to pay the tax suggests that the Christians were not part of the Jewish community with responsibilities for the Temple. But the instruction to Peter to pay it all the same has the implication that Matt.'s Church should show loyalty to the empire.

Stephanas The first person to be converted by Paul at *Corinth (1 Cor. 16: 15).

Stephen The most notable of the Seven (Acts 6: 5), and the Church's first *martyr. Sometimes called a '*deacon' on the basis of Acts 6: 1–2, though the noun is not used of him in the NT and any link with the later office in the Church of deacon is unlikely. When there was a dispute between 'Hellenists' (Jewish Christians from abroad who also spoke Greek) and the Hebrews (Palestinian Jewish Christians who spoke Aramaic) about fair shares in charitable gifts, the apostles appointed seven Hellenists to serve or organize (Greek, *diakonein*) the distribution, *Stephen among them. However, so far from undertaking any subordinate job, Stephen instantly became a Christian propagandist —though the *Twelve had appointed the *Seven precisely in order that they themselves should have more time for preaching. Stephen's active role brought him into conflict with the authorities; apparently he had claimed that Jesus as *Messiah had rendered the Temple and its worship obsolete. He is credited with a speech accusing the Jews of constantly rejecting those whom God had sent to them, and the mob set upon him and lynched him. The author of Acts describes Stephen's death in terms recalling that of Jesus; he died, with words of forgiveness on his lips. The immediate result was the dispersion of disciples to evangelize elsewhere; this included *Philip, the only other well-known member of the team of Seven.

steward Someone who looks after property in business (1 Kgs. 16: 9) or a city treasury (Rom. 16: 23). Jesus told a *parable (Luke 16: 1–8) of the Unjust Steward ('manager', NRSV). The word is also used in a metaphorical sense; a steward of divine mysteries (1 Cor. 4: 1) is one who has a divine commission in charge of the revelation. So Paul saw himself as a steward (1 Cor. 9: 17) who would one day have to give an account of his stewardship (1 Cor. 4: 3–4).

stocks A public instrument of *punishment in which the victim was pinned down by the legs (Job 13: 27) and sometimes fastened with shackles (Acts 16: 24, 26) which caused considerable pain.

Stoics Followers of the philosopher Zeno (335–263 BCE), who lived in *Athens and taught within a colonnade (Greek, *stoa*). They were numerous in NT times among Roman politicians and orators and emphasized the cosmopolitan nature of mankind, teaching that all are brothers and sisters in the world, but that the lot here of everyone is predetermined. The most effective means of dealing with this fate was to control one's own passions and ambitions so that inevitable external events had the least painful results. Stoicism was a philosophy of life which called for inner discipline, and had points of contact both with Jewish Wisdom literature (e.g. Wisd. of Sol. 7: 22–6) and with Christian teaching. For example, Stoics held that the universe was inspired by a divine **Logos* or Word (cf. John 1: 1–3), and there are parallels between the *Household Codes of the epistles and ethical teaching of the Stoics. The difference is that the ideal for Christians was to include a warmth and love which the Stoics would have regarded as rather reprehensible. Christian *martyrdom out of loyalty and love for Jesus was different from the steely courage to endure of the Stoics. According to Acts 17: 22–31 Paul encountered Stoic teachers at *Athens, and he made use of popular Stoic philosophy in his sermon there.

stone(s) Used in Palestine for building, for fighting, for executions, and even as a pillow (for Jacob, Gen. 28: 18). Knives and tools were made of stone. Stones were also placed to mark boundaries and landmarks (Josh. 15: 6; Deut. 19: 14) and to serve as *altars (Gen. 28: 18). Metaphorically, hard hearts are described as stones (Ezek. 36: 26). 'He that is wounded in the stones' (Deut. 23: 1, AV, RV; 'testicles', NRSV, REB, NJB) was excluded from the assembly.

stoning Capital *punishment for *blasphemy (Lev. 24: 15–16) or *idolatry (Deut. 17: 2–7) and other crimes. It took place outside a city and the victim was pelted with stones by the whole community (Deut. 13: 10). The death of *Stephen was organized in this way (Acts 7: 58), the Jews possibly taking the law into their own hands during the temporary absence of the Roman governor. Paul was stoned and left for dead at *Lystra (Acts 14: 19). It is recorded by *Josephus and Eusebius the Church historian that shortly before the fall of Jerusalem in 70 CE *scribes and *Pharisees hurled *James the Lord's brother from a pinnacle of the Temple and dropped heavy stones on him. It is disputed whether these stonings were a form of lynch law or whether the Roman authorities tolerated the killing of those who had broken the religious laws of Judaism.

storms About twice a year fierce rainstorms are expected in Palestine, with thunder in November–December. The Sea of *Galilee is subject to violent squalls (Mark 4: 37 f.).

stoves Metal braziers used by the wealthy, such as the one in the court of the *high priest, by which Peter warmed himself (Mark 14: 54).

stranger The stranger, or *sojourner (AV) of OT times, was a resident foreigner, who was given legal protection (Deut. 1: 16) but required to observe Israelite law (Lev. 16: 29). By the 1st cent. CE the term and the regulations were applied to proselytes, *Gentiles converted to full acceptance of *Judaism (Acts 2: 10; cf. Matt. 23: 15), or sometimes just to a person unknown (John 10: 5).

strangled animals Jewish *Law does not allow animals to be choked to death and requires slaughtering by cutting the throat and draining the blood, which is not to be consumed (for 'the blood is the life', Lev. 17: 14) even by *Gentiles resident within the *community (Lev. 17: 10–14). This regulation became part of the decree issued at the *Council of Jerusalem (Acts 15: 20) in order not to impair table fellowship between Jewish and Gentile Christians.

Strauss, David Friedrich (1808–74) German theologian, pupil of F. C. *Baur in Tübingen. His *Life of Jesus*, translated into English by the novelist George Eliot, suggested that there was little of historical value in the gospels, which were the product of the creative mind of the primitive Church working on narratives of the OT.

straw Used by the Israelites enslaved in *Egypt for binding mud bricks, but to make conditions harder for the *slaves, straw was later withheld (Exod. 5: 7–18). Straw was also valuable as fodder (1 Kgs. 4: 28) and the chaff (husks) was separated by winnowing from the edible corn and blown away. Using harvest as a symbol of the final judgment, *John the Baptist says that the 'chaff' will be destroyed (Matt. 3: 12), adding the method, that it will be by 'fire' (a hint of *Gehenna, Matt. 18: 9) which was not part of the agricultural process in that age.

Streeter, Bernett Hillman (1874–1937). English NT scholar, Provost of the Queen's College, Oxford, killed in an air crash, 1937. His *Four Gospels* (1924) remained for a long time the classical exposition of the view that Mark was the earliest gospel, that Matthew and Luke both made use of a source designated by the symbol *Q, and that in the case of Luke this source was first combined with his own material (a first draft of the gospel, '*Proto-Luke') and later supplemented by the use of Mark.

structuralism One facet of a larger movement in modern literary criticism, associated particularly with Claude Lévi-Strauss, a philosophical anthropologist, and now applied to biblical studies. It suggests that critical attention be focused less on evidence about authors or editors of the text but more on the text as it stands, and what it conveys to the readers. For texts have only a relational, not an essential, meaning. Every word fits into a complex pattern of binary oppositions, and it is this structure which gives it meaning. Structuralism is indifferent to historical investigation but, assuming the integrity of the text, analyses the structure of the human mind and group which have produced it. For basic human patterns form themselves into structures and are expressed in texts. A biblical text is therefore to be read and grasped as a whole and is more than a compilation of parts or sources. By it the basic convictions and conditions of the author are transmitted to the reader.

Critics of the movement argue that some knowledge of the authors of the biblical books is always helpful; and because most of the books are composite, a knowledge of the history of their composition remains indispensable.

stumbling-block The Greek word is *skandalon* (1 Cor. 1: 23) and means an obstacle which could cause a fall (Lev. 19: 14) or a fundamental difficulty for belief (Gal. 5: 11). It is wrong for convinced believers to put a stumbling-block in the way of less assured or hesitant believers (Rom. 14: 13). Some modern theologians have urged that elements in traditional Christian faith which are incredible to the modern educated mind should be reinterpreted, so that the real *skandalon* to belief in Christ should be made plain, rather than antiquated irrelevancies.

succession narrative Chs. 9–20 of 2 Sam. and chs. 1 and 2 of 1 Kgs. A historical document centred on the court of King *David. The narrative has non-historical elements when the author includes lively personal dialogues, but his main interest is political. He demonstrates the legitimacy of David's regime, defends its politics, and is exercised about who should succeed him.

Succoth (1) A place where *Jacob built a house; it was east of the *Jordan (Gen. 33: 17). (2) A place where the Israelites stayed after journeying from *Rameses after leaving *Egypt (Exod. 12: 37).

The word means 'booths' and gave its name to the autumn feast known as *Tabernacles.

suffering For both *Judaism and Christianity belief in the goodness of God has made the universal experience of suffering the supreme problem for theology. Broadly, two kinds of suffering are recognized in the Bible: that which comes upon us because of our humanity; and that which is visited upon the people of God because of their *faith (as Jews or Christians). The first category embraces illness, bereavement, anxiety, depression; the second is caused by persecution or by personal discipline. Several OT texts suggest that some human suffering comes as a result of *sin (Gen. 3: 14–19); the Exile is interpreted as punishment for the nation's sins (2 Kgs. 17). Job's 'comforters' are persuaded that he must have done some great wickedness to have deserved such suffering; *Job rejects that solution (cf. John 9: 3), and in Job 38–41 he receives an answer directly from God which is overwhelming in its power: the problem is beyond human understanding, and Job must relax and leave it to the *wisdom of God. *Apocalyptic writers took the view that undeserved suffering would receive its compensation in a state after death (Dan. 12: 2).

The second category is of suffering accepted for the sake of good, or even embraced, as did the Suffering Servant of *Deutero-Isaiah. In the NT followers of Christ are warned to expect to take up a *cross 'daily' (Luke 9: 23), and to suffer hostility and persecution. It is the message especially in 1 Pet. Such suffering is not only an imitation of Jesus; it is said to be a means of completing what is lacking in the Church's burden of suffering (Col. 1: 24).

Sumer Modern S. Iraq; *Ur was a principal city. The Sumerians' myths of creation and human civilization including a story of a *Flood are related to those of Gen. 1 to 11. The Sumerians developed a system of law and government and devised measurements of time and distance and their own calligraphy. They invented *cuneiform writing, and many Sumerian tablets have survived. The victories of King *Hammurabi of Babylon in about 1750 BCE brought Sumerian predominance to an end.

sun References to the sun in the *Bible include its beneficial effects in agriculture (Deut. 33: 14) and also the unpleasant injuries it causes (Ps. 121: 6). It symbolizes the *glory of God (Rev. 1: 16) but on the *day of the Lord the sun will be eclipsed (Joel 2: 10; Rev. 6: 12).

Worship of the sun was common in the ancient world and was strictly forbidden to Israelites.

Sunday For Christians, the first day of the week. The Greeks and Romans knew of seven planets, and named the days of the week after them: *dies solis* was one day, as was *dies martis*, etc. Christians at first used the same names as the Jews, so that the last day of the week was the Sabbath and Friday was the eve of the *Sabbath. From the 4th cent. CE the Church adopted the names of the planetary week and the rich symbolism surrounding the sun and its light could be appropriate to the Church which worshipped Christ, the true sun, on 'the Lord's Day', which had been the name used in the NT period (Rev. 1: 10). The *Easter narratives of the gospels (Mark 16: 2; Matt. 28: 1; Luke 24: 1; John 20: 1) all speak of the 'first day of the week'. A week later, the eighth day (John 20: 26) was the next Sunday.

Christians first began to worship on Sunday rather than on the Sabbath quite soon. This is suggested by the description of Paul at Troas (Acts 20: 7–12), confirmed by the *Didache (14: 1) and by *Ignatius. The communities probably met in the evening, following the Easter traditions (e.g. Luke 24: 29).

superscription The title which *Pilate ordered to be attached to the *cross of Jesus (Luke 23: 38; John 19: 19) or fastened round his neck. The word is also used for the words written on a coin (Mark 12: 16).

superstitious The Greek word in Acts 17: 22 can have a good or bad sense. AV 'too superstitious' and RV 'somewhat superstitious' imply the latter, but modern translations prefer 'extremely religious' or 'extremely scrupulous'.

Susanna, book of An addition to Daniel written just before 100 BCE in Hebrew or Aramaic but only known in the *LXX version. It is regarded as deutero-canonical scripture by Catholics and attached to Daniel as ch. 13 (so NJB). By Protestants the book is placed in the *Apocrypha. The story is of the virtuous Susanna who refuses the advances of two *elders; they in revenge bring her to trial in court on a false charge. Susanna is condemned to death, but in response to her prayer, God sends the young *Daniel to her aid. In separate cross-examinations, he shows that the elders' evidence is contradictory and unreliable. So she is released; they are sentenced to death. The purpose of the story is to illustrate how God sends help to his own.

symbols A word, name, person, or action which is related to what it symbolizes; it is not an arbitrary or conventional sign (as are, for example, the colours of traffic lights). It is not itself a biblical word, but symbols proliferate in the Bible. A symbol points to a reality which is not directly knowable and therefore has to be revealed indirectly by the symbol. In the OT God's presence is symbolized by fire and *cloud (Exod. 13: 21–2) and Israel's *covenant relationship with God is symbolized by the institution of the *Sabbath and *circumcision.

A clutch of symbols was associated with the feast of *Passover; unleavened bread was eaten, as a symbol of the *Exodus, reminiscent of the haste of the Israelites' departure; they did not have time even to collect food for the journey or wait for the dough to rise. The prophets used symbolic actions to convey their message—*Isaiah walked barefoot (Isa. 20) and *Hosea gave his children symbolic names (Hos. 1: 6, 9).

Places were often symbols—the Temple and its furnishings were symbols of holiness and the *ephod and the *breastpiece symbolized the dignity of the priesthood (Exod. 28–9). Certain numbers had acquired symbolic significance: *twelve, because of the number of tribes; *seven, as being the days of creation, indicating completeness; forty (Gen. 7: 12) became a round figure for a period of time. In the NT one and eight, being the days of resurrection, are symbols, and three, seven, and twelve seem to be important and to convey something more than the literal: hence the 'twelve apostles' was a symbol that a new people of God had been formed. In Rev. (13: 18) the famous number 666 on the beast is a symbol of *Nero. *Baptism with water is a symbol of newness of life in Christ and the *Lord's Supper with bread and wine both of the death of Christ in the past and the Messianic Banquet in the future Kingdom.

Modern scholarship is divided about what is strictly factual and what is mainly symbolical in the NT. Some who stress the historical value of the narratives would, however, deny some of the miraculous elements (e.g. in the story of the multiplication of the loaves, Mark 6: 30–44). Other scholars who have argued for a strong element of symbolism (e.g. Augustine in the 5th cent. and Austin *Farrer in the 20th cent.) suggested that the first narrative of the loaves (Mark 6) implied a message for Jews and the second (Mark 8) sanctioned the mission to the Gentiles but nevertheless held strongly to the credibility of miracles in the gospels as supernaturally caused.

synagogue Jewish synagogues are not mentioned in the OT but were well established by the NT era. They were possibly founded in *Babylon during the Exile in default of the Temple, though many scholars believe they originated in the Greek-speaking *Diaspora. They may have been at first community centres for Jews, to which regular worship was later added, and they were therefore a revolutionary concept. Sacrificial worship gave way to prayer, homilies, and education—worship in which priests were less essential than the readings from the *Torah. Synagogues were increasingly important during the Maccabean revolt (1 Macc. 3: 48), but it is probable that meetings took place in private houses rather than in designated buildings. Even in and around Jerusalem synagogues existed, but they probably had fewer functions than those in more remote places. In the capital, religious needs were still mainly catered for by the Temple.

After the fall of Jerusalem (70 CE) rabbinic Judaism required buildings for *teaching and *worship, sacred meals and legal proceedings, and by the 3rd cent. CE these were plentiful. They were often decorated with mosaics and reliefs. Many remains have been discovered by archaeologists, especially in *Galilee. A few may be of the 1st cent. Worship included scripture-reading, especially from the *Pentateuch, though the prophets were not omitted (Luke 4: 17); interpretation might be given; psalms were recited. Some scholars have held that the regular synagogue lectionary was already in existence in NT times, and may have influenced the structure of the gospels. Synagogues had a presiding minister and sometimes priests held office, though laymen were permitted to speak if qualified, by being able, for example, to translate from Hebrew into the vernacular.

The congregation sat on benches round the walls, a position which facilitated dialogue.

The word 'synagogue' is used once in the NT (Jas. 2: 2; NJB) for a Christian assembly or meeting (REB).

synchronic As opposed to *diachronic exegesis, synchronic is interested in complete biblical texts as they exist now and in the response of the reader to that text. This response may be from a position of faith or of non-faith.

syncretism A combination of two religions, as in Israel when the worship of *Yahweh, who had led them through the *wilderness, became infiltrated, according to the biblical narratives, with beliefs and practices of the *Canaanites, who worshipped the gods **El* and **Baal*. Such mixtures were sternly resisted by those prophets who looked back to the wilderness period as an ideal (Hos. 13: 4–5; Amos 2: 10–12). Foreign cults were introduced in the reign of Solomon, but from the time of the *Exile such influences were excluded; Jewish identity was safeguarded; syncretism was unacceptable, though Hellenization in the last two centuries BCE threatened the traditions and integrity of *Judaism.

Christianity related itself to its surroundings but did no more than borrow terminology from *Gnosticism, *Mithraism, and Platonic philosophy. The theory that Paul blended primitive Christianity with pagan mystery religions of a dying and rising god has had powerful and attractive advocates but is difficult to sustain.

synopsis Term used in NT studies for a book showing the three synoptic gospels in parallel columns, thus making it clear exactly where they overlap and where they diverge.

synoptists The generic name given to writers of the three gospels, of Matthew, Mark, and Luke, since these gospels can be printed in three parallel columns and compared, whereas the gospel of John stands apart in chronology, content, and theology. Because the three synoptists have so much in common, it has been a major task of NT critics to try to solve the 'synoptic problem' and to determine their relationships. Who borrowed from whom? The most widely accepted theory is that Mark was written first and that Matt. and Luke reproduced Mark in their gospels. This still leaves a great deal of material in Matt. and Luke which is not in Mark, and the majority view is that both evangelists had access to a body of Jesus' teaching ('*Q'), which they added to the material they had taken from Mark; this theory is called the Two Document Hypothesis. Finally, both Matt. and Luke had independent traditions of Jesus' birth and infancy which they prefaced to their gospels and material about the *resurrection at the end, as well as some other items (e.g. in Matt.'s Passion Narrative). The examination of the different orders and the verbal alterations in the common material enables *Redaction Critics to formulate the theological stance of each evangelist. Scholars who reject the Q hypothesis propose what they maintain is a simpler solution, that Luke knew and made use of both Mark and Matthew. The *Griesbach hypothesis, that Mark is an abbreviation of Matt. and Luke, has also won a few modern advocates.

Syntyche A believer in *Philippi who was in dispute with Euodia; Paul urges a reconciliation (Phil. 4: 2–3).

Syracuse Originally a city founded by *Corinth, but taken over by the Romans in 212 BCE. It lies on the SE coast of Sicily and the harbour gave Paul's ship, with its figurehead of *Castor and Pollux, three days' break on the journey to *Rome.

Syria In biblical history the term applies to the region round *Damascus, north of Israel, with the Meditarranean Sea on the west, known in the OT as Aram. *David conquered it (2 Sam. 8: 5 f.) but when Israel and *Judah split the territory became independent and repeatedly harassed Israel; on one occasion, however, Syria and Israel united in order to coerce Judah (Isa. 7). Assyria sacked Damascus in 732 BCE. Syria then ceased to be of importance until the time of the Greek (and later the Roman) empires. Its leading city, *Antioch, on the River Orontes, became the capital of the Seleucid Empire against which the *Maccabeans revolted in 165 BCE. The Romans made Syria into a province, and Christianity soon spread to Antioch (Acts 11: 19–26). Here members of the Church were first called Christians, and here Peter and Paul had a painful confrontation (Gal. 2: 11). Early in the 2nd cent. its bishop was *Ignatius.

Syrophoenician A *Gentile woman who asked Jesus to heal her daughter (Mark 7: 26); she came from the area round *Tyre and *Sidon and in Matt. (15: 22) she is called a Canaanite, which fitted the location. It is the only explicit healing in the gospel of Mark of a Gentile (though the demoniac Legion was roaming in Gentile country, Mark 5: 9), and could have been regarded by the early Church as Jesus' implicit authorization of its mission to the Gentiles, even though Jesus' own *ministry was restricted to Jews (Matt. 10: 5). The woman was probably a well-to-do Hellenized Phoenician, a member of the more privileged social group: her daughter lay on a decent 'bed' (Mark 7: 30), not a miserable 'mat' (Mark 2: 4).

Syrtis Two great gulfs on the north coast of *Africa, south of Sicily, much feared by sailors on account of treacherous sand banks (Acts 27: 17) and strong winds.

T

tabernacle A portable *sanctuary. There are instructions to build a tabernacle in Exod. (25: 8 ff.) and it is duly built (Exod. 35: 10 ff.). When completed it consisted of two compartments, the inner being the *Holy of Holies containing the *Ark. The narrative puts the episode, improbably, into the wilderness period, and describes how *Levites had to take down and reassemble this complicated structure every time the people moved on (Num. 1: 51). The amounts of gold and silver used sound incredible; and surely when the people are said to cross the River *Jordan (Josh. 3: 17), there would have been mention of the tabernacle if it was being carried.

Since this account forms part of the *P source, it seems likely that details of the Jerusalem Temple have been transposed into the description of the wilderness tabernacle. But not all the narrative is *retrojection. There is also a 'tent of meeting' described in Exod. (33: 7–11) which appears to be an alternative account of the tabernacle and is probably more in accordance with the facts. The Priestly source has perhaps coalesced the memories of a *wilderness sanctuary with a theological and idealized concept based on knowledge of the Jerusalem Temple. The 'tent of meeting' was a simple, easily movable, structure in which a single minister could operate. According to *P this is the location for the divine presence, or the **shekhinah*. (Exod. 40: 34–8).

In the NT there are references to the tabernacle in Heb. (8: 2, 5; 9: 21)—the true tabernacle is in *heaven—and in *Stephen's speech (Acts 7: 44–50), where Solomon's Temple is contrasted unfavourably with the ancient tabernacle made according to the pattern of the one in heaven. It is also possible that the Exod. narrative lies behind John 1: 14—where the Greek is literally 'the Word . . . tabernacled among us'.

Tabernacles, feast of An autumn festival of ingathering lasting eight days; people were instructed to 'dwell in *succoth*' (Lev. 23: 42; 'booths' is an alternative name for the feast) as a memorial and *thanksgiving for the safety of the tribes in the wilderness. The Court of Women in Herod's Temple was lit for seven days by giant candelabra, and this may have been the reason why John (8: 12 ff.) placed Jesus' discourse ('I am the light') in the context of the feast of Tabernacles.

Tabitha The Aramaic form of the Greek *Dorcas (Acts 9: 36), meaning 'gazelle', the name of a woman much given to works of charity. After illness and death, she was restored to life by the *prayer of Peter (Acts 9: 40).

tables (or tablets) (1) The 'tables of the law' (Exod. 24: 12) were the two stones on which the *Ten Commandments were carved on Mount *Sinai. (2) Writing tablets might be made of wood covered with wax (Isa. 30: 8; Luke 1: 63).

Tabor, Mount In the plain of *Esdraelon, and the scene in the OT of *Barak's assembly of his forces (Judg. 4: 6) and of the *death of *Gideon's brothers (Judg. 8: 18). Although not mentioned in the NT, it has traditionally, but improbably, been regarded as the site of Jesus' *transfiguration (Mark 9: 2). Tabor was not a 'high mountain'; Mount *Hermon, near *Caesarea Philippi, seems the more probable location.

tabret A musical instrument (1 Sam. 10: 5, AV), called a 'tambourine' (NRSV, NJB), a 'drum' (REB).

Tahpanhes A city in *Egypt to which Jews fled in 586 BCE from the wrath of the

Babylonians, taking *Jeremiah with them (Jer. 43: 7–8). In Greek the city was called Daphnae.

tale The word used as a technical term in *Form Criticism as the English translation of German *Novelle*—nine *miracle stories in Mark and five in John.

talent The Greek talent (Matt. 18: 24) was a sum of coins equivalent to 6,000 *drachmas (Luke 19: 13, NRSV; 'ten pounds', REB, NJB). 'Ten thousand talents' is a vast sum; the total taxation paid by the entire area of *Judaea, *Idumea, *Samaria, *Galilee, and *Perea in one year amounted only to 800 talents.

talitha cumi The command in Aramaic addressed to the 12-year-old daughter of *Jairus 'to get up'; for 'the child is not dead but sleeping' (Mark 5: 39).

Talmud Hebrew for '*teaching'; the name is given to two collections of Jewish commentaries on the *Mishnah, the Palestinian and the Babylonian. This commentary is called *Gemara* and is essentially analysis of the words of the Mishnah which is the central writing of *Judaism, dealing with agricultural and family life, mostly written in Aramaic. The Palestinian Talmud was completed by about 450 CE. The Babylonian, completed a century later, is very much longer and embodies many quotations from the OT.

Tammuz A Mesopotamian deity of the underworld, sometimes thought to have been the focus of a cult which worshipped him as a dying and rising god—mourned in autumn when vegetation seemed to die, and celebrated in spring as the crops began to grow. The cult sometimes endangered the purity of worship in Israel, as when *Ezekiel saw women weeping for Tammuz in the Temple (Ezek. 8: 14). However, this interpretation of Tammuz is speculative; the women may well have been engaged only in a mourning rite.

tanner One who prepared animals' hides for use as *leather for buckets and wineskins —and for the coverings for the tent of meeting (Exod. 36: 14). Some animals were 'unclean' for Jews, and this, added to the unpleasant smells of the business, forced tanners to live outside the main towns. Simon the tanner (Acts 9: 43) lived by the sea (Acts 10: 6), and Peter overcame his inhibitions to lodge with Simon.

tares Used in AV, RV for 'weeds' which resembled wheat in early growth. In the parable (Matt. 13: 36–43) the interpretation given regards the tares as unbelievers who may mix with believers but who will ultimately be judged by God.

targums Translations of the Hebrew OT into Aramaic to assist Jews in Palestine where Aramaic had become the language of daily speech; made just before and in the first three centuries of the CE. The translations often include interpretations. *Targums* are especially valuable in determining the original Hebrew text. They are also important for NT studies, since early *targums* may have been known to Jesus, Paul, and the evangelists, and could provide clues to the meaning of some NT texts, e.g. the exposition of Ps. 68: 18 in Eph. 4: 8.

Tarshish A port on the Mediterranean to which Jonah was fleeing (Jonah 1: 3), possibly in *Spain or *Cyprus (Isa. 23: 1)—though the cargo of *gold, *silver, *ivory, apes, and peacocks brought by the ships of Tarshish to King *Solomon (1 Kgs. 10: 22) argues a more exotic region. There is a dispute whether 'ships of Tarshish' (e.g. 1 Kgs. 22: 48) meant ships of a particular type or ships whose destination was Tarshish.

Tarsus 'No mean city' (Acts 21: 39), 16 km. (10 miles) up the River Cydnus in SE Asia Minor, the centre of cross-currents of commerce and culture and the residence of several distinguished Stoic philosophers. Paul is said (Acts 9: 11) to have been born in the Jewish community at Tarsus, but he spoke Greek like everyone else.

Tatian (120–73 CE) A Christian *Apologist in the east, who was not well received in the west. He was educated in *Greek philosophy but regarded Greek civilization as *evil. He compiled a life of Christ by weaving together the four gospels as used in the Syrian Church; this is called the Diatessaron. Its publication in the 2nd cent. (about 160 CE) is evidence of the widespread authority of the four gospels in the Church at that time.

taw The last letter of the Hebrew alphabet, and written as X in the old *Canaanite script (Ezek. 9: 4). It is the mark of the redeemed in Rev. 7: 3–4, and so became a special symbol in Christianity.

Taverns, Three A village on the *Appian Way about 53 km. (33 miles) south of *Rome, where Paul was met by Christians from the capital (Acts 28: 15).

taxes In ancient Israel the only taxes imposed on people mentioned in the Bible were for the maintenance of the *tabernacle and its officiants (Deut. 18: 1 ff.) but kings exacted much more in order to pay the army and the court (1 Kgs. 12: 11). In Palestine under the Romans a poll tax had to be paid to *Caesar (Matt. 22: 17). Jews paid a temple tax (Matt. 17: 24) of a half-*shekel every year, and also had to pay customs duties (Mark 2: 14) at *Jericho, since a frontier post which lay between the Roman province of *Judah and *Perea, part of the tetrarchy of Herod *Antipas. It provided an important source of his revenue. It was at *Jericho that Jesus encountered *Zacchaeus (Luke 19: 1–10), an official hugely unpopular because, as a subcontractor, he took the *money and remitted it to his employer, who in turn sent it on to *Rome. This procedure allowed much scope for greed and dishonesty. The collectors were known as *publicani*, hence the English 'publicans' of AV, RV.

teacher of righteousness The leader of the *Qumran community who interpreted scripture and was opposed by a 'wicked priest'. The identity and history of the teacher can only be surmised: possibly he was once a *high priest in Jerusalem and was opposed to the Hellenizing party.

teaching Most religious, moral, and cultural *education in Israel took place in the home, or in *synagogues when they had been established; but for a few there may have been schools. Preparation for civil servants was provided by *scribes attached to the royal court. Jesus taught by word of mouth (John 3: 2), as did Paul (Acts 18: 4–7). Teaching is mentioned as one of the variety of gifts in the Church (1 Cor. 12: 28).

Teman One of the sons of *Eliphaz, a son of *Esau, regarded as ancestor of the *Edomites (Gen. 36: 11). The tribe was sufficiently important to stand for the whole of Edom (Amos 1: 12), though it apparently occupied only the northern section of the country (Ezek. 25: 13). Its inhabitants had a reputation for *wisdom (Jer. 49: 7; Job 2: 11).

Temple The central *sanctuary. Plans for a permanent site for national *worship in Jerusalem were mooted by *David and executed by his son *Solomon (2 Sam. 24: 18 ff.; 1 Kgs. 6–7). The Temple was rectangular in shape and it had two courtyards : the inner one contained a bronze *altar (2 Chron. 4: 1). There was an immense basin, or laver, for ritual washings; two detached pillars called *Jachin and *Boaz stood at the entrance to the building itself. The inmost part of the building was the '*holy of holies'. Treasure accumulated in the Temple, and it was raided by *foreigners (1 Kgs. 14: 26) and by the kings of *Judah themselves (2 Kgs. 16: 8). Josiah repaired the building; the Babylonians destroyed it (586 BCE). *Ezekiel (40–2) had a *vision of a new Temple (571 BCE) based on the plan of Solomon's Temple, but it was never built. The exiles who returned to Jerusalem did manage to build the second Temple, which was smaller than Solomon's but lasted for 500 years. What *Herod began to do in 19 BCE was a grandiose development of the site during which the second Temple was not destroyed and *sacrifices were never interrupted. The reconstruction was still in process during the life of Jesus. It was finished in 64 CE only to be destroyed at the end of the Jewish Revolt, in 70 CE. (The foundations alone were left, since these could have been useful if the Romans had decided to erect a pagan temple on the site.)

Herod's Temple was enclosed by massive walls. An outer court was open for teachers, for public debate, and for the business of the money changers (for the Temple coinage, Matt. 21: 12). The Temple treasury was kept in the coinage of *Tyre, which was one of the most stable currencies of the time. This is a surprising custom, since the coins of Tyre bore a representation of the god Melkart. Notices in Latin and Greek warned *Gentiles not to venture beyond this outer court on pain of summary execution.

The next court was the Court of Women (Mark 12: 41), then the Court of Israel (for men) and the Priests' Court, with an *altar. Inside the Temple building itself a curtain (Mark 15: 38) separated the holy place from the 'Holy of Holies' which the high priests entered alone on the Day of *Atonement. The whole building complex occupied no less than a quarter of the area of the city.

In the *infancy narratives Jesus was brought to the Temple as a child (Luke 2: 22) and again in adolescence (Luke 2: 42). At the end of his ministry, he cleansed the outer court of commercial activities, thereby removing from it obstructions that made *prayer and *worship impossible for Gentiles (Mark 11: 15–19). This was in the prophetic tradition of symbolic actions, and John (2: 19–22) regards the cleansing as a 'sign' of Jesus' coming death. Jesus' prophecy of the destruction of the Temple (Mark 13: 2) seems to be accurately reported, since in the event the Temple was destroyed by the Romans by *fire*, and there are still many stones remaining on one another. The prophecy (also Matt. 24: 2) has not therefore been influenced by what happened. The destruction of the Temple would open the way for the revelation of God to all nations, and the Jewish Temple would be replaced by the Church as the new Temple (1 Cor. 3: 16 f.; Eph. 2: 19 ff.) The Church is a sanctuary and those who worship as Christians could be called its '*priests' (1 Pet. 2: 4 ff.).

temptation The primary meaning in the Bible is of testing. God tested the reality of *Abraham's belief by the command to *sacrifice his son *Isaac (Gen. 22: 1–9). But, secondly, the word acquires its more usual modern meaning of enticement to *sin (which does not come from God, Jas. 1: 12–15). The petition in the *Lord's Prayer that we be not led into temptation is a case of the primary meaning: that we may be spared the ultimate test of *apostasy. It is also the meaning of Jesus' temptations in the *wilderness, when he was put to the test after his *baptism: was he really prepared to be a *Messiah—*Son of God in the terms of the Suffering Servant? The brief narrative in Mark (1: 12 f.) with a reference to wild *beasts coming to Jesus (paradise regained?) is expanded by Matt. (4: 1–11) and Luke (4: 1–13) into a dramatic and pictorial story; there is no mountain in existence from which all the kingdoms of the world can be surveyed (Matt. 4: 8)! It is possible that the narratives in Matt. and Luke had the emperor Caligula in mind as a model for *Satan. Caligula did in fact grant a kingdom to a Jewish ruler, *Agrippa I, and at his court the etiquette of prostration was initiated—much to Jewish consternation. Cf. Matt. 4: 9. The narratives specify that the temptations came after forty days of fasting, and, supported by the quotations from Deut., it is plain that the evangelists interpret Jesus' temptations as a reversal of Israel's temptations in the wilderness after the *Exodus. There the Israelites repeatedly put God to the test (e.g. Deut. 6: 16). 'If he brought us out of Egypt, why does he let us suffer these hardships?' But the temptations to which they succumbed in forty years are now successfully resisted by one individual, and God's purpose is being fulfilled. Jesus refuses to use his status as Son of God, recently established at his baptism (Matt. 3: 17), to exercise a worldly sovereignty or to save his own life (Matt. 4: 5). Cf. Matt. 27: 42–3.

Ten Commandments The *Law given to Moses according to the OT at Exod. 20: 1–17 and Deut. 5: 6–21, though in fact developed over centuries. It served as the foundation principles for the *community's life. The Commandments, also known as the *Decalogue, prohibit idolatry and call for a total cessation of work on the Sabbath. They demand respect for individuals and for the life of the nation. They provide an ethical basis for a secure existence. Their influence on both Jewish and Christian communities has been enormous.

tents Made of goats' hair, supported by poles, and fastened to the ground by pegs (Judg. 4: 21). Whole households might occupy a single large tent (Exod. 33: 8), but the tents of kings could be magnificent (Dan. 11: 45).

Paul had been trained to make tents, probably out of *leather, and when he met *Aquila and Priscilla, who had the same occupation, he stayed with them in *Corinth (Acts 18: 3).

teraphim Statues of household gods; legal as well as religious importance was attached to possessing them, hence the fury of *Laban when he discovered that his daughter *Rachel had stolen his gods (Gen. 31: 19). They were part of the equipment of the personal shrine of *Micaiah, who was much put out when certain Danites persuaded his Levite chaplain to transfer to their shrine at Laish, bringing Micaiah's *teraphim* with him (Judg. 18: 24). The statues were man-size and, put inside a bed, they effectively deceived *Saul's servants who were pursuing *David (1 Sam. 19: 13–16). In the reforms of *Josiah, *teraphim* were among the cultic objects banned.

terebinth A large *tree with a long life, much esteemed in Israel (Isa. 6: 13), offering a shaded site suitable as a burial ground (1 Chron. 10: 12; 'oak tree', NRSV, REB; 'tamarisk', NJB). Its graceful branches are praised by Ben Sirach (Ecclus. [= Sir.] 24: 16).

Tertullus The prosecuting counsel on behalf of the Jews who accused Paul of disturbing the peace (Acts 24: 1–8). The name is *Latin but he may have been a Jew. His advocacy was only partly successful; Paul's trial was adjourned by *Felix the procurator (Acts 24: 22).

testament A *covenant, as at *Sinai, and used of OT by Paul (Gal. 3: 15–18). It is also used of a speech or *blessing given to children or followers, as by Moses (Deut. 33); this served as the model for such works as the *Testaments of the Twelve Patriarchs.

Testaments of the Twelve Patriarchs A Christian work of the 2nd cent. CE, though the core of it is a Jewish document containing doctrines found also in the *Dead Sea scrolls (e.g. a belief in two *Messiahs) and therefore in existence before 70 CE. The content, exhortations supposedly given by *Jacob to his sons on his death-bed, is a compendium of ethical instructions, but the Jewish traditions have been reinterpreted in the Christian interest.

testimony books A possible collection of testimonies to Christ assembled before the gospels were compiled. It is suggested that there was an urgent need in the Church for such a collection of testimonies (Latin, *testimonia*). These were OT texts supporting the Christian claims about Jesus' Messiahship, and in particular the concept of a suffering *Messiah. Christian *teachers searched the OT for anything that might lend credibility to their controversies with the Jews; it was, for example, necessary to explain how one of the Twelve could have betrayed Jesus, as well as other historical details which were an embarrassment. Books of testimonies were certainly available to the Fathers (e.g. Cyprian, in 248 CE) and similar collections were in use to support doctrine in the *Qumran community. However, because NT quotations of the same OT text sometimes differ in the wording, it may be that there was a body of oral testimony tradition used in the Church and that the hypothetical books did not exist.

tetrarch The title accorded by the Romans to the ruler of a dependent state, inferior to an ethnarch. Herod *Antipas and *Philip, son of *Herod the Great, and also *Lysanias of *Abilene, are given the title in Luke 3: 1, which is more accurate than 'king' in Mark (6: 14), since Herod's application at *Rome to be given the title of king was turned down and led to his banishment.

tetragrammaton Greek for 'four letters', that is the sacred Hebrew name for God, the consonants YHWH (Exod. 3: 15); but because it was considered too holy to pronounce, Adonai ('Lord') was substituted by readers of the text. When the vowels of Adonai were inserted into YHWH the artificial name *Jehovah was produced and established for generations by the AV. English translations (except NJB) adopt the convention LORD for the Hebrew *Yahweh*. (*Elohim* is rendered 'God'.)

text of the Bible It is the purpose of textual criticism (formerly known as 'Lower Criticism') to try to recover as far as possible the authentic words of the books of the Bible as they were originally penned. It is an enormous task. Nothing at all remains of any of the authors' own writing. What we do have is a vast number of texts copied by *scribes, versions translated into ancient languages, and a host of quotations in the Fathers which act as checks on the various MSS. It was so easy for mistakes to creep in: a word or two could be omitted by mistake, or repeated; if scribes were writing from dictation words of similar sound could be misheard. Sometimes a scribe could produce an alteration for doctrinal reasons.

1. *The OT*. The basic text of the Hebrew OT is that provided by the Masoretes around 500–1000 CE, on which very few variants exist. Before this time, the Hebrew script consisted only of consonants. The Masoretes then introduced pointing (vowel signs) and they were scrupulous about accuracy. Their work can be tested by the Hebrew of some of the *Dead Sea scrolls, MSS older by 1,000 years than any other extant Hebrew MS. The Hebrew of the OT can be checked by translations, especially those in *Greek, notably the *LXX of the 3rd cent. BCE, translated for the use of Jews in *Egypt. Other Greek translations were made, and in 245 CE Origen arranged the OT (in a work called the *Hexapla*) in six parallel columns: Hebrew, a Greek transliteration, three Greek translations, and Origen's own translation. The Greek translation included the deuterocanonical books of the *Apocrypha which were not in the Hebrew OT, and it was the *LXX version of the OT that was used by the first Christians. A Latin translation from the Hebrew was made by Jerome; it is called the *Vulgate. The Hebrew OT was translated into Syriac early in the 1st cent. CE; it was adopted by the Christian Syriac Church, and known as the *Peshitta (which also contains most of the NT).

2. *The NT*. Thousands of Greek MSS exist, of which the oldest almost complete MSS go back to the 4th and 5th cents. CE. Some 2nd- and 3rd-cent. fragments are on *papyrus; other MSS are on parchment; all were bound in the form of books (Latin *codices*). They have a continuous text which may be in uncials (capitals) or minuscules (lower-case letters); some are organized as lectionaries, with portions for the daily services. The words do not have breathings or accents, and there are no spaces between words, no punctuation marks, no divisions into chapters and verses. The most important uncials are Codex *Sinaiticus of the 4th cent., *Codex Vaticanus, and *Codex Alexandrinus. The Codex *Bezae contains both Greek and Latin texts of the gospels and Acts. The earliest fragment of a papyrus MS is of John 18: 31–4, 37–8, discovered in Egypt and published by C. H. Roberts in 1935, dated about 130 CE, and now in Manchester.

textus receptus Latin for 'the received text' and applied to the 1550 Greek edition of the NT, largely the work of Erasmus of Rotterdam (1516). It was the text in general use until the 19th cent. and underlay the AV of 1611. However, the wealth of discoveries and the development of textual criticism made other editions of the Greek NT imperative. *Westcott and *Hort provided an updated Greek NT on which the RV of 1881 was based. The *textus receptus* of the Hebrew OT refers to the Second Rabbinic Bible published in Venice in 1524 CE.

Thaddaeus One of the twelve *apostles in Mark 3: 18 and Matt. 10: 3 but there are variations in the names of the twelve in each of the four lists in the NT. In Luke 6: 16 and Acts 1: 13 'Thaddaeus' is replaced by '*Judas', son of *James. He is also sometimes known as Lebbaeus (Matt. 10: 3 NRSV marg.) Such variations are due to the traditions being transmitted through and by different Christian *communities.

thanksgiving sacrifice (NRSV) **/thank offering** (REB) An animal *sacrifice offered in gratitude for some favour received; the meat was shared between the donor and the priest, who had laid his hand on the victim's head. They ate round the *altar at a kind of *communion meal (hence NJB 'communion sacrifice' at Lev. 7: 13) which included unleavened cakes (Lev. 7: 12). In *Judaism a thanksgiving (*berakah*) was offered at meals, and Jesus followed this practice at the *Lord's Supper. He gave thanks (Greek, *eucharistesas*, 1 Cor. 11: 24), which became a technical term by the 2nd cent. for the Breaking of *Bread.

theatre An area designed for public recitations and performances. Greek cities had outdoor theatres not only for the classical tragedies and comedies but also for official assemblies. They were often large and were constructed with remarkably good acoustics. Paul was minded to enter the theatre at *Ephesus (Acts 19: 30–1) but was dissuaded because of the crowd's hostility. The Jews put forward a spokesman, *Alexander, perhaps to disclaim responsibility for Paul and Christianity, but he was shouted down.

Thebes A leading city of *Egypt, called No in *Hebrew, which was conquered by the *Assyrians in 663 BCE (Nahum 3: 8–10).

Thebez A city near *Shechem (Judg. 9: 49) where *Abimelech was struck on the head by a millstone thrown by a woman (Judg. 9: 50, 53).

theologoumenon A way of speaking of God or divine things without using the name of God—e.g. the '*cloud' which accompanied the Israelites in the *wilderness.

theology In the narrowest sense, the study of the knowledge of God; but more widely, in modern usage, the rational account of a religion as serviced by a range of subdisciplines which include the study of sacred texts, *ethics, doctrine, history, and liturgy. The word does not occur in the Bible but the practice of theology began as early as Paul, who had himself studied Jewish theology (Phil. 3: 5). Theological controversy is found in most of the epistles, and the disputants argued about truth and error, including or excluding rival views about the gospel (Gal. 1: 6) or Christ (1 John 4: 2–3).

Attempts have been made to bring the diverse materials of the NT under one umbrella as a 'theology of the NT', argued, for example, under dogmatic principles, such as Church, *ministry, and *sacraments. In Germany, the best-known recent NT theology was that of Rudolf *Bultmann who, after a short introduction on Jesus' message, then described the theologies of Paul and John from an existentialist point of view, well adapted for the modern Christian. Since then it has been argued that the NT data are far too variegated to be brought into a single system. Round a core of consensus the remainder has to be recognized as attempts to express its significance for very different communities. The core is the salvation-event of Jesus Christ himself, especially the *cross and *resurrection and his preaching of the Kingdom, and the apostolic proclamation of this event.

OT scholars in the 19th cent. shrank from the task of ordering the extraordinary variety of OT narratives into a coherent systematic theology, and produced instead histories of the religion of Israel. More recently, scholars have turned away from a purely descriptive account of the OT to recording it as a body of literature in which history and *faith are inextricably combined and which in the Christian community of faith is linked to the NT. Through the OT interpretation of Israel's history, this offers a perspective for understanding events in our own day. For example, *liberation and *feminist theologians can approach the history of the Exodus with their particular human concerns: OT history is a model for regarding contemporary movements as the setting for God's continuing work of *salvation.

theophany An appearance or revelation of God, often in the OT in connection with a *sanctuary (Exod. 40: 34–8); as to Isaiah in the Temple (Isa. 6: 1); God's presence is indicated by fire or wind and *clouds (Deut. 33: 2; Ps. 68: 33; 104: 3). According to Exod. 33: 40, no one can see God and survive. He is never seen in the theophany at Mount *Sinai except in the older narrative of Exod. 24: 9–11, where Moses occupies a merely subordinate position; in the later narratives he is supreme. Rather different was the revelation to *Elijah in 'a still, small voice' (1 Kgs. 19: 12, AV, RV; 'a sound of sheer silence', NRSV; 'a faint, murmuring sound', REB; 'a light murmuring sound', NJB). In the NT Jesus saw the *Spirit of God descending like a *dove at his *baptism (Matt. 3: 16) and at the *Transfiguration there came a voice from the cloud (Matt. 17: 5).

Theophilus Greek for 'lover of God'; the person to whom the third gospel (Luke 1: 3) and the Acts (1: 1) are dedicated; someone of distinction who might have been the author's patron, who helped him with the publication of the two volumes. Some scholars have thought the name was merely a *pseudonym.

Thessalonians, Paul's epistles to the Arranged in the NT as the eighth and ninth of the epistles. From *Philippi, Paul came to *Thessalonica with *Timothy and *Silas and spent about a month there (Acts 17: 1–9) before disturbances obliged them to make an unscheduled departure for *Beroea, and thence to *Athens, without Timothy and Silas; there Paul waited for them to catch him up (Acts 17: 16), as Timothy, at any rate, did (1 Thess. 3: 1–3), having been previously dispatched from Athens to Thessalonica to support the Church there. Paul wrote 1 Thess. after Timothy had returned to him—but by this time Paul had moved on from a rather indifferent reception at Athens to the volatile city of *Corinth (Acts 18: 5). Here, in 51 CE, Paul wrote 1 Thess. and (unless *Gal. can claim precedence) it was the first of all his epistles; it is also rather special in having no dramatic, controversial theme.

The Church at Thessalonica comprised both Jews and *Gentiles, and the latter were probably perplexed by the *apocalyptic teaching concerning the **parousia* and the *Judgement. Paul's Christian belief was that Christ's *resurrection was the first stage in an eschatological series of events, whereas the Thessalonian Gentiles may once have been initiates in the cult of Serapis with the promise of an agreeable afterlife, and so were now as Christians expecting very soon to enjoy a painless share in a general resurrection.

Because of Paul's hasty exit from Thessalonica, he now could say he was grateful for the warmth of those who did accept him. His motives were entirely honourable. Their response was one of joy (1: 6). Yet he was anxious about the state of the young Church there (1 Thess. 3: 5) and urged the Christians to have *hope*. True, some of the members had already died, but they would not be at a disadvantage when Christ returned: Christians in this life already shared in the new life in Christ.

Much of the message of 1 Thess. is repeated in 2 Thess. which creates a problem of authorship. Could Paul have written so similar a letter to the same Church so soon? And yet with the similarity go some differences: a word like 'calling' (2 Thess. 1: 11) does not bear its normal Pauline meaning, (1 Thess. 2: 12), and imitation (2: 14) is used differently in 2 Thess. (3: 9). The same Greek word is indeed translated differently in each place by REB. The *eschatology of 2 Thess. 2 urges the recipient to look out for the signs of the *End, whereas 1 Thess. 5: 4–5 warns them that the End will come quite unexpectedly. It would also seem possible that the reference to 'the man of lawlessness' occupying the Temple (2 Thess. 2: 3–4) depends on Mark 13: 14–17, which is later than anything Paul could have written. The language does not refer to a historical event, such as Caligula's intention to erect a statue of himself in the Temple in 40 CE.

The transparent effort of 2 Thess. to gain credibility as a Pauline letter is apparent in 2: 1–2, where there is mention of a letter purporting to come from Paul, teaching that the *day of the Lord had already come. And 3: 17 rather overdoes the appeal to his own handwriting, for if the Thessalonians had just read 1 Thess. it is odd that they should need to have the letter authenticated in this way. It is better to assume that a letter had gone round in Paul's name asserting that the Day had already come, and that a disciple of Paul wrote 2 Thess. in about 80 CE, explaining that the events to precede that day had not yet taken place. The letter is an encouragement to the Christians in Thessalonica to be firm in the faith. The expected return of Christ is no excuse for refusing to work now for a living (2 Thess. 3: 11–12).

Thessalonica The capital of the Roman province of *Macedonia and situated on the Via Egnatia; it enjoyed much prosperity and was an important city for Paul's missionary work. Religious cults flourished, both pagan and Jewish, and divine honours were accorded the goddess Roma—evidence of indebtedness to Roman benefactors. An inscription has been found which is evidence that the unusual term 'politarchs' was used for the rulers of the city, as correctly given in Acts 17: 6 (NRSV marg.).

Theudas A Messianic pretender who promised to lead his followers in revolt against *Rome; he was killed according to *Josephus in 44 CE. In Acts 5: 36–7 Theudas is regarded as preceding another revolutionary, *Judas, who died in 6 CE, so it would seem that either Josephus or the Acts has made a mistake. The time when *Gamaliel was speaking (Acts 5: 34) was about 30 CE, so that Theudas' rebellion had not then taken place.

However, since there were numerous other disorders, it is just possible that there was another leader called Theudas who lived at an earlier date.

Thomas One of the Twelve, called Didymus ('the twin', John 11: 16). He is named in the lists of *apostles (e.g. Luke 6: 15) but only becomes important in the gospel of John. He has earned the title 'doubting Thomas' because in John's narrative he declined to believe in the *resurrection of Jesus until he was given tangible evidence (John 20: 24–9). When Jesus made himself known, Thomas exclaimed, 'My Lord and my God' (20: 28). Such an address to Jesus shows that by the time John was written, the rigid monotheism of *Judaism was beginning to have to take on board the Christian veneration of Jesus. It implies a belief that Jesus is exalted to the throne of God and seems to echo John 10: 30 ('I and the Father are one').

Thomas, gospel of Among the Greek papyri discovered at *Oxyrhynchus were three fragments containing sayings of Jesus. It turned out that these formed part of a complete work, the gospel of Thomas, which in a Coptic text of about 350 CE was among the important writings found at *Nag Hammadi in 1945. The original fragments had generated discussion about the history of the transmission of Jesus' sayings in the oral tradition. Then the publication of the 114 sayings of Jesus (without any narrative continuum) in the complete gospel stimulated lively scholarly controversy. Some of the sayings have close resemblances to those in the canonical synoptic gospels; others are quite different, of which five have been reliably claimed as authentic.

It is disputed whether the sayings resembling those in the canonical gospels come from an *independent* oral tradition (and therefore are perhaps closer to what Jesus in fact said) or whether they show knowledge of those gospels (and so must be later). The first view is taken by those who believe that the Thomas sayings represent a collection which is 'very, very early' and remarkably free of any ecclesiastical interests. On the other hand, Thomas is without doubt a *Gnostic work. In that system Christ shares with true believers a secret knowledge (*gnosis*) which enables them to be liberated from the evils of this world. The sayings in Thomas conform to this system; it is the words which offer the means of salvation—not, as in mainstream Christianity, Jesus' *death and *resurrection.

Although some of Thomas' parallels with sayings in the canonical gospels are shorter, that is not a certain indication of being earlier and independent; after all, Matt., following Mark, often abbreviated that gospel in order to accommodate more of his own special material. Evidence of Gnostic adaptation of Jesus' parables has been shown to be in such a way that they have become vehicles of esoteric Gnostic teaching. And the tradition which Thomas was adapting was probably not an oral tradition but the written canonical gospels, as indicated when one of the Greek fragments from Oxyrhynchus presupposes Luke 8: 17, which has altered Mark 4: 22.

It would seem that Thomas is not a 'fifth gospel' offering an additional and faithful reproduction of Jesus' sayings, but a work designed for its own clientele with very few sayings of Jesus of demonstrable authenticity.

thorns, crown of A crown pressed into the head of Jesus (Matt. 27: 29). Briers and brambles were common and grew in thickets in Palestine, and could well have supplied the thorns. The acanthus wreath, together with the purple cloak, or cape, and a reed as a *sceptre, was part of the soldiers' mockery, representing Jesus as a king.

thousand Used either as a literal number in the OT (2 Kgs. 19: 35) or to express a long time (Ps. 90: 4) or a military unit (1 Sam. 8: 12). In the NT large numbers are reckoned in thousands (Mark 6: 44; Acts 2: 41; Rev. 21: 16, NRSV marg., NJB).

three Like other numbers three had a symbolic significance in the biblical tradition; it suggested completeness: *heaven, earth, and the underworld, for example. Three major feasts are mentioned (Exod. 23: 14–19); the *sanctuary was divided into three parts (1 Kgs. 6: 2–22); *Jonah was in the belly of the fish for three days and nights (Jonah 1: 17); and in the NT Jesus was raised from the dead on the third day, or 'after three days' (Matt. 12: 40; 27: 63; Luke 24: 46; 1 Cor. 15: 4). In Christian theology, the Godhead is a *Trinity—Father, Son, and Holy Spirit.

throne The symbol both of earthly and of divine *judgement (1 Kgs. 7: 7; Ps. 47: 8). It symbolized the power of *Satan (Rev. 2: 13; 13: 2; 16: 10).

In another metaphorical sense, Jesus inherits the throne of *David (Luke 1: 32) and is exalted to the right hand of God's throne (Heb. 8: 1).

Thummim Devices to ascertain the will of God. **Urim* and *Thummim* were consulted by the chief priest (Num. 27: 21; 1 Sam. 14: 41; 28: 6), but they fell into disuse. They were worn in a pouch on the *ephod garment (Exod. 28: 30).

thunder Used metaphorically for the voice of God (Job 37: 2, 5; Ps. 29: 3; John 12: 28–9) and the awesome holiness of God (Isa. 33: 3, REB).

Thunder, Sons of 'Boanerges'—the name given by Jesus to *James and *John (Mark 3: 17), probably on account of their stormy tempers.

Thyatira A city 56 km. (35 miles) NE of *Smyrna, in modern Turkey, with valuable trading contacts in the 1st cent. CE. When Paul visited *Philippi he met and converted *Lydia from Thyatira, a seller of purple goods (Acts 16: 14–15). The fourth of the seven letters in Rev. (2: 18–29) was addressed to the Church in Thyatira.

Tiamat In the *Babylonian creation *myth she is the raging female monster of chaos or the sea; she is represented variously as a cow, goat, dragon, or woman, and is slain by Marduk ('Merodach', Jer. 50: 2, NRSV). Out of the two halves of her body emerged heaven and earth (cf. Gen. 1: 6). The primeval flood (the 'deep' of Gen. 1: 2; 7: 11, and 8: 2), *tehom* in Hebrew, is a term originating in the mythology of Babylonian Tiamat.

Tiberias A city founded by Herod *Antipas (20 CE) as his new capital to replace Sepphoris, in honour of the emperor Tiberius on the western shore of the Sea of *Galilee (also called the Sea of Tiberias, John 6: 1; 21: 1), on an important trade route. It had a good deal of independence and because it was the power base of the despised Herodian family, Jesus seems to have avoided it and also Sepphoris. It is mentioned only in passing by John (6: 23), and never in the synoptic gospels. *Josephus was in command of forces which captured the city early in the Jewish War of 66–70 CE and he surrendered it to the Roman forces in 67. In the following centuries it had a strong Jewish cultural influence, and the *Mishnah and the Palestinian *Talmud emerged from Tiberias—even though, strictly, it was unclean according to Jewish *Law, for it was built on the site of a cemetery.

Tiberius Roman Emperor, 14–37 CE, and the '*Caesar' of Jesus' *ministry and *crucifixion. The coin shown to Jesus when he was replying to a question about taxation bore this emperor's portrait (Mark 12: 13–17).

Tiglath-Pileser King of *Assyria 745–727 BCE, who greatly expanded the empire, until it almost reached the border of Egypt. After the rebellion of Israel and other states, these areas became Assyrian provinces (2 Kgs. 15: 29); but *Judah under King *Ahaz did not join the alliance and was spared annihilation, though not taxation (2 Kgs. 16: 17–18). It was alleged that Ahaz introduced Assyrian cultic practices into Jerusalem (2 Kgs. 16: 10–18).

Tigris One of the two major rivers that made possible the civilization of the ancient Near East and on which *Nineveh and *Assher stood.

tile Writing or drawing (Ezek. 4: 1) was impressed on clay and then baked hard into tiles. In the account of the healing of the paralytic, Luke 5: 19 changes Mark 2: 4: the earlier gospel described a poor Palestinian house with a *roof of straw but Luke's more sophisticated readers would expect a house to be roofed with tiling.

time For the Hebrews, time was duration: it went on for a certain measurable length, short (1 Sam. 9: 16), or longer (Eccles. 3: 1), or indefinite (Mic. 4: 7).

In the NT the Greek *chronos* is used for a length of time (John 7: 33) and *kairos* for an opportune time or moment of decision (John 7: 8). A third word, *aion*, broadly indicates an era and is used to make the distinction between 'this present era' or 'age' and 'the coming era', as in Eph. 1: 21.

Timon One of the Seven appointed by the apostles to perform administrative duties in the Church of Jerusalem (Acts 6: 5).

Timothy (1) An Ammonite officer with the Syrian army who opposed *Judas Maccabaeus (164 BCE) and was killed (1 Macc. 5: 6–8). (2) A devoted younger assistant to Paul mentioned in the epistles of Paul and in Acts. He was much involved with the Thessalonian mission and when Paul was alone in* Athens he sent Timothy to *Thessalonica to give that Church support in a time of persecution (1 Thess. 3: 1–5); on rejoining Paul, now at *Corinth, he was able to bring reassuring news. Next, from *Ephesus, Timothy was dispatched to bring some order to the Corinthian Church (1 Cor. 16: 10–11). There was also a plan for Timothy to go to *Philippi (Phil. 2: 19–24). According to the Acts, Timothy was the son of a Jewish mother and Gentile father, and evidently his mother (Eunice, according to 2 Tim. 1: 5) was not able to prevail in the household at *Lystra to bring up Timothy as a Jew (Acts 16: 3). So Paul circumcised Timothy, probably in order to make him as much a Jewish Christian as he and Silas were, thereby smoothing relations with adherents of *synagogues. With pure Gentiles (such as *Titus, Gal. 2: 3) it was a matter of principle for Paul that they should *not* submit to the requirements of the Jewish Law.

Timothy, epistles to Together with the epistle to Titus, these letters are called *'Pastoral' because they deal primarily with the concerns of members of the Churches. Traditionally, it has been held that after Paul was released from prison in *Rome he continued his evangelistic ministry in *Crete, where he was re-arrested (cf. Tit. 1: 5). In this event, Paul's theology and ethical teaching had developed since writing to the Corinthians. But is this plausible? Fathers from the time of *Irenaeus regard the epistles as authentically Pauline, it is true, but arguments against this view are strong: the heresy denounced is more Gnostic than anything otherwise known in Paul's lifetime, but early in the 2nd cent. such teaching was common. The organization of the Church has become a dominant issue, requiring much advice in the epistles. The vocabulary and literary style of the epistles seem, even in translation, to be different from Paul's. However, the majority view is that the epistles were written later in the 1st or early in the 2nd cent. by a pseudonymous author, who sincerely believed that he was writing what would have been Paul's own words in the new situation.

The letters consist of a charge to Timothy (2: 1–6: 19) outlining the essential qualities for Church leadership and deploring excessive asceticism, as proposed by heretics. The second epistle commends the experience of suffering as being evidence of true vocation (2 Tim. 1: 8–2: 26).

Tirhakah An Egyptian *Pharaoh who reigned 690–664 BCE and who led a campaign against Assyria (2 Kgs. 19: 9; Isa. 37: 9). Assyrian annals record the Egyptian defeat in about 701 BCE, when Tirhakah would have been a young commander in the field.

Tirshatha Used in AV (Ezra 2: 63; Neh. 8: 9) as the Persian title for the 'governor' (NRSV, REB) of *Judah.

Tischendorf, Constantin von (1815–74) NT professor at Leipzig. He made a remarkable and unexpected discovery of the Codex *Sinaiticus when on visits to the monastery of St Catherine on Mount *Sinai in 1844 and 1859.

tithes A tenth of annual income, set aside for religious purposes and validated by *Jacob's vow at *Bethel (Gen. 28: 22). In the second Temple the tithe was collected in the Temple to maintain the priesthood (Neh. 10: 37–8), but it was not always paid willingly or in full (Mal. 3: 8, 10). Before the Exile, however, kings had collected the tithe in kind (fruits and flocks) and stored some of it in the Temple (2 Chron. 31: 5–6). It would seem that the tithe was then a royal tax but that the kings used it to maintain the Temple. Jesus does not condemn the tithing practice of his own day, which *Judaism continued, but he was censorious, as was Amos (4: 4) before him of those who put the duty of tithing over and above compassion and faith (Matt. 23: 23).

Titius Justus A Gentile sympathetic to *Judaism in *Corinth who gave accommodation to Paul in a house-church after he had been evicted from the local *synagogue (Acts 18: 7). Titius Justus was possibly a Roman citizen. Some Greek MSS read the name as Titus Justus.

tittle A small mark used in Hebrew script to distinguish letters otherwise similar (Matt. 5: 18, AV; 'dot', REB; 'little stroke', NJB); so comes to signify a minute point of legalism.

Titus Paul's Gentile assistant (2 Cor. 8: 23) who travelled with him to Jerusalem (Gal. 2: 1–10) and whom Paul refused to circumcise; it would have constituted a damaging precedent. Titus was probably Paul's emissary to the difficult Corinthians, taking the 'severe' letter (2 Cor. 2: 3–9). Following a delay in reaching Paul after this visit (Titus did not get to *Troas as planned), they eventually met in *Macedonia and Titus reported favourably—so much so, that Paul felt able to ask him to go back to *Corinth again.

The epistle to Titus (one of the

'*Pastorals') contains advice on Church and household organization but it is widely held to come from a later date for the same reasons as apply to *1 and 2 Timothy.

Tobit, book of A book included in the *Apocrypha by Protestants and by Roman Catholics in the deuterocanonical literature of the OT. It was in the Greek *LXX, although probably it was originally composed in Hebrew or Aramaic. Fragments of the book in both these languages are among the *Dead Sea scrolls. The author of the book is unknown, but lived in the *Diaspora among Gentiles, rather than in Palestine. The suggested place is *Antioch in *Syria, where extensive commercial activities provided a link with the area (ancient Assyria) where the story of the book is placed. Another suggestion is that *Babylon could have been the author's home, in view of the fascination with *angels—who had been introduced into Judaism under Persian influence.

Although in the story Tobit is an official in the Assyrian government of *Sennacherib, there are too many historical errors to regard it as genuinely autobiographical, though it pretends to be (Tob. 1: 3–3: 6). The reference to the Temple (14: 4–5) has no hint about the desecration by *Antiochus Epiphanes in 165 BCE; but its ethical and theological ideas resemble those of Ecclus. [= Sir.] (180 BCE).

The Aramaic of Tob. in the *Qumran scrolls is said to be more typical of the age of Ezra (4th cent. BCE) than of Daniel (2nd cent.), so possibly the date of composition was about 250 BCE.

The story's hero, Tobit, is a devout Jew who suffered on account of his zeal in giving burial to executed Jews. But he has deposited a large sum of money with a relative in Rages in Media and he asks his son Tobias to fetch it. Tobias does so in company with a guide called Raphael, who turns out to be an angel with a gift for extracting magical spells from the organs of fish. This he does, but in addition Raphael organizes a marriage for Tobias with one Sarah, daughter of a kinsman of Tobit, who had been widowed seven times.

The book teaches the importance of burying the dead and of avoiding marriage with non-Jews by Sarah's marrying within her own family. The author expresses great reverence for Jerusalem (1: 6–8; 14: 5) and in general indicates the very best ethical traditions of Judaism—integrity, honesty, charity. He is influenced by *Deuteronomic theology, that *righteousness is rewarded with prosperity (Tob. 14: 2) and wickedness in the end with disaster (Tob. 13: 12).

The delightful tale of Tobit has inspired some superb paintings—by Botticelli (who provides *three* angels!), Rembrandt, and Titian.

tombs Jewish dead were always buried, and skeletons of the 1st cent. CE of crucified victims have been discovered in a cemetery in Jerusalem. If burial took place in a grave, it would be covered with stones after the interment. Larger tombs, or sepulchres, could be cut into limestone rock and a stone door tightly fastened, forming an imposing mausoleum. The doors were shaped like wheels and were rolled up a track when necessary and held by a block, which was simply moved out of the way when the door was to return to seal the tomb. Probably this is the kind of tomb which *Joseph of Arimathaea owned (John 20: 12), but its exact location is unidentifiable.

tongues, speaking in The act of speaking, or singing, in a language incomprehensible to the speaker and the audience, familiar in times of intense religious emotion in Christian and non-Christian communities. It was experienced at *Corinth (1 Cor. 12: 10) but Paul advised the Church not to stress this particular *gift, also known by the term **glossolalia*. *Love was the greatest gift, not the display of ecstatic utterance.

The events in the Upper Room at *Pentecost (Acts 2) when the *Spirit empowered the apostles, as also perhaps at *Ephesus when Paul laid his hands on some converts (Acts 19: 6), suggest that the utterances were in foreign speech intelligible to the natives of those regions. There is evidence that at the end of the 17th cent. CE persecuted

Huguenots in France spoke in foreign languages; but modern linguistic scholars have shown that the utterances spoken in tongues do not meet the criteria of what constitutes a natural human language.

topaz The ninth jewel in the wall of the New Jerusalem (Rev. 21: 20).

Topheth In the valley of Hinnom, south of Jerusalem. The place had an evil reputation because children had been sacrificed there to the gods *Baal and Molech (Jer. 7: 31) and King *Ahaz and *Manasseh are alleged to have so offered their own sons (2 Chr. 28: 3; 33: 6). King *Josiah forbade the practice (2 Kgs. 23: 10).

Torah Hebrew for 'instruction' or 'teaching'. The translation '*Law' is too restrictive: there *is* a body of legal enactments but it is within a corpus of divine revelation and teaching—the whole being known as Torah. In addition to the statutes (Exod. 18: 16) the Torah contains instructions for *worship (Lev. 6: 14), sacrificial procedure, right conduct, and precautions to preserve purity (Lev. 10: 10; 14: 57; Deut. 4–7; 11–15; 23). In Deut. the Torah refers to the whole book, all that went to maintain Israel's cultural and religious identity, based on the law book that was allegedly found in the Temple (2 Kgs. 22: 8). Josiah's contemporary, Jeremiah, uses Torah in a wide sense as in Deut., and in the psalms (e.g. Ps. 1: 2) the study of Torah is praised as being the whole point and purpose of existence. Disobedience or neglect would bring exile, disaster, and death, even for kings (Deut. 17: 18–20). In later Judaism the *Pentateuch came to be known as the written Torah, while the oral Torah consisted of the traditions that were eventually written in the *Mishnah, forming the basis of the *Talmud.

Moses is credited with enunciating the Torah (Deut. 4: 44) but subsequently Torah in its widest sense was promulgated by kings, priests, and wise men (Prov. 6: 20). It was a divine revelation and responding to it was the devout Israelite's greatest *joy (Ps. 119). In the NT Jesus warns his followers that the precepts of the Law take them only to the threshold of the kingdom: the Law forbade murder; within the kingdom there cannot even exist anger (Matt. 5: 22). Paul regards observance of the Law as the badge of Judaism, from which Christians are released, for Christ alone is the agent of *salvation.

tosefta Aramaic for 'addition'; consisting of commentary on the *Mishnah compiled in the 3rd cent. CE.

town clerk An official of cities in the Roman Empire with responsibility for public order. At *Ephesus (Acts 19: 35–41) he intervened to quell the riot instigated by the silversmiths who feared that Paul's preaching would jeopardize their businesses.

towns It is not easy to discover the criteria for distinguishing 'cities', 'towns', and 'villages' in the Bible, especially as English translations vary. *Bethlehem is a 'town' (Judg. 17: 8, NRSV, NJB) but a city in AV, RV, REB. Perhaps, in general, cities were fortified and larger. However, urban centres as big as *Ephesus had officials translated as 'town clerk' (Acts 19: 35). Jesus is said to have gone through both towns and villages (Luke 13: 22); but Paul, townsman *par excellence*, does not seem to have been interested in rural ministry.

Trachonitis Part of the territory of Herod Philip (Luke 3: 1), later governed by Herod *Agrippa I. Situated in the NE of Palestine, its OT name is Bashan. *Caesarea Philippi and *Bethsaida are two NT cities of the region.

trade The Bible is full of references to trade, from the time of *Jacob (Gen. 34: 10) to the merchants and sailors of Babylon (Rev. 18: 15, 17). Palestine was criss-crossed by trade routes—especially north to south, but also west to east: great powers to the south and the NE inevitably used the roads of Palestine in war and peace. Isaiah refers to a route that ran along the coast (Isa. 9: 1); there was a route through the *Transjordanian hills of *Edom and *Moab to *Damascus; and a third route to Israel linked Jerusalem, *Bethlehem, *Hebron, and *Beersheba in the south, and proceeded north through *Shechem and Bethshan towards the Sea of *Galilee. From east to west, there were routes through the passes at Shechem and Beersheba. The routes were not necessarily safe (Luke 10: 29–37) but they had staging-posts and also customs posts where taxes were levied on goods passing through.

With such facilities for transport, and the availability of donkeys and camels, trade flourished (Ezek. 27: 12–25) bringing wealth (Ezek. 28: 5) but also vandalism (Ezek. 28: 16, 18) and financial scandals (Hos. 12: 7), as well as financial rewards (Luke 19: 20–6).

tradent One who is responsible for preserving and handing on the oral tradition, such as a *teacher or preacher or missionary, in the form of *apophthegms or similar *pericopae.

tradition Much OT and NT scholarship in the 20th cent. has investigated the social contexts in which the narratives were transmitted orally before they were encapsulated in writing. A *community or nation in constant change adapts its religious practices accordingly, and critics have been interested to go behind the literary sources of the *Pentateuch, for example, to discern the process by which those sources attained their final form. Thus, attention was given to the life and aims of the Priestly editors from whom emerged the *P source of the *Pentateuch.

In NT studies work has been devoted to the process of transmitting the various items in the gospels which were identified by the *Form Critics. There are two significant passages in 1 Cor. (11: 23–5 and 15: 3–5) in which Paul explains that what he passes on to his converts he had himself received and checked out at Jerusalem; the traditions about the *Eucharist and the *Resurrection as recorded by Paul therefore go back to the pre-Pauline era—the short period of time between the *crucifixion and Paul's first visit to the leaders of the Church in Jerusalem (Gal. 1: 18). Paul asserts that he received the tradition 'from the Lord', and this might refer either to a direct revelation, as on the Damascus road, or, more probably, to the Lord (Jesus) as the source of a tradition that was passed on by a human chain of believers.

In Roman Catholic theology 'tradition', as defined at the Council of Trent in 1546, has a special meaning as a body of belief equal in authority with scripture. It has been held to consist of teaching given by the risen Jesus to the *apostles during the forty days after the *Resurrection (Acts 1: 3). Some Catholic theologians, however, regard tradition as the ongoing interpretation, within the Church, of revelation which is either explicitly or implicitly contained in scripture.

Trajan Roman emperor, 98–117 CE, and a successful soldier. Correspondence between him and *Pliny the Younger has provided interesting evidence about the state of the Church in Asia Minor (the provinces of Bithynia and Pontus) in 111–12 CE. Pliny asks the emperor's advice on whether Christians are to be punished solely on grounds of professing the faith or whether only offences connected with religion are punishable; and whether continuing obstinacy merited execution. These enquiries correspond closely to the substance of 1 Peter 4: 12–16, and have persuaded some scholars that this epistle is to be given a date early in the 2nd cent. CE.

trance A period when the normal faculties of perception and recognition are in suspense. Some visions recorded in the NT (e.g. of Peter, Acts 10: 10) were received during a trance.

transfiguration The account in the *synoptic gospels (Mark 9: 2–8; Matt. 17: 1–8; Luke 9: 28–36) of the supernatural transformation into *glory of the appearance of Jesus. There is no mention of the event in the gospel of John, but there is a reference at 2 Pet. 1: 18. It has been held that the incident has been wrongly placed in the gospels in the ministry of Jesus, and that it was originally the account of a post-resurrection appearance. This, however, is unlikely; the form of the narrative is too different from that of the *Easter narratives. It is possible to regard the event as a paranormal experience of luminosity, such as is recorded of some of the saints, which has been given an interpretation by the *evangelists by means of the mention of Moses and *Elijah (representing the *Law and the *Prophets), who bear witness to Jesus; the *cloud is a symbol for the divine presence (Exod. 33: 7–11). The glory of the Lord once settled on Mount *Sinai (Exod. 24: 16) and the cloud covered it for six days. Mark says that Jesus took the three disciples up the mount 'six days' after the events of *Caesarea Philippi, though Luke (9: 28) alters the date to 'about eight days after', possibly to make the theological point that the Transfiguration was a preview or anticipation of Christ's Resurrection glory—for 'eight' pointed to Sunday, the first (or eighth) day of the week.

Peter is recorded as thinking it would be appropriate to build three dwellings (or *tents, *booths), for Jesus, Moses, and Elijah (Mark 9: 5)—a reference to the feast of *Tabernacles when people lived as once the Israelites were believed to have done in the *wilderness (Lev. 23: 42) and would do so again in the Messianic age. But Peter is rebuked for putting Jesus on a level with Moses and Elijah—'this is my unique Son' (Mark 9: 7).

The Transfiguration episode must therefore be understood against its OT background, especially with regard to Moses, whose face once shone with the reflected glory of God (Exod. 34: 29); on the mount Jesus' figure shone with its own glory, and his clothes were 'dazzling white' (Mark 9: 3).

Transjordan The part of *Palestine to the east of the River *Jordan; mountainous, with good annual rainfall. It comprises the OT areas of *Gilead, *Moab (east of the *Dead Sea), and *Edom.

translations of the Bible into modern English What has been known in Britain since 1611 as the 'Authorized Version'—though it was never formally authorized (or in the USA as the 'King James Version') was a revision of earlier Bibles by six companies: three groups worked on the OT, two on the NT, and one on the Apocrypha. Fifty-four scholars were enlisted, of whom the names of forty-seven are known, and they met in Oxford, Cambridge, and London (Westminster). King James I approved the proposal in 1604 and was always much interested in its progress. It became the most read and most loved Bible in English. Its phrasing was often dependent on *Tyndale's version. But by the 19th cent. it was being regarded as unsatisfactory in the light of much new knowledge through the discovery of great MSS like Codex *Sinaiticus. This led to a revision, initiated by the Church of England, by a committee of scholars from several Protestant Churches and the publication of the Revised Version—NT in 1881, OT in 1885. Although there was extensive correspondence with an American revision committee, a joint version was not published, but the American variations were printed in an appendix to the English RV, and in 1900–1 the American committees published their Standard Version, with the English variations in an appendix. The RV was disliked for its pedantic literalism, but it was the Bible used by students until the American Revised Standard Version was published (NT 1946, OT 1952).

A host of new translations appeared throughout the 20th cent. in Britain and America. The RSV was kept under scrutiny and a major revision was published in 1989 and called the New Revised Standard Version; it adopts 'inclusive language', avoiding masculine-oriented terms. The NRSV thus stands in a succession which goes back beyond the AV.

A wholly new English translation was the New English Bible (NEB; NT 1961, OT 1970), which was revised in 1989 as the Revised English Bible; it is more dignified and more suitable for liturgical use than NEB, as well as, like NRSV, adopting inclusive terminology except where scholarly integrity would be compromised.

Roman Catholics have shared in the preparation of NRSV and REB, but they have also had their own independent translations, notably that of Ronald Knox in 1945–9, and the Jerusalem Bible (1966), which owed much to the French translation (1956) published by the Jerusalem Dominicans; the New Jerusalem Bible appeared in 1985 and achieved considerable acclaim, though its use of 'Yahweh' in the OT is controversial. The titles of OT books follow the traditional English usage (e.g. 1 and 2 Samuel). The New American Bible, by Roman Catholic translators, was completed in 1970. It uses the best available Hebrew and Greek texts. A revision of the NT was published in 1989. A number of translations of markedly Protestant-Evangelical flavour have been published: the Living Bible in 1971, the Good News Bible (Today's English Version) in 1966–79. The New International Version (1973–8) was the work of translators 'committed to the authority and infallibility of the Bible as God's Word in written form'. Sometimes in this version a doctrinal consideration seems to be in control, as when Hebrew *almah* (Isa. 7: 14), which means 'a young woman', is rendered 'virgin'; it does so because it is translated thus by the *LXX of Isa. 7: 14 and this verse of Isa. is quoted in Matt. 1: 23 in connection with the conception of Jesus in Mary: the translation (which follows AV of Isa. 7: 14) is unreliable.

See also Bible, English translations of.

travel (1) OT: People did not travel very far unless they were traders or soldiers. Sea journeys were repugnant to Israelites and roads were often no more than paths trodden down by cattle and pedestrians. Egyptians had wagons (drawn by oxen? Gen. 46: 5). Horses were available for armies (2 Kgs. 18: 23) and donkeys and camels for civilians. (2) NT: By the 1st cent. CE Roman government had transformed the opportunities for travel. Pirates had been driven off the seas, and roads had been constructed to link the main cities. Hence the journeys which Paul was able to plan (Rom. 15: 24), and the pilgrimages on foot to Jerusalem (Luke 2: 41 f.) Roman roads were paved and milestones (a godsend to modern archaeologists) recorded distances.

treasure *Gold and *silver were kept in a safe store in a *sanctuary (1 Kgs. 7: 51). In Herod's Temple alms boxes shaped like trumpets were placed in the Court of Women (Mark 12: 41). The chief priests declined to put there the thirty pieces of silver which *Judas rejected (Matt. 27: 6).

trees Israel was not well supplied with trees, but timber was constantly needed for buildings and on farms. Wood for the Temple was imported (1 Kgs. 5: 8). Among the trees mentioned in the Bible are cedars (abundant on Mount *Lebanon), cypress, palm (which has a great capacity for survival), and sycamine (mulberry).

trial(s) of Jesus Long before the end of Jesus' life there had been conflicts with Jewish religious leaders—over *sabbath observance and in connection with *miracles of *healing (Mark 3: 6). Traps for him were set and it is not therefore a surprise to read that this opposition came to a head in Jerusalem after Jesus had entered the city with a measure of popular support (Mark 11: 9) and caused chaos in the outer court of the Temple (Mark 11: 15). According to the gospels, after his whereabouts had been betrayed by *Judas, Jesus was arrested, brought to trial before the *Sanhedrin, and condemned to death. A second trial then took place before the Roman Governor since the Jews, as a subject people, had no legal right to inflict capital *punishment, and *Pontius Pilate acceded to the wishes of the Jewish leaders, and condemned Jesus to death. He died by the Roman method of execution—*crucifixion (Mark 15: 24, 37).

The historicity of the gospel narratives has been questioned, especially by Jewish scholars, who maintain that the *evangelists are guilty of putting back into the time of Jesus the antagonism of the Church for the Synagogue. It is argued that the trial before the Sanhedrin is a fiction designed to fasten the blame for Jesus' death on the Jews; and that the historical fact is that Jesus was condemned by the Roman authorities as a dangerous disturber of the peace. The story of the nocturnal trial (Mark 14: 53–64) is not credible—even Luke recognized the improbability and recorded it as happening in daylight (Luke 22: 66); possibly Luke is using a different tradition. The assertion is disputable that the Jews were obliged to hand Jesus over to the Romans because they had been deprived of the right to execute (John 18: 31).

In favour of the gospels' accounts it is held that a preliminary trial before the Sanhedrin accords with Roman procedures elsewhere; it was the diplomatic skill of the high priest to unite the several factions in the Sanhedrin to obtain a unanimous verdict of guilt. The charge of blasphemy could have been based on Deut. 13: 5; a false prophet is to be executed. This was then rephrased in political terms (Luke 23: 2), and Pilate gave way to the pressure. That the Jewish authorities could not themselves inflict the death penalty (John 18: 31) under the Romans is corroborated in the *Mishnah. This is not overturned by the death of *Stephen (Acts 7: 57), which reads more like a lynching than a judicial procedure.

It is undeniable that the gospels from Mark through Matthew to John show a tendency to exonerate Pilate and maximize the responsibilities of the Jews (e.g. Matt. 27: 24), but the essential facts of trials before the Sanhedrin and before Pilate are defensible.

The gospel of John replaces the trial before the Sanhedrin with an investigation before Annas, high priest until deposed in 15 CE, and his son-in-law Caiaphas, the reigning high priest (John 18: 12–28). Peter's threefold denials are recorded in this section. Cf. Mark 14: 66–72. John's account of this trial is historically improbable. There is no reason why the ex-high priest *Annas should have acted as prosecutor. And 18: 20 f. assumes that there are no witnesses to testify, which would be illegal. Possibly John has carefully transposed dialogue of the Sanhedrin trial into John 10 to mark the climax of Jesus' self-revelation.

It is reported in Luke (23: 6–16) that Pilate referred Jesus' case for Herod's opinion, but Herod had no legal right to be consulted. If therefore this scene in the Passion narrative is historical, it was perhaps no more than Pilate's crude attempt to make amends for a slight inflicted on Herod (referred to in Luke 13: 1).

tribes of Israel The nation of Israel was divided into *twelve tribes, recognizably related but each having a specific character dependent on its experience of migration, war, and economic fortunes. In the tradition all the tribes were in *Egypt and all came out at the *Exodus (Gen. 49), though it is possible that in fact certain tribes were not in Egypt at all. Many modern scholars are sceptical about the antiquity of the tribal system itself and hold that the tribal names refer to the geographical areas where they lived, without any necessary family linkage. In their own traditions, the tribes traced their corporate existence to the eponymous ancestor *Jacob or Israel (Gen. 32: 28; 35: 10). Six tribes were regarded as descended from Jacob's wife *Leah: *Reuben, *Simeon, *Levi, *Judah, *Issachar, and *Zebulon. Two tribes came from Rachel: *Joseph (subdivided into Ephraim and Manasseh) and *Benjamin. On the fringe of these tribes were those alleged to be descendants of the concubines Zilpah (*Gad and *Ashur) and Bilhah (*Dan and *Naphtali).

The twelve tribes are not identical in each of the OT lists: Levi is omitted in Num. 1: 20–43 and 26: 5–50, but the total of twelve is maintained by regarding Ephraim and Manasseh as separate tribes. Levi possessed no land but members of the tribe were distributed amongst the others. In 722 BCE the northern tribes (called Ephraim or Israel) were taken to *Assyria; the southern tribes (called Judah) were exiled to *Babylon in the 6th cent.

After the Exile tribal distinctions were lost, but the number twelve retained a symbolic importance: the NT Church claims to be the successor of the Jews by the use of the number twelve—the twelve apostles will judge the twelve tribes (Matt. 19: 28). James addressed Christians as the twelve tribes in the dispersion (Jas. 1: 1); and just as in the OT lists there is a certain discrepancy so that by combining them there become thirteen tribes, so the lists of the apostles as given in the four gospels and the Acts are not identical and could be made into thirteen names.

tribunal The rostrum occupied by judges in the Roman Empire. Paul was several times brought before tribunals and accused by opponents. The proconsul *Gallio sat in the court on a judge's bench at *Corinth (Acts 18: 12–17) and dismissed the case. Paul was also tried before the tribunal (judgement-seat) in *Caesarea (Acts 25: 6–21); and it happens that at *Philippi excavations have revealed what may have been the tribunal before which Paul and *Silas were accused by the owner of a slave girl with a spirit of divination (Acts 16: 16). The apostles were imprisoned but soon released and given an apology (Acts 16: 39).

tribute It was common during the history of the OT for a conquering nation to impose a tax as tribute on its subject peoples, thus increasing its prosperity and self-esteem. David exacted tribute from neighbouring states which he conquered (2 Sam. 8: 2–10), and *Solomon demanded taxation and imposed conscription on his own people, with the help of a professional civil service (1 Kgs, 4: 7). One-off payments of tribute were made by *Ahaz (2 Kgs. 16: 8) and *Hezekiah (2 Kgs. 18: 14) to Assyria. But when annual payments were demanded the subject state was made responsible for collecting and delivering it, as when Judah paid tribute to Persia (Ezra 4: 20). In the NT era Herod paid tribute to Rome and assessed his own taxation requirements accordingly.

Trinity The Christian doctrine of the triune God. The Trinity is not explicitly mentioned in the NT: it was defined as a result of continuous exploration of the biblical data. The one God is made known as Father, Son, and *Holy Spirit; each is God, but each is a distinct 'Person'. The gospel of John provides most of the NT foundations for the doctrine; Jesus repeatedly refers to his Father who sent him and is one with him (John 14: 7–10) and the Holy Spirit (John 16: 13–15). The final commission of Jesus is reported in trinitarian terms (Matt. 28: 19) and Paul uses similar language of the gifts of the Spirit (1 Cor. 12: 4 ff.) and in his benediction (2 Cor. 13: 13). This experience of the Trinity was thrashed out in debate and treatises, and was formulated at the Council of Nicaea in 325 CE.

Trito-Isaiah The name often applied to chs. 56–66 of the book of Isaiah, or their author, dated later than chs. 40–55.

Troas A port on the NW coast of Asia Minor which in NT times had the status of a Roman colony. Strictly, the place was Alexandria Troas, the second name being added because it was near the famous ancient city of Troy (Ilium of the Homeric era). It was vital for communication and was even twice considered as a possible capital city for the empire. Paul had a *vision at Troas in which a man from *Macedonia invited him to come over to Europe (Acts 16: 8–10); and on a second visit Paul restored *Eutychus to health after he had fallen from a third-storey window (Acts 20: 5–12).

Trogyllium A peninsula on the west coast of Asia Minor, between *Samos and *Miletus (Acts 20: 15, AV, NJB, NRSV marg.)

Trophimus One of Paul's companions from *Ephesus who travelled with him to Jerusalem (Acts 20: 4–5) and who was suspected by Jews of taking him, a *Gentile, into the Temple (Acts 21: 29). A Trophimus is also mentioned at a later stage as being ill in *Miletus (2 Tim. 4: 20).

trumpets The Israelites used long rams' horns as trumpets, and these may still be heard in modern *synagogues. Moses made two silver trumpets (Num. 10: 2) which were blown by the priests (Num. 10: 8). There was a feast of Trumpets on the first day of the seventh month (Num. 29: 1), which was kept as a day of rest.

truth In the OT the 'truth' of God means his reliability (Isa. 65: 16); and reports which are personally verified are established as true and reliable (1 Kgs. 10: 6–7). In the NT truth is correct knowledge (1 Tim. 4: 3; 2 Tim. 2: 25), which is a Greek concept, but it also has the OT sense of reliability (2 Cor. 7: 14) and is to be obeyed (Gal. 5: 7). Much of the gospel of John is an exposition of the truth of God (John 3: 33) which is revealed by Christ (John 8: 26), who is himself the truth (John 14: 6). The *Holy Spirit of truth will come (John 15: 26) and the Church will be led into all truth by him (John 16: 13).

Tryphena and Tryphosa Greetings were sent to these two sisters(?) in Rom. 16: 12; the names were fairly common.

Tubalcain A forger of instruments of copper and iron (Gen. 4: 22) and a descendant of *Cain. The *Kenites were believed to derive their origins from him.

Tübingen critics A group of scholars round F. C. *Baur (1792–1860) at Tübingen University who took a radical view of the NT: only Rom., 1 and 2 Cor., and Gal. were accepted as authentically Pauline, and these epistles revealed a deep opposition between Jewish and *Gentile Christianity. At a later date the historically unreliable Acts represented a Gentile attempt to present a harmonious picture of the early Church, and the gospel of John was the crown of a theological development rather than an account of the Jesus of history. Baur's view that much of the NT was to be dated in the 2nd cent. depended on the so-called epistles of *Ignatius being even later, but the painstaking work of J. B. *Lightfoot of Cambridge demonstrated that the seven epistles attributed to Ignatius (107 CE) were authentic. This partly undermined the Tübingen thesis. There was the further weakness that Baur assumed that Matthew was the first gospel to be written and that Mark's theological motifs argue a later date—a view generally now discarded. Nevertheless, it would be widely agreed today that the harmonious picture of the Acts distorts the facts. Conflicts were passionate and intense, as between Paul and the Jerusalem leaders. In the end, Pauline principles prevailed and we only have the victor's side of the struggle.

Twelve, the Jesus chose a group of twelve (usually called '*apostles' by Luke, e.g. 6: 13) 'to be with him' (on his travels). The Twelve symbolized the new Israel since the old Israel was regarded as made up of twelve *tribes (Matt. 19: 28). The Twelve bore witness to Jesus' *ministry and to his *resurrection. The defection of *Judas Iscariot necessitated that the number twelve be made good by the election of *Matthias (Acts 1: 15–26). Subsequent deaths did not create vacancies that needed to be filled.

twin brothers In Greek mythology, *Castor and *Pollux, sons of *Zeus. They were regarded as protectors of seamen and ships often had their figures on the bows, as did Paul's on the journey from *Malta to Italy (Acts 28: 11).

Tychicus One of Paul's companions on his last visit to Jerusalem (Acts 20: 4–5). A Tychicus is mentioned as taking letters to the Churches of *Colossae and *Ephesus (Col. 4: 7–9; Eph. 6: 21–2).

Tyndale, William Born in England, 1494; executed for heresy near Brussels, 1536. He made the first English translation of the Greek NT and later of the Hebrew *Pentateuch. It greatly influenced the phraseology of the AV. A complete copy of the first edition of Tyndale's NT (printed in Worms in 1526) was acquired by the British Library in London, 1994.

types/typology A recognition that in the Bible there is discernible a pattern of *salvation events. Some of what happened in the OT is seen to be anticipations of events recorded in the NT, and some of the narratives in the gospels seem to be reflected in the Acts. The anticipations are called 'types' and the fulfilments are the 'antitypes'. Thus the story of the *Exodus is repeated in the synoptic gospels; the Israelites cross the *Red Sea, yield to temptations of doubt and disillusionment for forty years in the *wilderness, and then Moses on Mount *Sinai presents the people with the *Law. In the gospels Jesus is *baptized in the water by *John, is tempted for forty days in the wilderness, and then gives the *Sermon on the Mount (Matt. 5–7). The difference is that where Israel failed, by repeatedly grumbling and doubting God's determination, Jesus succeeded. The gospels are, as it were, retelling the story of Israel, but giving the events of Jesus as its climax and rationale. It could be shown that Matt.'s account of Jesus recapitulates that of Israel with Gen. (Matt. 1: 1), Exodus (2: 15), Deut. (5–7); a ministry about kingship and prophecy; an exile (Calvary) and restoration (the Resurrection). The principle behind such *exegesis is that God had the same purpose in the NT as he always had (cf. Heb. 13: 8). He is consistent. Though his plan failed because of Israel's weakness, he did not change his plan but brought it to completion through Jesus.

It is possible that there is a typological parallelism between the Passion narrative in Luke and the later chapters of Acts. In the gospel, the *Lord's Supper and teaching about ministry is followed by four trials (before the *high priest, the *Sanhedrin, *Pilate, and *Herod); the *crucifixion; three days in the grave; the *resurrection. In the Acts, there is a meal at *Troas (Acts 20: 7), a discourse about ministry (20: 28 ff.); four trials, before the Sanhedrin, *Felix, *Festus, and Herod *Agrippa II; the shipwreck, followed by three months with *Publius. And the climax of the book is Paul's arrival in *Rome.

If such a theory is plausible, the purpose of the literary device is to show how the disciple must be as his master; that if the way of Jesus was in humiliation, there can be no other way for the Church (Luke 9: 23).

The story of the *Flood and *Noah's *ark is treated typologically by 1 Pet. 19–21.

Tyre An important city on an island off the Phoenician coast, often associated with *Sidon. Tyre enjoyed close relations with Israel when *David and *Solomon reigned, and cedars of Tyre were supplied for the building of the Temple (1 Kgs 5; 7: 13–46). Later, there were troubles in Israel (the northern kingdom) when *Baal worship from Tyre (which controlled Sidon) was introduced by Queen *Jezebel (1 Kgs. 16: 31; 2 Kgs. 9: 10).

Some of those who listened to Jesus came from Tyre (Mark 3: 8), and he is once reported to have travelled into that district (Mark 7: 24). He had a better opinion of Tyre and Sidon than of some of the Galilean cities which rejected his message (Matt. 11: 21–2).

U

Ugarit The ancient name of the modern *Ras Shamra on the Mediterranean coast in *Syria. Excavations started in 1929 by French archaeologists revealed large deposits of *cuneiform texts in Akkadian and in Ugaritic and give much information about the social organization and *religion of the city. The importance of these texts for OT studies is the light they shine on the cultures surrounding Israel and the legends of their deities, such as *Baal, who is a storm-god. (Ps. 29 portrays Yahweh in very similar terms.) Their language can also sometimes shed light on obscure Hebrew passages in the Bible.

unbelief In the Bible, more a matter of wilful disobedience than intellectual doubting. The 'fool' who denies the existence of God (Ps. 14: 1–4) does so not by argumentation but by his perverse way of living. He refuses to seek the Lord and so has no knowledge of him. This is the major *sin in the world, sin which the *Spirit condemns (John 16: 9). Denial of the work of the *Son of God is tantamount to making God a liar (1 John 5: 10).

unclean There is a comprehensive list in the *Pentateuch of creatures that are regarded as unclean and therefore forbidden as food, or even to be touched (Lev. 11: 8); those animals that ruminate (chew the cud) but do not have cloven hooves, such as camels and hares, or vice versa in the case of pigs. Other unclean creatures include rats and mice, eels and crabs, insects (but not locusts), and some birds.

The attitude to unclean food caused bitter discussion in the early Church. Jewish Christians wished to continue the observance of the Law and demanded that Gentile converts should do so too. Paul resisted the demand, and denounced his opponents in his letter to the Galatians. The conflict is also reflected in the gospels (e.g. Mark 7: 19). Cf. Acts 10. *See also* clean and unclean.

unction An anointing with oil: kings and priests were to be *consecrated with oil (1 Sam. 10: 1 ff.; Lev. 8: 12). *Healing with the use of oil is commended by James (5: 14–15). *Messiah ('Christ') means 'anointed' and the fact that Christ was sent to be 'anointed with the oil of gladness' (Heb. 1: 9) for his Messianic work, and his reception of the *Holy Spirit (Luke 4: 18), provided a model for the subsequent work of the Church which used rites of anointing for bestowal of the Spirit, as at *ordinations.

universalism The belief that all human beings will ultimately share in the grace of God's *salvation. The NT evidence is unclear. Paul maintains that faithful Christians are preordained to salvation (Rom. 8: 29) but does not assert that others are preordained to damnation. He offers the hope not only that the 'full number of the *Gentiles' will be saved (Rom. 11: 25) but also that ultimately 'all Israel will be saved' too (Rom. 11: 26; cf. Rom. 5: 19). The NT evidence against a *doctrine of universalism is in 1 John (5: 12) and Matt. 25: 41, which assert that the *cross is not only a revelation of the *love of God but also a *judgement on all who will not accept that love. It could then be argued that, because God will never *compel* someone to embrace the offer of salvation, therefore an absolutely critical choice must in the last resort rest with the individual. Although there is no limit to the resources of divine love, yet it is conceivable that even the tiniest dying ember of individual responsibility could be so dead that it could not be fanned by God into a flicker of response. This is the meaning of the imagery of judgement.

unknown God Paul had noticed an *altar in *Athens, dedicated to 'an unknown god' (Acts 17: 23). Although such an altar has not been found by archaeologists, inscriptions are known of altars dedicated 'to unknown gods' (plural). The reasoning was that one or other of the gods might show anger at having been overlooked. Such an inscription was designed to cover all such contingencies. At any rate Paul is described as making use of current religious and cultural concepts rather than demonizing them. It represents an attitude of measured appreciation.

unleavened bread Made without yeast, and ordered for cereal offerings in the Temple (Lev. 2: 4, 11). The theory was that the process of fermentation was a kind of corruption from a previously pure condition (1 Cor. 5: 8).

During the festival of Unleavened Bread at the beginning of *harvest only unleavened bread was eaten. This custom was grafted on to Israelite history by being associated with the *Exodus when, according to the tradition, the people left *Egypt in such haste that the women did not even have time to let the dough rise (Exod. 12: 14–20). Thus Unleavened Bread and *Passover came to be regarded as one festival.

Upper Room The scene of the *Lord's Supper (Mark 14: 15) and of the outpouring of the *Spirit on the *disciples at *Pentecost (Acts 1: 13). The room seems to have been reasonably spacious and this suggests that its owner must have been both comfortably off and friendly to Jesus, like the man who lent his colt (Mark 11: 3).

Ur A city situated in antiquity on the River *Euphrates; it had a sophisticated culture and a history of prosperity. It has been thoroughly excavated, and the evidence is that the name 'Ur of the Chaldeans' (Gen. 11: 28) is an anachronism imported by the *Priestly editor during the era of Babylonian (Chaldean) imperialism around the 7th cent. BCE. It was the home of *Abraham before his migration to *Canaan. He left the amenities of the city for the precarious and harsher life of a nomad. There are plenty of modern parallels.

Uriah A *Hittite proselyte to Yahwism (as indicated by his Yahwistic name), who served as an officer in *David's army. David arranged to have him killed in battle in order that he could marry his wife *Bathsheba (2 Sam. 11).

Urim* and *Thummim A means by which leaders of the Israelites believed they could receive divine guidance (Deut. 33: 8). They had disappeared from use by the time of the Exile. They may have been objects such as pebbles, or two stones (one white, one black) held by priests in their garments, and thrown to give answers (1 Sam. 14: 41).

Ur-Markus The name given by several German NT scholars for a supposed first edition of the gospel of Mark. The evidence used to support the theory is the existence of some phrases in our gospel of Mark which do not seem to have been known by Matt. and Luke, and of doublets such as the Feedings in chs. 6 and 8, which seem to be two traditions of the same event.

Ussher, James (d. 1656) Irish Anglican archbishop, who produced a scheme of biblical chronology, beginning with *Creation in 4004 BCE, a date long perpetuated in editions of AV.

usury Interest charged on a loan; *Hebrews were not to charge interest on loans to needy Israelites (Exod. 22: 25) but it was permissible to take it from non-Israelites (Deut. 23: 20). Various humane regulations limited the rights of creditors: bailiffs could not thrust open the door and enter a house (Deut. 24: 10–11). But it is also true that legal safeguards were often ignored or violated (2 Kgs. 4: 1; Isa. 50: 1; Matt. 18: 25). The NT accepts the practice of banking as an ordinary, not in itself a reprehensible, part of life (Matt. 25: 27).

Uzziah King of *Judah jointly with his father from 791 BCE, solo from 767 BCE; he waged successful *wars, but for presuming to take on priestly duties in the Temple he was stated by the Chronicler to be infected with *leprosy (2 Chr. 26: 19). In the year of Uzziah's *death (734 BCE), *Isaiah had his inaugural *vision (Isa. 6). In 2 Kgs. 15 he is called 'Azariah'.

V

valleys Numerous among the mountains of Palestine. They were locations for grazing cattle (1 Chron. 27: 29) and growing crops (1 Sam. 6: 13), but also for battles (Gen. 14: 8).

vanity A familiar refrain in Eccles. (e.g. 1: 2), asserting that life has no meaning. It is a theme in Isa. (57: 13) that idols are nothing but vanity (cf. Acts 14: 15). In the OT and NT 'vanity' does not mean empty pride or conceit.

vaticinium ex eventu (or ***post eventum***) The term applied to a passage in the prophets or the gospels which has the form of a prediction but is in fact written in the knowledge of the event having occurred (e.g. probably Luke 21: 20).

veil (1) Worn by *Hebrew women at the time of *marriage (Gen. 24: 65–7). Also by Moses after he had been 'conversing' with God (Exod. 34: 33): the story was taken up by Paul (2 Cor. 3: 13), who compared to Moses' disadvantage the action of veiling his face so that the people should not realize that the reflected *glory had come to an end with the complete openness he (Paul) displayed towards the Corinthians. (2) The *curtain which separated the *Holy of Holies in the Temple from the Holy Place (2 Chron. 3: 14), through which only the High Priest passed (Lev. 16: 12). The gospel writers claim that it was torn at Jesus' death (Mark 15: 38), symbolizing the access to God which Jesus has opened for everybody. According to *Josephus the veil was a tapestry on which the heavens were depicted. Mark's readers might have recalled that at Jesus' *baptism the heavens were rent (Mark 1: 10). At the baptism, Jesus' role was signified; at the *crucifixion it was consummated.

verbal inspiration The theory that God so inspired the authors of the biblical texts that they had to be written precisely in those words and no others. It has been held in most eras of Jewish and Christian interpretation of the scriptures. It is therefore to be distinguished from modern theories of inspiration which place the intervention of God in the events of history (**Heilsgeschichte*) rather than in the narratives. Verbal inspiration is also different from theories which locate the work of the Spirit in the creative imaginations of the individual authors. Verbal inspiration emphasizes the uniqueness of scripture ('the oracles of God', Rom. 3: 2), divinely given (2 Pet. 1: 21) and accepted as such by Jesus (John 10: 35) and by the rabbis. Traditional Jewish interpretation regards the text as dictated by God and therefore without error, inconsistency, or obsolescence.

Christian interpretation in the Fathers and the Middle Ages held that beyond the literal meaning of the text there were others, especially the *allegorical, but the literal sense was fundamental, and was interpreted in a conservative way. The Reformers rejected allegorical interpretation but put great stress on the authority of the Bible in their controversies with Rome. But the doctrinal clash with Rome, and serious discrepancies amongst the Reformers themselves, led to rival interpretations of the biblical text, and to a new freedom in exploring its meaning.

In 17th-cent. England Moses' authorship of the *Pentateuch was being questioned. In Germany there were soon systematic examinations of language, history, and authorship of the biblical books. Biblical criticism had arrived. Against it, those who upheld a doctrine of verbal inspiration conceded that individual writers had their own styles, and the books of the Bible represented different *genres. But conservative Evangelicals maintain that, essentially, inspiration must entail the total accuracy of the historical records and the truth of its doctrines. The First Vatican Council (1870) took a similar view that inspiration 'is incompatible with error... It is impossible that God himself can err', a view reinforced when the authorities crushed Roman Catholic Modernism in 1907. However, the papal encyclical *Divino Afflante Spiritus* of September 1943 allowed a new and more open approach into Roman Catholic biblical scholarship, strongly encouraged by the Pontifical Biblical Commission in a document called *The Interpretation of the Bible in the Church* (1995).

Most modern scholars accept that the biblical documents must be rigorously examined with all the available resources of linguistics, archaeology, and historical enquiry. For them, any theory of verbal inspiration, however eroded or redefined, is incompatible with scholarly freedom. It is on such a basis that allegiance to Jesus must be either offered or withdrawn.

versions Early translations of the *Hebrew OT and *Greek NT. They were made into other languages of the ancient Near East and also into *Latin. These are among the aids available to a textual critic in the never-ending search for a text as close as possible to the original. *Tatian, in the 2nd cent., compiled his Diatessaron in Syriac, and this was itself put into Arabic, Latin, and other languages. Other Syriac translations of parts of the NT have been discovered; the standard version of the whole Bible, apart from five books of the NT, used in the Syriac Church is called the *Peshitta (= 'simple'). The five missing NT books were added in the 7th cent. CE. Other versions were the Coptic, Armenian, and Ethiopic. A Latin version (the 'Old Latin') became necessary with the growth of the use of this language in the West; and the *Vulgate (= 'popular') was Jerome's Latin version, which superseded the Old Latin and became the official Bible of the Roman Catholic Church. English translations of the Bible are called versions when they are revisions rather than new translations—hence NRSV.

Vespasian The Roman general in the Jewish War of 66–70 CE; but before Jerusalem was captured he was nominated emperor and returned via *Alexandria to *Rome, leaving his son *Titus to complete the siege. He had previously (43 CE) been in command of forces in Britain, and he subdued the Isle of Wight. He died in 79 CE after a reign of conspicuous modesty and affability, and was succeeded by Titus.

Via Dolorosa The route followed by *Christian pilgrims in Jerusalem commemorating the journey taken by Jesus carrying his *cross (Mark 15: 20–2), and punctuated with nine stations for *prayer. The present route was fixed in the 18th cent. and does not correspond with historical probability, but is convenient for the fourteen 'stops' as finally settled for Roman Catholics in the 19th cent.

victory The triumph of God, e.g. over *death (1 Cor. 15: 54–5). Ultimately the victory will be on the *day of the Lord (Ezek. 39: 7–8), when all creation will be made new. In Eph. (6: 11–17) there is a picture of the victory of the person who has *faith couched in terms of the accoutrements of *war.

villages Although not easy to differentiate the terms 'village', 'town', and 'city' in the Bible (*Bethlehem is both a *city in Luke 2: 4 and a village in John 7: 42), a village perhaps resembled a hamlet or satellite grouping adjacent to an enclosed city. Thus the word translated 'dependencies' (NRSV) at Josh. 15: 45 is in Hebrew 'daughters', indicating the dependence of the outlying villages on the mother city.

vine Vineyards and their produce were of great economic, social, and religious importance in Palestine. When the Israelites after the Settlement in *Canaan began to cultivate them, it signalled their intention to stay, for the vine needs long-term and intensive care. The climate was favourable; the hilly terrain suitable—more so than for cereals. Vineyards needed protection from thieves and booths were erected at the time of *harvest (Isa. 1: 8). Isaiah also describes (Isa. 5: 1–7) the different stages required to secure a successful harvest.

*Wine was used for barter (2 Chron. 2: 10) as well as for *feasts (Dan. 1: 5, 8, 16) and in *worship (Jer. 51: 7; Isa. 28: 7–8; cf. Acts 2: 13). Early Christian leaders were advised to be temperate (1 Tim. 3: 3, 8), though not to be total abstainers, like the OT Rechabites (Jer. 35: 7–9). *Wine was drunk at the *Passover festival, and Jesus blessed it at the *Lord's Supper (Mark 14: 23, 25) and the rite became universal in the Church (1 Cor. 11: 25). The vine is used as a symbol of Jesus by the gospel of John (15: 1) and Christians are its branches.

vinegar Sour wine; offered to Jesus on the *cross in a pathetic attempt to diminish the pain (Mark 15: 36) but perhaps understood later as a further infliction of suffering (Luke 23: 36) if it was seen as a fulfilment of the suggestion of vinegar as poison (Ps. 69: 21).

virgin birth The assertion in the gospels of Matthew and Luke that Jesus was born of Mary without the intervention of a human partner. The pregnancy was initiated by the *Holy Spirit (Matt. 1: 18; Luke 1: 35). The earliest writers in the NT (Mark and Paul) show no knowledge of such a virginal conception, and it is suggested that the narratives are a **midrash* on the *LXX of Isa. 7: 14 which prophesies a birth from a 'virgin' (in the Greek). The Greek *parthenos* was used to translate the Hebrew *almah*, which means a 'young woman', and is so translated at Isa. 7: 14 by NRSV, REB, NJB, thus correcting the 'virgin' of AV, RV. But modern translations of Matt. 1: 23 correctly translate the Greek *parthenos* by 'virgin' because Matt. is quoting the LXX. As early as the 2nd cent. CE the Jewish controversialist Trypho was pointing out that the Hebrew did not mean a virgin but that Isa. 7: 14 was referring to the natural birth of *Hezekiah. Indeed the LXX went out of favour with Greek-speaking Jews, who opted for the literal translation by *Aquila in the early 2nd cent. CE. Later in the 2nd cent. the pagan philosopher Celsus claimed that Jesus' father was Panthera, a Roman soldier.

Since historical enquiries cannot settle the truth or otherwise of the gospels' accounts, theological arguments are brought in. In favour of the traditional *doctrine it is argued that it is appropriate as a means by which God made a decisive break with the old, sinful humanity and inaugurated in Jesus a new, untainted humanity. The miraculous birth came to be regarded as a sign of the divine nature of Jesus. On the other hand, some modern theologians argue that the doctrine must imply that the humanity of Jesus was somehow impaired, since with only one parent he could not have been fully human. If Jesus was provided miraculously with chromosomes, specially created by God, with no human ancestry, how did he have a human inheritance, of the house of *David (Matt. 1: 17, 20; Rom. 1: 3)?

virtue Lists of virtues in the NT correspond closely with those in the philosophical tradition of the *Stoics (e.g. Phil. 4: 8). There is an overlap between Christian and natural *ethics, though the contexts differ. In the NT virtue is said to be the result of God's *grace (2 Pet. 1: 3) and is to be in imitation of Christ himself (1 Cor. 11: 1).

vision Paranormal experience interpreted by the subject as revelation, and usually of some object or scene, either heavenly or earthly, and either of the present or of the future. *Isaiah saw God on his throne (Isa. 6); *Amos had visions of locusts (Amos 7: 1–3), of fire (drought; 7: 4–6) and of a plumb line (7: 7–8) which he interpreted as a message of doom for Israel.

vision of God Described as the supreme goal of the Christian life, based on one of the beatitudes, 'Blessed are the pure in heart, for they shall see God' (Matt. 5: 8). It is attainable only by being detached from self-centred desires of the senses and by being on fire with *love.

Visitation, the The name given to the visit of *Mary, the mother of Jesus, to her cousin *Elizabeth, according to Luke (1: 39–55). It is commemorated in the liturgy of the East and in the Book of Common Prayer on 2 July and in the Roman Catholic and modern Anglican calendars on 31 May.

vocation Calling. The modern usage of a vocation to a profession is not found in the NT, though it was developed in the 16th cent. by Luther and Calvin as part of their assault on Roman Catholicism, which had regarded vocation as primarily a call to enter a religious order. In the NT people are called to a new life in Christ (1 Cor. 1: 26) and *apostles have a vocation to encourage them (1 Cor. 1: 1).

vow A solemn promise made with a religious sanction. *Jacob vowed that if God protected him he would give him a tenth of what he earned (Gen. 28: 20–2). There were regulations about vows (Mal. 1: 14) and a tariff of payments to accompany and validate them (Lev. 27: 8). Paul cut his hair in fulfilment of a vow (Acts 18: 18) and paid the expenses of other vow-makers (Acts 21: 24).

Vulgate The translation of the Bible from the original languages into Latin by *Jerome (from 383 to 405 CE) undertaken at the request of Pope Damasus to bring order into the various existing versions. It became the official Bible of the medieval Church, though after Jerome there were some subsequent revisions. Jerome himself made some additions to the text; they were of dubious authenticity and had doctrinal implications.

vulture The king of birds, according to the *Hebrews (Lam. 4: 19, where English versions translate 'eagles') but considered *unclean (Lev. 11: 13). Of the species known in Palestine, probably that referred to by Jesus (Matt. 24: 28), is the griffon vulture, as in Prov. 30: 17, translated 'eagles' by AV, but 'vultures' in modern translations.

wadi Streams that flow only irregularly in Palestine and are therefore unreliable for water supplies and symbolical of deceit (Jer. 15: 18). The word is Arabic and is frequently used in modern commentaries.

wages Payment for *labour in Israel was made at the end of each day (Deut. 24: 14), or year (Isa. 21: 16), to shepherds (Gen. 29: 15), nurses (Exod. 2: 9), prophets (Neh. 6: 10–13), and craftsmen (Isa. 46: 6). The amounts are not normally mentioned, but according to Jesus' *parable an employee contracted with an employer in a vineyard for a *denarius a day (Matt. 20: 2).

wailing *Death was accompanied with ritual mourning and professional wailers were employed (2 Sam. 3: 31; Matt. 9: 23; Acts 9: 39). The length of time for wailing depended on the social position of the deceased.

walls It was essential in ancient times to protect vineyards with walls of unhewn stones and *cities with quite massive fortifications. The earliest walls of *Jericho consisted of solid stone 2 m. (6 feet) thick. From the walls defenders of a city could engage an enemy (2 Sam. 11: 20–1) and down a wall fugitives could make their escape (Josh. 2: 15). Paul was let down the wall of *Damascus in a basket (Acts 9: 25); he later uses a metaphor to describe the high priest *Ananias as a 'whitewashed wall' (Acts 23: 3).

war War was a frequent fact of life throughout the OT era, and the geographical situation of Palestine between great powers inevitably dragged Israel and *Judah into their conflicts. The *Hebrews themselves, in coming out of *Egypt, fought the existing inhabitants of *Canaan according to Josh. (though there is no archaeological corroboration) in order to secure territory for themselves, and the kings from *Saul to the time of the Exile (586 BCE) were engaged in constant warfare to maintain or extend territory. After the Exile the Jews were in conflict with neighbours as well as with the great empires, and there were popular uprisings led by the *Maccabees in the 2nd cent. BCE and Simon *Bar-Kochba in the 2nd cent. CE. With such a history of war it is not surprising that Israel's God is often described as a warrior (Isa. 42: 13) who helps his people against their enemies (1 Sam. 30: 26). When the nation experienced a disaster it was interpreted as the withdrawal of God's help (Ezek. 23: 24–5). After the Exile war was seen as a possible means of winning freedom from foreign oppressors, possibly under the leadership of a Messianic Son of *David (Pss. of Sol.) or by direct divine intervention. War had become an *eschatological concept. In the NT the use of military force is disavowed (Acts 1: 6; John 18: 36) though expected to continue (Luke 14: 31 f.), and war is used in the sense of spiritual conflict (Eph. 6: 12).

watches In NT times the day was divided into four watches of three hours each (Luke 12: 38), which was the Roman system. The command of Jesus to keep awake (Mark 13: 35) refers to the four watches of the night, and these correspond to four dramatic events in the subsequent passion narrative (Mark 14–15)—the *Lord's Supper, *Gethsemane and Jesus' arrest, Peter's denials, and the handing over of Jesus to *Pilate.

water Immensely important in the economy, religion, and imagery of Israel; rainfall determined the pattern of life, where settlements could be established and the sort of cultivation that was practicable. To offer a cup of water was a gesture of *hospitality (Matt. 10: 42), as was the provision of a basin of water to wash guests' feet (Luke 7: 44). Washing with water was prescribed to clean away ritual impurities (Lev. 11: 29–38) as well as to sterilize skin diseases (Lev. 14: 8). *Naaman was cleansed of *leprosy by washing in the River *Jordan (2 Kgs. 5: 14), and a blind man received his sight after washing in the pool of *Siloam (John 9: 7).

Water has a place in the *myths of Genesis: there was a primeval ocean which God divided into the upper waters (which provided rain) and the lower waters (which formed the sea). The latter came to represent chaos—the sea was always thought of as turbulent—but God sat triumphantly over it (Ps. 29: 10). To be submerged in the ocean (Ps. 69: 1–2) was equivalent to being in anguish of spirit, and *salvation was the experience of being rescued from the water (of death; Ps. 18: 16). It is in accord with this concept of water that Jesus described his coming *crucifixion as '*baptism' from which his human spirit shrank (Mark 10: 38), and in the early Church baptism was understood to be a participation in the *death of Jesus (Rom. 6: 3); the baptized are those whom God has saved out of the *flood (1 Pet. 3: 20).

wave offering Part of the ritual of *sacrifices. The person making the offering extended his hands on which the priest placed the offering and then put his hands underneath, and together they were raised up to the *Lord (Exod. 29: 24). The breast of the animal was waved (swung) to and fro in the direction of the *altar (Exod. 29: 26).

Similar rituals were practised by Nazirites when they made their vow of abstinence. The offering, in this case, was a ram's shoulder.

The priests received as their dues the more generous share of the wave offering (Lev. 7: 34).

wax Tablets of beeswax were used for writing purposes by both *Assyrians and *Romans, and by *Zechariah for writing the name of *John the Baptist (Luke 1: 63).

way The literal meaning of a path came to be used metaphorically in both OT and NT of the pattern of divine activity (Isa. 55: 8–9). In the NT Jesus is himself identified with the Way in as much as he opens up a new relationship with God (John 14: 4–6). The early *disciples were known as followers of the Way (Acts 9: 2). The expression in Isa. 40: 3 may have been the origin of the term in the *Qumran Manual of Discipline, where the monastic life of the community is described as 'the Way'.

waymark Like a milestone, only to indicate the course of a track rather than its distance (Jer. 31: 21, AV; 'road markers', NRSV; 'signposts', REB, NJB).

wealth In OT times wealth was measured in terms of cattle or sheep and *silver and *gold (Gen. 24: 35). It is recognized that these possessions came from God, who had ordained that material blessings are given to some so that they can help the poor (Deut. 15: 10).

While the NT also accepted that riches are given in order to assist the poor (2 Cor. 8: 8–10), the possession of wealth is viewed more negatively, especially in the gospel of Luke (6: 20–1, 24–9). Any display of luxury is condemned (Luke 16: 19–31) because of the probability of injustice (Jas. 5: 1–6) and because it is a distraction from the basic purpose of life (Matt. 6: 21).

weapons The Israelites were equipped with both offensive and defensive weapons for their incessant warfare: those used in hand-to-hand fighting consisted of clubs (Prov. 25: 18) and hammers (Jer. 51: 20), swords and daggers (Judg. 3: 16–22; 1 Sam. 17: 51), spears (1 Sam. 17: 7), and the rod and staff familiar in the pastoral Ps. 23: 4; these were certainly weapons of war according to Isa. 10: 5. For attacking a distant enemy the Israelites used slings (1 Sam. 17: 40) and bows and arrows (Jer. 50: 14). For defence, there were shields (1 Chron. 12: 24) and breastplates (1 Kgs. 22: 34) and greaves (1 Sam. 17: 6). Chariots were sometimes in use (1 Kgs. 10: 26; Ps. 68: 17).

Many biblical weapons are introduced in a metaphorical sense in the NT (2 Cor. 10: 4; Eph. 6: 11–17).

weather *Rain normally falls in the winter in *Palestine—some in October–November and the heaviest in January to March, with lighter rain in April or May. The western slopes of the hills receive the largest amounts. Temperatures are not extreme, except for the great heat of the *Jordan valley.

wedding There is no mention in the Bible of weddings as religious ceremonies; they were a legal and social occasion marked by customs developed over the ages. There was a procession apparently from the home of the bride, who was veiled (S. of S. 6: 7), to that of the bridegroom (Matt. 25: 6), but the details in Jesus' parable are confusing—would the procession be at night? And would a shop selling oil be open after midnight (Matt. 25: 9)? There followed a lengthy banquet (Matt. 22: 2). The *parables of Jesus which centre on weddings indicate something of the *joy as the *community celebrate a new family. A wedding reception at *Cana attended by Jesus and his mother is described in John 2: 1–11. Guests at a wedding reception were expected, just as they are today, to wear their best clothes (Matt. 22: 11–12).

weeds In the AV *'tares', which grow along with wheat and are difficult to separate because the roots become intertwined. In Jesus' *parable (Matt. 13: 36–43) an interpretation regards the weeds as unbelievers, whose *judgement at the end of the day lies solely in the hand of God.

week Jews and Christians reckoned time in periods of weeks each lasting seven days, but whereas Jews observed Saturday, the *Sabbath, as a day of rest (cf. Gen. 2: 2–3), Christians began to keep the first day special as being that of the Lord's *Resurrection (Acts 20: 7).

Weeks, feast of Celebrated at the beginning of the wheat *harvest (Exod. 34: 22), seven weeks after the presentation of the barley omer (= the first sheaf; Exod. 23: 16) during the feast of Unleavened Bread. It was also called *Pentecost (Lev. 23: 16) and by NT times was kept as the anniversary of the giving of the *Law on Mount *Sinai, and by the Church as the day on which the *Holy Spirit empowered the *apostles (Acts 2: 1–4).

weights and measures The *law in OT times demanded honest weights and measures (Lev. 19: 35 f.), though this was not always observed (Amos 8: 4–5). There was no unified system covering the whole country. Metals were weighed in talents; one was divisible into 60 minas of 60 *shekels each, but the weight of a shekel varied.

Height and length were measured in the OT in cubits; dry capacity in homers, which could be loaded on a donkey, or ephahs, containers which could hold a person (Zech. 5: 7). Liquids were measured by a bath (72 litres) and a hin (1/6 of a bath).

In the NT talents are mentioned in the *parable in Matt. 18: 23 ff. A talent weighed perhaps 36 kg. or a hundredweight (Rev. 16: 21); a pound (John 12: 3) was the Roman libra—about 317 g. or 12 ounces.

Distance and length and height in the NT are measured in terms of cubits (Matt. 6: 27, AV, NRSV marg.), and a boat 'two hundred cubits from the shore' (John 21: 8, AV, NRSV marg.) would have been about half a stadion or 91.4 m. (100 yards) away (REB, NJB). The 'mile' of the *Sermon on the Mount (Matt. 5: 41) would have been the Roman mile of 5,000 Roman feet (1,480 m. or 1,618 yards). Capacity is measured in the NT by Roman measures, so that a 'firkin' (AV, RV, John 2: 6) would have been from 91 to 136 litres or about 20 to 30 gallons (NRSV, REB, NJB).

Weiss, Johannes (1863–1914) German NT scholar at Heidelberg. He strongly emphasized the *eschatological teaching of Jesus.

well The means of tapping underground water or springs (Gen. 21: 25). Because water was sometimes scarce, wells, artificially dug, were extremely important and often fiercely fought over for possession. They could also be in the form of deep cisterns for storing water or even immuring prisoners (Jer. 38: 6). The top of wells might be covered by a circular stone (Gen. 29: 2; Exod. 21: 33) and the water drawn up by means of a vessel attached to a rope.

Wellhausen, Julius (1844–1918) German OT scholar at Marburg. He was responsible for refining the theory that the *Pentateuch was an edited composition of four sources each deriving from a different historical period. They were given the symbols *J, *E, *D, *P.

Westcott, Brooke Foss (1825–1901) English NT scholar. As a professor in Cambridge, he wrote three commentaries, notably one on the gospel of John, and together with Fenton *Hort published an edition of the Greek NT taking into account all the available evidence; this then served as the basis for the *English Revised Version of 1881. Westcott became bishop of Durham and evinced a deep concern for the welfare of miners.

Western Text A family of NT MSS, including *D and the Old Latin version, which contain interesting variations from other principal MSS, usually in the form of additions (e.g. Acts 15: 29, REB marg.).

whale The whale was not known to the *Hebrews, and the word translated 'whale' (Ezek. 32: 2, AV, and more famously Jonah 1: 17 and Matt. 12: 40) refers to a large sea creature or 'sea monster' (NRSV, REB, NJB).

wheat Various types of wheat existed in Palestine. The quantity produced depended on the vagaries of the *weather. Crops could be plentiful (1 Kgs. 5: 10–11), or negligible (Gen. 43: 1–2). The translation 'corn' (e.g. Mark 2: 23) is adopted by AV, REB, and NJB, but NRSV has 'grain', as is preferred in America.

wheel Wheels were used in OT times on chariots and wagons and on the lavers in the Temple court (1 Kgs. 7: 30–3). *Ezekiel's *vision of God's throne includes wheels with eyes (Ezek. 1: 15–28). Wheels were part of the equipment for a potter, and were also found at the head of a well for guiding the rope.

whirlwind A swirling wind or tornado was rare in Israel but violent *storms were common and are frequently described as 'whirlwinds', e.g. Job 37: 9; Jer. 23: 19.

They are vivid metaphors for a sudden attack (Jer. 4: 13) or divine vengeance or anger (Hos. 8: 7).

wicked Both OT and NT anticipate certain *punishment for the wicked (Ps. 9: 17; Matt. 13: 49). It was a constant puzzle that the wicked prospered (Ps. 37: 36; 69: 4; 73: 3–12) and frequently there is the refrain that sooner or later they will be punished. Wickedness is a deep-seated *evil in the heart (Jer. 17: 9) and is inspired by the *devil (1 John 3: 12).

widows In a patriarchal society in which women 'belonged' to a man, a widow was in a very exposed and vulnerable position. She was on her own—outside normal relationships. Hence in the OT widows (and orphans) were an especial charge on the community's care (Exod. 22: 22 ff.) and of concern to God (Ps. 68: 5). The OT prophets frequently appealed to the nation's conscience on their behalf (e.g. Isa. 1: 23). The *apostles arranged for charitable distributions to widows (Acts 6: 1 ff.). Some of the younger widows took advantage of this (1 Tim. 5: 9 ff.) and restrictions had to be imposed. Paul, though reluctantly (1 Cor. 7: 40), thought it reasonable that marriageable widows who had the opportunity might remarry and later this liberty seems to have been positively accepted (1 Tim. 5: 14) but with a sly innuendo (5: 11).

wilderness Desert area of sand or rock—but also sloping ground unsuitable for cereals but possible for grazing cattle or sheep (Exod. 3: 1). It was a much treasured tradition of the Hebrews that their forefathers had wandered for 'forty years' following the *Exodus from *Egypt, after complaining in the wastes of *Sinai (Deut. 8: 15). In retrospect this became idealized as a time of incorruption (Hos. 2: 14–15). Jesus spent 'forty days' in the wilderness (Matt. 4: 1–11) tempted by *Satan to doubt his mission and visited by the wild *beasts (Mark 1: 13) who roamed the wildernesses (Ezek. 34: 25). The *Qumran monastery was established in the wilderness in the light of Isa. 40: 3; members of the Community were preparing 'a highway for their God'.

will Used in Gal. 3: 15 and Heb. 9: 16 of the last testament made against *death and in illustration of the idea of a *covenant or agreement or arrangement. There is a finality about a will.

wind Regarded by the *Hebrews as a more than a natural phenomenon; Hebrew *ruach* means not only wind but also 'breath' and 'spirit'. Hence the play on the words in John 3: 8. In the literal sense wind is mentioned in Luke 12: 55; a tempestuous wind called Euroclydon (Acts 27: 14, AV; 'Euraquilo', RV; 'the northeaster', NRSV, REB, NJB) was a sudden typhoon which blew down from the mountains of *Crete.

window Few Palestinian *houses had windows, but the house at *Troas, the Roman colony in Asia Minor where Paul preached until midnight, was better equipped. A young man, *Eutychus, overwhelmed by sleep and perhaps by the heat of the oil lamps, fell from an unglazed third-storey window (Acts 20: 9).

wine In the OT wine, fermented grape juice, is associated with festivals and occasions of *joy, and with *covenants. It also symbolizes blood, being derived from the red juice of crushed grapes (Isa. 63: 2). Similarly, in the NT new wine (Matt. 9: 17) stands for the new *covenant or new *faith which cannot be contained within the framework of the old. And at *Cana the significance for the evangelist of his account of Jesus changing a great quantity of *water into wine (John 2: 6–10) is an indication of the liberating joy of the Gospel for all mankind. Wine was used at the *Lord's Supper when Jesus inaugurated the new covenant in his blood (Mark 14: 24), and it has subsequently been used at the Christian *Eucharist or *Communion (1 Cor. 11: 25).

wine-bibbers Jesus was accused of being a wine-bibber (Matt. 11: 19, AV; a 'drunkard' NRSV, NJB, a 'drinker', REB) by hostile critics of his way of life. They contrasted it with the austerity of *John the Baptist—though the same people had criticized John because *he* was austere!

winnowing The process of hurling threshed grain into the air with a kind of fan; the wind separated the valuable grain from the chaff (Ruth 3: 2) so that it could be stored. It was *John the Baptist's metaphor for the divine *judgement (Luke 3: 17).

wisdom Not so much an intellectual gift in the OT as a technical skill (Exod. 36: 8) or ability to live satisfactorily (Prov. 1: 5–12). Above all wisdom is a quality or attribute of God, who created the world (Prov. 8: 22–31) and gave Israel the *Law (Ecclus. [= Sir.] 24: 1, 23). The noun *hokma* in Hebrew is feminine, and wisdom is often personified as a woman or a 'sister' (Prov. 7: 4), active with God in *creation (Job 28: 25–7). The latter concept is veering dangerously near the *apostasy condemned by *Jeremiah (44: 17) which provided *Yahweh with a female consort.

Wisdom literature in the OT consists of Proverbs, Job, and Ecclesiastes, together with Ecclesiasticus, the Wisdom of Solomon, and several of the Psalms (e.g. 37). The teaching of these books is restrained and reasonable, based on human experience rather than the early history of Israel—though the book of Wisdom (10–12) alludes to events from *Adam to *Sinai. Because Solomon had achieved a reputation for wisdom (1 Kgs. 4: 29–34) the books of Prov., Eccles., and Wisd. came to be ascribed to his authorship. These books, however, were compiled by groups trained in traditional wisdom-culture, such as existed also in *Egypt and other countries in the Near East. There are parallels between the Egyptian Wisdom of Amenemope and the OT Prov., and also between other Egyptian literature and several of the Pss. All over the Near East divisions between the powerful and the disadvantaged were remarked, and the sages who wrote the books of wisdom made their pithy observations about the reasons for such injustices. Nevertheless much of the literature was probably instruction for the sons of the wealthy given by scholars. The books of Job and Eccles. have a stronger theological content than the other books. These may well have been produced to fill a theological vacuum when the old structures of cult, court, and Temple had broken down at the time of the Exile. The Wisdom literature has a kind of universal relevance and practical role. Moreover the personification of Wisdom (a feminine noun in both Hebrew and Greek) in Prov. where Wisdom is closely associated with God in creation (Prov. 8: 22–31) in due course offered theological ideas to writers of the NT and afterwards. They had to grapple with a doctrine of Christ which did justice both to their monotheism and to their worship of Jesus. So Wisdom thinking lies behind Matt. 11: 27, but the parallels which might be cited (e.g. Wisd. 2: 13, 16) are with the disciple of Wisdom, as the 'Son of God' not with Wisdom herself. But in Matt. 11: 28–30 and 23: 34–6 Jesus himself is presented as Wisdom who speaks. Paul (1 Cor. 1: 24) asserts that Christ is the Wisdom of God, and the OT personification of Wisdom is the background of the language about Christ in John 1: 1–18; Eph. 3: 8–10, and Col. 1: 15–20.

Wisdom of Solomon A Greek book composed in *Alexandria in the latter part of the 1st cent. BCE or the early part of the 1st cent. CE. It contains a warning to the *wicked of a future *judgement and an affirmation of the doctrine of *immortality; it derides the cults of idols and suggests that they are made in the likeness of human beings to meet human needs—grieving parents who want to remember their children, and kings who find it convenient to invoke a supernatural authority to support their powers. Possibly the aim of the writer is to urge Jews in *Alexandria to be vigilant against the wiles of *idolatry. The book is not quoted in the NT, though possible allusions to Wis. 7: 26 have been detected in *Heb. (1: 3), an epistle which has other links with Alexandria, and Paul is often held to be dependent on the book, especially in his letter to the Romans.

The pseudonymous author uses the name *Solomon, whose wisdom was renowned, and in 6: 12–10: 21 there is a kind of Solomonic autobiography in which the king praises wisdom, God's partner, and prays for her continuing support (9: 10).

wise men A term which embraces sages of Israel as well as those in *Egypt and *Babylonia, and in particular the *Magi, astrologers from the east, who visited the infant Jesus (Matt. 2: 1–18). In Israel from the time of *Solomon there may have existed wise men who taught at the king's court in order to preserve the nation's traditions (Jer. 8: 8–9; 18: 18).

witchcraft Forbidden in the *Deuteronomist Code (Deut. 18: 10–11) but licensed by *Manasseh (2 Kgs. 21: 6). A woman who practises witchcraft is condemned to death (Exod. 22: 18) in the ancient Book of the Covenant; this verse was until 1722 cited by English lawyers as authority for burning witches. The 'witch of Endor' (1 Sam. 28) was a medium who indulged in necromancy, and was able to call up *Samuel from the dead, even though her practice was officially banned; but the condemnation of the *prophets (Jer. 27: 9; Mal. 3: 5) shows that sorcerers and mediums continued to answer popular demand.

withered hand A condition in which the muscles are shrunken and the limb is shorter than it should properly be; the cause may have been infantile *paralysis. Jesus healed such a sufferer on a *Sabbath (Matt. 12: 13).

witness At a trial two or three witnesses are required to corroborate evidence (Matt. 26: 60; Heb. 10: 28). They can initiate sentence of *death (Deut. 17: 7; Acts 7: 58). False testimony is severely punished (Deut. 19: 16–21). In the NT a witness is one who can testify to the deed of Jesus in his ministry, death, and *resurrection (Acts 1: 22). The word has come to have a specialized meaning when applied to someone who has witnessed to Jesus even to the extent of dying for him—hence a '*martyr' (derived from the Greek *martus*, 'witness'). The word is on the way to this meaning in Heb. 12: 1, where the 'cloud of witnesses' who in spite of suffering have retained the *faith are compared to cheering spectators in a stadium.

wizard Male and female practitioners in sorcery or divination who were unacceptable in Israel (Lev. 19: 31). *Saul's visit to *Endor to persuade a woman to bring up *Samuel from the dead (1 Sam. 28: 9) went against his own legislation. Sorcerers were banned again under *Josiah (639–609 BCE; 2 Kgs. 23: 24). *See* witchcraft.

woe The antithesis of 'blessed' or 'happy', as in Luke 6: 24 ff. where Jesus laments the unhappy condition before God of those who are rich, complacent, and self-satisfied, compared with the poor and the persecuted (Luke 6: 20–3) who are blessed. Two Jewish towns, *Chorazin and *Bethsaida, are condemned as miserable and doomed (Matt. 11: 21) and contrasted with the potential for repentance in the Gentile cities *Tyre and *Sidon. Three woes (disasters) are predicted for the inhabitants of the earth (Rev. 8: 13); this is a theme frequently met in Jewish writings of this period, that there will be a time of bitter suffering, the 'Messianic woes', sometimes expressed as birth-pangs, preceding the dawn of *salvation, a belief expressed by Paul in terms of the cross (2 Cor. 5: 14).

wolf A fierce creature which hunted by itself, rather than in packs, in Palestine (John 10: 12) at night (Hab. 1: 8). Its vicious habits were so greatly feared (Ezek. 22: 27; Zeph. 3: 3; Acts 20: 29) that its powers of destruction were proverbial (Jer. 5: 6).

woman In Israel in the early period the *man was the absolute ruler of the extended *family; if a husband died, the widow was given to the nearest brother of the deceased (Deut. 25: 5–10). Women had no powers, could make no decisions, though they could sometimes engage in inspired trickery (Gen. 27 and 31). The subordination of women was maintained during the monarchical period; *divorce was exclusively open to a husband, and a woman's *adultery was, according to the *law, a capital offence (Lev. 20: 10). Women were not entitled to own property and were kept in states of impurity—e.g. during and after menstruation and after childbirth (the length of purdah was doubled if a baby was female). In practice, however, a measure of humanity possibly prevailed, and women were not excluded from worship ceremonies (Deut. 16: 13–14). Women are to be honoured as parents equally with men (Exod. 20: 12). Women could seize opportunities: *Abigail, wife of Nabal, loaded a great quantity of her husband's most valued goods on to donkeys and delivered them to *David (1 Sam. 25: 23), and Queen *Jezebel certainly exercised power (1 Kgs. 21: 7). There had also been women prophets in Israel—*Miriam and *Deborah, and 'wise women' (2 Sam. 20: 16–22). Huldah interpreted *Deuteronomy for King *Josiah (2 Kgs. 22: 14–20). There is a picture of the ideal domesticated woman in Prov. 31: 10 ff.—but this perhaps is an example of how the generally subordinate condition of women was due to the tradition being shaped by men. Legislation after the Exile imposed more restrictions on women. They were no longer able to participate with men in worship, but in the Second Temple were relegated to an outer court. Their testimony was not accepted in courts, and they could not teach the *Torah. By the NT era there had been a few changes: it is assumed that women in the Graeco-Roman world of the gospels may legally divorce their husbands (Mark 10: 12); but there is evidence that in post-70 CE Judaism male Jews still gave thanks to God that they were not born female (Tosefta Berakot, 7: 18).

In the gospels there is no indication that Jesus in his teaching or actions displayed a biased, masculine ('sexist') attitude to women, and his being called '*Son of God' by the evangelists was more a theological recognition of his close relationship to God than an assertion of maleness. The birth of Jesus gives *Mary a prominent role in the scheme of *salvation, and women are also prominent in the resurrection narratives: it is they who receive the first revelation that Jesus has been raised. Between the *birth and *resurrection there are notable healing *miracles for women—the distressing case of menorrhagia (Mark 5: 24–34) and the Gentile *Syro-Phoenician girl (Mark 7: 24–30). Women anoint Jesus (e.g. Luke 7: 36–50), and Mary and *Martha are described, along with their brother *Lazarus, as 'loved' by Jesus (John 11: 5). Jesus' relationships with women accord with his teaching that the kingdom of God implies a new community of love which embraces all mankind (Luke 13: 10–17). People are welcomed by Jesus irrespective of race, status, or gender, and those who are called to leadership are chosen on the basis of God's gracious spirit not on accidents of birth.

In the Church there were undoubtedly tendencies to retain traditional masculine superiority (1 Tim. 2: 11–12). When, however, Paul urges women in the Corinthian Church to ask their husbands at home about matters discussed rather than in the public assembly (1 Cor. 14: 35) he is making a local regulation for a local problem, in the interests of good order, just as he instructed male *prophets to be silent if their speech proved unedifying. Paul certainly did expect women normally to speak to the whole congregation (1 Cor. 11: 5).

The tenor of the Christian gospel is in favour of a true evaluation of women (Gal. 3: 28), though the Church down the ages has been slow and reluctant to incorporate this belief into its institutions and rituals. See *feminism.

wonders Paul in Jerusalem claimed that the truth about Jesus was confirmed by the signs and wonders that God had done through him (Acts 2: 22) and of which he assumed many in the audience must have been aware. Yet there was a warning that such phenomena could be the work of false Christs (Mark 13: 22); they are not trustworthy (John 4: 48).

word Revelation: a central and dynamic concept in both OT and NT. (1) It was by God's 'word' that the *heaven and *earth were made (Gen. 1: 3, 6, 9). (2) The *Ten Commandments are his words (Exod. 20: 1). They are often referred to as the 'Ten Words'. *Prophets speak the Lord's word (Jer. 1: 2), which is the revelation of his will and purpose. (3) In the NT Jesus preaches the word (Mark 2: 2). (4) In the gospel of John Jesus is identified with God's word—his revelation or his meaning (John 1: 1, 14). The word in this sense is not identified with the written words of any scripture. (5) In Paul and Acts (e.g. 6: 2), the word means the Christian proclamation.

work Work was natural to mankind from the beginning (Gen. 2: 15) but is not a *punishment. It was only *sin that turned it into a wearisome and interminable drudgery (Gen. 3: 16 ff.). However, it can be given a positive value—not for the mere accumulation of wealth (Prov. 23)—but as a means of sharing in the creative process of God (Col. 3: 23). So refusing to work, even if for the religious reason of awaiting the return of the Lord (1 Thess. 4: 11; 2 Thess. 3: 10) is reprehensible. Paul worked at his craft in order to maintain himself without being a charge on his converts (Acts 18: 3; 1 Cor. 9: 1–7). Work is a service to the *community, and when an individual suffers through lack of work the whole Body suffers, for the body consists of many related members (1 Cor. 12).

works Used in the NT of human behaviour. 'Good works' are the human response to the *grace of God; but they do not earn the grace of God. Much controversy since the Reformation has been caused by the Reformers' alleging that Catholics sought to earn *salvation by good works, whereas in their view believers are justified only by Christ through *faith. The epistle of James (2: 14) was not well liked by Martin Luther, who suspected a *doctrine of salvation by works in it, and thus a regression to

*Judaism—which was itself a misunderstanding of Judaism.

world No single or unitary concept in the Bible. There is in the OT a strong sense of the created order, from which God as its maker is entirely distinct, but there is no conception of the universe as a system to be explored, explained, and loved. The NT has several words in *Greek translated 'world' of which *kosmos* is the most frequent, meaning the universe (Acts 17: 24). A second word is *oikoumene* (e.g. Heb. 1: 6), meaning the inhabited world; and, thirdly, *aion* used for the contrast between 'this world' and 'the world to come'. Paul exhorts his readers not to be conformed to this world, *aion*, or age (Rom. 12: 2). According to *John, the world was made through Christ (John 1: 3), taking up the OT idea of the word in Prov. 8: 22, but yet the world apart from Christ stands under *judgement (John 16: 8–11). In this derogatory sense, the world is to be renounced (1 John 2: 15–17), though Jesus did not want the *disciples to withdraw from the world (John 17: 15) in the sense of society. Nor did Paul (1 Cor. 5: 10).

worm Worms in the Bible may be either maggots or larvae of various kinds of insect. It is a term of contempt when applied to human beings (Job 25: 6) and is particularly associated with *death and decay (Isa. 14: 11; Acts 12: 23). It was a worm that attacked the bush under which *Jonah was sheltering (Jonah 4: 7) outside *Nineveh.

wormwood A plant with a bitter taste, and therefore a metaphor for bitterness (Prov. 5: 4) and sorrow (Jer. 23: 15). A star which fell from *heaven and poisoned the waters is called Wormwood (Rev. 8: 11).

worship The homage rendered to God (Exod. 20: 1–6) in gestures and appropriate words, but also, the *prophets insisted, in ethical conduct (Amos 5: 21–2). *Sacrifices were offered to God not as food but as *gifts that were costly to the offerer; animals for sacrifice had to be unblemished.

The Temple in Jerusalem was the centre of the nation's worship from the time of *Solomon. There was a daily sacrifice of a lamb in the morning, and a second lamb, together with cereal offerings, was offered by the *high priest on the *Sabbath. After the Exile the Day of *Atonement was added to the ritual in the Second Temple. There were in addition voluntary offerings made by lay people; the main condition for participation in this service was to be in a state of ritual purity.

*Prayers and singing were part of the Temple worship (1 Chron. 16: 4–6), and people used many of the psalms. They were also used in worship in the *synagogues which began to spring up in the *Dispersion as centres for Jewish immigrants and were very soon places of religious instruction. Early Christian worship took over features of Jewish synagogue worship, including appropriate furnishings, though according to Acts the early followers of Jesus joined in Temple worship (Acts 3: 1). But former rituals were soon given radically new meanings: sacrifice was no longer bloody—the whole Temple institution was obsolete (Heb. 9: 13–14). *Circumcision was replaced by *baptism, *laying on of hands was associated with special commissions (Acts 13: 2–3). Above all, there was a new form of sacramental worship with *bread and *wine in commemoration of the sacrifice of Jesus and in anticipation of the joyous *Messianic banquet in the age to come (1 Cor. 11: 26). The first day of the week was set aside for this communion, which was accompanied by prophesying, reading, singing, and prayers, all in a spirit of joy and *thanksgiving (1 Cor. 14: 26; Col. 4: 16; 1 Thess. 5: 16–18). Justin Martyr in the early 2nd cent. says that on Sundays 'memoirs of the apostles' (i.e. a gospel) were read also.

wrath The attitude of God towards *sin and *evil, the inevitable response of his holiness to human wickedness. In the OT sudden *death or disaster was interpreted as the manifestation of God's wrath and reasons sought for it, as when Uzzah touched the *Ark with a profane hand (2 Sam. 6: 7). God's wrath is not arbitrary or unjustifiable; always he is gracious and loving (Ps. 103: 8), but his love sometimes seems like anger to those who experience it. Paul maintains belief in the wrath of God (Rom. 1: 18; Eph. 2: 3; 1 Thess. 1: 10) but the NT emphasizes that 'God is love' (1 John 4: 8). God's forbearance to the sinner is no less than his wrath against sin, though it is not to be evaded that the NT also speaks of a Day of Wrath, of irrevocable *judgement (Matt. 25: 31–45).

Wrede, William (1859–1906) German NT scholar born in Hanover, professor at Breslau. His book on the **Messianic Secret in the Gospels* (1901) suggested that Jesus' instructions to keep quiet about his work (e.g. Mark 1: 44) were not spoken by him but were later apologetic devices of the Church to account for the failure of Jesus' contemporaries to acclaim him as *Messiah. Wrede's theory was accepted in the next generation by the *Form Critics, as a presupposition of their discipline, and also foreshadowed *Redaction Criticism.

writing Spread from southern *Mesopotamia to *Egypt in the 3rd millennium BCE in the form of *cuneiform (wedge-shaped) signs; this was largely replaced by an alphabetical system in Palestine early in the 2nd millennium BCE, based on consonants (vowels borrowed from the Phoenicians were added by the Greeks at some time before the 8th cent. BCE). Inscriptions, to be permanent, were incised in stone, as (it is recorded) on the tables of the *Law brought by Moses from *Sinai (Exod. 31: 18). But a prayer scroll (pre-exilic) has been discovered in Jerusalem, and one of the *Qumran scrolls has the writing on sheets of copper. For everyday use *leather and *papyrus were available as writing material, and both took ink made from soot and applied with a brush or *pen.

The OT records exchanges of written communications. Letters were frequently sent and received at the royal courts of Israel and *Judah, from *David's time (2 Sam. 11: 14) and in the reign of *Hezekiah (2 Chron. 30: 1), and also to and from the Persian court (Ezra 4: 7).

It was fairly common for people in Israel to be able to read—even for children (Isa. 10: 19), but when *Baruch wrote down the words of *Jeremiah, and the king required them to be read to him by an official (Jer. 36: 21), it may have been that the king could not himself read. According to the NT, Paul also dictated letters (Rom. 16: 22), but it is likely that he could read both *Hebrew and *Greek. Certainly he signed his letters with his own hand (Col. 4: 18). The letters of Paul were read aloud in the Church meetings, and sent on to other Churches (Col. 4: 16). When *Philemon received his private letter, he would have read it aloud, since silent reading was generally unknown in antiquity.

Writings, the The third of the three divisions of the *Hebrew Bible, established as a separate category at the end of the 1st cent. CE and consisting of Pss., Prov., Job, S. of S., Ruth, Lam., Eccles., Esther, Dan., Ezra–Neh., and Chron.

Xerxes Ruler of *Persia from 486 to 465 BCE; called *Ahasuerus in Ezra 4: 6. His defeat by the Greeks at the naval battle of *Salamis in 480 BCE is described by the Greek historian Herodotus. There are problems about the biblical chronology, since the complaint against the returning Jews (Ezra 4: 6) must have been made just before 445 BCE. However, the problems may be easily solved by detaching 4: 6 from 4: 7, which belongs in the reign of a different ruler.

Ahasuerus (Xerxes I) is the Persian ruler in the book of Esther, but this book is more in the nature of legend than history. In Dan. 9: 1 Ahasuerus is said to be the father of *Darius the Mede, though in fact he was the son of Darius the Great.

Xystus Mentioned by *Josephus as a place in Jerusalem surrounded by porticoes and used by gymnasts and for public assemblies; it was designed by *Herod the Great. It has sometimes been suggested that it was the scene of the trial of Jesus before *Pilate.

Yahweh Occurs nearly 6,000 times in the OT as the name for God, but increasing reverence caused it to be replaced in public reading by '*Adonai*' ('My great Lord'). When vowels were added to the Hebrew text, those of *Adonai* were combined with YHWH to warn readers to substitute *Adonai* for Yahweh. So '*Jehovah' is an incorrect hybrid of the Christian Middle Ages. In modern Jewish writing the same respect is maintained by writing 'G—d'.

The Pentateuchal source *J (from German Jahveh) traces the name Yahweh to the earliest ages (Gen. 4: 26) but *P attributes the first use to Moses (Exod. 6: 2–3). It is suggested that the name means 'He causes to be what exists', i.e. Yahweh is the creator—but it is much disputed whether the divine name 'means' anything.

year In order to synchronize lunar and solar years, Jews added an extra *month to twelve lunar months, as required. Before the Exile, the new year began in autumn when agricultural work was finished; but in the Exile the Jews took over the *Babylonian year, which began in the spring, as the civil year, while retaining the former New Year for religious use.

The observance of Rosh Hashanah ('beginning [literally 'head'] of the year') still takes place in September–October (the first day of the month Tishri). According to Jewish tradition, there was a period of God's judgement which began on this day and lasted until **Yom Kippur* ten days later.

yod The tenth and smallest letter of the Hebrew alphabet; rendered *iota* (NRSV marg.; *jot in AV) in Matt. 5: 18, as being the smallest letter in the Greek alphabet.

yoke A frame of wood with vertical pins which effectively separated the *oxen so that together they could pull heavy loads. The yoke was a single cross-bar with rope nooses that were fastened round the animals' necks. The cross-bar was attached to a shaft, and so the wagon was drawn.

Figuratively, the yoke was used for hardship or submission. In typical prophetic style, *Jeremiah wore a yoke round his neck to symbolize his message that *Judah should submit to *Babylon (Jer. 27: 2). Jesus spoke of his 'easy yoke' (Matt. 11: 29–30), one that did not chafe. *Rabbis spoke of taking the yoke of the kingdom of heaven, by which they meant submitting to the sovereignty of God's *will.

Yom Kippur Hebrew for the Day of *Atonement, described in Lev. 16, and observed as a fast on the tenth day of the seventh *month (Lev. 16: 29), and a day still very widely observed by Jews. It was the one day of the year when the *high priest entered the *Holy of Holies in the inmost court of the Temple to offer incense. The ritual of this day is used in Heb. 8–9 to explain the sacrificial work of Christ.

YRSLM Hebrew letters for Jerusalem, found on coins of the Jewish rebellions of 66–70 CE and 132–5 CE.

Z

Zacchaeus A tax collector who climbed a tree to get a better view of Jesus in *Jericho (Luke 19: 1–10).

Zacharias *See* Zechariah.

Zadok A priest in the time of *David (2 Sam. 20: 25). He supported David during the revolt of his son *Absalom and urged him to appoint *Solomon as his successor (1 Kgs. 1: 8; 1: 39–45). According to 1 Chron. (6: 1–8; 49–53) he was descended from *Aaron, so was an Aaronite priest, but it has been widely held by modern scholars that he was of Jebusite origin, that is, one of the inhabitants of the city conquered by David. His name shows similarity with that of *Melchizedek, described in Gen. 14 as priest of Jerusalem.

The descendants of Zadok controlled the Jerusalem priesthood (Ezek. 40: 46) until 171 BCE when the office passed to the *Hasmoneans. The Zadokite priests were the first leaders of the community established at *Qumran. The name *Sadducees may be connected with Zadok.

Zadokite fragments Also known as the *Damascus Document, similar to the Manual of Discipline from *Qumran, containing the rules of the *community. This document was discovered in the *geniza* attached to an ancient *synagogue in Cairo in 1896, and is now in Cambridge.

Zarephath A port on the coast of *Phoenicia where *Elijah was given accommodation by a widow; he supplied her with food and restored her dying son to life (1 Kgs. 17: 8–24). The event is recalled by Jesus (Luke 4: 25–7 where AV uses 'Sarepta').

zeal Used in NT in both a bad sense—translated 'jealousy' (Acts 5: 17, NRSV, REB, NJB)—and a good sense (2 Cor. 7: 7). Paul had a zeal for *Judaism before he became an *apostle for Christ (Gal. 1: 14).

Zealots Those prepared to take up arms against the Romans occupying Palestine. As a group they came into existence at the beginning of the Jewish rebellion in 66 CE, but there was much social unrest as early as 6 CE, when Judas the Galilean led a revolt against the census of Quirinius (Acts 5: 37). Jesus had a sympathizer with this movement among the *Twelve (Luke 6: 15). By 66 CE, however, the name came to signify an organization or coalition of bandits, terrorists, and the unemployed to fight the Romans. The identification of Jesus with a popular movement of rebellion has been argued on the ground of his suffering crucifixion, the form of execution for a terrorist or rebel. The theory, however, is unlikely, since Jesus' teaching about the *kingdom of God was not given in language typical of the later Zealotry (cf. Mark 12: 17). It could be that *Simon, called the Zealot by *Luke but 'the Cananaean' by Mark (3: 18) which means in Aramaic a zealot in the sense of an enthusiast, gets this name because of his personal character. Or perhaps he was an enthusiast for God (like Paul, Acts 22: 3).

Zebedee The father and partner of *James and *John (Mark 1: 19–20) and a fisherman who employed staff (Mark 1: 20). His wife left home and followed Jesus (Matt. 27: 56), at least at the end of his life.

Zeboim (1) A valley near *Michmash (1 Sam. 13: 17–18). (2) A town near *Lydda where some of the returning exiles settled (Neh. 11: 34).

Zebulun Son of *Jacob, by *Leah (Gen. 30: 19–20); he gave his name to the tribe who were established near the *Phoenicians along the coast (Gen. 49: 13) and in central *Galilee (Josh. 19: 10–16), a region served by passable roads. *Nazareth was in this area (Matt. 4: 15).

Zechariah A common name in the Bible (Zacharias in AV). (1) The son of Jehoiada, stoned by the populace (2 Chron. 24: 20–3) and probably referred to in Matt. 23: 35. (2) The Minor Prophet after the Exile (Zech. 1: 1–7). (3) The father of *John the Baptist (Luke 1: 5) to whom an *angel appeared to announce the forthcoming birth to his wife *Elizabeth. When he asked for a confirmatory sign he was struck dumb—until the child was born, when he then blessed God and spoke of the hope for a *Messiah for whom John would prepare the way (Luke 1: 76). (4) Son of Baris, murdered by *Zealots shortly before the fall of Jerusalem in 70 CE according to *Josephus.

Matt. 23: 35 is mistaken in regarding Zechariah (1, above) as 'son of Barachiah': possibly this was an attempt to harmonize the reference with Zech. 1: 1 or even also with Isa. 8: 2.

Zechariah (4, above) would have been an anachronism on the lips of Jesus, but it is not impossible that recollection of this recent case of innocently shed blood was an influence on Matt.

Zechariah, book of The penultimate book of the OT. Chs. 1–8 are the work of Zechariah, son of *Iddo (according to Ezra 5: 1; grandson according to Zech. 1: 1), and describe the harsh conditions of life in *Judah after the Return from Exile. He followed *Haggai in urging the rebuilding of the Temple (about 518 BCE). There are eight visions in these chapters, linked by the theme of the consecration of *Joshua as *high priest. The land is to be cleansed of all who do not observe the *Law and the *visions promise God's continuing protection of Jerusalem.

The remaining chapters of Zech. (and Mal.) were added in the 5th and 4th cents. BCE and the authors seem dissatisfied with the nation's leadership in their time and use their visions of an *apocalyptic character to express hopes that in the end there will be a final vindication for the nation. Gentiles too will be included in the final act of *salvation (Zech. 14: 16).

Zedekiah The last king of *Judah, placed on the throne by the Babylonian conqueror *Nebuchadnezzar in 597 BCE (2 Kgs. 24: 17). Jeremiah advised him to live and reign under Babylonian rule—but he rebelled; Jerusalem was again besieged and captured, and Zedekiah was taken in chains to *Babylon (2 Kgs. 25: 6–7), where he was forced to witness the murder of his sons before his eyes were put out.

Zephaniah, book of Zephaniah was one of the twelve Minor *Prophets of the OT. He lived in the reign of *Josiah (646–609 BCE) and may have been descended from King *Hezekiah. He was one of those who were appalled by the apostasies of *Manasseh (697–642 BCE). The message of his book is that Jerusalem should repent, urgently, and forsake idolatrous rites (Zeph. 1: 12). The coming *day of the Lord will bring vengeance (1: 14–16), verses taken up in the 13th-cent. Christian hymn *Dies irae, dies illa*. It will not be a day of defeat for Judah's enemies but of war against the covenant people. Beginning in the city itself, battles will spread everywhere, devastating the people of Judah and foreign countries equally. The only hope lies in *repentance (2: 3) and the small number who do will indeed be saved (2: 7, 9); Jerusalem will rejoice again and the Lord will exult as on a day of festival (3: 17).

Zerubbabel A member of the house of *David who is mentioned in Haggai (1: 1; 2: 2) along with the *high priest *Joshua. He was charged to rebuild the Temple. In Zech. (4: 6–10) he is to do the work assisted only by the *Spirit of the Lord of Hosts. He refuses any bogus help from self-proclaimed worshippers of Yahweh (Ezra 4: 1–3). It may be that Zerubbabel owed his position as governor (Hag. 1: 1) to *Darius I of *Persia.

Zeus In Greek mythology, the chief of the gods, who struck people in anger with thunderbolts from Mount Olympus. By NT times Zeus was identified with the chief god of any of the religions in the Hellenistic region. *Barnabas is taken to be Zeus (Acts 14: 12; 'Jupiter', AV, RV) at *Lystra.

Ziba A servant of *Saul (2 Sam. 9: 9), who remained loyal to *David during the revolt of *Absalom (2 Sam. 16: 1–4).

ziggurat A temple built on a series of stages from a square base on which there was a *sanctuary. The god was supposed to reside at the top. There were ziggurats at *Nineveh, *Babylon (built between 1900 and 1600 BCE), and *Ur, and it is thought that the narrative of the tower of *Babel (Gen. 11: 1–9) may have been inspired by a neighbouring ziggurat in *Mesopotamia. *Jacob's dream of a staircase (Gen. 28: 12–15) may have been triggered by seeing such a building.

Zion ('Sion' in AV); used for Jerusalem (as 2 Sam. 5: 6–10) or for part of Jerusalem, such as the Temple mount (Ps. 2: 6; 48: 11), though the description (1 Kgs. 8: 1) of *Solomon bringing the *Ark to the Temple from Zion suggests they were once distinct areas, Zion being to the south. Zion came to be described in elaborate imagery in both Jewish and Christian *worship. It is a sacred place (Ps. 48: 1–2) and a secure mountain that cannot be moved (Ps. 125: 1–2). It was thought to be unassailable (Mic. 3: 9–12), as Isaiah (10: 24) promised when the Assyrians approached. It was destroyed, but it will be restored (Isa. 51: 1–6). Zion becomes 'the heavenly Jerusalem' (Heb. 12: 22), the destiny of Christian pilgrims. In 1 Pet. (2: 6) there is a quotation from Isa. (28: 16) where it is said that God has placed a *corner-stone in Zion. This is now applied, in the epistle, to Christ.

Zipporah Wife of Moses, daughter of Reuel (*Jethro), priest of Midian (Exod. 2: 16–22). When Moses returned to *Egypt to lead out the people of Israel in the Exodus, Zipporah and her sons stayed in Midian (Exod. 18: 2–5).

ziv The second *month of the Jewish *year (1 Kgs. 6: 1).

Zoan A city in *Egypt where Moses and *Pharaoh met (Ps. 78: 12, 43). For a time it was the capital of Egypt (Isa. 19: 11).

Zophar One of Job's 'comforters' (Job 2: 11). He contends that Job's misfortunes must be due to *wickedness. If only Job could see it all from God's point of view, he would realize that he has actually got off quite lightly. But in the prose conclusion to the book, his advice, and that of his friends, is derided by God (Job 42: 7–8). Still, Job interceded for him and for Eliphaz and Bildad (Job 42: 9).

Zoroaster A prophet in *Persia thought to have died in 551 BCE, though some scholars date him very much earlier, who worshipped the god Ahura-Mazda. The religion he founded was dualistic, with a strong belief in *Satan, an *evil power, and in *angels. It is suggested that the religion influenced *Judaism in the last two centuries BCE, which is reflected in *Daniel and certain writings of the *Apocrypha and *Pseudepigrapha. The *Magi who are said to have come from the east to visit the infant Jesus (Matt. 2: 1) could have been astrologers with Zoroastrian connections.

Zuzim A people said to have been conquered by Chedorlaomer, one of the four kings (Gen. 14: 1) who fought in the valley of the *Dead Sea (Gen. 14: 3–5). They may have existed to the north of Mount *Gilboa.

MEASURES, WEIGHTS, AND VALUES

A TABLE reproduced by permission of Oxford University Press from the Revised English Bible.

No precise modern equivalents can be given for the units of measurement, weight, and value used in the ancient world, which themselves varied at different times, in different places, and in different contexts of use. The approximate equivalents given below may be helpful as an indication of the order of magnitude implied by a particular term.

length

Unit	*Approx. equivalent in metres*	*As read at*
hand's breadth	0.075	Ezek. 40: 5
span	0.225	1 Sam. 17: 4 (REB marg.)
cubit (short) = 6 hand's breadths	0.45	Judg. 3: 16 (REB marg.)
cubit (long) = 7 hand's breadths	0.525	2 Chr. 3: 3

Weights and Values

Unit	*Approx. equivalent in grammes*	*As read at*
gerah	0.6	Ezek. 45: 12
shekel (sacred) = 20 gerahs	12	Lev. 27: 25
mina = 50 shekels	600	1 Kgs. 10: 17
mina = 60 shekels	720	Ezek. 45: 12
talent = 3,000 shekels	36,000	Exod. 38: 25

Mention is made (Gen. 23: 16) of a shekel of 'the standard recognized by merchants'; its relationship to the sacred standard is uncertain.

The 'pound' of the New Testament (John 12: 3) may be referred to the Roman standard of about 317 grammes.

Related to gold or silver, the weights tabulated above are frequently used as measures of value. In the Old Testament 'beka' (*lit.* half) is used to signify a half-shekel (Exod. 38: 26). The 'talent' of the New Testament (Matt. 18: 24) evidently signifies a large but not precise monetary value.

Coins

The 'daric' (1 Chr. 29: 7) was a gold coin weighing just over 8 grammes, said to have been equivalent to a month's pay for a soldier in the Persian army. What is referred to as a 'drachma' (Neh. 7: 70) may have been a silver coin of about 4.4 grammes.

The 'denarius' of the New Testament (Mark 14: 5) is said to have been the equivalent of a day's wage for a labourer.

Measures of Capacity: Dry Measures

Unit	*Approx. equivalent in litres*	*As read at*
kab	2.5	2 Kgs. 6: 25
omer	4.5	Exod. 16: 32
seah	15	1 Sam. 25: 18 (REB marg.)
ephah = 10 omers	45	Exod. 16: 36
kor = 10 ephahs	450	1 Kgs. 4: 22
homer = 10 ephahs	450	Ezek. 45: 11

Liquid Measures

Unit	*Approx. equivalent in litres*	*As read at*
log	1	Lev. 14: 10
hin = 12 log	12	Num. 15: 7
bath = 6 hin	72	Ezek. 45: 14
kor = 10 bath	720	Ezek. 45: 14

IMPORTANT DATES IN BIBLICAL HISTORY

BCE

c.1720	Hammurabi ruler of the first Babylonian Empire
c.1700	Abraham
c.1250	The Exodus
c.1200–1020	The Judges
c.1020	Saul unites the tribes as first king
c.1000	David is anointed king, and later captures Jerusalem
c.962	Accession of Solomon; building of the Temple is begun
c.924	Rehoboam succeeds Solomon; the kingdom is divided when the northern tribes choose Jeroboam as king
c.880	Omri establishes Samaria as capital of Israel
865	Activity of Elijah
853	Shalmanezer III of Assyria records battle with Ahab of Israel
842	Revolt of Jehu
754/3	Foundation of the city of Rome
734	Syro-Ephraimite conspiracy
727–698	Hezekiah king of Judah
722	Destruction of city of Samaria and end of the independent northern kingdom. Partial deportation
701	Assyrian record of the siege of Jerusalem
698	Manasseh king of Judah
639	Josiah king of Judah; killed in 609
597	Jehoiachin becomes king; Nebuchadrezzar besieges Jerusalem. Partial deportation of inhabitants
586	Jerusalem captured; many inhabitants exiled
586–538	The Babylonian Exile
538–520	The Return from Exile and building of the Second Temple
490	Persians defeated by Athenians at Marathon
445	Nehemiah in Jerusalem
332	Alexander the Great conquers Palestine
323	Empire divided amongst Alexander's generals. Seleucus gains Babylonia
175	Antiochus IV Epiphanes in control
167	Desecration of the Temple and the Maccabean War
164	Judas Maccabaeus defeats Lysias, enters Jerusalem, and rededicates the Temple (14 December)
142–63	Judaean independence
63	Capture of Jerusalem by Pompey
37–4	Herod the Great
20	Reconstruction of the Temple begins
c.4	Birth of Jesus

CE

6	Judaea placed under control of Roman prefect by Emperor Augustus
14	Tiberius Caesar succeeds Augustus
c.25	Philo of Alexandria expounds Judaism
26–36	Pontius Pilate prefect of Judaea
c.30	Crucifixion of Jesus
c.34	'Conversion' of Paul
41	Execution of James by Herod Agrippa I
49	Expulsion of Jews from Rome by Emperor Claudius
51	Paul before Gallio in Corinth
54	Nero becomes emperor
c.60	Paul in prison
64	The great fire in Rome
66–73	The Jewish War
70	Jerusalem captured by Titus; the Temple destroyed
73	Jewish rebels commit suicide after last stand at Masada
81	Domitian becomes emperor
93	Josephus writes his *Antiquities of the Jewish People*
98	Trajan becomes emperor
132–5	Second Jewish revolt under Simon Bar-Kochba put down by Romans
c.135	Aelia Capitolina built on site of Jerusalem by Emperor Hadrian

SELECT BIBLIOGRAPHY

MANY of the themes in this dictionary can be followed up in the numerous commentaries on individual books of the Bible that are available, and in the *Oxford Companion to the Bible* (OUP, 1993) and in *A Dictionary of Biblical Interpretation* (SCM Press, 1990). A collection taking a broad look at books of the Bible more generally has been published in the Oxford Bible Series (OUP) where, in addition to two volumes of an introductory character to both testaments, there are seven others—on the Old Testament, on the Gospels, and on Pauline Christianity. A useful introductory book is John Barton's *What is the Bible?* (SPCK, 1991).

For further reading:

Old Testament

John Barton, *Reading the Old Testament* (DLT, 1984).
N. L. Crenshaw, *Old Testament Wisdom* (SCM Press, 1981).
B. Otzen *et al.*, *Myths in the Old Testament* (SCM Press, 1980).
G. Vermes, *The Dead Sea Scrolls* (SCM Press, 1994).

Old Testament Guides on individual books in course of publication since 1984 by the Sheffield Academic Press.

New Testament

J. D. G. Dunn, *Unity and Diversity in the New Testament* (SCM Press, 1977).
J. L. Houlden, *Ethics and the New Testament* (T. & T. Clark, new edn., 1993).
E. P. Sanders, *Paul* (OUP, 1991).
E. P. Sanders, *The Historical Figure of Jesus* (Allen Lane, 1993).
Graham Stanton, *Gospel Truth?* (HarperCollins, 1995).
N. T. Wright, *The New Testament and the People of God* (SPCK, 1992).

New Testament Guides on individual books in course of publication since 1992 by the Sheffield University Press.

New Testament Environment

W. A. Meeks, *The First Urban Christians* (Yale University Press, 1983).
John Riches, *The World of Jesus* (CUP, 1990).
J. Stambaugh and D. Balch, *The Social World of the First Christians* (SPCK, 1986).

MAPS

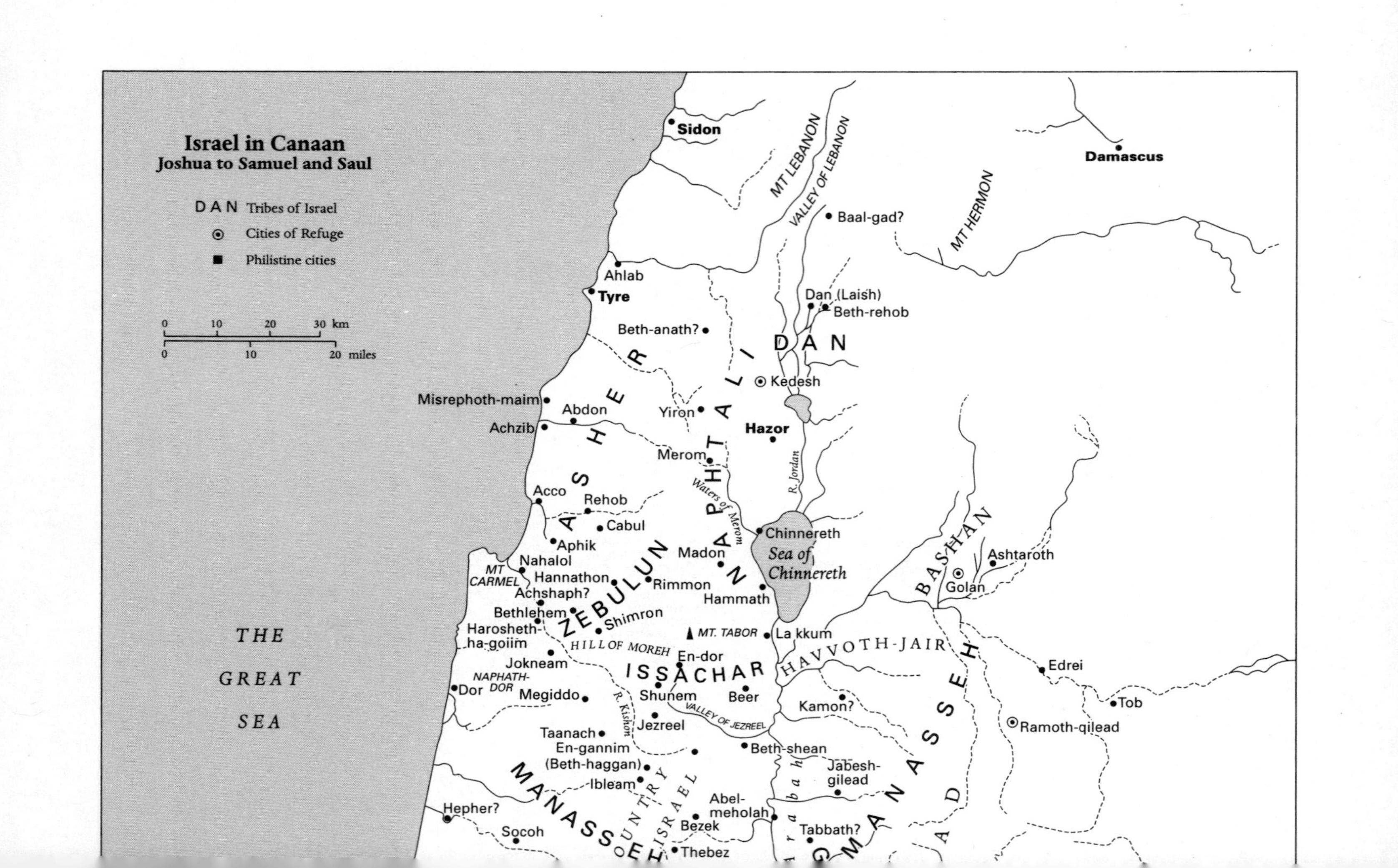
Israel in Canaan
Joshua to Samuel and Saul
DAN Tribes of Israel
Cities of Refuge
Philistine cities
0 10 20 30 km
0 10 20 miles
THE GREAT SEA
Sidon
Tyre
Ahlab
Damascus
MT LEBANON
VALLEY OF LEBANON
MT HERMON
Baal-gad?
Dan (Laish)
Beth-rehob
DAN
Beth-anath?
Kedesh
Hazor
Misrephoth-maim
Abdon
Achzib
Yiron
Merom
Waters of Merom
R. Jordan
ASHER
NAPHTALI
Acco
Rehob
Cabul
Aphik
Nahalol
MT CARMEL
Hannathon
Achshaph?
Rimmon
Madon
Chinnereth
Sea of Chinnereth
Hammath
ZEBULUN
Bethlehem
Shimron
Harosheth-ha-goiim
HILL OF MOREH
MT. TABOR
La kkum
En-dor
Jokneam
ISSACHAR
NAPHATH-DOR
Dor
Megiddo
Shunem
Beer
R. Kishon
VALLEY OF JEZREEL
Jezreel
Taanach
En-gannim (Beth-haggan)
Ibleam
Beth-shean
MANASSEH
HILL COUNTRY OF ISRAEL
Hepher?
Socoh
Thebez
Bezek
Abel-meholah
Arabah
Jabesh-gilead
Kamon?
Tabbath?
HAVVOTH-JAIR
BASHAN
Golan
Ashtaroth
Edrei
Tob
Ramoth-gilead
MANASSEH
GAD

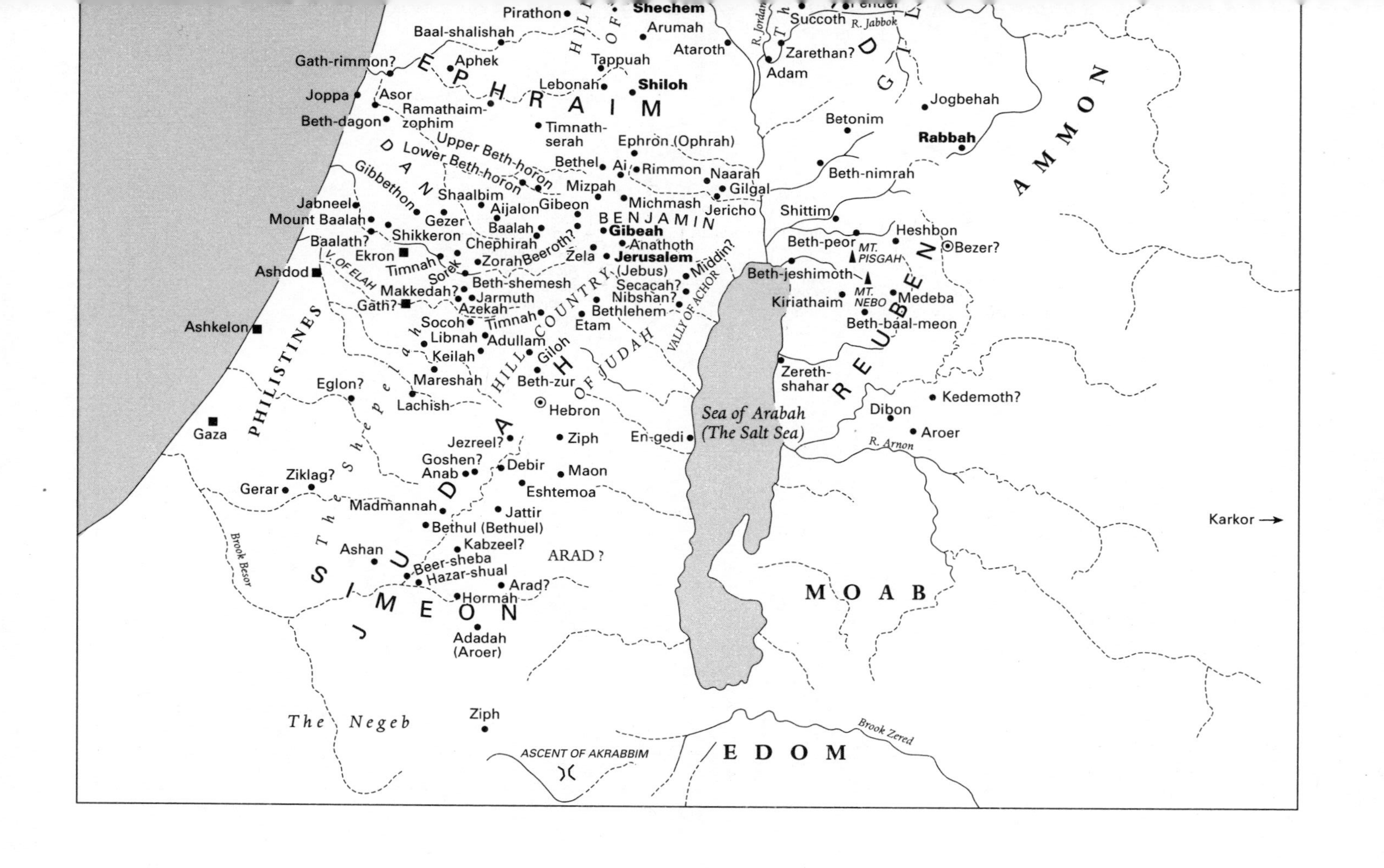

Pirathon
Shechem
Succoth
R. Jabbok
R. Jordan
Baal-shalishah
Arumah
Ataroth
Zarethan?
Adam
Gath-rimmon?
EPHRAIM
Aphek
Tappuah
Lebonah
Shiloh
Joppa
Asor
Ramathaim-zophim
Beth-dagon
Jogbehah
Betonim
Rabbah
AMMON
Timnath-serah
Ephron (Ophrah)
Upper Beth-horon
Lower Beth-horon
DAN
Bethel
Ai
Rimmon
Naarah
Gilgal
Beth-nimrah
Gibbethon
Shaalbim
Mizpah
Michmash
Jericho
Jabneel
Mount Baalah
Gezer
Aijalon
Gibeon
BENJAMIN
Shittim
Shikkeron
Baalah
Gibeah
Heshbon
Baalath?
Chephirah
Beeroth?
Anathoth
Beth-peor
MT. PISGAH
Bezer?
Ekron
Timnah
Zorah
Zela
Jerusalem
(Jebus)
Middin?
Ashdod
V. OF ELAH
Sorek
Beth-shemesh
Beth-jeshimoth
MT. NEBO
Makkedah?
Jarmuth
Secacah?
Nibshan?
Kiriathaim
Medeba
Gath?
Azekah
Timnah
Bethlehem
VALLY OF ACHOR
Beth-baal-meon
Ashkelon
Socoh
Etam
HILL COUNTRY OF JUDAH
Libnah
Adullam
REUBEN
PHILISTINES
Keilah
Giloh
Mareshah
Beth-zur
Zereth-shahar
Eglon?
Lachish
Hebron
Kedemoth?
Sea of Arabah
(The Salt Sea)
Dibon
Gaza
The Shephelah
Jezreel?
Ziph
En-gedi
Aroer
R. Arnon
Goshen?
Anab
Debir
Maon
Ziklag?
Gerar
JUDAH
Eshtemoa
Madmannah
Jattir
Karkor
Bethul (Bethuel)
Brook Besor
Kabzeel?
Ashan
ARAD?
Beer-sheba
Hazar-shual
SIMEON
Arad?
Hormah
MOAB
Adadah
(Aroer)
The Negeb
Ziph
Brook Zered
ASCENT OF AKRABBIM
EDOM

Black Sea
R. Sangarius
PHRYGIA
Gordion
Gomer
(Gimarrai)
Meshech
(Mushki)
R. Halys
Usiana
R. Hermus
LYDIA
Sardis
(Sepharad)
R. Maeander
JAVAN
CILICIA
(KHILAKKU)
Kue
Musri
Rhodes
Crete
(Caphtor)
Cyprus
(Iadanna)
Amanus Mts
HATTINA
R. Orontes
Arvad
Lebanon
Gebal (Byblos)
Berytus
Riblah
Lebo-
Hamath
Helbon
Sidon
Tyre
Damascus
Acco
Ushu
Hauran
ISRAEL
Samaria
AMMON
Jerusalem
Gaza
JUDAH
MOAB
Raphia
Sela
EDOM
The Great Sea
(The Upper Sea, the Western Sea)
Libya
Pelusium
Sais
Zoan
(Tanis)
Migdol
Tahpannes
Athribis
Heliopolis
(On)
Memphis
(Noph)
R. Nile
Sinai
Ezion-geber
(Elath)
EGYPT
Hermopolis
Lycopolis
(Siut)
Thebes
Red Sea
Syene
ETHIOPIA

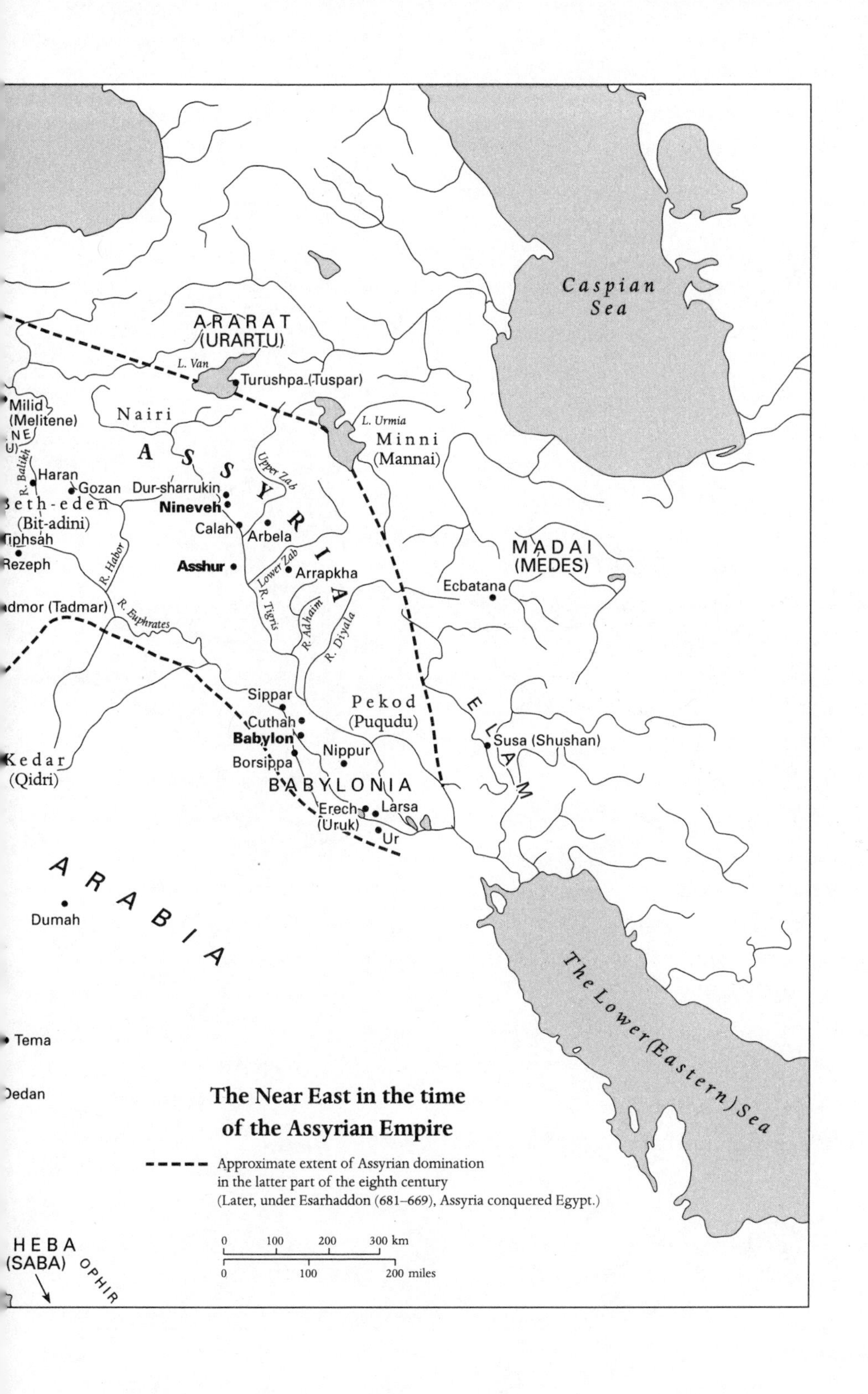

The Near East in the time of the Assyrian Empire

Approximate extent of Assyrian domination in the latter part of the eighth century
(Later, under Esarhaddon (681–669), Assyria conquered Egypt.)

0 100 200 300 km
0 100 200 miles

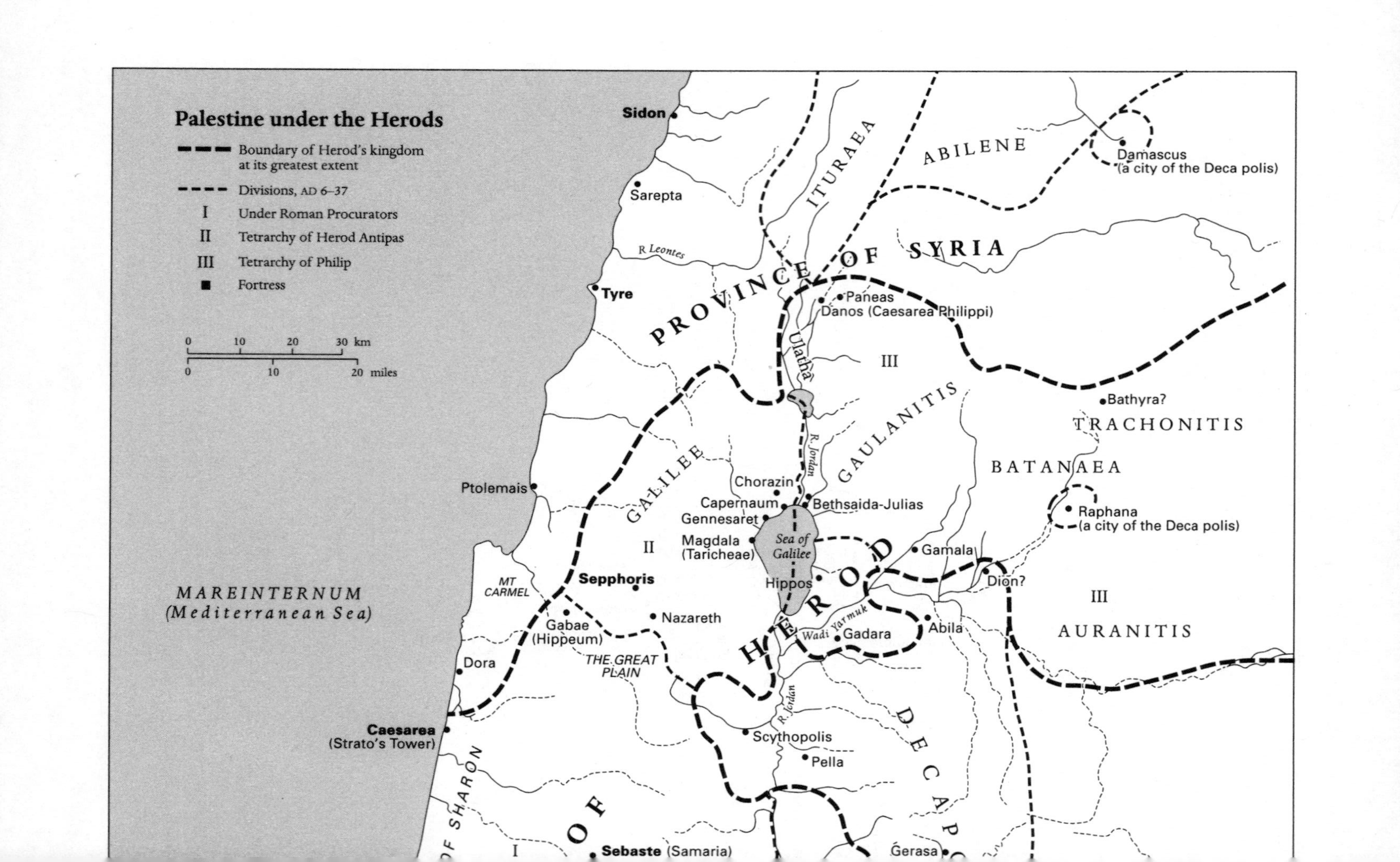

Palestine under the Herods
Boundary of Herod's kingdom at its greatest extent
Divisions, AD 6–37
I Under Roman Procurators
II Tetrarchy of Herod Antipas
III Tetrarchy of Philip
Fortress
0 10 20 30 km
0 10 20 miles
MAREINTERNUM
(Mediterranean Sea)
Sidon
Sarepta
R Leontes
Tyre
Ptolemais
MT CARMEL
Dora
Caesarea
(Strato's Tower)
SHARON
ITURAEA
ABILENE
Damascus
(a city of the Deca polis)
PROVINCE OF SYRIA
Paneas
Danos (Caesarea Philippi)
Ulatha
III
GAULANITIS
Bathyra?
TRACHONITIS
BATANAEA
Raphana
(a city of the Deca polis)
GALILEE
R. Jordan
Chorazin
Capernaum
Gennesaret
Bethsaida-Julias
Magdala
(Taricheae)
Sea of Galilee
II
Sepphoris
Hippos
Gamala
Dion?
III
AURANITIS
Nazareth
Gabae
(Hippeum)
THE GREAT PLAIN
HEROD
Wadi Yarmuk
Gadara
Abila
Scythopolis
Pella
DECAPO
OF
I
Sebaste (Samaria)
Gerasa

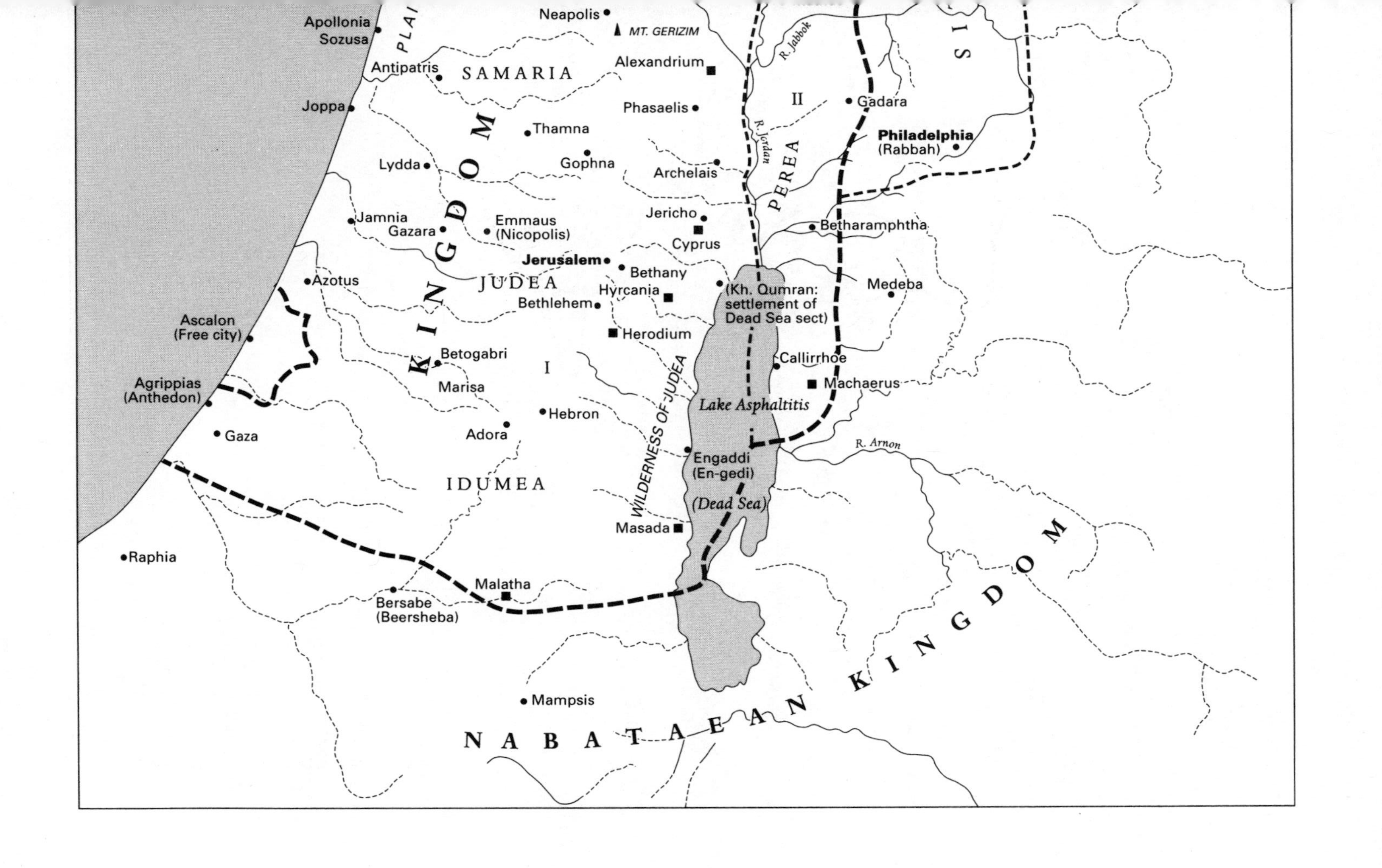

Apollonia
Sozusa
PLAI
Neapolis
MT. GERIZIM
Antipatris
SAMARIA
Alexandrium
Joppa
Phasaelis
KINGDOM
Thamna
Lydda
Gophna
Archelais
Jamnia
Gazara
Emmaus
(Nicopolis)
Jericho
Cyprus
Jerusalem
Bethany
Azotus
JUDEA
Hyrcania
Bethlehem
Ascalon
(Free city)
Herodium
Betogabri
I
Marisa
Agrippias
(Anthedon)
Hebron
Adora
Gaza
IDUMEA
WILDERNESS OF JUDEA
Masada
Raphia
Malatha
Bersabe
(Beersheba)
Mampsis
R. Jabbok
R. Jordan
II
PEREA
Gadara
Philadelphia
(Rabbah)
IS
Betharamphtha
(Kh. Qumran:
settlement of
Dead Sea sect)
Medeba
Callirrhoe
Machaerus
Lake Asphaltitis
Engaddi
(En-gedi)
(Dead Sea)
R. Arnon
NABATAEAN KINGDOM

R. Po
PANNONIA
Siscia
DACIA
Pola
Ravenna
Ariminum
Apennine Mts
Sirmium
Singidunum
Perusia
Ancona
ILLYRICUM (DALMATIA)
Viminaccium
ITALY
Salonae
R. Tiber
ROME
Sea of Adria
R. Danube
Oescus
Istros
Tomi
Ostia
Three Taverns
Forum of Appius (Appii Forum)
Naissus
MOESIA
Novae
Antium
Odessus
Tarracina
Capua
Scodra
Scupi
Sardica
Puteoli
Beneventum
Mesembria
Neapolis
Pompeii
Canusium
Philippopolis
Capreae (Capri)
Paestum
Dyrrhachium
THRACE
Brundisium
Appollonia
Tarentum
MACEDONIA
Philippi
Sybaris
Beroea
Neapolis
Bosphorus
Thessalonica
Byzantium
ROMAN EMPIRE
Samothrace
Corcyra
THESSALY
Nicomedia
Panormus
Croton
Larissa
Cyzicus
Nicaea
Nicopolis
Troas (Alexandria)
Assos
Prusa
Messana
Rhegium
Actium
Adramyttium
SICILY
Lesbos
Mitylene
Catana
Aegean Sea
Pergamum
ASIA
Corinth
Thyatira
Syracuse
ACHAIA
Cenchreae
Chios
Olympia
Athens
Smyrna
Sardis
Philadelphia
Ephesus
Sparta (Lacedaemon)
Malta (Melita)
Laodicea
Samos
Miletus
Colossae
PISIDIA
Cos
Cnidus
Rhodes
LYCIA
Perga
Attalia
Paul's Journey to Rome
CRETE
Patara
Myra
Phoenix
Salmone
Cauda
Fair Havens
Lasea
Lepcis Magna
Mediterranean Sea
(Mare Internum, Mare Nostrum)
Cyrene
Greater Syrtis
CYRENAICA
AFRICA
LIBYA
Canopus
Alexandria
Sais
Naucratis
Heliopolis
Memphis
EGYPT
Oxyrhynchus
R. Nile
Paul's 1st journey
Antioch
Iconium
Lystra
Perga
Attalia
Derbe
Antioch
Salamis
Paphos
Paul's 2nd journey
Philippi
Neapolis
Thessalonica
Beroea
Troas
Corinth
Antioch
Iconium
Athens
Ephesus
Lystra
Derbe
Antioch
Cenchreae
Caesarea
Paul's 3rd journey
Philippi
Troas
Assos
Mitylene
Chios
Samos
Ephesus
Miletus
Antioch
Corinth
Cos
Patara
Rhodes
Tyre
Ptolemais
Caesarea
Jerusalem

The Background of the New Testament
Rome and the East
(including Paul's Journeys)